Developments in Fish Telemetry

Developments in Hydrobiology 195

Series editor
K. Martens

Developments in Fish Telemetry

Proceedings of the Sixth Conference on Fish Telemetry held in Europe

Edited by

Pedro R. Almeida[1,2], Bernardo R. Quintella[2], Maria J. Costa[2] & Andrew Moore[3]

[1] *Department of Biology, University of Évora, Portugal*
[2] *Institute of Oceanography, Faculty of Sciences University of Lisbon, Portugal*
[3] *Centre for Environment, Fisheries & Aquaculture Science, Lowestoft, United Kingdom*

Reprinted from Hydrobiologia, Volume 582 (2007)

Organisation:

Sponsors:

Springer

Library of Congress Cataloging-in-Publication Data

A C.I.P. Catalogue record for this book is available from the Library of Congress.

ISBN-13: 978-1-4020-6236-0

Published by Springer,
P.O. Box 17, 3300 AA Dordrecht, The Netherlands

Cite this publication as Hydrobiologia vol. 582 (2007).

Cover illustration: Yagi antenna used to radio track adult sea lamprey during the upriver spawning migration in River Mondego (Portugal). Photo: I. Póvoa.

Printed on acid-free paper

Printed in the Netherlands

TABLE OF CONTENTS

Hydrobiologia (2007) 582:1–4
DOI 10.1007/s10750-006-0564-7

FISH TELEMETRY

Introduction

Pedro R. Almeida · Bernardo R. Quintella · Maria J. Costa · Andrew Moore

Freshwater and marine ecosystems continue to be the focus of significant environmental change associated with an expanding human population and a rapidly changing climate. Many organisms within these environments, and fish in particular, are susceptible to increasing pressures, ranging from exploitation to modifications and loss of habitat. In order to manage and conserve these populations and protect biodiversity, these resources must be managed in a sympathetic and responsible manner.

One biological tool that is progressively at the cutting edge of aquatic conservation is telemetry, which is increasingly being used by scientists to investigate fundamental aspects of animal biology in order to conserve and manage natural resources.

Guest editors: P. R. Almeida, B. R. Quintella, M. J. Costa and A. Moore
Developments in Fish Telemetry

P. R. Almeida (✉)
Department of Biology, University of Evora, 7002-554 Évora, Portugal
e-mail: pmraposo@mail.telepac.pt

P. R. Almeida · B. R. Quintella · M. J. Costa
Institute of Oceanography, Faculty of Sciences
University of Lisbon, Campo Grande, 1749-016 Lisboa, Portugal

A. Moore
Centre for Environment, Fisheries & Aquaculture Science, Lowestoft Laboratory, Pakefield Road, Lowestoft Suffolk NR33 0HT, UK

The Sixth Conference on Fish Telemetry held in Europe was organised by the University of Evora and the Institute of Oceanography of the Faculty of Sciences of the University of Lisbon in June 2005. The Conference attracted 162 delegates from 25 countries stimulating a multidisciplinary approach to problem solving and further promoting technological development and innovation in the field of wildlife telemetry. The Conference provided an excellent forum for scientists and engineers to exchange views and to disseminate information on recent developments in order to support and provide the scientific basis for the sustainable management of aquatic resources.

At the conference 74 oral and 56 poster contributions presented, and after review by members of the editorial board, 26 manuscripts were selected for inclusion within these proceedings. The selected papers are organised into four sections: (1) Human Impacts and Fisheries; (2) Migration and Behaviour; (3) Species Conservation and Habitat Rehabilitation; (4) Methodology and New Technology.

We would like to thank the support of two institutions for the organization of the Conference, the University of Evora and the Institute of Oceanography (Faculty of Sciences, University of

Lisbon), as well as the sponsorship of Administração do Porto de Sines, Ambassade de France au Portugal, Bacalhôa wines of Portugal, British Council (Portugal), Câmara Municipal de Sesimbra, Direcção-Geral das Pescas e Aquicultura, Fundação Luso-Americana, Fundação para a Ciência e a Tecnologia, Instituto da Água, Instituto de Turismo de Portugal, Instituto Italiano di Cultura Lisboa, Oceanário de Lisboa, and Região de Turismo da Costa Azul. We would also like to thank the members of the Conference Secretariat, Scientific and Organizing Committees, and all the attendees for contributing directly or indirectly to the success of this event. Special thanks go to 68 independent referees for helping to peer-review and select the manuscripts for the proceedings.

Scientific committee

Pedro Raposo Almeida
(Univ. of Evora/Inst. of Oceanography, Portugal)
Marie-Laure Begout Anras
(CREMA, France)
Niels Jepsen
(Danish Inst. for Fisheries Research, Denmark)
Robert Scott McKinley
(University of British Columbia, Canada)
Andrew Moore
(CEFAS Lowestoft Laboratory, UK)
Maria Teresa Spedicato
(COISPA Tecnologia & Ricerca, Italy)
Eva Thorstad
(NINA, Norway)
Hiroshi Ueda
(Hokkaido University, Japan)

Organising committee

Pedro Raposo Almeida
(Univ. of Évora/Inst. of Oceanography, Portugal)
Nuno Andrade
(Institute of Oceanography, Portugal)
Bernardo Quintella
(Institute of Oceanography, Portugal)

Conference secretariat

Carlos Alexandre
(Institute of Oceanography, Portugal)
Elsa Cabral
(Institute of Oceanography, Portugal)
João Ferreira
(Institute of Oceanography, Portugal)
Sílvia Pedro
(Institute of Oceanography, Portugal)
Sara Pinela
(Institute of Oceanography, Portugal)
Inês Póvoa
(Institute of Oceanography, Portugal)

Sponsors

Administração do Porto de Sines
Ambassade de France au Portugal
Bacalhôa, wines of Portugal
British Council (Portugal)
Câmara Municipal de Sesimbra
Direcção-Geral das Pescas e Aquicultura
Fundação Luso-Americana
Fundação para a Ciência e a Tecnologia
Instituto da Água
Instituto de Turismo de Portugal
Instituto Italiano di Cultura Lisboa
Oceanário de Lisboa
Região de Turismo da Costa Azul

Commercial exhibitors

Advanced Telemetry Systems, Inc., U.S.A.
Biotrack Ltd, UK
CEFAS Business Development Unit, UK
ENSID Technologies Ltd, New Zealand
Hydroacoustic Technology, Inc., U.S.A.
Lotek Wireless, Inc., Canada
VEMCO (AMIRIX Systems Inc.), Canada

List of participants

Surname	First name	Country
Aarestrup	Kim	Denmark
Abecassis	David	Portugal
Afonso	Pedro	Portugal
Alcocer-Peñas	Alex	Portugal
Alexandre	Carlos	Portugal
Alfredsen	Jo Arve	Norway
Almeida	Pedro	Portugal
Andrade	Nuno	Portugal
Arai	Nobuaki	Japan
Argyle	Ray	USA
Arnekleiv	Jo Vegar	Norway
Assis	Carlos	Portugal
Azeda	Carla	Portugal
Baeta	Filipa	Portugal
Barroso	Pedro	Portugal
Batista	Marisa	Portugal
Beaumont	William	UK
Beeman	John	USA
Bégout	Marie-L.	France
Belpaire	Claude	Belgium
Bendall	Barry	UK
Bentes	Luís	Portugal
Berg	Søren	Denmark
Bjørnsen	Jan Eyolf	Norway
Breukelaar	André	The Netherlandss
Butterworth	Kevin	Canada
Buysse	David	Belgium
Cabral	Henrique	Portugal
Caffrey	Joe	Ireland
Calado	João	Portugal
Carbonara	Pierluigi	Italy
Cardoso	Inês	Portugal
Carol	Joaquim	Spain
Chittenden	Cedar	Canada
Coeck	Johan	Belgium
Cooke	Steven	Canada
Cortes	Rui	Portugal
Costa	José L.	Portugal
Costa	Maria J.	Portugal
Cowley	Paul	South Africa
Crossin	Glenn T.	Canada
Croze	Olivier	France
Dagorn	Laurent	Seychelles
Debowski	Piotr	Poland
Dedual	Michel	New Zealand
Domingos	Isabel	Portugal
Dupont	Etienne	Belgium
Egan	Lawrence	Canada
Enders	Eva	Canada
Ferreira	João	Portugal
Ferreira	Telma	Portugal
Finstad	Anders G.	Norway
Fonseca	Vanessa	Portugal
França	Susana	Portugal
Fredrich	Frank	Germany
Geeraerts	Caroline	Belgium

continued

Surname	First name	Country
Greer	Sandra	Canada
Hahn	Lisiane	Brazil
Hinch	Scott	Canada
Holand	Bård	Norway
Horký	Pavel	Czech Republic
Hunter	Ewan	UK
Huovinen	Teemu	Finland
Hyvärinen	Pekka	Finland
Jacobsen	Lene	Denmark
Jepsen	Niels	Denmark
Jonsson	Sigurdur T.	Iceland
Kenward	Robert	UK
Kerwath	Sven	South Africa
Keskinen	Tapio	Finland
Knight	Carolyn	UK
Koed	Anders	Denmark
Komeyama	Kazuyoshi	Japan
Kubo	Toshihiko	Japan
Lembo	Giuseppe	Italy
Lino	Pedro	Portugal
Løkkeborg	Svein	Norway
Lopes	Marta	Portugal
Lucas	Martyn	UK
Lund	Vikebø E.	Norway
Maia	Anabela	Portugal
Makiguchi	Yuya	Japan
Marmulla	Gerd	Italia
Marques	Joana F.	Portugal
Matthews	Kathleen	USA
McKinley	R. Scott	Canada
Melnychuk	Michael	Canada
Michalsen	Kathrine	Norway
Midttun	Lars	Norway
Mitamura	Hiromichi	Japan
Mitsunaga	Yasushi	Japan
Moore	Andrew	UK
Naesje	Tor F.	Norway
Niezgoda	George	Canada
O'Dor	Ronald	USA
Økland	Finn	Norway
Östergren	Johan	Sweden
Ovidio	Michael	Belgium
Pálsson	Ólafur K.	Iceland
Parkinson	Derrick	New Zealand
Pedersen	Michael	Denmark
Pedro	Silvia	Portugal
Pereira	Nuno	Portugal
Pereira	Tadeu	Portugal
Peter	Armin	Switzerland
Pincock	Douglas	Canada
Pinela	Sara	Portugal
Pinheiro	Paulo	Portugal
Plantalech	Núria M.	Spain
Potin	Gaël	France
Póvoa	Inês	Portugal
Prista	Nuno M.	Portugal
Quintella	Bernardo	Portugal

continued

Surname	First name	Country
Ransom	Bruce	USA
Rego	Ana	Portugal
Reid	Andy	UK
Righton	David	UK
Rikardsen	Audun H.	Norway
Rivinoja	Peter	Sweden
Rogado	Leonor	Portugal
Rogan	Ger	Ireland
Rooney	Sean	Ireland
Sakamoto	Wataru	Japan
Santos	Carmen	Portugal
Santos	Patrick	Portugal
Saraniemi	Maare	Finland
Schulz	Uwe Horst	Brazil
Scolamacchia	Maria	Italy
Scruton	David	Canada
Semmens	Jayson	Australia
Silva	Gilda	Portugal
Soria	Marc	France
South	Andy	UK
Spedicato	Maria T.	Italy
Stakenas	Saulius	UK
Steig	Tracey	USA
Stickler	Morten	Norway
Suíssas	Cláudia	Portugal

continued

Surname	First name	Country
Svendsen	Jon	Denmark
Takashi	Toshinori	Japan
Takeshi	Yamane	Japan
Taquet	Marc	France
Tavner	Dave	UK
Teixeira	Amílcar	Portugal
Teixeira	Célia	Portugal
Thorstad	Eva B.	Norway
Thorsteinsson	Vilhjalmur	Iceland
Tranmæl	Even	Norway
Tsang	Peggy	Canada
Tsuda	Yuichi	Japan
Ueda	Hiroshi	Japan
Uglem	Ingebrigt	Norway
Valente	Alexandre	Portugal
Vasconcelos	Rita	Portugal
Vinagre	Catarina	Portugal
Walker	Alan M.	UK
Walls	Sean	UK
Ward	Bruce	Canada
Webber	Dale	Canada
Welch	David	Canada
Winter	Erwin	The Netherlands
Yamagishi	Yukiko	Japan
Zamora-Hernandez	Lluís	Spain

Hydrobiologia (2007) 582:5–15
DOI 10.1007/s10750-006-0547-8

FISH TELEMETRY

Efforts to aid downstream migrating brown trout (*Salmo trutta* L.) kelts and smolts passing a hydroelectric dam and a spillway

Jo Vegar Arnekleiv · Morten Kraabøl · Jon Museth

Abstract The autumn and spring descent of 41 brown trout kelts (spent trout) (average total length L_t 75.9 cm) and the spring descent of 27 hatchery-reared smolts (average L_t 26.2 cm) were studied by radiotelemetry in 1993/1994 (kelts) and 1997/1998 (kelts and smolts) at the Hunderfossen dam and power plant, south-east Norway. In 1999 we studied spring descent of 48 untagged kelts by visual observations and video-monitoring at one spillway. In autumn 1993 and 1997, 62.4% and 44.0% of the tagged kelts migrated downstream to the dam at water temperatures between 0.1 and 0.8°C. During release of spillwater, the kelts gathered along the dam with limited movements. Neither smolts nor kelts used the 2 m submerged turbine shafts as a pathway to migrate downstream Hunderfossen dam. Nor did release of large amounts of deep water through spillways provide downstream migration possibilities for kelts and possibly for smolts. The majority of both smolts and kelts migrated downstream at short periods of surface water release through the spillways, indicating high importance of surface water release. The threshold value of descend of kelts at surface release was between 1 and 4 m^3 s^{-1} which correspond to a water column between 12 and 36 cm. These findings are highly relevant regarding hydroelectric development in river systems containing iteroparous salmonid species.

Keywords Brown trout · Kelts · Downstream migration · Regulated river · Spillway · Water release

Guest editors: P. R. Almeida, B. R. Quintella, M. J. Costa and A. Moore
Developments in Fish Telemetry

J. V. Arnekleiv (✉) · M. Kraabøl
Norwegian University of Science and Technology, Museum of Natural History and Archaeology, 7491 Trondheim, Norway
e-mail: jo.arnekleiv@vm.ntnu.no

J. Museth
Norwegian Institute for Nature Research, 2624 Lillehammer, Norway

Introduction

On a worldwide scale probably thousands of hydropower dams, impoundments, tunnel outlets, reduced water flows in rivers, channelizations and other physical obstructions have interrupted the movements of fish and prevented natural reproduction and migration between separate habitats (Jungwirth, 1998; Northcote, 1998; Rivinoja et al., 2001; Carlsson et al., 2004). Construction of fish passages (fishways) is a common solution to this problem. Although many facilities, especially for upstream movements of salmonids, have been reported well functioning, the efficiency of many fishways have been questioned (Linløkken, 1993;

Larinier, 1998) and even newly designed fish passages may have low efficiency (Aarestrup et al., 2003). However, facilities and guidance to secure downstream migration for post-spawners and their progeny have been given less attention. For example, the main focus in Scandinavian fisheries management is to safeguard only the upstream spawning migration of salmonids. However, recent studies have shown that intra- and instream migrations are common in landlocked brown trout (*Salmo trutta* L.) populations, and that the current regime with fish passages may have negative effects on the populations (Carlsson et al., 2004).

Lake Mjøsa, the largest lake in Norway, contains populations of fast-growing and piscivorous brown trout with considerable commercial and recreational interests (Aass & Kraabøl, 1999). The Hunder trout population is a migratory, fast-growing strain using the Gudbrandsdalslågen River for reproduction. This trout strain is known for particularly large individuals (maximum body mass of 15–18 kg). They migrate between their feeding areas in the lake where they feed on smelt (*Osmerus eperlanus* L.), vendace (*Coregonus albula* L.) and small whitefish (*C. lavaretus* L.) (Aass et al., 1989) and the spawning grounds in the Gudbrandsdalslågen River (Arnekleiv & Kraabøl, 1996). Much of the spawning migration stops at the Hunderfossen dam (Aass et al., 1989). The Hunderfossen power plant severely interferes with the upstream spawning migrations of adult brown trout as well as the lake-ward migrations of kelts (spent trout) and smolts (Aass et al., 1989; Arnekleiv & Kraabøl, 1996, 1999). To maintain the ecological connection eliminated by hydropower impoundment in 1964, a fishway was built contemporaneously with the power station. Each year, 75–600 brown trout pass the fish ladder, and the ascent has been studied in relation to water discharge and water temperature (Jensen & Aass, 1995). Brown trout that succeeded to pass the fish ladder migrated rapidly to their spawning sites 4–63 km above Hunderfossen dam (Arnekleiv & Kraabøl, 1996; Kraabøl & Arnekleiv, 1998). However, less attention has been given to the downstream migration of kelts and smolts. These fish normally spawn repeatedly during their life span, and they probably require access to feeding habitats in the lake Mjøsa below the hydropower dam after spawning since there is little fish food available in the river.

Hydroelectric dams, however, provide different migrating pathways such as turbines, sluiceways and fishways (Skalski et al., 2002; Scruton et al., 2003a, b; Rivinoja 2005) with various mortality depending on physical properties of the technical installations (Montèn, 1985; Jepsen et al., 1998; Coutant & Whitney, 2000; Muir et al., 2001). The sluiceways at Hunderfossen dam release water from both deep and surface layers of the reservoir. Both the turbines and the fishway release water from submerged gateways. The aims of this study was (1) to investigate which of these waterways were used by descending smolts and kelts, and (2) to analyze if the descend of smolt and kelts were dependent on surface water release, and (3) to identify threshold values for descending kelts.

Study site

River Gudbrandsdalslågen drains a 17,000 km^2 catchment area and mean water discharge in the lower parts of the river is 60–100 m^3 s^{-1} in winter and 350–500 m^3 s^{-1} in summer. Maximum water discharge during floods in spring and summer is 1000–2500 m^3 s^{-1}. The Hunderfossen power station exploits a 46 m high waterfall, about 15 km upstream the outlet to Lake Mjøsa. The power generating water is abstracted from a river reservoir through two turbines and led back into the river through a tunnel about 4.4 km downstream from the dam (Fig. 1). The minimum discharge between the dam and the tunnel outlet is 1.8 m^3 s^{-1} in winter and spring (October–July), and varies between 5 and 20 m^3 s^{-1} the rest of the year. In late autumn and winter, the minimum water discharge is released in the fish ladder and this is normally the only water release passing the dam in this period. When the total river discharge exceeds the 300 m^3 s^{-1} capacity of the two Kaplan turbines, the surplus spillwater is released through seven spillways (Fig. 2). Normally, excess water is released through the

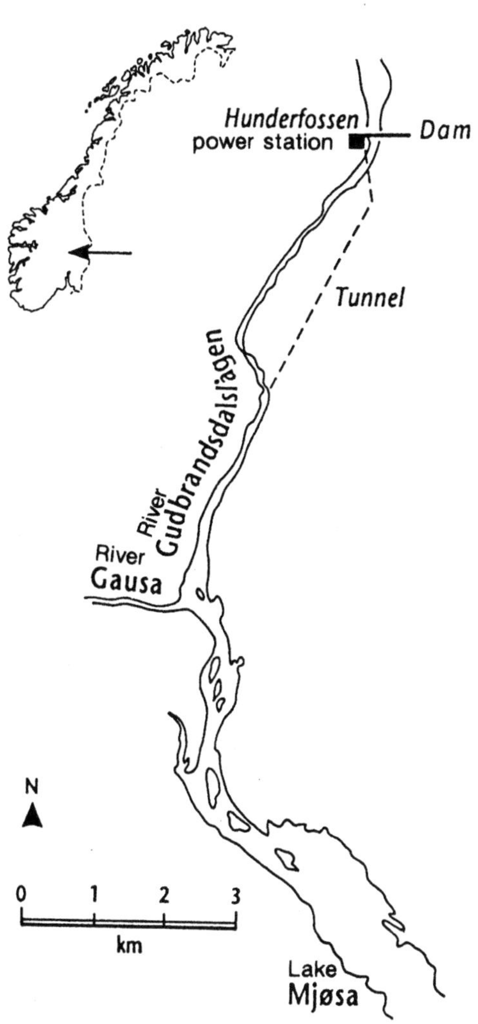

Fig. 1 Location of the Gudbrandsdalen River and the Hunderfossen power station, Norway

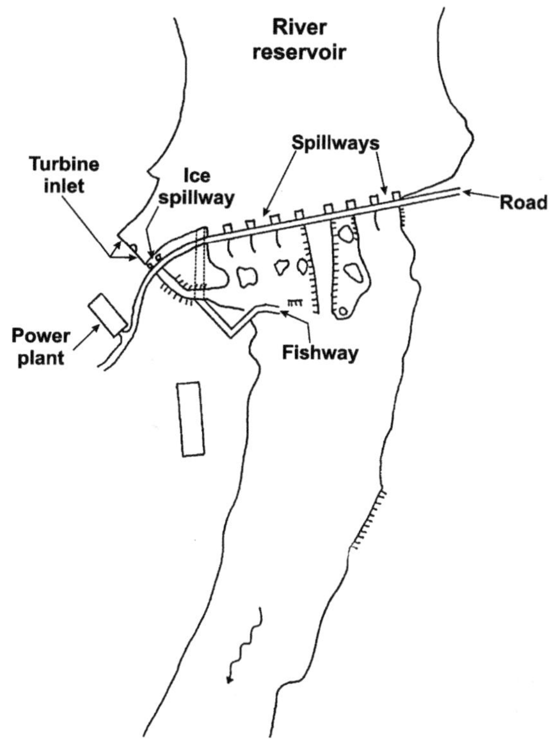

Fig. 2 Schematic diagram of the Hunderfossen dam, showing the different spillways, fishway and turbine inlet

bottom of the six spillways (bottom water release). Each spillway has a capacity of 267 $m^3 s^{-1}$ when fully opened and thus provide release of surface water similar to a natural fall. However, these spillways are seldom fully opened, and the water discharge is divided between the spillways providing bottom water release. Another spillway is draining floating ice and litter away from the turbine inlet (Fig. 2). This ice spillway is 6 m wide and has a capacity of surface water release of 1–40 $m^3 s^{-1}$, which correspond to a water column between 12 and 198 cm.

Materials and methods

Radio tracking and visual observations of kelts

The descent of brown trout kelts at the Hunderfossen dam and power station, was studied in two periods; 1993–1994 and 1997–1998, and visual observations were carried out in spring 1999. Forty one brown trout (BL range = 60–91 cm, BL mean = 75.9 cm) were caught in the fishway and radiotagged from July to late September in 1993 and 1997 (Table 1). They were released upstream of the dam. During tagging, the fish were placed in a partially covered cylindrical tank filled with well-oxygenated water. An ATS' radio transmitter (Models 16M, 7PN or 3PN Eiler activity, 142 MHz) (Eiler, 1990), was externally attached to each fish below the dorsal fin (Mellas & Haynes, 1985; Thorstad et al., 2000). The transmitter weighed between 16 and 27 g in air (0.5–1.5% of the mass of the fish). Fish were tracked manually each day, or one to two days a week in

Table 1 Number and length of resident brown trout studied by radio tracking at the Hunderfossen dam in 1993/1994 and 1997/1998

Year	Sex	n	Average total body length (cm)	Range	Dates of capture (range)
1993	Males	5	70.6	65–80	09.07–24.09
	Females	11	71.1	63–79	09.07–24.09
1997	Males	11	77.2	61–91	29.07–09.10
	Females	14	74.8	60–89	29.07–09.10

winter and early spring. Due to a screen at the turbine inlet, adult brown trout could not pass through the turbines.

In spring 1998, telemetry experiments were undertaken to try to guide the downstream migration of kelts and smolts passing the dam, by using surface water release of 1–40 $m^3 s^{-1}$ at the ice spillway. In 1999, descending kelts were visually observed from a bridge 5 m from the spillway during the first hours of spillwater release. Observing personell noted the number of kelts observed at the spillway entrance and number of kelts passing the spillway during trials of five different discharges of surface water release of 1, 4, 6, 15 and 25 $m^3 s^{-1}$. In addition, a videocamera was used to inspect the detailed behaviour of descending kelts.

Telemetry studies of smolt migration

To study the downstream movements of smolts passing the dam, we used 27 smolts (BL range = 22–30 cm, BL mean = 26.2 cm) from the local hatchery. All smolts (2 years old and F1-generation of native, wild stock) were tagged with external attached radio-transmitters (Model TXP-1, Televilt AB, Sweden, 142 MHz). The transmitters weighed 2.6 g in the air (1.3% of the mass of the smolt). The smolts were released in the reservoir above the dam on 14 May 1998 and tracked manually every 4 h each day and night during the experiments of surface water release (14–25 May, 1–6 June, 9–25 June). In the remaining periods between 25 May and 5 August, we manually tracked the fish every second day. Smolt had three possible routes to pass the dam: the turbines, the fishway or the spillways. By manual trackings, we assumed to determine which of the pathways used by descending smolts.

Statistical analyses

We used logistic regression models to analyse the relationship between the probability of observing descending kelts and smolts (separate tests), and the explanatory variables "water temperature", "total spillwater discharge", and "surface water discharge through the spillways". The test was carried out in Minitab (Release 13.0). To test whether the positions of kelts differed between periods before and during spillwater release (not normally distributed data), a non-parametric Wilcoxon Sign Ranks test was used.

Results

Descending kelts in 1993–1994

In November–December 1993 (period 1), 10 kelts descended from the spawning grounds and approached the dam. At water temperatures ranging between 0.1 and 0.8°C, they displayed a restless behaviour ranging up to 3.2 km upstream the dam. In early May 1994 (period 2), the additional 6 overwintering kelts migrated downstream to the dam and the restless behaviour continued until spillwater was released through the spillgates. During release of spillwater, the kelts assembled at the dam with limited movements (Fig. 3). The positions in period 1 were not significantly different from period 2 (Wilcoxon Signed Ranks test, $Z = -0.391$, $P = 0.693$) despite the higher water temperature range (5–6°C). The positions in period 1 were significantly different from period 3 (Wilcoxon Signed Ranks test, $Z = -4.301$, $P < 0.001$) and positions at period 2 were also significantly different from period 3 (Wilcoxon Signed Ranks test, $Z = 0.344$, $P = 0.001$).

Logistic regression model revealed no significant effect of total spillwater ($Z = 0.43$, $P = 0.668$), discharge in spillgate 1 ($Z = 0.02$, $P = 0.983$) or discharge in the ice spillway ($Z = 1.18$, $P = 0.238$) on the probability of observing descending kelts during spring 1994. However, all 16 kelts descended during three short periods of surface water release of 267 $m^3 s^{-1}$ through spillgate 1. During these three periods, seven, six and three kelts descended, respectively. No kelts descended during intermediate periods with bottom water release up to 469 $m^3 s^{-1}$ from the reservoir (Fig. 4).

Descending kelts in 1997–1998

In May 1998, 19 kelts descended through the spillgates. The logistic regression model revealed that there were a significant effect of water discharge in the ice spillway ($Z = 4.13$, $P < 0.001$) on the probability of observing descending kelts, and no significant effect of total spillwater discharge ($Z = 0.980$, $P = 0.327$) and water temperature ($Z = -1.15$, $P = 0.248$) (Fig. 5).

Visual observations of descending kelts in 1999

This 3.75 h observation study revealed that kelts passed the ice spillway at 4 $m^3 s^{-1}$ but not at 1 $m^3 s^{-1}$ (Table 2). In total, 48 kelts descended at water discharges between 4 and 25 $m^3 s^{-1}$. However, the degree of hesitation observed (expressed as percent descending kelts compared to total number observed at the sluiceway) revealed that only 39% of the observed kelts at the sluiceway descended at 4 $m^3 s^{-1}$, whilst 62% descended at 6 $m^3 s^{-1}$, 80% at 15 $m^3 s^{-1}$ and 100% at 25 $m^3 s^{-1}$ (Table 2). Despite this, the highest descending rate was observed at 4 $m^3 s^{-1}$. These results outpoint a descending threshold value between 1 and 4 $m^3 s^{-1}$, which correspond to a water column of 12 and 36 cm. Some hesitation occurred between 4 and 15 $m^3 s^{-1}$, and at 25 $m^3 s^{-1}$ kelts descended directly when approaching the sluiceway. The video recordings revealed that kelts descended the sluiceway with positive rheotaxi and propulsive flexations with their caudal fin, which provided them a slower falling velocity compared to the water column.

Descending smolts in 1998

In May-July 1998, all 27 smolts descended through the spillgates. The logistic regression model revealed that there were a significant effects of water discharge through the ice spillway ($Z = 4.94$, $P < 0.001$) and total spillwater discharge ($Z = 2.79$, $P = 0.005$) on the probability of observing descending smolts. Further, no significant effect of water temperature ($Z = -1.66$, $P = 0.097$) was found. These findings implies that

Fig. 3 Median positions (•) of ten radio-tagged kelts above the Hunderfossen dam in three different periods; (**a**) November 24–December 22 in 1993, (**b**) May 1–8 in 1994 and (**c**) May 12–June 9 in 1994. The vertical lines (whiskers) show the 10 (⊥) and 90 (⊤) percentiles

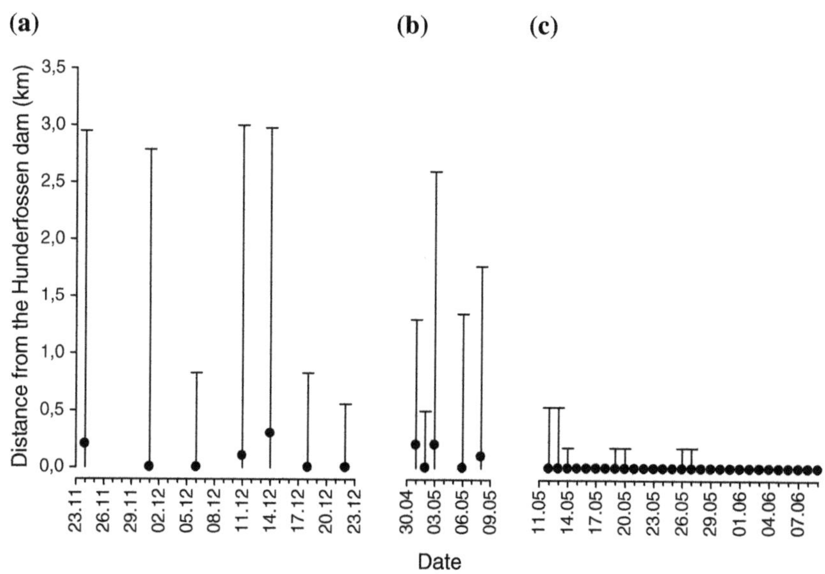

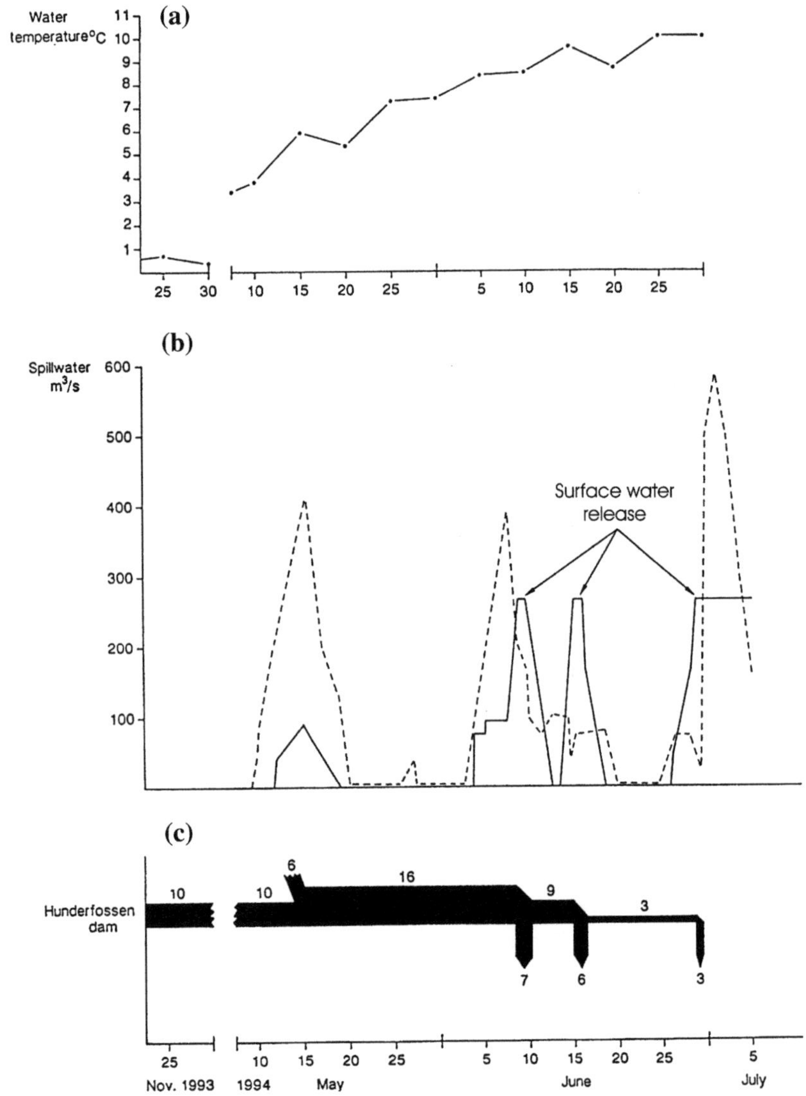

Fig. 4 (**a**) Water temperature, (**b**) water discharge passing the dam (total discharge minus the discharge at spillway 1 – stippled line, discharge at spillway 1 – solid line) and (**c**) downstream migration of radiotagged kelts (number of kelts shown) at Hunderfossen in 1993/1994

descending smolts use both the surface and bottom released spillwater as migratory pathways through the dam.

Discussion

This study demonstrated that neither smolts nor kelts used the 2 m deep submerged turbine shafts as a pathway to migrate downstream Hunderfossen dam. A few fish descended through the fishway entrance which is about 0.5 m submerged. Nor did release of large amounts of deep water through spillways provide downstream migration possibilities for kelts. We were not able to distinguish whether the smolts migrated by surface or bottom water release through the spillway. The majority of both smolts and kelts migrated downstream at short periods of surface water release through the spillways, indicating high importance of surface water release. The threshold value of descend of kelts at surface release was between 1 and 4 $m^3 s^{-1}$ which correspond to a water column between 12 and 36 cm. These findings are highly relevant regarding hydroelectric

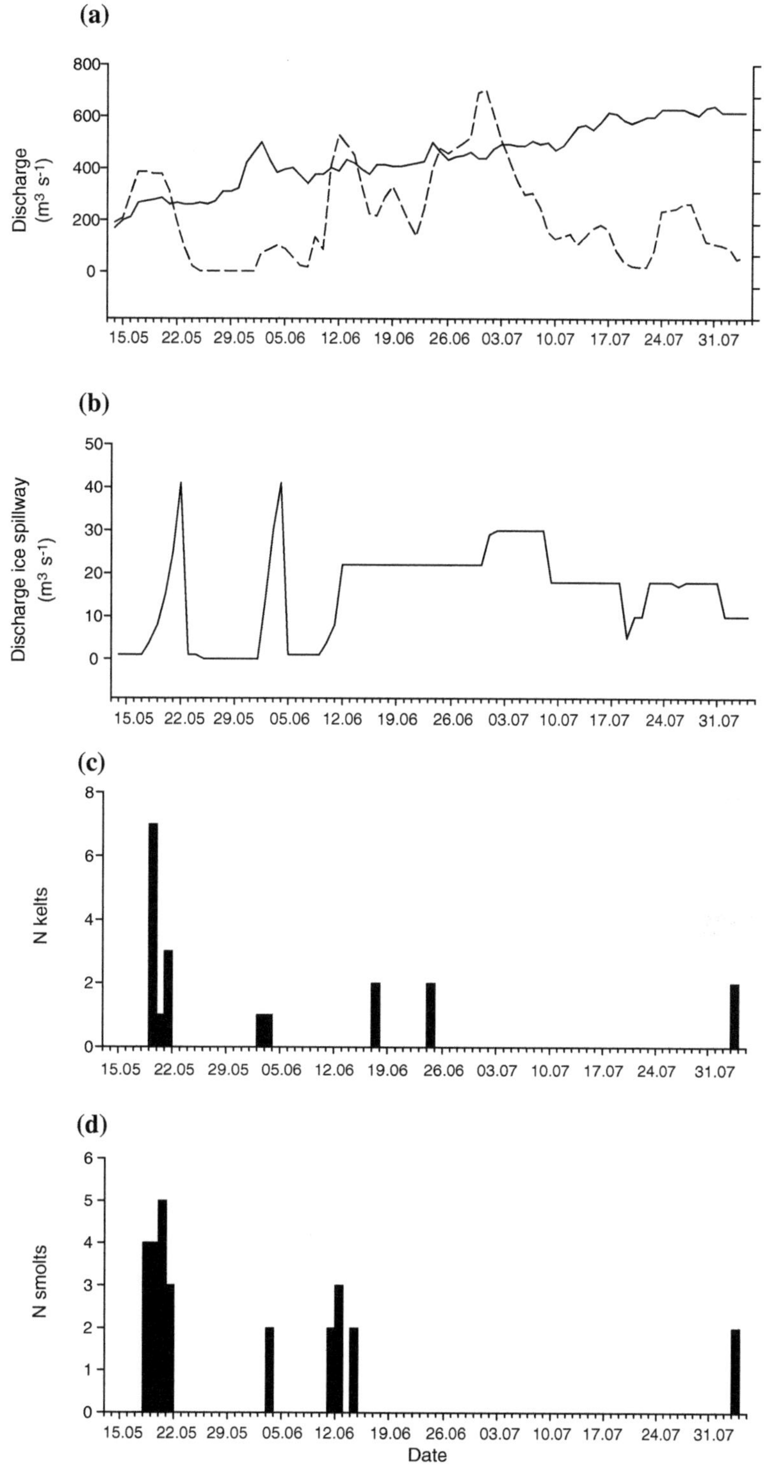

Fig. 5 (**a**) Water temperature (solid line) and total discharge passing the dam (stippled line), (**b**) surface water discharge through the ice spillway, (**c**) number of descending radio-tagged kelts and (**d**) number of descending radio-tagged smolts at Hunderfossen dam and power station, May 14–August 4 in 1997

development in river systems containing iteroparous salmonid species. Paradoxically, discharges ranging from 300 to 500 $m^3 s^{-1}$ released through submerged openings at the turbine intake or spillgates acted as a downstream migration barrier, whilst surface water

Table 2 Summary of the observations of kelts at the ice spillway at different water discharge during the experiment 21 May 1999

Discharge at the ice spillway $m^3 s^{-1}$	Time	Minutes of observation	Number of kelts observed	Number of descending kelts	Descending kelts per minute	Percent descending kelts of number observed
4	0830–1015	62	59	23	0.37	39
6	1015–1145	71	21	13	0.18	62
15	1145–1240	43	5	4	0.09	80
1	1240–1300	19	0	0	0	0
25	1300–1330	30	8	8	0.27	100
Sum		225	93	48		

release of 4 $m^3 s^{-1}$ provide a highly effective pathway for kelts and probably smolts.

The turbine pathways seems to be a common rout for salmonid smolts passing hydroelectric facilities (Montèn, 1985; Hvidsten & Johnsen, 1997; Jepsen et al., 1998; Coutant & Whitney, 2000; Muir et al., 2001; Rivinoja, 2005). In our study, none of the smolts descended through the turbines. This may be explained by the large smolt size of Hunder trout (mean BL = 26.2 cm, range = 22–30 cm) compared to other salmonid smolts which are considerably smaller. Rivinoja (2005) and Kemp et al. (2005a) reported that the downstream movements of some larger smolt was altered in the vicinity of a power station similar to Hunderfossen Power Plant. These large smolts avoided turbine entrance and did not continue downstream migration. Kemp et al. (2005a, b) observed that downstream migrants hesitated to descend into dark spaces with accelerating water velocities. We suggests that there is some body size selectivity regarding turbine entrance of migrating smolts which may be explained by swimming performance positively correlated to BL.

Downstream migrating kelts possibly avoided the turbine entrance of the same reason. In addition, the grid in front of the turbine shaft may also act as an obstacle considering the rapid and accelerating water velocity at this intake area. Kelt may find it difficult to maneuver with precision under such conditions and may therefore neglect this pathways of several reasons. In 1994, none of the kelts used the ice spillway. The water discharge was 8 m^3s^{-1} during most of the time. However, the release of spillwater through spillgate 1 on the opposite end of the dam during the study period possibly attracted kelts away from the ice sluiceway and thereby prevented them from using this pathway. In 1998, surface water was released through the ice spillway and adjacent spillgates. These findings demonstrate the potential of guiding kelts by manipulating release points through the dam.

After spawning, 44–62% of the trout moved downstream to the reservoir above the dam in late autumn, and by early May, 100% of the surviving tagged kelts had moved downstream to the dam. A post-reproduction migration back to the lake in autumn or in early spring is also observed in other lake-migrating brown trout strains (Arnekleiv & Rønning, 2004; Rustadbakken et al., 2004). At the first release of bottom water at the spillways, kelts assembled at the dam, and tracking observation showed that they were swimming back and forth along the dam. We suggest that this behaviour was a searching behaviour to find their way downstream. The praxis to only open for bottom water release in spring and early summer obviously prevented kelts from passing the dam on their post-spawning migration in these periods. Because Hunder trout often spawn repeatedly, usually with an intermediate year in lake Mjøsa (Aass et al., 1989), it is vital that brown trout are given free passage on their downstream migration. During temperature increase in spring and early summer, large, piscivory brown trout probably have no prey fish to feed on in the river, and migration in salmonids may be triggered by the amount or the lack of available food (Northcote, 1992). Obstacles that are preventing, or delaying fish in reaching its main food source of smelt, vendace and small whitefish in the lake (Aass et al., 1989), probably put considerable constraints and stresses on kelts and may cause increased mortality.

Fish often depend on special hydraulic conditions to pass physical obstacles. Cold water increases problems of passing obstacles for ascending fish (Jonsson, 1991) and a temperature of 7–9°C for brown trout to negotiate obstacles are found in several rivers (Ovidio & Phillipart, 2002; Rustadbakken et al., 2004). However, temperature requirements to swim downstream are less, and obstacles are obviously more easily cleared going downstream than upstream. At Hunderfossen, kelts were active during late autumn and winter at temperatures 0.1–0.8°C, and five kelts passed the fish ladder or spillway at 1.5°C in late autumn. This is in accordance with findings for descending anadromous brown trout in Imsa River (Jonsson & Jonsson, 2002). In autumn, the daily number of descending brown trout in Imsa River is positively correlated with discharge and negatively with water temperature, whereas in spring, high water temperature appeared to positively influence the descent (Jonsson & Jonsson, 2002). At Hunderfossen, the downstream migration of kelts occurred during increased water discharge and temperature. However, the maneuvering of the spill gates, giving either bottom water release or surface water release, probably overrode the effect of water temperature. The experiments with surface water release showed that kelts were concentrated in the upper water column and did not pass the dam (with one exception in autumn) at bottom water release or in the fishway. The upper entrance of the fishway is about 0.5 m below surface, and this is probably the reason for its low efficiency for downstream migration. Numbers of kelts were often observed in the surface water layer at the turbine inlet and spillways in April and early May, before any spillwater release occurred (O. Caspersen, pers. comm.). In 1998, 253 brown trout ascended the fish ladder at their spawning migration. In the visual observation study, with surface water release in May 1999, 48 kelts passed the dam in 3.75 h of ≥ 4 m^3 s^{-1} surface water release on their post-spawning downstream migration. This means that the entire spawning populations from previous season are able to descend this spillway within 20 h of ≥ 4 m^3 s^{-1} surface water release. At this particular site, a discharge of only 4–8 m^3 s^{-1} (36–64 cm water depth, 1.3–2.7% of the total turbine discharge) may be enough to secure the descent of kelts.

Brown trout exhibit great variation in life history traits (Klemetsen et al., 2003). Ontogenetic habitat shifts occur in populations performing migrations between spawning and nursery areas in rivers and feeding areas in lakes or at sea. Correspondingly, smolt migration from nursery areas to feeding areas in lakes or at sea is common. Also in inland waters with migratory lake-run brown trout populations, a smolt migration similar to that of anadromous brown trout occurs. In anadromous brown trout, smolt is migrating in spring/early summer in the upper water layer at night in southern Norway (Jonsson, 1985, Hembre et al., 2001). The radio tagged smolt at Hunderfossen passed the dam through the water release in the spillway and not through the turbines. However, experiments were conducted with hatchery smolt and there is a possibility that wild smolt will behave in a different way and may pass through the turbines. Catches of adult trout from smolt stocked above and below the dam, gave greater catches of the stocked fish from below the dam, indicating a mortality of smolt passing the turbines or dam (P. Aass, pers. comm.).

This study revealed that neither kelts, nor smolts entered the turbine pathway during their downstream migration. Kelts almost exclusively used surface water released from two spillways as a migration route, whereas smolts probably also descended through deep water release from the spillways. The threshold value of downstream migrating kelts corresponds to a water column between 12 and 36 cm released as surface water. Further, these results demonstrate the importance of significant effort providing surface water releases at hydroelectric facilities which interrupt iteroparous and/or diadromous salmonid migratory systems.

Acknowledgments Financial support was given by the Norwegian Electricity Federation. We thank Lars Rønning for technical assistance and two anonymous referees for improving the manuscript.

References

Aarestrup, K., M. C. Lucas & J. A. Hansen, 2003. Efficiency of a nature-like bypass channel for sea trout (*Salmo trutta*) ascending a small Danish stream studied by PIT telemetry. Ecology of Freshwater Fish 12: 160–168.

Aass, P., P. Sondrup Nielsen & Å. Brabrand, 1989. Effects of river regulation on the structure of a fast-growing brown trout (*Salmo trutta* L.) population. Regulated Rivers: Research and Management 3: 225–266.

Aass, P. & M. Kraabøl, 1999. The exploitation of a migrating brown trout (*Salmo trutta* L.) population; change of fishing methods due to river regulation. Regulated Rivers; Research & Management 15: 211–219.

Arnekleiv, J. V. & M. Kraabøl, 1996. Migratory behaviour of adult fast-growing brown trout (*Salmo trutta*, L.) in relation to water flow in a regulated Norwegian river. Regulated Rivers: Research and Management 12: 39–49.

Arnekleiv, J. V. & M. Kraabøl, 1999. Upstream and downstream migrations of brown trout at the Hunderfossen power plant. pp. 49–55. In Kamula, R & A. Laine, (eds), Proceedings of the Nordic Conference on Fish Passage. DN-notat1999-1: 169 pp.

Arnekleiv, J. V. & L. Rønning, 2004. Migratory patterns and return to the catch site of adult brown trout (*Salmo trutta* L.) in a regulated river. River Research and Applications 20: 929–942.

Carlsson, J., K. Aarestrup, F. Nordwall, I. Näslund, T. Eriksson & J. E. L. Carlsson, 2004. Migration of landlocked brown trout in two Scandinavian streams as revealed from trap data. Ecology of Freshwater Fish 13: 161–167.

Coutant, C. C. & R. R. Whitney, 2000. Fish behavior in relation to passage through hydropower turbines: A review. Transactions of the American Fisheries Society 129: 351–380.

Eiler, J. H., 1990. Radio transmitters used to study salmon in glacial rivers. American Fisheries Society Symposium 7: 364–369.

Hembre, B., J. V. Arnekleiv & J. H. L'Abée-Lund, (2001). Effect of water discharge and temperature on the seaward migration of anadromous brown trout, *Salmo trutta*, smolts. Ecology of Freshwater Fish 10: 61–64.

Hvidsten, N. A. & B. O. Johnsen, 1997. Screening of descending Atlantic salmon (Salmo salar L.) smolts from a hydropower intake in the river Orkla, Norway. Nordic Journal of Freshwater Research 73: 44–49.

Jensen, A. J. & P. Aass, 1995. Migration of a fast-growing population of brown trout (*Salmo trutta* L.) through a fish ladder in relation to water flow and water temperature. Regulated Rivers: Research and Management 10: 217–228.

Jepsen, N., K. Aarestrup, G. Rasmussen & F. Økland, 1998. Survival of radio-tagged Atlantic salmon (*Salmo salar*) and trout (*S. trutta*) smolts passing a reservoir during seaward migration. Hydrobiologia 371–372: 347–353.

Jonsson, B., 1985. Life history patterns of freshwater resident and sea-run migrant brown trout in Norway. Transactions of the American Fisheries Society 114: 182–194.

Jonsson, N., 1991. Influence of water flow, water temperature and light on fish migration in rivers. Nordic Journal of Freshwater Research 66: 20–35.

Jonsson, B. & N. Jonsson, 2002. Migration of anadromous brown trout *Salmo trutta* in a Norwegian river. Freshwater Biology 47: 1391–1401.

Jungwirth, M., 1998. River continuum and fish migration – Going beyond the longitudinal river corridor in understanding ecological integrity. In Jungwirth, M., M. S. Schmutz & S. Weiss (eds), Fish migration and fish bypasses. Fishing News Books. Blackwell Science Ltd., Oxford, 127–145.

Kemp, P. S., M. H. Gessel & J. G. Williams, 2005a. Fine-scale behavioural responses of Pacific salmonid smolts as they encounter divergence and acceleration of flow. Transactions of the American Fisheries Society 134: 390–398.

Kemp, P. S., M. H. Gessel & J. G. Williams, 2005b. Seaward migrating subyearling chinook salmon avoid overhead cover. Journal of Fish Biology 67: 1381–1391.

Klemetsen, A., P.-A. Amundsen, J. B. Dempson, B. Jonsson, N. Jonsson, M. F. O'Connell & E. Mortensen, 2003. Atlantic salmon *Salmo salar* L., brown trout *Salmo trutta* L. and Arctic charr *Salvelinus alpinus* (L.): a review of aspects of their life histories. Ecology of Freshwater Fish 12: 1–59.

Kraabøl, M. & J. V. Arnekleiv, 1998. Records of spawning locations in fast-growing brown trout populations in the rivers Gudbrandsdalslågen and Gausa. Vitenskapsmuseet Rapp. Zool. Ser. 2: 1–28 (in Norwegian, English summary).

Larinier, M., 1998. Upstream and downstream fish passage experience in France. In Jungwirth, M., M. S. Schmutz & S. Weiss (eds), Fish migration and fish bypasses. Fishing News Books. Blackwell Science Ltd., Oxford: 127–145.

Linlokken, A., 1993. Efficiency of fishways and impact of dams on the migration of grayling and brown trout in the Glomma river system, South-eastern Norway. Regulated Rivers: Research and Management 8: 145–153.

Mellas, E. J. & J. M. Haynes, 1985. Swimming performance and behaviour of rainbow trout (*Salmo gairdneri*) and white perch (*Morone americana*): effect of attaching telemetry transmitters. Canadian Journal of Fisheries and Aquatic Sciences 42: 488–493.

Montèn, E., 1985. Fish and turbines; fish injuries during passage through power station turbines. Vattenfall AB, Stockholm, Sweden.

Muir, W. D., S. G. Smith & B. P. Sandfjord, 2001. Survival of juvenile salmonids passing through bypass systems, turbines and spillways with and without flow deflectors at Snake river dams. North American Journal of Fisheries Management 21: 135–146.

Northcote, T. G., 1992. Migration and residency in stream salmonids – some ecological considerations and evolutionary consequences. Nordic Journal of Freshwater Research 67: 5–17.

Northcote, T. G., 1998. Migratory behaviour of fish and its significance to movement through riverine fish passage facilities. In Jungwirth, M., M. S. Schmutz & S. Weiss (eds), Fish migration and fish bypasses. Fishing News Books. Blackwell Science Ltd., Oxford, 3–18.

Ovidio, M. & J. C. Philippart, 2002. The impact of small physical obstacles on upstream movements of six species of fish. Hydrobiologia 483: 55–69.

Rivinoja, P., 2005. Migration problems of Atlantic salmon (*Salmo salar* L.) in flow regulated rivers. Doctoral thesis No. 2005:114. Sweedish University of Agricultural Aciences, Umeå.

Rivinoja, P., S. McKinnell & H. Lundqvist, 2001. Hindrances to upstream migration of Atlantic salmon (*Salmo salar*) in northern Swedish river caused by a hydroelectric power-station. Regulated Rivers: Research & Management 17: 101–115.

Rustadbakken, A., J. H. L'Abée-Lund, J. V. Arnekleiv & M. Kraabøl, 2004. Reproductive migration of brown trout in a small Norwegian river studied by telemetry. Journal of Fish Biology 64: 2–15.

Scruton, D. A., R. S. McKinley, N. Kuowen, W. Eddy & R. K. Booth, 2003a. Improvement and optimisation of fish guidance efficiency (FGE) at a behavioural fish protection system for downstream migrating Atlantic salmon (*Salmo salar*) smolts. River Research & Applications 19: 605–617.DOI:10.1002/rra.735.

Scruton, D. A., C. J. Pennell, M. J. Robertson, E. W. Clarke & R. S. McKinley, 2003b. Telemetry studies of the entrainment and fate of downstream migrating Atlantic salmon (*Salmo salar*) smolts at hydroelectric installations on the Expoits River, Newfoundland, Canada. Proceedings from the Fifth conference on fish telemetry held in Europe 9–13 June 2003, Ustica, Palmero, Italy.

Skalski, J. R., R. Townsend, J. Lady, A. E. Giorgi, J. R. Stevenson & R. D. McDonald, 2002. Estimating route-specific passage and survival probabilities at a hydroelectric project from smolt radiotelemetry studies. Canadian Journal of Fisheries and Aquatic Sciences 59: 1385–1393. DOI: 10.1139/F02-094.

Thorstad, E. B., F. Økland & B. Finstad, 2000. Effects of telemetry transmitters on swimming performance of adult Atlantic salmon. Journal of Fish Biology 57: 531–535.

Hydrobiologia (2007) 582:17–24
DOI 10.1007/s10750-006-0542-0

FISH TELEMETRY

Walking speed and area utilization of red king crab (*Paralithodes camtschaticus*) introduced to the Barents Sea coastal ecosystem

Terje Jørgensen · Svein Løkkeborg · Anders Fernö · Marianne Hufthammer

Abstract The red king crab (*Paralithodes camtschaticus*) was introduced in the Barents Sea in the 1960s and soon established a viable population. Proper management and exploitation of the Barents Sea king crab stock require better understanding of the spatial dynamics at different scales. This study examines the small-scale movement patterns of seven adult male crabs tracked for a period of up to one month from mid July to mid August at 150 m depth in a semi-enclosed fjord on the Russo-Norwegian border. The crabs were tagged with acoustic transmitters and their movements monitored with an acoustic positioning system. Low walking speeds (<0.01 m s^{-1}) were most frequent but the crabs could move at a maximum speed of 0.15 m s^{-1} and walk an actual distance of up to 270 m over a period of one hour. However, the crabs usually moved within a relatively restricted area with mean hourly longest rectilinear distance varying from 26 to 64 m. The crabs alternated between periods of low and high activity, which could reflect feeding in and movements between food patches. The lack of a diel activity rhythm may be due to high light levels during the polar summer night, or a chemically mediated food search strategy.

Keywords Red king crab · Telemetry · Walking speed · Area utilization · Barents Sea

Guest editors: P. R. Almeida, B. R. Quintella, M. J. Costa and A. Moore
Developments in Fish Telemetry

T. Jørgensen (✉) · S. Løkkeborg · A. Fernö
Institute of Marine Research, P.O. Box 1870, Nordnes, 5817 Bergen, Norway
e-mail: terjej@imr.no

S. Løkkeborg
e-mail: svein.lokkeborg@imr.no

A. Fernö
Department of Biology, University of Bergen, P.O. Box 7800, 5020 Bergen, Norway
e-mail: anders.ferno@bio.uib.no

M. Hufthammer
Directorate of Fisheries, Region West, P.O. Box 233, 5804 Bergen, Norway
e-mail: marianne.hufthammer@fiskeridir.no

Introduction

Red king crabs (*Paralithodes camtschaticus*) were transferred from the north-western Pacific and introduced at the Murman coast (Russia) of the Barents Sea in the 1960s (Orlov & Ivanov, 1978). Today, the distribution of the king crab stock in the Barents Sea ranges from Cape Kanin (Russia) in the east to Sørøya (Norway) in the west, and crabs are numerous in coastal areas. The management and exploitation of the Barents Sea king crab stock face several challenges related to assessment, harvest and bycatch reduction in gears targeting groundfish. Animal behaviour is central to understanding many aspects of marine fisheries, and is of particular importance in relation to resource assessment and harvest (Fernö & Olsen, 1994). Movements and changes in the distribution of target species in time and space influence their availability to sampling and commercial fishing gears.

Knowledge of the behavioural ecology of this introduced species in its new environment is scarce, and information on small-scale movements, migration and distribution is therefore needed. Walking speed is crucial for the range of the area crabs cover during food search, for the type of potential prey species and for migration capability. Use of space provides information on mobility and home range, and the feeding strategy in a patchy environment.

Acoustic telemetry is a commonly used technique to investigate spatial dynamics of individuals (e.g. Moore & Russell, 2000; Thorstad et al., 2002). Several studies have used ultrasonic telemetry to track movements of crabs (e.g. Wolcott & Hines, 1990; Shirley & Wolcott, 1991; González-Gurriarán & Freire, 1994; Hines et al., 1995); but few on a high-resolution (minutes or a few hours) time scale (but see Hill, 1978; Skajaa et al., 1998). In this study we investigated the spatial and temporal dynamics of the red king crab. The aims of the study given the constraints of a stationary positioning system were thus to investigate (i) the dynamics of walking speed, including diel rhythm and (ii) the small-scale spatial dynamics. Seven individuals were tagged with acoustic transmitters in a fjord in northern Norway and tracked for a period of up to one month. To the best of our knowledge, this is the first high-resolution time scale tracking study of the red king crab.

Materials and methods

The experiment was carried out from 12 July to 13 August 1997 in the Kobbholmfjord, close to the Russo-Norwegian border. The fjord is connected to the Barents Sea through a narrow inlet with a sill of about 35 m depth. The main fjord basin is 155 m deep with mainly clay/mud sediment.

Table 1 The size of the tagged crabs, the operating frequencies of the transmitters and the total number of positional fixes during the study. CL and CW are carapace length and width

Crab	Size (mm)		Tag freq (kHz)	No. positional fixes
	CL	CW		
1	124	148	72	1,940
2	127	140	75	1,961
3	134	157	63	2,667
4	190	223	65.5	1,651
5	149	176	69	2,693
6	145	168	54	2,584
7	175	182	78	1,204

Crabs to be tagged were caught with conical pots baited with herring (*Clupea harengus*) and soaked for two days. Seven crabs (Table 1), all mature males (only males are allowed to be caught in the commercial fishery), were tagged with VEMCO VR 16-4H acoustic transmitters (56 mm long and 16 mm in diameter and weighed approx. 10 g in water). The transmitters had a pulse period of 1,500 ms and each transmitter operated on a unique frequency. The transmitter was attached to the merus of the fourth periopod.

The stationary telemetry positioning system (VRAP, Vemco Ltd., Halifax, Canada) consisted of three hydrophone buoys anchored to the seabed in a triangular configuration. The capture/release site of the crabs was at the centre of the triangle of buoys. The distances between the buoys were between 460 and 704 m. The buoys communicated via radio transmission with a base station onboard a research vessel. The hydrophones were hung at a depth of 100 m (i.e. 55 m above the seabed). The VRAP system used was capable of tracking up to ten transmitters operating at different frequencies. However, the system monitored the frequencies sequentially, 34 s for each frequency on each scan, and the elapsed time between successive positional fixes thus varied with the number of animals being tracked. In this study the location of each crab was monitored at 5 min intervals. The walking speed estimates are the average speed across the time interval between two consecutive positional fixes. Under optimal acoustic conditions a transmitter can be localized to an accuracy of about one metre within the triangle of buoys. When a transmitter is outside the triangle the accuracy decreases.

Data collected during the first 12 h following release of the crabs were excluded from the analysis to reduce the possible influence of capture/release on behaviour. Spurious positional fixes due to background noise or reflected signals were manually removed before data analysis. A positional fix was regarded as spurious if it, when previous and successive fixes were close to each other, deviated markedly from the track followed by the crab. Speed, distance and direction of movement between positional fixes were calculated for all fixes where the time since the last positional fix did not exceed 12 min. These data were used to calculate mean walking speed and longest rectilinear distance (LRD) between any two fixes during periods of one hour. LRD can be considered as the diameter of a circle circumscribing the animal's locations during the specified period and provides a rough measure of the area the animal has covered during the specified time.

To study diel rhythms in walking speed, time series of hourly mean walking speed were analysed by calculating the autocorrelation coefficient (Box et al., 1994) for time lags ranging from one to 24 h. This analysis does not allow for missing values. To establish time series of sufficient length for analysis (arbitrarily set at minimum 48 h), missing values in series, with at the most one missing observation per 24 h, were replaced with the arithmetic mean of the immediately preceding and following observations.

To explore evidence of directionality in the walking directions of the crabs, the set of movement directions between consecutive positional fixes were plotted as a rose diagram for each crab. To test the null hypothesis of uniformity on the circumference of the circle, a Kuiper test was used (Jammalamadaka & SenGupta, 2001).

Results

All tagged king crabs showed large fluctuations in their walking speed during the observational period (Fig. 1). Low speeds were most frequent in that 20–50% of the observations were below 0.01 m s^{-1}. The median walking speed for the whole observational period varied from 0.01 to 0.03 m s^{-1} for the individual crabs (Fig. 2). The maximum speed of individual crabs ranged from 0.08 to 0.15 m s^{-1}, while the maximum median speed over 1 h intervals ranged from 0.06 to 0.11 m s^{-1}.

The crabs usually moved within a relatively restricted area with mean hourly LRD varying from 26 to 64 m. The size of this area for individual crabs ranged from a minimum of 0–5 m to a maximum of 89–216 m. The longest distance a crab actually walked over a period of one hour was 269 m. Although the movements of the crabs did not show any marked directionality (Fig. 3), the distributions for all crabs deviated significantly from uniformity ($P < 0.01$).

Patterns of alternating periods of low and high walking speed were frequently recorded (Fig. 4). The length of these periods varied and they were observed both at day and night. Autocorrelation analysis of hourly walking speed showed significant and consistent correlations for time lags of 1–2 h. Although the data for one of the crabs (crab 6) indicated a significant correlation at a lag of

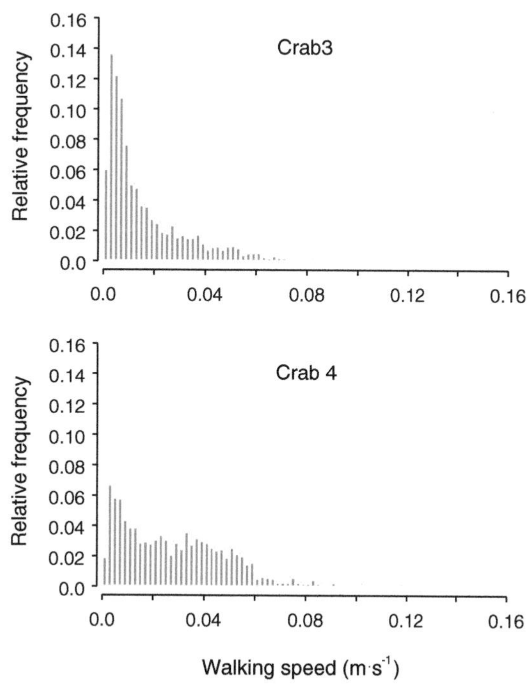

Fig. 1 Examples of frequency distributions of the recorded walking speed observations for the tagged red king crabs (crabs 3 and 4)

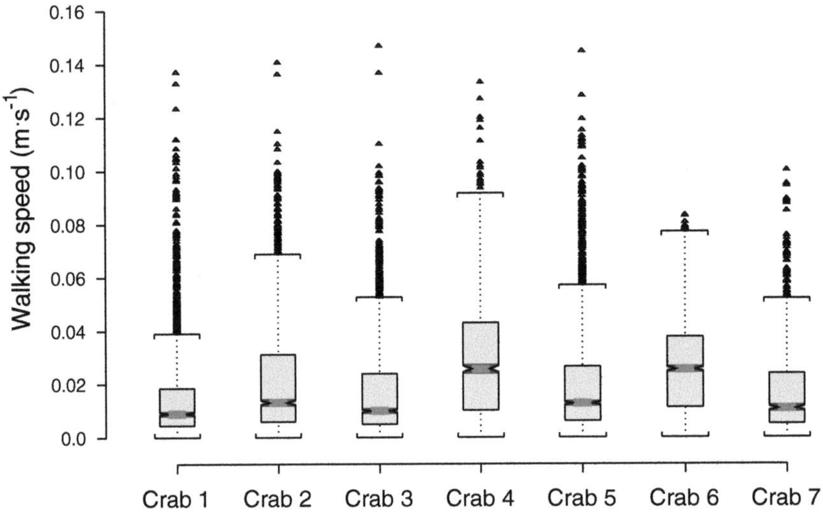

Fig. 2 Box plots of the individual walking speed recordings for the entire experiment. Notches show approximate 95% confidence intervals of the median. Whiskers confine observations not more than 1.5 times the interquartile range from the quartiles

10–14 h, the autocorrelation analyses revealed no consistent diel pattern in walking speed (Fig. 5).

Discussion

This tracking study was conducted several months after the mating season, and adult king crabs of the size tagged presumably experience low predation risk. We studied crabs that had been caught with baited pots indicating that they were actively feeding. Thus the movement pattern observed should primarily reflect feeding behaviour and foraging strategy. Hines et al. (1995) also used optimization of energy needs to explain movement patterns of tracked adult blue crab (*Callinectes sapidus*) and spider crab (*Maja squinado*).

The crabs were often stationary or moved slowly (<0.01 m s^{-1}), and this movement pattern may reflect active feeding. Red king crab is a generalist preying mainly on sedentary prey such as polychaetes and bivalves (Tarverdieva, 1976; Rafter et al., 1996; Gerasimova, 1997). The crab is thus likely to remain stationary when handling and ingesting such prey items. Periods when crabs were stationary or moved slowly may also reflect periods of inactivity. After feeding to satiation and in periods of low feeding motivation, the crab is likely to stay inactive to minimise energy costs of moving. Edible crab (*Cancer pagurus*) has been shown to exhibit long periods of inactivity after periods of feeding activity (Skajaa et al., 1998).

Red king crabs are also capable of moving at high speed. The highest walking speeds observed in our study may reflect crabs moving between food patches. We observed movement patterns that indicate a foraging strategy with alternating staying in and moving between patches (Fig. 4). With a patchy prey distribution, the crab may stay within a limited area until the patch is grazed down. The time a crab stayed varied between one and several hours and autocorrelation analysis of hourly walking speed showed consistent correlations for time lags of 1–2 h.

The highest walking speed recorded was 0.15 m s^{-1} while the highest hourly median speed was 0.11 m s^{-1}. The actual speed is presumably even higher because the true path of the crab between successive fixes likely deviates from a straight line (see Løkkeborg et al., 2002). A speed of 0.11 m s^{-1} indicates that king crabs could walk about 400 m in one hour, and the longest recorded distance in this experiment was 269 m. These observations coincide well with the results of tagging studies of red king crabs in the northeastern Pacific, showing a maximum rate of migration for males of 13.1 km/day (0.15 m s^{-1}) (Marukawa, 1933; cited by Stone et al., 1992).

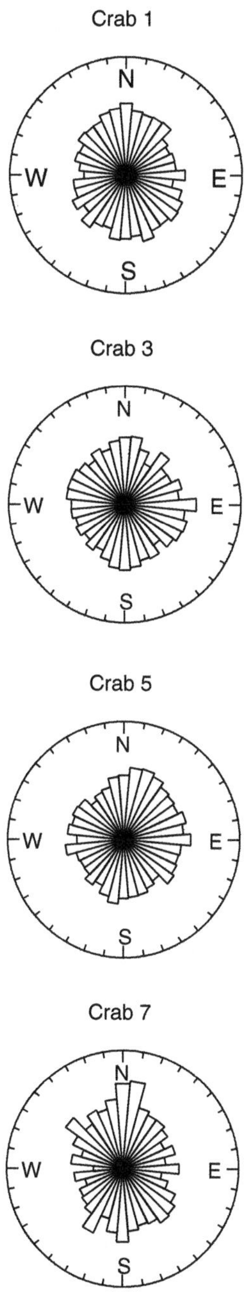

Fig. 3 Distribution of the set of recorded walking directions for each of the tagged crabs. The rose diagrams use a binwidth of 10 degrees

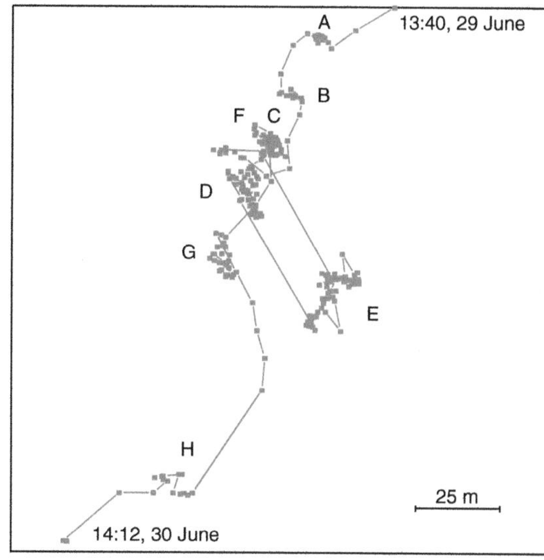

Fig. 4 The recorded track of crab 3 from 13:12 h 29 June to 11:03 h 30 June. Note the pattern of shifts between rapid movement and periods of slow or no movement. The letters A–G show the chronological order of locations where the crab stayed for extended periods; (A) 1 h 6 min, (B): 55 min, (C) 2 h 14 min, (D) 4 h 2 min, (E) 5 h 28 min, (F) 1 h 31 min, (G) 2 h 3 min. Individual position fixes are plotted as squares. Time interval between neighbouring position fixes was between 5 and 6 min for all but two fixes for which it was 10 min (the latter connected by hatched lines)

Walking speeds exceeding 0.08 m s^{-1} have been observed in the laboratory for king crabs following bait odours (Zhou & Kruse, 2000). For smaller crab species, Hill (1978) observed *Scylla serrata* to move at speeds up to 0.06 m s^{-1} (66 m in 30 min) and Skajaa et al. (1998) reported a mean speed of 0.05 m s^{-1} for *Cancer pagurus* during reactions to odour plumes.

The autocorrelation analysis presented no evidence of a consistent diel rhythm in activity. The observations were conducted during the polar summer when light levels are relative high at night, and a clear activity rhythm associated with a cyclic change in light level is thus less likely. Cod showed less pronounced rhythm in swimming activity during late spring with higher light levels at night compared to fall (Løkkeborg, 1998; Løkkeborg & Fernö, 1999). Furthermore, chemical senses play an important role in food search of crustaceans (Hazlett, 1968; Fuzessery & Childress, 1975; Ache et al., 1976; Rittschof, 1992), and king crab feeds mainly on sedentary prey that should be easy to locate and catch even at night. Several decapod crustaceans have been shown to exhibit a nocturnal activity rhythm (e.g. Atkinson, 1974; Hill, 1976, 1978; Rebach, 1985; Bjordal,

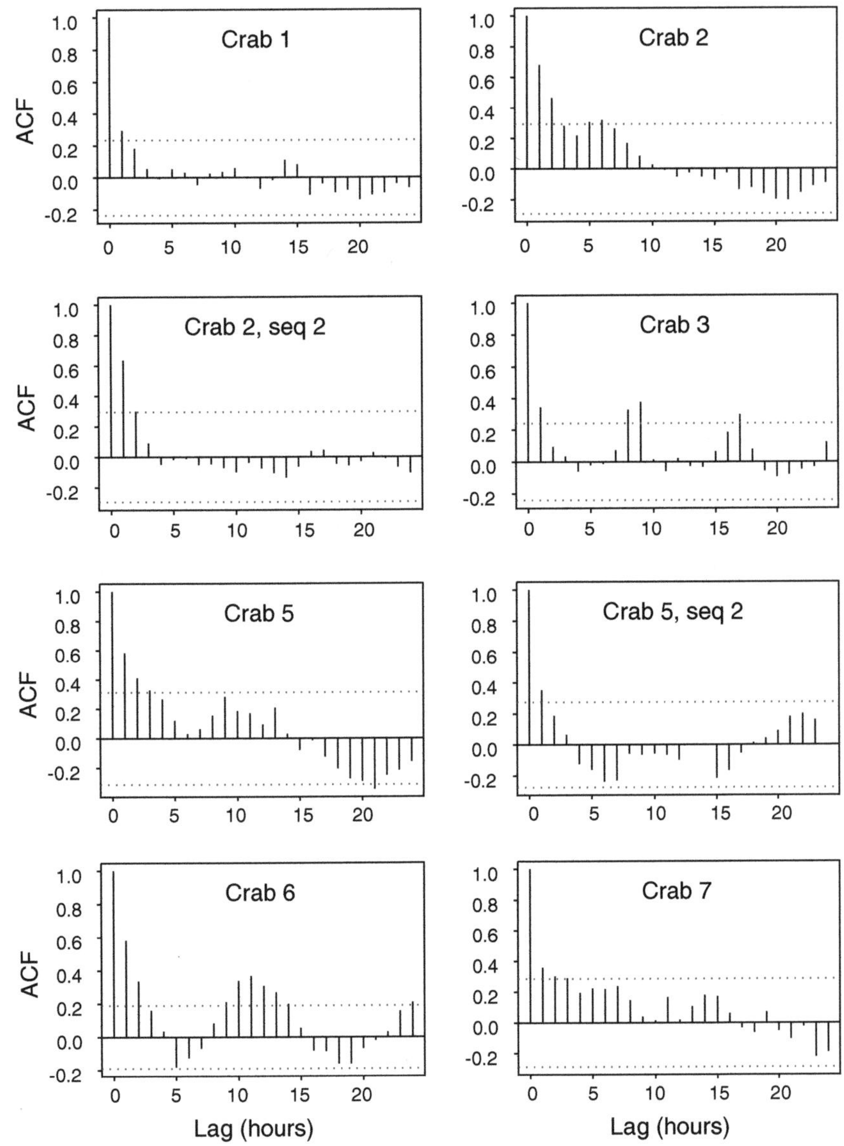

Fig. 5 Plots of the autocorrelation function (ACF) of median hourly swimming speed for time lags of 1–24 h. The horizontal dotted line shows correlations that would be significant at 5‰. For each crab, the analysis is based the longest continuous time series of recorded hourly walking speed (see text for details). The time series varied between 48 and 70 h. No analysis was made for crab 4 with the longest time series of only 35 h

1986; Jernakoff, 1987; Chatterton & Williams, 1994), and the nocturnal rhythm of the edible crab was explained by reduced risk of predation from visual predators (Skajaa et al., 1998). Adult king crab of the size tracked in this study should experience low predation, and no potential predators were observed at the experimental site. Thus, the king crabs tracked in this study were not likely to gain increased feeding success or decreased predation risk by exhibiting a diel activity rhythm.

The diameter of the area within which the individual crabs moved during a period of one hour ranged from a few to about 200 m. Crabs could occasionally move and stay out of range of the positioning system for periods of several days. The range of this system is about one kilometre, and the crabs may thus have moved over a considerably larger area during the observational period of one month. Although the movement directions deviated significantly from uniformity for all crabs, no marked and consistent directionality was seen. This is to be expected from the topographic and oceanographic conditions; the study area was located at the centre of a fjord bottom plain and currents were below the recordable minimum value (0.02 m s^{-1}) for the current meter used (SD6000, www.sendordata.net).

The findings of this study have implications for king crab stock assessment and measures to mitigate bycatches in demersal fisheries targeting ground fish. The tagged king crabs were active throughout the 24-h cycle and the probability of encountering both active and passive gears should be independent of time of day. The wide range of the area occupied by the crabs and their capability of walking long distances within relatively short periods of time mean that encounter rates with sampling gears used in assessment surveys are high. On the other hand, because the king crabs did not exhibit a diel rhythm and altered between relatively short periods of low and high activity, there is no time window in summer when the probability of encounters with commercial gears is low. Hence, efforts to reduce interactions between king crab and fishing gears should focus on designing gears with low catching efficiency for crab, e.g. floating pots and gillnets off the bottom (Godøy et al., 2003a, b).

Acknowledgements The authors are indebted to Bente K. H. Ulvestad for skilful assistance in the screening and analysis of the data. We also thank the two anonymous referees for helpful comments.

References

Ache, B. W., Z. M. Fuzessery & W. E. S. Carr, 1976. Antennular chemosensitivity in the spiny lobster, *Panulirus argus*: comparative tests of high and low molecular weight stimulants. Biology Bulletin 151: 273–282.

Atkinson, R. J. A., 1974. The activity rhythm of *Goneplax rhomboids* (L.). Marine Behaviour and Physiology 2: 325–335.

Bjordal, Å., 1986. The behaviour of Norway lobster towards baited creels and size selectivity of creels and trawls. Fiskeridirektoratets Skrifter Serie Havundersøkelser 18: 131–137.

Box, G. E. P., G. M. Jenkins & G. C. Reinsel, 1994. Time Series Analysis: Forecasting and Control, 3rd edn. Holden-Day.

Chatterton, T. D. & B. G. Williams, 1994. Activity patterns of the New Zealand cancrid crab *Cancer novaezelandiae* (Jacquirot) in the field and laboratory. Journal of Experimental Marine Biology and Ecology 178: 261–274.

Fernö, A. & S. Olsen, 1994. Marine fish behaviour in capture and abundance estimation, Fishing News Books, 221 pp.

Fuzessery, Z. M. & J. J. Childress, 1975. Comparative chemosensitivity to amino acids and their role in the feeding activity of bathypelagic and littoral crustaceans. Biology Bulletin 149: 522–538.

Gerasimova, O. V., 1997. Analysis of king crab (*Paralithodes camtschatica*) trophic links in the Barents Sea. International Council for the Exploration of the Sea Council Meeting 1997/GG:3.

Godøy, H., D. Furevik & S. Løkkeborg, 2003a. Reduced bycatch of red king crab (*Paralithodes camtschaticus*) in the gillnet fishery for cod (*Gadus morhua*) in northern Norway. Fisheries Research 62: 377–384.

Godøy, H., D. Furevik & S. Stansen, 2003b. Unaccounted mortality of red king crab (*Paralithodes camtschaticus*) in deliberately lost pots off Northern Norway. Fisheries Research 64: 171–177.

González-Gurriarán, E. & J. Freire, 1994. Movement patterns and habitat utilization in the spider crab *Maja squinado* (Herbst) (Decapoda, Majidae) measured by ultrasonic telemetry. Journal of Experimental Marine Biology and Ecology 184: 269–291.

Hazlett, B. A., 1968. Effects of crowing on agonistic behavior of hermit crab *Pagurus bernhardus*. Ecology 49: 573–575.

Hill, B. J., 1976. Natural food, foregut clearance rate and activity of the crab *Scylla serrata*. Marine Biology 34: 109–116.

Hill, B. J., 1978. Activity, track and speed of movement of the crab *Scylla serrata* in an estuary. Marine Biology 47: 135–141.

Hines, A. H., T. G. Wolcott, E. González-Gurriarán, J. L. González-Escalante & J. Freire, 1995. Movement patterns and migrations in crabs: telemetry of juvenile and adult behaviour in *Callinectes sapidus* and *Maja squinado*. Journal of the Marine Biological Association of the United Kingdom 75: 27–42.

Jammalamadaka, S. Rao & A. SenGupta, 2001. Topics in Circular Statistics. Series on Multivariate Analysis, 5. World Scientific Press, Singapore 322 pp.

Jernakoff, P., 1987. Foraging patterns of juvenile western rock lobster *Panulirus Cygnus* George. Journal of Experimental Marine Biology and Ecology 113: 125–144.

Løkkeborg, S., 1998. Feeding behaviour of cod, *Gadus morhua*: Activity rhythm and chemically mediated food search. Animal Behaviour 56: 371–378.

Løkkeborg, S. & A. Fernö, 1999. Activity pattern and food search behaviour in cod, *Gadus morhua*. Environmental Biology of Fishes 54: 345–353.

Løkkeborg, S, A. Fernö & T. Jørgensen, 2002. Effect of position-fixing interval on the estimated swimming speed and movement pattern on fish tracked with a stationary positioning system. Hydrobiologia 482: 259–264.

Moore, A. & I. Russell, 2000. Advances in Fish Telemetry. Proceedings of the Third Conference on Fish Telemetry in Europe, Held in Norwich, England, 20–25 June 1999. CEFAS, Lowestoft. 264 pp.

Orlov, Y. I. & B. G. Ivanov, 1978. On the introduction of the Kamchatka king crab *Paralithodes camtschatica* (Decapoda: Anomura: Lithodidae) into the Barents Sea. Marine Biology 48: 373–375.

Rafter, E., E. M. Nilssen & J. H. Sundet, 1996. Stomach content, life history, maturation and morphometric parameters of red king crab, *Paralithodes camschaticus*, from Varangerfjord area, North Norway. International Council for the Exploration of the Sea Council Meeting 1996/K:10.

Rebach, S., 1985. Rhytmicity under constant conditions in the rock crab, *Cancer irrotatus*. Bulletin of Marine Science 36: 454–466.

Rittschof, D., 1992. Chemosensation in the daily life of crabs. American Zoologist 32: 363–369.

Shirely, M. A. & T. G. Wolcott, 1991. A telemetry study of microhabitat selection by premolt and molting blue crabs, *Callinectes sapidus* (Rathbun), within a subestuary of the Pamlico River, North Carolina. Marine Behaviour and Physiology 19(2): 133–148.

Skajaa, K., A. Fernö, S. Løkkeborg & E. K. Haugland, 1998. Basic movement pattern and chemo-oriented search towards baited pots in edible crab (*Cancer pagurus* L.). Hydrobiologia 372: 143–153.

Stone, R. P., C. E. O'Clair & T. C. Shirley, 1992. Seasonal migration of female red king crabs in a southeast Alaskan estuary. Journal of Crustacean Biology 12(4): 546–560.

Tarverdieva, M. I., 1976. Feeding of the Kamchatka king crab *Paralithodes camtschatica* and tanner crabs *Chionoecetes bairdi* and *Chionoecetes opilio* in the southeastern part of the Bering Sea. Soviet Journal of Marine Biology 2: 34–39.

Thorstad E. B., I. A. Fleming & T. F. Næsje, 2002. Aquatic telemetry. Proceedings of the Fourth Conference on Fish Telemetry in Europe. Hydrobiologia 483, 292 pp.

Wolcott, T. G. & A. N. Hines, 1990. Ultrasonic telemetry of small-scale movements and microhabitat selection by molting blue crabs (*Callinectes sapidus*). Bulletin of Marine Science 46(1): 83–94.

Zhou, S. & G. H. Kruse, 2000. Capture efficiency and size selectivity of two types of pots for red king crabs in the Bering Sea. Alaska Fisheries Research Bulletin 6: 94–103.

Hydrobiologia (2007) 582:25–34
DOI 10.1007/s10750-006-0563-8

FISH TELEMETRY

Movements of undersized spotted grunter (*Pomadasys commersonnii*) in the Great Fish Estuary, South Africa: implications for fisheries management

Tor F. Næsje · Amber-Robyn Childs · Paul D. Cowley · Warren M. Potts · Eva B. Thorstad · Finn Økland

Abstract The purpose of this study was to test the hypothesis that an estuarine species as the spotted grunter *Pomadasys commersonnii* is evenly distributed within the estuary and uniformly exploited by different groups of fishers. This was done by investigating the movements and area use of spotted grunter in relation to the fisheries. The position of 20 acoustically tagged fish was recorded during 36 days in February and March 2003, by manual tracking and automated data logging receivers. Information on the fisheries in the estuary was collected through visual registration of fishing effort and interviews. The spotted grunter moved on average 1.0 km (SD ± 0.7) between positional fixes. Most of the fish were positioned in the lower part of the 12 km long estuary, as 70% of the positional fixes were within the first 3 km, and 89% within the first 6 km. Approximately half (49%) of all the fixes were between 1.0 and 1.5 km from the estuary mouth. The spotted grunter used on average 4.9 km (SD ± 4.9) of the length of the estuary, and there was no significant relationship between the length of the estuary used and the body length of the fish (26–39 cm TL). Subsistence fishers accounted for 73% of fishing lines in the water, while recreational fishers accounted for the rest. Ninety-three percent of the lines were recorded within the first 6 km from the estuary mouth, of which 80% were recorded within the first 3 km. Almost 1/3 of the fishing effort was recorded between 1.0 and 1.5 km from the mouth. The hypothesis that the estuarine dependent species spotted grunter was evenly distributed within the estuary was rejected. However, there was a significant relationship between the distribution of the fishing effort of the subsistence fishers and the fish, indicating that the spotted grunter was uniformly exploited within the estuary by this group of fishers. In contrast, there was no relationship between the distribution of fish and recreational anglers.

Keywords Movement behavior · *Pomadasys commersonnii* · Area use · Fisher distribution · Catch · Fisheries management

Guest editors: P. R. Almeida, B. R. Quintella, M. J. Costa and A. Moore
Developments in Fish Telemetry

T. F. Næsje (✉) · E. B. Thorstad · F. Økland
Norwegian Institute for Nature Research (NINA), Tungasletta 2, NO-7485 Trondheim, Norway
e-mail: tor.naesje@nina.no

A.-R. Childs
Department of Ichthyology and Fisheries Science (DIFS), Rhodes University, P.O. Box 94, Grahamstown 6140, South Africa

P. D. Cowley · W. M. Potts
South African Institute for Aquatic Biodiversity (SAIAB), Private Bag 1015, Grahamstown, South Africa

Introduction

Estuaries are productive ecosystems that provide numerous fishery opportunities and food for people. In South Africa, temporarily resident estuarine fish populations are exploited for both food, mainly subsistence, and recreation. Little is known about the status of estuarine fisheries, but it is widely believed that they have contributed to the over-exploitation of several species, which are also important components of other coastal fisheries (Cowley et al., 2004; Griffiths, 1997). Due to the paucity of information regarding estuarine fishing and its impacts, sustainable utilization and effective management of the resources are currently difficult (Boyd et al., 2000).

The Great Fish Estuary in the Eastern Cape Province of South Africa is situated in an economically depressed rural area with high unemployment rates, resulting in a high dependence on estuarine fishery resources by subsistence fishers. This sector was formally recognized for the first time in 1998 (Marine Living Resources Act). Inhabitants generally consist of the historically disadvantaged people who use low technology gear to catch fish, predominantly for their own consumption. Several of the primary targeted fishery species are also caught by recreational anglers. Consequently, the potential for growth and development of the subsistence sector appears to be limited (Lamberth & Turpie, 2003). According to these authors, 14 of the 80 utilized estuary-associated species found along the South African coastline are considered overexploited, including the spotted grunter (*Pomadasys commersonnii*, Lacepede 1801). Since 1992, spotted grunter has been categorized as a recreational species (i.e., may not be sold). This is primarily because of its inshore distribution and estuarine dependence, hence its greater vulnerability to exploitation relative to other, more marine species (Fennessy, 2000). The species is currently managed by means of a daily bag limit (five fish per person per day) and a minimum legal size limit (40 cm total length (TL)). The effectiveness of these fisheries management measures has never been evaluated, nor have alternative measures such as closed seasons or area closure (e.g., protected areas) been considered.

The purpose of this study was to test the hypothesis that an estuarine species as the spotted grunter *Pomadasys commersonnii* is evenly distributed within the estuary and uniformly exploited by different groups of fishers. This was done by investigating the movement behavior of legally undersized (<40 cm) spotted grunter by telemetry methods, and by testing the effects of the distribution of the fish on the spatial distribution of fishing effort.

Study area

Great Fish River catchment

The 650 km long Great Fish River enters the Indian Ocean at 33°29′28″ S and 27°13′06″ E (Fig. 1). The river system has a catchment area of approximately 30,300 km^2, and a mean annual runoff of 525×10^6 m^3 yr^{-1} (Vorwerk et al., 2003).

The estuary mouth is permanently open and maintained by enhanced freshwater inputs from an inter-basin transfer system located on the Orange River (Vorwerk et al., 2003), accounting for continuous nutrient inputs and, hence, elevated phytoplankton production. The main channel in the mouth region of the estuary is approximately 30 m wide and is restricted by the presence of extensive sand banks (Cowley &

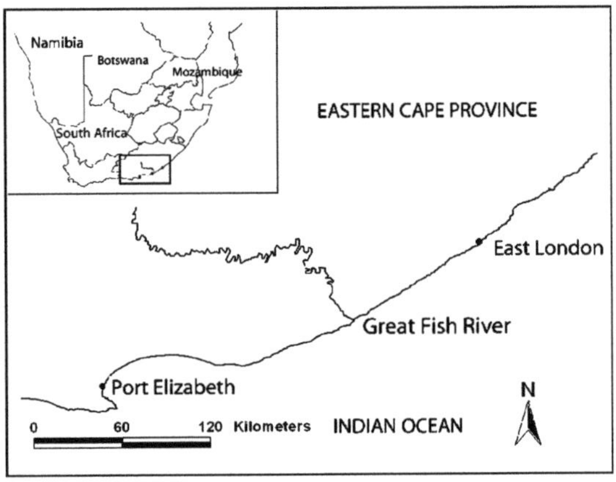

Fig. 1 Map of South Africa, with insert showing the location of the Great Fish River in the Eastern Cape Province

Daniel, 2001). Following flood events, the main channel can be up to 200 m wide.

The estuary is mostly shallow, ranging between 1 m and 2 m (mean 1.4 m), except for some areas in the lower and upper reaches that have depths of up to 3 and 6 m, respectively (Cowley & Daniel, 2001). The estuary is highly influenced by the tide with a tidal prism of 1.6×10^6 m^3 (Allanson & Read, 1987). The surface area of the estuary is approximately 1.95 km^2.

The Great Fish Estuary fisheries

A study on the catches of subsistence and recreational fishers was performed from October 2003 to March 2004 in the Great Fish Estuary (Potts et al., 2005). Spotted grunter dominated the catches in terms of number and mass, followed by dusky kob (*Argyrosomus japonicus*) and white seabarbel (*Galeichthys feliceps*).

Of the spotted grunter captured, 85% were retained. Subsistence fishers captured most spotted grunter both in terms of numbers (67%) and mass (74%) (Potts et al., 2005). A relative large proportion of the landed catch comprised small individuals. Of the retained catch, 29% of the spotted grunter was below the legal size limit of 40 cm TL.

Angler bag frequencies (number of fish caught per person per day) varied largely between fishers (Potts et al., 2005). The maximum daily catch of spotted grunter by an individual angler was 14 fish. Of the 1,157 fishermen interviewed, most did not catch a spotted grunter (70%) during a daily outing. Of the fishers that did capture spotted grunter, a catch rate of only one fish per angler per day was most frequently observed. The bag limit of five per day was exceeded on only 3% of fisher outings (Potts et al., 2005).

Materials and methods

Tagging of fish

The transmitters used (V8SC-2L-R256 coded pingers, VEMCO Ltd, Halifax, Canada), with an expected battery life of 112 days, were 8.5 mm in diameter, 28 mm in length and weighted approximately 3.1 g in water being on average 0.75% of the fish body weight (min.: 0.5%, max.: 1.6%). These pingers (69 kHz) emitted acoustic pulse trains randomly every 5–15 s.

Twenty spotted grunter (mean = 336 mm ± 32.5 SD TL, range 263–387 mm) were tagged in the Great Fish Estuary between 21 January and 1 February 2003 (Table 1, Fig. 2). The fish were caught with rod and line using mud prawn (*Upogebia africana*) or sand prawn (*Callianassa kraussi*) as bait. Captured fish were immediately placed in a 50 l tank with estuary water containing 2-phenoxy ethanol (approximately 1.0 ml l^{-1}). During surgery, the gills were continuously flushed with estuarine water without anesthetics. The incision in the body cavity was closed using two independent silk sutures (2/0 Ethicon). The duration of the surgical procedure was on average 2 min 48 s, after which the fish was placed in a recovery bath, and released at the catch site once they had recovered. The time from the fish was placed in anesthetic bath to it was released was on average 17 min (range 8 min–41 min), with the exception of one fish that was kept longer in the recovery bath (11 h 27 min).

Spotted grunter from KwaZulu-Natal attain 50% sexual maturity at lengths of 300 mm (standard lengths (SL)) for males and 360 mm (SL) for females (Wallace, 1975). Similarly, Webb (2002) found that males from the South Eastern Cape attained 50% maturity at 305 mm TL. It is not possible to externally sex spotted grunter, but according to their body length, the tagged individuals might have consisted of both mature and immature fish. All the fish studied, however, were below the legal size limit for fisheries (<40 cm TL).

Tracking of fish

Two methods were used to monitor spotted grunter movements. Fish positions were recorded by manual tracking from a boat using a VEMCO VR60 and a directional hydrophone. Additionally, four automated data logging receivers, VEMCO VR2 were used to monitor the presence or absence of individual fish within an omnidirectional range of the receivers.

Table 1 Transmitter code, body length (fork and total length), tagging date, number of positions manually recorded, average number of days between fixes, and last date manually tracked for the spotted grunter tagged with acoustic transmitters in the Great Fish River estuary in January and February 2003

Fish code	Fork length (mm)	Total length (mm)	Tagging date	Date last manually recorded	Number of fixes manual tracking	Average number of days between fixes
20	297	317	21/01/03	16/03/03	25	1.5
21	307	334	21/01/03	23/03/03	34	1.3
22	271	297	01/02/03	06/03/03	20	1.4
23	354	380	01/02/03	24/03/03	36	1.3
24	304	330	01/02/03	28/02/03	18	1.2
25	284	313	01/02/03	03/03/03	19	1.3
26	291	314	01/02/03	24/03/03	35	1.3
27	300	328	01/02/03	18/02/03	12	1.0
28	354	382	01/02/03	24/03/03	34	1.4
29	346	377	27/01/03	24/03/03	36	1.3
30	282	308	01/02/03	24/03/03	36	1.3
31	330	357	27/01/03	09/03/03	12	2.7
32	293	318	01/02/03	24/03/03	36	1.3
33	300	329	27/01/03	09/03/03	13	2.5
34	256	263	21/01/03	17/02/03	8	1.4
35	330	357	27/01/03	17/02/03	8	1.4
36	358	387	27/01/03	28/02/03	15	1.4
37	344	363	21/01/03	23/03/03	23	1.8
38	296	319	26/01/03	21/02/03	12	1.3
39	328	355	01/02/03	24/03/03	36	1.3

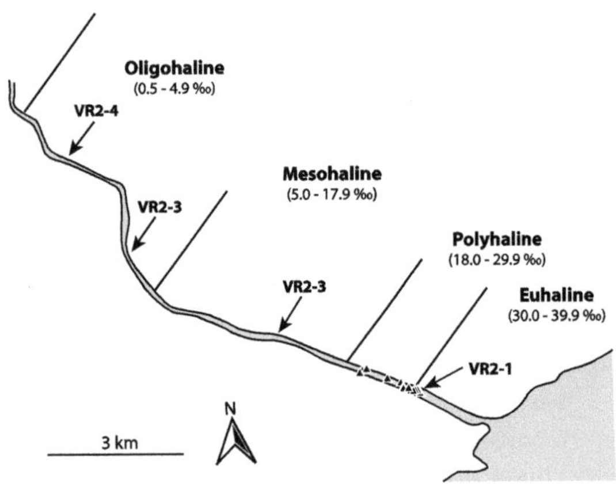

Fig. 2 Great Fish Estuary divided into salinity regions according to Venice system based on mean bottom salinities. Arrows indicate the positions of the automated data-logging receivers (VR2 1–4). Triangles indicate the catch and release sites of the 20 tagged spotted grunters. The total length of estuary is 12 km

Manual tracking

Tracking took place between January and March 2003, in the form of two 16 day sampling periods with daily tracking, and an interim period of 16 days, during which fish were tracked every third day. The sampling periods were 7–22 February (daily), 25 February–9 March (every third day) and 9–24 March (daily); a total of 36 days. The daily route tracked was standardized doing a return trip that started at the mouth of the estuary and turning at the head region of the estuary (12 km from the mouth). Tracking started at the same time every day (9 h) and continued until all fish were found, or, if not, until 18 h. When a tagged fish was located, positional fixes were taken using a GPS. The reception range of the VEMCO VR60 was approximately the same as for the VEMCO VR2, and under normal tracking conditions the minimum range exceeded the width of the estuary.

Automated data-logging receivers

The four automated data-logging receivers (VEMCO VR2) were situated along the estuary 1.0, 3.7, 7.6 and 10.3 km from the mouth (Fig. 2). The reception range of each logging station varied between 310 and 600 m, but at times and certain directions might have been considerably less due to variation in wave action, salinity, depth, and physical obstructions such as the road bridge pylons, large rocks and deep holes. The lowermost receiver (VR2-1, Fig. 2) could not be situated in the estuary mouth due to the strong currents and wave actions in this area. Since the transmitters had a life expectancy of 112 days, the automated receivers were removed from the estuary on 16 April 2003.

Independence of positional fixes obtained by manual tracking

There was no significant correlation between distance moved between consecutive positional fixes by individual fish and the number of days between the fish was positioned at these fixes for 17 of the 20 fish tagged. For the three remaining individuals, there was a significant correlation between distance moved and the numbers of days between consecutive fixes. However, the coefficient of determination was weak for two of these fish (linear regression, $P = 0.04$, $R^2 = 0.13$; $P = 0.05$, $R^2 = 0.11$, respectively). For the third fish, the distances between fixes were longer when it was positioned every third day than daily ($P = 0.004$, $R^2 = 0.33$). For the pooled data set there was no significant relationship between number of days between fixes and distance moved between the corresponding positional fixes (log transformed) (linear regression, $P = 0.26$, $R^2 = 0.003$). Based on this, we pooled the data of individual positioning within the estuary.

Fishery data

Aspects of the recreational and subsistence fishery on the estuary were visually recorded from the manual tracking boat between January and March 2003. The following data were collected: number of lines in the water, number of fishers, fishery sector (recreational or subsistence), whether they were fishing from shore or boat, and their location on the estuary (using a GPS). As individual fishers often used more than one rod and/or hand line, fishing effort was recorded as number of lines used as each line only had one hook. Classification of fishers to the recreational or subsistence sector was based on interviews or personal knowledge of the fishers. The distribution of fishing effort of the different groups of fishers was sorted in 500 m zones from the mouth of the Great Fish Estuary. The effects of the distribution of the fish on the fishing effort of the different groups of fishers was tested.

Results

Movements between river, estuary and sea

Six fish (30%) were resident in the estuary and remained in the estuary every day during manual tracking (35 or 36 fixes, Table 1). Two fish (10%) were mainly resident in the estuarine environment, but undertook short sea trips during the manual tracking period, and each trip lasted for only a day. Eleven fish (55%), however, left the estuary during manual tracking and remained in the marine environment. Ten of the 11 fish left the estuary between 11 and 29 days after being tagged, and one fish left the estuary after 45 days. One individual moved between the marine, estuarine and riverine environments and finally disappeared after being recorded on the uppermost receiver.

Proportion of time spent in the estuary

Altogether, 50% of the fish made return-trips to the sea, and the number of return trips for individual fish varied between zero and five (mean 0.8 ± 1.27 SD). Nine fish made return trips before permanently leaving the estuary, while four fish left the estuary for good the first time they went to sea. The spotted grunter spent on average 66% of the period with manual tracking (36 days) in the estuary, being in the marine environment for the rest of the time.

There was no relationship between the fish length (mm) and total number of fish positions in the estuary ($P = 0.30$; $R^2 = 0.06$). Hence, movements to sea could not be linked to fish length.

Although the duration of the primary study lasted 48 days, information obtained at a later date from (i) fish recaptured in the estuary, and (ii) fish registered during a later telemetry study, indicated that fish return to, or use, the estuary for prolonged periods. Two of the individuals that migrated to sea were caught by anglers in the estuary 11 and 17 months after tagging. Further, three other fish that were resident in the estuary were recorded in the estuary during another study seven months after tagging. Two of these were recorded up to 24 months after tagging.

Movements relative to catch site

The average distance from the initial catch site to each positional fix in the estuary was 1.49 km (individual means from 0.07 to 6.45 km, SD = ±2.32 km). More than fifty percent of the fish (11 individuals) were on average located less than 1 km from their respective catch sites. There was no significant relationship between the average distance between each fix and the catch site (log transformed) and fish length ($P = 0.37$, $R^2 = 0.05$).

Movements between positional fixes

The average distance moved between positional fixes was 1.01 km (individual means from 0.06 to 2.67 km, SD = ±0.65). There was no significant relationship between fish length (mm) and the average distance moved between positional fixes (linear regression: $P = 0.87$; $R^2 = 0.002$).

Distribution in the estuary

Most of the tagged fish were positioned in the lower part of the estuary, as 70% of the positional fixes were located within the first 3 km of the estuary, and 89% of the positions were within the first 6 km (Fig. 3). Approximately half (49%) of the total observations were recorded between 1.0 and 1.5 km from the estuary mouth.

The uppermost automated data logging receiver (VR2-4) situated 10.3 km from the mouth confirmed that very few fish ever entered the riverine environment. Only four individuals (20%) were recorded on this receiver, while only two of these were manually tracked above the receiver.

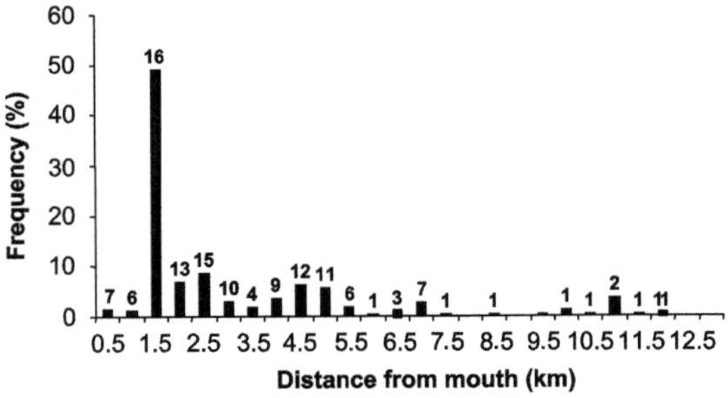

Fig. 3 Distribution of positional fixes ($n = 468$) of tagged spotted grunters per 500 m zones within the Great Fish Estuary between February and March 2003. Numbers above the bars indicate number of individuals recorded within each zone

Estuary stretch use

The total length of the estuary used by individual fish can be described by the distance between the furthermost positions registered during manual tracking. The length of estuary used was on average 4.93 km (individual means from 0.15 to 11.99 km, SD = ±2.70 km). The majority of fish (15 individuals) utilized between 3.2 and 7.1 km of the estuary, while three fish used less and two fish more. There was no significant relationship between the fish length and length of estuary used (linear regression: $P = 0.91$, $R^2 = 0.001$).

Fisher composition and effort

The total number of fishing lines recorded over the study period was 1441. Subsistence fishers, all fishing from shore, accounted for most lines (73%). Recreational fishers accounted for the remaining 27%, of which 17% were fishing from the shore and 10% from boats.

Distribution of fishing effort

Within the estuary, 93% of the lines were recorded within the first 6 km, of which 80% were recorded within the first 3 km. The highest concentration of effort (28%) was recorded between 1.0 and 1.5 km from the mouth.

Recreational boat anglers (57 boats) were recorded from the mouth to 7.5 km up the estuary (Fig. 4). Within this area, 99% of boat fishing effort occurred within the first 6 km, of which 67% was within the first 3 km. Although more than a third of the boat fishing effort occurred within the first 500 m, the remaining effort was more evenly distributed along the estuary.

Recreational shore anglers were mainly recorded (94%) within 1.5 km of the estuary mouth with little effort ($n = 14$ lines) recorded further up the estuary (Fig. 4). Most of the effort was recorded around the road bridge approximately 1 km from the estuary mouth as 39% of the lines were observed within the 500 m zone

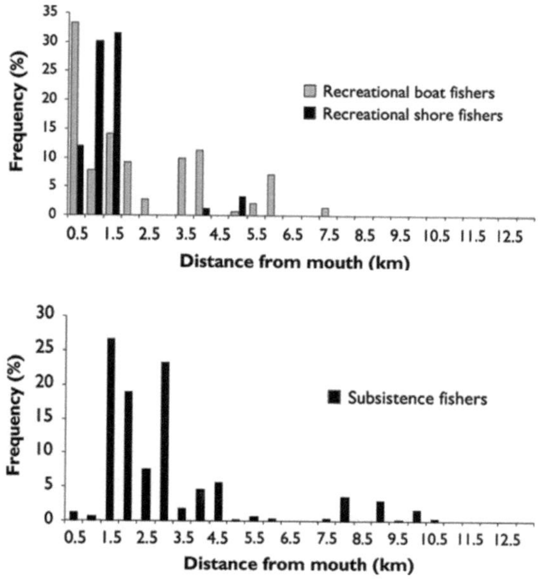

Fig. 4 Distribution of fishing effort (total number lines in the water) divided into groups of recreational boat fishers and recreational shore fishers (top), and subsistence fishers all fishing from the shore (bottom), in February and March 2003. The effort is sorted in 500 m zones from the mouth of the Great Fish Estuary

below the bridge, and 40% within the 500 m zone above the bridge.

The shore-based subsistence fishing effort was recorded from the mouth to 10.5 km upriver (Fig. 4). Ninety-one percent of all lines recorded were within the first 6 km of the estuary, of which 78% were recorded within the first 3 km.

Distribution of spotted grunter and fishers

There was a significant relationship between the distribution of fish (expressed as the number of positional fixes per 2.5 km length of estuary) and the distribution of subsistence fishers along the estuary (linear regression, $P = 0.013$, $R^2 = 0.21$). There was, however, no relationship between the distribution of fish and (i) recreational shore anglers ($P = 0.60$, $R^2 = 0.041$) and (ii) recreational boat anglers ($P = 0.94$, $R^2 = 0.001$).

Discussion

Spotted grunters are wholly dependent on estuaries as juveniles and are thought to vacate these nursery habitats with the onset of sexual maturity (Whitfield, 1994a). The findings of this study revealed that spotted grunter made extensive use of both the estuarine and marine environments. Almost half (45%) of the fish made return trips to the sea during the study and 55% of the tracked fish were at sea at the end of the 48 day study period.

Spotted grunters are known to spawn off the KwaZulu-Natal coastline from September to December, extending until February (Harris & Cyrus, 1997, 1999; Wallace, 1975; Wallace & van der Elst, 1975). Due to the lack of evidence that spawning occurs outside of KwaZulu-Natal it has been suggested that all South African spotted grunters comprise a single stock (Webb, 2002). The movement of fish into the marine environment four to five months prior to the start of the spawning season may be linked to an ontogenetic shift in habitat in connection with sexual maturation, and may represent the beginning of a more marine phase in the life history of spotted grunter. An alternative explanation, however, may be that the individuals that left the estuary were sexually mature and the reason for departure was to initiate their spawning migration.

Spotted grunter may stay in estuaries for prolonged periods and return after some time at sea. The recorded presence of 25% of the tagged fish in the Great Fish Estuary 12 to 24 months after being tagged, suggests that spotted grunter display fidelity towards their nursery estuaries. These findings are supported by conventional tagging studies compiled by Bullen & Mann (2000, 2004) indicating that tagged spotted grunters are largely resident to certain estuaries. The majority of the recaptured fish (>80%) were caught in close proximity to the initial release site (<3 km). However, some recaptured fish showed fairly extensive movements in excess of 50 km (Bullen & Mann, 2000, 2004).

Based on this study it is evident that a significant proportion of the fish in the studied size category (26–39 cm TL) spent time at sea during late summer and autumn. This information may have implications for the management of this important fishery species, for example, for the estimation of the estuarine stock size.

The distribution and movements of fish within an area exposed to extensive fisheries, as the Great

Fish Estuary, have implications for the choice of management regulations to ensure sustainable resource utilization. Movements of spotted grunter within the estuary were characterized by high variability, and there was no significant relationship between the fish length and average distance moved between fixes. While being in the estuary, the maximum distance between positional fixes varied considerably among individuals. The majority of the fish (75%), however, utilized between 3.2 and 7.1 km of the estuary. Despite this, most of the positional fixes were in the lower part of the estuary.

When capturing fish for this study, the fishing effort was spread over the estuary. Most fish, however, were caught and tagged in the lower reaches where catch rates were highest. We therefore do not attribute the river stretch use of the fish to fidelity to the catch site, but rather to the elevated abundance of fish in the lower reaches. This might be due to a more favorable environment caused by external factors as for example turbidity (Cyrus & Blaber, 1987), salinity (Whitfield, 1994b), temperature (Marshall & Elliot, 1998), availability of food resources (Marais, 1984; Whitfield, 1988) and/or reduction of predation risk (Blaber, 1973; Whitfield & Blaber, 1978). The influence of these factors, however, was not evaluated for the purposes of this study.

The degree of overlap of fishing effort and fish distribution has implications for the implementation of corrective fishery management measures. The larger proportion of fish that are vulnerable to exploitation, the more important it may be to improve fishery control measures. All fishing activities on the Great Fish Estuary were conducted either from shore or from boats by people fishing with rod and line or handline. Subsistence fishers accounted for 73% of the fishing effort, and recreational fishers for the rest. The distribution of subsistence fishers overlapped largely with the distribution of the fish. There was no such correlation, however, between the distributions of the two groups of recreational fishers. The reason for this might be that the subsistence fishers are temporarily resident on the banks of the estuary and possibly have better knowledge of the best fishing sites. In addition, being more mobile than the two other groups, recreational boat anglers move more frequently around and were more evenly distributed in the lower and middle part of the estuary.

The stock status of the spotted grunter is considered to be slightly over-exploited (Fennessy, 2000). However, when compared with other estuarine-dependent angling species in South Africa, such as dusky kob and white steenbras (*Lithognathus lithognathus*), the state of spotted grunter is relatively healthy (Mann, 2000). The most probable explanation for this is that the exploitation rate of spotted grunter in the marine environment is low (Childs, 2005). However, estuarine degradation due to freshwater abstraction (Cyrus, 2000), increased fishing pressure in estuarine systems (Lamberth & Turpie, 2003), and lack of compliance to fisheries regulations (Potts et al., 2005) may cause spotted grunter populations to decline in the future.

A reduction in the bag limit, which is currently set at five fish per person per day, would need to be substantial to have any effect on the fish population. Most fishers (83%) did not catch any spotted grunter or only one fish during a daily outing, and only 3% of the fishers exceeded the legislated daily bag limit (Potts et al., 2005). In addition, a large proportion (29%) of caught and retained spotted grunter was below the legal size limit of 40 cm TL (Potts et al., 2005). Consequently, effective management of spotted grunter on the Great Fish Estuary by way of bag limits and size restrictions would require improved law enforcement and/or better compliance by fishers. However, such changes would be difficult considering that the Great Fish Estuary is situated in a rural area and that there is a high dependence on fish catch by a large subsistence sector. Alternative measures include closed seasons and protected areas. However, a closed season will have negative impacts on the subsistence fishers who might be deprived of food and income. Therefore, over-exploitation of undersized spotted grunter might be best controlled by establishing a protected area. Such a no-fishing zone on the Great Fish Estuary should be established in the lower reaches (the area between one and two kilometers from the estuary mouth) as this area represented a highly used zone by undersized spotted grunter. However, when the stock status is uncertain and life history knowledge poor, as for the spotted grunter, complementary measures acting at the same time (for example legal size and protected area) might be the best management option.

In conclusion, the hypothesis that the estuarine dependent species spotted grunter was evenly distributed within the estuary was rejected. However, there was a significant relationship between the distribution of the fishing effort of the subsistence fishers and the fish, indicating that the spotted grunter was uniformly exploited within the estuary by this group of fishers. In contrast, there was no such relationship between the distribution of fish and recreational anglers.

Acknowledgements We would like to express our sincere thanks to Rupert Harvey for his dedicated assistance in the field. We are also grateful to Don Reynolds, Simon Daniel, Aidan Wood, Brad Ripley and Sherman Ripley for help with catching and tagging of the fish and various forms of assistance with field work, and to Thia and Hendrik Swart for access to the caravan park and its facilities during the study. Alan Whitfield is acknowledged for support and advice on various aspect of this project. The project was funded by the South Africa / Norway Programme on Research Cooperation (National Research Foundation of South Africa, and the Research Council of Norway), the South African Institute for Aquatic Biodiversity (SAIAB), and the Norwegian Institute for Nature Research (NINA). We are grateful to these institutions for their financial support.

References

Allanson, B. R. & F. H. L. Read, 1987. The response of estuaries along the south eastern coast of southern Africa to marked variation in freshwater inflow. Institute for Freshwater Studies Special Report No. 2/87. Rhodes University, Grahamstown, South Africa.

Blaber, S. J. M., 1973. Population size and mortality of the marine teleost *Rhabdosargus holubi* (Pisces: Sparidae) in a closed estuary. Marine Biology 21: 219–225.

Boyd, A. J., I. Barwell & S. Taljaard, 2000. Report on the National Estuaries Workshop. 3–5 May 2000, Port Elizabeth, South Africa. Report No. 2, Marine and Coastal Management Implementation Workshops.

Bullen, E. & B. Mann, 2000. Sedgwick's/ORI/WWF tagging programme: spotted grunter (*Pomadasys commersonnii*). Data report, Oceanographic Research Institute 2000/5, 7 pp.

Bullen, E. & B. Mann, 2004. Sedgwick's/ORI/WWF tagging programme: Summary of spotted grunter (*Pomadasys commersonnii*) tagged in South Africa. Data report, Oceanographic Research Institute 2004/11, 2 pp.

Childs, A. -R., 2005. Movement patterns of spotted grunter, *Pomadasys commersonnii* (Haemulidae), in the Great Fish Estuary, South Africa. MSc Thesis, Rhodes University, Grahamstown, South Africa, 151 pp.

Cowley, P. D. & C. Daniel, 2001. Estuaries of the Ndlambe municipality (EC 105). Great Fish, Klein Palmiet (Brak), East Kleinemonde, West Kleinemonde, Riet, Rufanes, Kowie, Kasouga, Kariega, Bushmans and Boknes estuaries. Special report prepared for the Institute of Natural Resources, 63 pp.

Cowley, P. D., A. D. Wood, B. Corroyer, Y. Nsubuga & R. Chalmers, 2004. A survey of fishery resource utilization on four Eastern Cape estuaries (Great Fish, West Kleinemeonde, East Kleinemonde and Kowie). *In* Protocols Contributing to the Management of Estuaries in South Africa, with a Particular Emphasis on the Eastern Cape Province. Volume III. Project C, Supplementary Report C5, 126–165.

Cyrus, D., 2000. A preliminary assessment of impacts on estuarine associated fauna resulting from an intrabasin transfer and fresh water abstraction from aquatic systems in the Richards Bay area of KwaZulu Natal, South Africa. African Journal of Aquatic Sciences 26: 115–120.

Cyrus, D. P. & S. J. M. Blaber, 1987. The influence of turbidity on juvenile marine fish in the estuaries of Natal, South Africa. Continental Shelf Research 7: 1411–1416.

Fennessy, S. T., 2000. Stock assessment and management of spotted grunter in KwaZulu-Natal (Final report). Oceanographic Research Institute. 9 pp.

Griffiths, M. H., 1997. Influence of prey availability on the distribution of dusky kob *Argyrosomus japonicus* (Sciaenidae) in the Great Fish River estuary, with notes on the diet of early juveniles from three other estuarine systems. South African Journal of Marine Science 18: 137–145.

Harris, S. A. & D. P. Cyrus, 1997. Composition, abundance, and seasonality of larval fish in Richards Bay Harbour, KwaZulu-Natal, South Africa. South African Journal of Aquatic Science 23: 56–78.

Harris, S. A. & D. P. Cyrus, 1999. Composition, abundance and seasonality of fish larvae in the mouth of Durban Harbour, KwaZulu-Natal, South Africa. South African Journal of Marine Science 21: 19–39.

Lamberth, S. & J. Turpie, 2003. The role of estuaries in South African fisheries: economic importance and management implications. African Journal of Marine Science 25: 131–157.

Mann, B. Q. 2000. Southern African marine linefish status reports. Oceanographic Research Institute special publication, 257 pp.

Marais, J. F. K., 1984. Feeding ecology of major carnivorous fish from Eastern Cape estuaries. South African Journal of Zoology 19: 211–223.

Marshall, S. & M. Elliot, 1998. Environmental influences on the fish assemblage of the Humber Estuary, UK. Estuarine, Coastal and Shelf Science 46: 175–184.

Potts, W. M., P. D. Cowley, B. Corroyer & T. F. Næsje, 2005. Trends in resource utilisation on the Great Fish Estuary. NINA Report 50, 36 pp.

Vorwerk, P. D., A. K. Whitfield, P. D. Cowley & A. W. Paterson, 2003. The influence of selected environmental variables on fish assemblage structure in a range of southeast African estuaries. Environmental Biology of Fishes 66: 237–247.

Wallace, J. H. 1975. The estuarine fishes of the east coast of South Africa. III. Reproduction. Investigational Report Oceanographic Research Institute 41, 51 pp.

Wallace, J. H. & R. P. van der Elst, 1975. The estuarine fishes of the east coast of South Africa. Part IV. Occurrence of juveniles in estuaries. Part V. Ecology, estuarine dependence and status. Investigational Report of the Oceanographic Research Institute 42, 63 pp.

Webb, G. A., 2002. Biology and demography of the spotted grunter *Pomadasys commersonnii* (Haemulidae) in South African Waters. MSc Thesis. Rhodes University, Grahamstown, 135 pp.

Whitfield, A. K., 1988. The fish community of the Swartvlei estuary and the influence of food availability on resource utilisation. Estuaries 11: 160–170.

Whitfield, A. K., 1994a. Abundance of 0+ juvenile marine fishes in the lower reaches of three South African estuaries with differing freshwater inputs. Marine Ecology Progress Series 105: 257–267.

Whitfield, A. K., 1994b. An estuary-association classification for the fishes of southern Africa. South African Journal of Science 90: 411–417.

Whitfield, A. K. & S. J. M. Blaber, 1978. Food and feeding ecology of piscivorous fishes at Lake St Lucia, Zululand. Journal of Fish Biology 13: 675–691.

Hydrobiologia (2007) 582:35–42
DOI 10.1007/s10750-006-0545-x

FISH TELEMETRY

Effects of salmon lice infection and salmon lice protection on fjord migrating Atlantic salmon and brown trout post-smolts

Rolf Sivertsgård · Eva B. Thorstad · Finn Økland · Bengt Finstad · Pål Arne Bjørn · Niels Jepsen · Trude Nordal · R. Scott McKinley

Abstract Effects of artificial salmon lice infection and pharmaceutical salmon lice prophylaxis on survival and rate of progression of Atlantic salmon (n = 72) and brown trout post-smolts (n = 72) during their fjord migration, were studied by telemetry. The infected groups were artificially exposed to infective salmon lice larvae in the laboratory immediately before release in the inner part of the fjord to simulate a naturally high infection pressure. Groups of infected Atlantic salmon (n = 20) and brown trout (n = 12) were also retained in the hatchery to control the infection intensity and lice development during the study period. Neither salmon lice infection nor pharmaceutical prophylaxis had any effects on survival and rate of progression of fjord migrating Atlantic salmon post-smolts compared to control fish. Atlantic salmon spent on average only 151.2 h (maximum 207.3 h) in passing the 80 km fjord system and had, thus, entered the ocean when the more pathogenic pre-adult and adult lice stages developed. The brown trout, in comparison to Atlantic salmon, remained to a larger extent than Atlantic salmon in the inner part of the fjord system. No effect of salmon lice infection, or protection, was found in brown trout during the first weeks of their fjord migration. Brown trout will, to a larger extent than Atlantic salmon, stay in the fjord areas when salmon lice infections reach the more pathogenic pre-adult and adult stages. In contrast to Atlantic salmon, they will thereby possess the practical capability of returning to freshwater when encountering severe salmon lice attacks.

Keywords *Salmo salar* · *Salmo trutta* · Telemetry · Salmon lice · Migration · Aquaculture

Guest editors: P. R. Almeida, B. R. Quintella, M. J. Costa and A. Moore
Developments in Fish Telemetry

R. Sivertsgård
Norwegian College of Fishery Science, University of Tromsø, 9037 Tromsø, Norway

E. B. Thorstad · F. Økland · B. Finstad (✉) · T. Nordal
Norwegian Institute for Nature Research (NINA), Tungasletta 2, 7485 Trondheim, Norway
e-mail: bengt.finstad@nina.no

P. A. Bjørn
Norwegian Institute of Fisheries and Aquaculture Research, Breivika, 9291 Tromsø, Norway

N. Jepsen
Department of Inland Fisheries, Danish Institute for Fisheries Research, Vejlsøvej 39, 8600 Silkeborg, Denmark

R. S. McKinley
West Vancouver Laboratory, The University of British Columbia, Vancouver, BC, Canada VTV 1N6

Introduction

Atlantic salmon, *Salmo salar* L., and brown trout, *Salmo trutta* L., smolts leave freshwater in the spring. Atlantic salmon migrate to the open ocean, whereas brown trout remain feeding in fjords and near coastal areas (e.g. Lyse et al., 1998; Hansen et al., 2003; Klemetsen et al., 2003; Rikardsen, 2004). During the last decades, the marine parasite salmon louse *Lepeophtheirus salmonis* (Krøyer) has become one of the most serious pathogens on outward migrating post-smolts (Tully & Nolan, 2002; Heuch et al., 2005). Fish farming is believed to contribute significantly to the elevated salmon lice levels in wild post-smolts (Bjørn et al., 2001; Butler, 2002; Gargan et al., 2003). Major physiological disturbances, such as high levels of plasma cortisol and reduced osmoregulatory ability, occur in both Atlantic salmon and brown trout when the lice develop into mobile stages (reviewed by Tully & Nolan, 2002), and may even kill heavily infected hosts (Finstad et al., 2000; Bjørn et al., 2001). Sub lethal effects, such as reduced growth and swimming performance, have also been reported (e.g. Bjørn & Finstad, 1997; Nolan et al., 1999; Wagner et al., 2003). This may altogether impair fish from reaching feeding areas, finding food and avoiding predators, but such behavioural effects of salmon lice infection has never been studied in free-swimming salmonid hosts.

The aim of this study was to examine the effects of salmon lice infection and salmon lice protection on survival and rate of progression of Atlantic salmon and brown trout post-smolts during the first stage of their marine migration. Groups of post-smolts were artificially infected with salmon lice (Bjørn & Finstad, 1997), given a prophylactic pharmaceutical protection against salmon lice, or served as untreated control groups. Post-smolts were followed through a fjord system by using acoustic telemetry and survival and rate of progression were compared between the groups. Similarly infected fish were coincidentally kept in the hatchery to control infection intensity and lice development in free-swimming hosts during the study period.

Materials and methods

The study was carried out in the Romsdalsfjord system, Norway (62°40′ N 8°10′ E) (Fig. 1). First-generation hatchery-reared offspring of Atlantic salmon and brown trout of the River Eira stock (age 2+) were tagged with acoustic

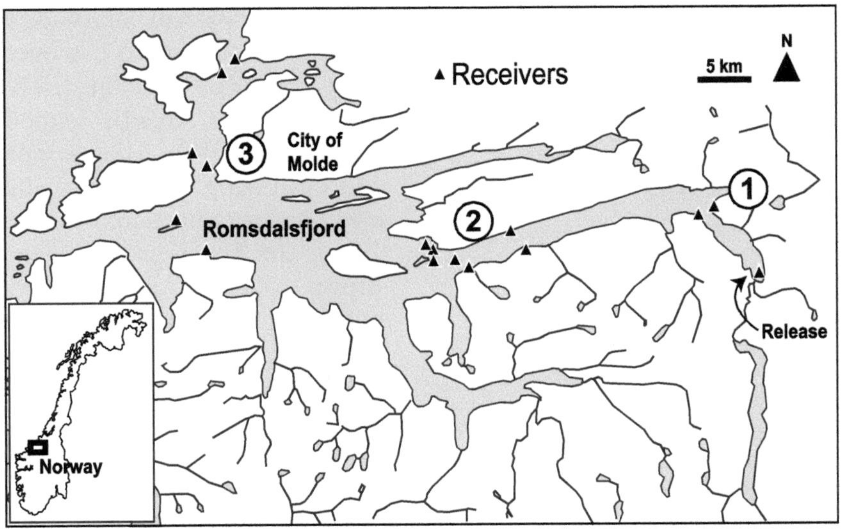

Fig. 1 The Romsdalsfjord system, Norway, with the release site for acoustically tagged Atlantic salmon and brown trout smolts in 2003. Automatic receiver sites 1–3 are indicated

transmitters (VEMCO Ltd., Canada, model V8SC-6L, 9 × 20 mm, weight in air 3.3 g, weight in water 2.0 g, life time 70 days) as described in Finstad et al. (2005). Individuals of each species were divided into groups (1) artificially infected with salmon lice, (2) protected against salmon lice infection with a prophylactic pharmaceutical product, substance EX (Pharmaq, Norway), and (3) untreated controls. The prophylactic substance protects the fish against salmon lice infection for up to 16 weeks, have low toxicity for warm-blooded animals and humans, and its function is to prevent synthesis of chitin in salmon lice (Bernt Martinsen, Pharmaq, pers. comm.).

For artificial salmon lice infection, infective salmon lice copepodids from Austevoll Research Station (Boxaspen & Næss, 2000) were added to the infection tank (34‰, 7°C) in accordance with the methods of Bjørn & Finstad (1997). Each fish in the tank was exposed to approximately 200 newly hatched copepodids (Glover et al., 2003). The tank was containing both trout and salmon, but the species was separated from each other. Control fish were sham infected. The protected group was treated with lice protection in a separate trial, and the treatment period was 30 min in oxygenated water with more than 7 mg/l O_2. In this period, 1,000 l of water was withheld and 1 g of active substance (10 ml/ 1,000 l) was added to the fish in each tank (one tank for each group) (See Table 1 for number of fish in each group).

The fish were allowed one day of recovery in tanks, before being transported to a nylon net recovery cage (volume 0.5 m^3) at the release site in the fjord (Fig. 1) from where they were released 12 h later at high tide during 10–14 May 2003. A total of 72 Atlantic salmon and 72 brown trout were tagged and released in the inner part of the fjord system near the mouth of the River Eira (Table 1). An equal number of fish from each group were released at the same time, together with 20–25 non-tagged smolts. The recovery cage was placed at salinity ≥30‰ at two meters depth because juvenile stages of salmon lice can be killed at ≤20‰ salinity (Johnson & Albright, 1991).

Tagged fish were automatically recorded when passing 15 VR2-receivers (VEMCO Ltd.) moored at three sites in the fjord system 7.1 km (site 1), 29–37 km (site 2) and 64–75 km (site 3) from the release site (Fig. 1). The range of the receivers did not cover the whole transect of the fjord (each receiver covered typically a radius of 200–260 m for transmitters at 0.5–3.0 m depth, but range varied with conditions). Hence, 'survival rates' are minimum estimates because some individuals may have passed without being recorded. However, standardised methods allow for comparison among groups. The release time for individual fish was based on recording by a receiver placed at the

Table 1 Number of Atlantic salmon and brown trout smolts tagged with acoustic transmitters and recorded at different receiver sites in the Romsdalsfjord system in 2003

Number tagged	Group	Mean total body length ± SD (mm)	Mean body mass ± SD (g)	Number (proportion) recorded at site 1	Number (proportion) recorded at site 2	Number (proportion) recorded at site 3
Atlantic salmon						
22 (16)	Lice infected	252 ± 2.7	147 ± 39	13 (59%)	10 (45%)	4 (18%)
25 (16)	Lice protected	254 ± 2.1	149 ± 31	13 (52%)	9 (36%)	3 (12%)
25 (17)	Control	254 ± 2.3	141 ± 32	11 (44%)	9 (36%)	3 (12%)
Brown trout						
22 (17)	Lice infected	235 ± 1.5	133 ± 25	7 (31%)	5 (23%)	2 (9%)
25 (19)	Lice protected	239 ± 1.6	121 ± 20	8 (32%)	3 (12%)	1 (4%)
25 (20)	Control	236 ± 1.5	112 ± 20	2 (8%)	2 (8%)	0 (0%)

The number tagged given in parentheses is the number of fish recorded by the receiver at the release site (see text). Total body length and body mass is given as mean values ± SD release site. After the fish was released, the fish had to swim a short distance from the recovery cage before it entered the detection range of this first receiver. The receiver was not able to pick up coded signals from the fish in the recovery cages. Results on rate of progression through the fjord system were, therefore, based on a smaller sample size than the total number of tagged fish (Table 1). All receivers were operating until the last week in August.

To control infection intensity and lice development of the acoustically tagged and released fish, 20 Atlantic salmon and 12 brown trout from the infection tank were retained in the hatchery (laboratory control groups). Between 5 and 10 fish of each species were examined for salmon lice development 15 and 25 days after infection, respectively. The saltwater in the hatchery was supplied from the fjord; a similar lice development was therefore expected for the laboratory control groups and the fish tracked during the fjord migration. The terms prevalence, intensity, abundance, median, intensity range and mean variance to mean ratio follow Bush et al. (1997).

Salinity was high (>30‰) at all depths in the inner part of the fjord system on 21 May. From the second week of June, the salinity decreased in the surface layer, with salinity lower than 23‰ down to 1 m depth. The salinity was still approximately 30‰ at all depths in the outer part of the system. The water temperature was around 7°C in the upper 5 m on 21 May. The surface temperature had increased to 10–14°C in the second week of June and to 16°C in mid-July.

Results

Survival

There was no difference between groups in proportions of fish detected at receiver site 1, 2 and 3, respectively, for neither Atlantic salmon (Pearson χ^2-tests; $\chi^2 = 1.1$, $P = 0.59$; $\chi^2 = 0.58$, $P = 0.75$; $\chi^2 = 0.49$, $P = 0.78$) nor brown trout ($\chi^2 = 4.6$, $P = 0.10$; $\chi^2 = 2.2$, $P = 0.33$; $\chi^2 = 2.4$, $P = 0.30$) (Table 1). Thus, no differences in survival between groups were detected based on proportions of post-smolts recorded at the different receiver sites.

Rate of progression

Atlantic salmon post-smolts spent on average 151.2 h (maximum 207.3 h) in passing the entire fjord system with the slowest migrating individual spending 336.3 h in the fjord system (Table 2). Brown trout remained to a larger extent within the fjord system and the last individual was recorded 31 days after release (Tables 1, 2). If including brown trout individuals for which the exact release time was not recorded (see Materials and methods), the last individual was recorded in the fjord system 11 weeks after tagging, but only three individuals (one protected and two control fish) were recorded more than 5 weeks after release.

Rate of progression through the fjord system, from release to last recording measured as body lengths per second, did not differ among the

Table 2 Time from release to last recording at the different receiver sites of acoustically tagged post-smolts in the Romsdalsfjord system in 2003

Salmon lice infected			Salmon lice protected		Control	
Site	*n*	Time (h)	*n*	Time (h)	*n*	Time (h)
Atlantic salmon						
1	3	11 (8–15, 4)	7	32 (7–56, 17)	7	40 (10–101, 29)
2	8	127 (80–365, 96)	8	92 (65–131, 24)	6	67 (34–105, 23)
3	4	137 (100–160, 26)	3	144 (135–158, 12)	2	186 (164–207, 31)
Brown trout						
1	5	286 (37–537, 194)	4	665 (25–2062, 655)	2	305 (94–516, 299)
2	3	390 (79–575, 222)	2	133 (106–160, 39)	2	643 (632–653, 15)
3	3	575 (413–736, 162)	1	636 (–)	0	

Time is given as mean with range and SD in parenthesis. *n* is number of fish groups in neither Atlantic salmon (Kruskal–Wallis test; $\chi^2 = 1.13$, $P = 0.57$) nor brown trout (Kruskal–Wallis test; $\chi^2 = 1.22$, $P = 0.55$). Rates of progression (mean ± SD) were for Atlantic salmon 0.41 ± 0.20 bl s^{-1} (protected, $n = 18$), 0.45 ± 0.30 bl s^{-1} (control, $n = 15$) and 0.48 ± 0,20 bl s^{-1} (infected, $n = 15$), and for brown trout 0.17 ± 0.14 bl s^{-1} (protected, $n = 7$), 0.06 ± 0.03 bl s^{-1} (control, $n = 4$) and 0.13 ± 0.15 bl s^{-1} (infected, $n = 11$).

Development of salmon lice in laboratory control groups

The infection intensity was high for both Atlantic salmon and brown trout; 100% of the fish in the experiment were infected (Table 3). There was no difference in intensity between 15 and 25 days post-infection neither in Atlantic salmon nor brown trout (Mann–Whitney *U*-test, salmon: $U = 49.0$, $P = 0.97$, brown trout $U = 22.0$, $P = 0.20$, Table 1). After 15 days, the infection was dominated by early chalimus stages (Fig. 2). After 25 days, the infection was dominated by mobile stages (Fig. 2). No significant differences in intensities were found between the species neither at day 15 nor at day 25 (Mann–Whitney *U*-tests, day 15: $U = 12.0$, $P = 0.1$, day 25: $U = 22.0$, $P = 0.20$, Table 3) (Fig. 2).

Discussion

Neither artificial salmon lice infection nor the pharmaceutical prophylaxis against salmon lice infection had any effects on survival or rate of progression in fjord migrating Atlantic salmon and brown trout post-smolts. Since the laboratory control fish had 100% prevalence and high intensity of salmon lice, the infected, tagged and released fish must also have been highly infected when released in the inner part of the fjord system.

Fast migration through the fjord system (average 151.2 h, maximum 207.3 h) might be the reason for no differences in survival and progression rates between infected and non-infected Atlantic salmon post-smolts. At 15 days post-infection, laboratory retained Atlantic salmon and brown trout were only infected with chalimus-larvae; developmental stages in the released fish were expected to be similar. A possible reinfection of salmon lice in released Atlantic salmon post-smolts was also expected to be dominated by chalimus larvae (Finstad et al., 2000). Heavy chalimus infection has in a series of laboratory experiments mainly been associated with stress responses whereas more severe alterations in physiological homeostasis and mortality mainly occur when pre-adult and adult stages appear (reviewed by Tully & Nolan 2002; Heuch et al., 2005). Thus, in the present study, the Atlantic salmon post-smolts had already left the 80 km long Romsdalsfjord system before pre-adult and adult stages of lice had developed and start to physiologically compromise the fish more severely (Tully & Nolan, 2002). Accordingly, positive effects of the pharmaceutical prophylaxis against salmon lice infection would not be expected until after the salmon had left the fjord system and entered the open ocean. However, more importantly, no adverse effects of lice protection were detected in survival or rate of progression on fjord migrating Atlantic salmon post-smolts.

The released brown trout remained to a larger extent than Atlantic salmon in the inner part of

Table 3 Recording of salmon lice on laboratory control groups 15 and 25 days after artificial infection

	Atlantic salmon		Brown trout	
Days after infection	15	25	15	25
Number of fish	10	10	5	7
Prevalence (%)	100	100	100	100
Intensity (mean ± SD)	43.1 ± 19.1	39.7 ± 6.9	24.4 ± 10.2	33.9 ± 7.4
Inter quartile range (IQR)	41.8	12.8	18	16
Median	41.5	39.5	30.0	34.0
Intensity range	21–70	31–51	9–33	24–44
Variance to mean ratio ($S^2\chi^{-1}$)	8.9	1.2	4.2	1.6

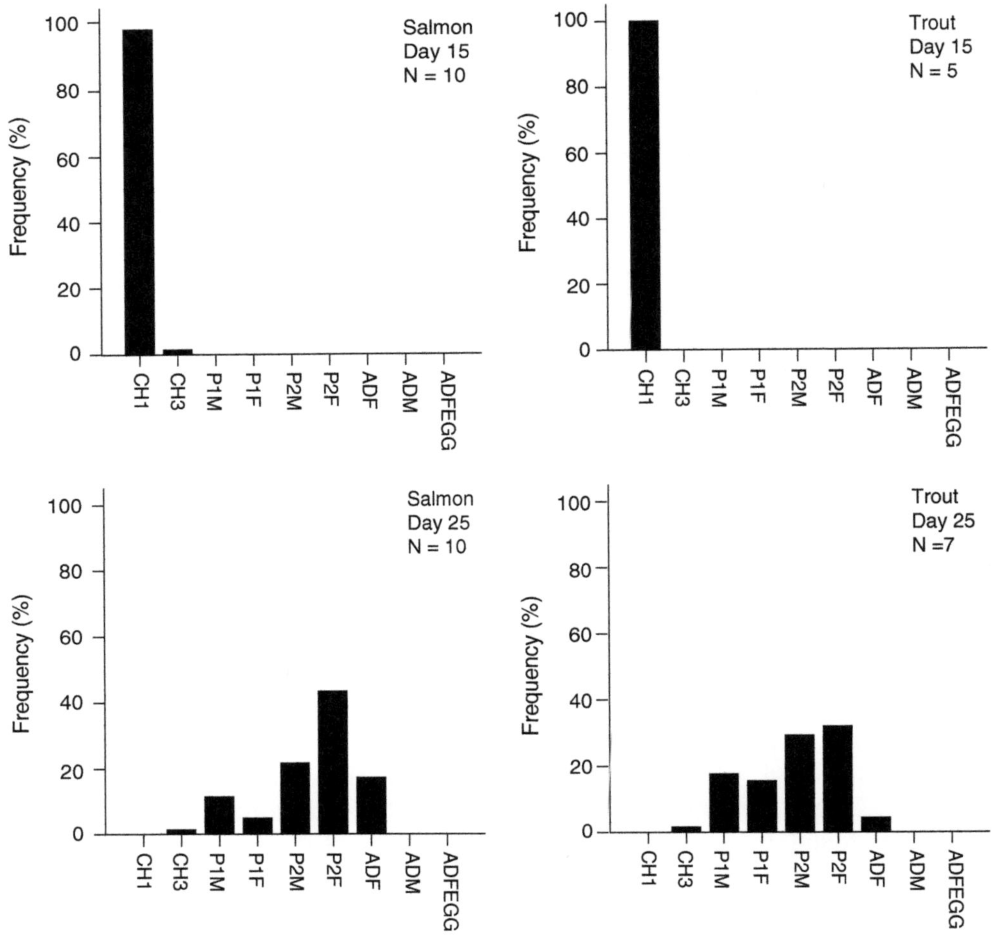

Fig. 2 Frequency distribution of developmental stages of salmon lice in laboratory control groups of Atlantic salmon and brown trout post-smolts 15 and 25 days after artificial infection. CH1 = first and second chalimus stage combined, CH3 = third and fourth chalimus stage combined, P1M = first pre-adult male, P1F = first pre-adult female, P2M = second pre-adult male, P2F = second pre-adult female, ADM = adult male, ADF = adult female, ADFEGG = adult female with egg strings

the fjord system, supporting previous telemetry studies (Finstad et al., 2005), as well as other studies of wild Norwegian sea trout populations (Berg & Berg, 1987; Lyse et al., 1998; Knutsen et al., 2001). As in Atlantic salmon, effects of salmon lice infection and protection were expected to increase a few weeks after release. By then, the artificial salmon lice infection would have developed into the more pathogenic pre-adult and adult stages. By this time, released fish in the infected and control group would also have been susceptible for reinfection by salmon lice copepodids. However, only three individual brown trout were recorded more than 5 weeks after release. This indicates that either all groups of brown trout suffered a high mortality in the fjord system, or that they stayed in parts of the system not covered by the listening stations. New results also indicate that predation risk can be very high in estuarine and inner fjord areas (Thorstad et al., 2007; Niels Jepsen, pers. comm.). Together, this might explain the unexpectedly low recovery rate in trout, which usually feed only within a short distance of their native river mouth (Berg & Berg, 1987; Berg & Jonsson, 1990). A few individuals may also have moved to the outer coastal areas, but the lower proportion recorded by the listening stations may indicate that they migrated to a lesser extent than salmon through the fjord system, concurring with earlier studies of marine sea trout ecology (see references above). Brown trout post-smolts also tend to move closer to shore than Atlantic salmon post-smolts (Thorstad et al., 2004) such that the likelihood of brown trout being recorded when passing the receivers should be higher than for Atlantic salmon.

In conclusion, the present study indicates that chalimus salmon lice infection have limited effects on Atlantic salmon and brown trout post-smolts fjord survival and migration. Thus, previous laboratory studies indicating only minor effects on post-smolts of chalimus larvae are confirmed under natural field conditions. When the infection had developed into mobile stages, Atlantic salmon post-smolts had most likely left the fjord and near-coastal areas. Unlike brown trout, heavily infected Atlantic salmon post-smolts therefore do not possess the practical capability of returning to estuarine and freshwater when encountering severe salmon lice attacks (Bjørn et al., 2001), and might die unnoticed at sea (Holst et al., 2003). Brown trout will to a greater extent stay in the fjord areas when severe infection loads are reached and may counteract this by returning to trout will also be more susceptible to adverse effects of continuous reinfection of salmon lice because they remain in the fjord and coastal areas with extensive fish farming activity and a high infection pressure (e.g. Tully et al., 1999; Bjørn et al., 2001; Heuch et al., 2005).

Acknowledgements We would like to thank Karin Boxaspen and Bernt Martinsen for applying salmon lice larvae and the lice protection, respectively. Further, Svein Erik Gaustad, Espen Holthe, Petter Sira, Bjørg Anne Vike and Torbjørn Utigard are thanked for assistance during the project. The study was financed by the European Commission contract no: Q5RS-2002-00730, AquaNet Canada (RSM), the Norwegian Institute for Nature Research and Statkraft Energy AS. The experimental procedures conform to Norwegian ethical requirements.

References

Berg, O. K. & M. Berg, 1987. Migrations of sea trout, *Salmo trutta* L., from the Vardnes river in northern Norway. Journal of Fish Biology 31: 113–121.

Berg, O. K. & B. Jonsson, 1990. Growth and survival rates of the anadromous trout, *Salmo trutta*, from the Vardnes River northern Norway. Environmental Biology of Fishes 29: 145–154.

Bjørn, P. A. & B. Finstad, 1997. The physiological effects of salmon lice infection on sea trout post smolts. Nordic Journal of Freshwater Research 73: 60–72.

Bjørn, P. A., B. Finstad & R. Kristoffersen, 2001. Salmon lice infection of wild sea trout an Artic char in marine and freshwaters: the effects of salmon farms. Aquaculture Research 32: 947–962.

Boxaspen, K. & T. Næss, 2000. Development of eggs and planktonic stages of salmon lice (*Lepeophtheirus salmonis*) at low temperatures. Contributions to Zoology 69: 51–55.

Bush, A. O., K. D. Lafferty, J. M. Lotz & A. Shostak, 1997. Parasitology meets ecology on its own terms: Margolis et al. revisited. Journal of Parasitology 83: 575–583.

Butler, J. R. A., 2002. Wild salmonids and sea lice infestation on the west coast of Scotland: sources of infection and implications for the management of marine salmon farms. Pest Management Science 58: 595–608.

Finstad, B., P. A. Bjørn, A. Grimnes & N. A. Hvidsten, 2000. Laboratory and field investigations of salmon lice [*Lepeophtheirus salmonis* (Krøyer)] infestation on Atlantic salmon (*Salmo salar* L.) post-smolts. Aquaculture Research 31: 795–803.

Finstad, B., F. Økland, E. B. Thorstad, P. A. Bjørn & R. S. McKinley, 2005. Migration of hatchery-reared Atlantic salmon and wild anadromous brown trout post-smolts in a Norwegian fjord system. Journal of Fish Biology 66: 86–96.

Gargan, P. G., O. Tully & W. R. Poole, 2003. Relationship between sea lice infestation, sea lice production and sea trout survival in Ireland, 1992–2001. In D. Mills (ed.), Salmon at the Edge. Blackwell Science, Oxford, pp. 119–135.

Glover, K. A., Ø. Skaala, F. Nilsen, R. Olsen, A. J. Teale & J. B. Taggart, 2003. Different susceptibility of anadromous brown trout (*Salmo trutta* L.) populations to salmon louse (*Lepeophtheirus salmonis* (Krøyer, 1837)) infection. ICES Journal of Marine Science 60: 1139–1148.

Hansen, L. P., M. Holm, J. C. Holst & J. A. Jacobsen, 2003. The ecology of post-smolts of Atlantic salmon. In D. Mills (ed.), Salmon at the Edge. Blackwell Science, Oxford, pp. 25–39.

Heuch, P. A., P. A. Bjørn, B. Finstad, J. C. Holst, L. Asplin & Nilsen, F., 2005. A review of the Norwegian 'National action plan against salmon lice on salmonids': the effect on wild salmonids. Aquaculture 246: 79–92.

Holst, J. C., P. Jacobsen, F. Nilsen, M. Holm, L. Asplin & J. Aure, 2003. Mortality of seaward-migrating post-smolts of Atlantic salmon due to salmon lice infection in Norwegian salmon stocks. In D. Mills (ed.), Salmon at the Edge. Blackwell Science, Oxford, pp. 136–137.

Johnson, S. C. & L. J. Albright, 1991. Development, growth and survival of *Lepeophtheirus salmonis* (Copepoda: caligidae) under laboratory conditions. Journal of the Marine Biological Association of the United Kingdom 71: 425–436.

Klemetsen, A., P.-A. Amundsen, J. B. Dempson, B. Jonsson, N. Jonsson, M. F. O'Connell & E.

Mortensen, 2003. Atlantic salmon *Salmo salar* L., brown trout *Salmo trutta* L. and Artic charr *Salvelinus alpinius* (L.): a review of aspects of their life histories. Ecology of Freshwater Fish 12: 1–59.

Knutsen, J. A., H. Knutsen, J. Gjøsæter & B. Jonsson, 2001. Food of anadromous brown trout at sea. Journal of Fish Biology 59: 533–543.

Lyse, A. A., S. O. Stefansson & A. Fernö, 1998. Behaviour and diet of sea trout post-smolts in a Norwegian fjord system. Journal of Fish Biology 52: 923–936.

Nolan, D. T., P. Reilly & S. E. Wendelaar Bonga, 1999. Infection with low numbers of the sea louse *Lepeophtheirus salmonis* induces stress-related effects in post-smolt Atlantic salmon (*Salmo salar*). Canadian Journal of Fisheries and Aquatic Sciences 56: 947–959.

Rikardsen, A. H., 2004. Seasonal occurrence of sea lice *Lepeophtheirus salmonis* on sea trout in two north Norwegian fjords. Journal of Fish Biology 65: 711–722.

Thorstad, E. B., F. Økland, B. Finstad, R. Sivertsgård, P. A. Bjørn & R. S. McKinley, 2004. Migration speeds and orientation of Atlantic salmon and sea trout post-smolts in a Norwegian fjord system. Environmental Biology of Fishes 71: 305–311.

Thorstad, E. B., F. Økland, B. Finstad, R. Sivertsgård, N. Plantalech, P. A. Bjørn & R. S. McKinley, 2007. Fjord migration and survival of wild and hatchery-reared Atlantic salmon and wild brown trout post-smolts. Hydrobiologia 582: 99–107.

Tully, O. & D. T. Nolan, 2002. A review of population biology and host-parasite interactions of the sea lice *Lepeophtheirus salmonis* (Copepoda: Caligidae). Parasitology 124: 165–182.

Tully, O., P. Gargan, W. R. Poole & K. F. Whelan, 1999. Spatial and temporal variation in the infestation of sea trout (*Salmo trutta* L.) by the caligid copepod *Lepeophtheirus salmonis* (Krøyer) in relation to sources of infection in Ireland. Parasitology 119: 41–51.

Wagner, G. N., R. S. McKinley, P. A. Bjørn & B. Finstad, 2003. Physiological impact of sea lice on swimming performance of Atlantic salmon. Journal of Fish Biology 62: 1000–1009.

Hydrobiologia (2007) 582:43–54
DOI 10.1007/s10750-006-0555-8

FISH TELEMETRY

Upstream migration of adult chum and pink salmon in the Shibetsu River

Y. Makiguchi · H. Nii · K. Nakao · H. Ueda

Abstract The migratory behavior and swimming patterns of anadromous upstream migratory fish have been poorly described in the Shibetsu River in eastern Hokkaido, Japan. In this 2004 study, we used electromyogram (EMG) transmitters and depth/ temperature (DT) loggers to compare the upstream migratory behavior of adult male chum salmon (*Oncorhynchus keta*) and pink salmon (*O. gorbuscha*) in the canalized and reconstructed segments of the Shibetsu River, where a part of canalized section was preliminary reconstructed meander to restore a more natural section. The EMG transmitter and DT logger were externally attached to the left side of the body, below the front edge of the dorsal fin. Fish of both species often migrated along the riverbanks and near the bottom of the water column, sometimes engaged in holding behavior, which was defined as cessation of swimming during their upstream migration for 5 minutes. Modal swimming depth calculated by DT loggers for chum salmon (0.2–0.4 m) was shallower than pink salmon (0.6–0.8 m). Further, modal swimming speeds measured by calibrated EMG for chum salmon (0.2–0.4 BL s^{-1}) were slower than pink salmon (1.2–1.4 BL s^{-1}). Pink salmon swam faster as well as in relatively deeper than chum salmon, suggesting that they expend more energy than chum salmon in the reconstructed segment. Based on these results, it seemed likely that the upstream migration behavior of chum and pink salmon was different with species-specific strategies.

Guest editors: P. R. Almeida, B. R. Quintella, M. J. Costa and A. Moore
Developments in Fish Telemetry

Y. Makiguchi (✉) · H. Ueda
Division of Environmental Science Development, Graduate School of Environmental Science, Hokkaido University, North 9 West 9, Kita-ku, Sapporo, Hokkaido 060-0809, Japan
e-mail: yuya-m@fsc.hokudai.ac.jp

H. Nii · K. Nakao
Hokkaido Farming Fisheries Promotion Corporation, North 3 West 7, Chuo-ku, Sapporo, Hokkaido 060-0003, Japan

H. Ueda
Laboratory of Aquatic Bioresources and Ecosystems, Field Science Center for Northern Biosphere, Hokkaido University, North 9 West 9, Kita-ku, Sapporo, Hokkaido 060-0809, Japan

Keywords Migratory behavior · Electromyogram (EMG) · DT logger · Meandering reconstruction

Introduction

It is well known that Pacific salmon (*Oncorhynchus* spp.) typically do not feed and must use their stored energy during upstream spawning migration. The upstream migration of anadromous fish can be energetically expensive, particularly when travel distances are long and river environmental conditions are adverse (Bernatchez & Dodson, 1987; Leonard & McCormick, 1999). Studies using electromyogram (EMG) telemetry found that sex, size and swimming speed patterns of salmon are extremely important factors affecting energy use during upstream migration (Hinch et al., 1996; Hinch & Rand, 1998, 2000). Hinch et al. (2002) found that sockeye salmon (*O. nerka*) were more variable in swimming speeds than pink salmon (*O. gorbuscha*) though the causes were not entirely clear. These studies on upstream migration behavior of salmon were conducted in long and turbulent rivers like the Fraser River in Canada. Bernatchez and Dodson (1987) compared energetic costs of migration for short- and long-distance upstream migrating anadromous fishes, and concluded that only the long-distance migrants (particularly some stocks of Pacific salmon) swam at optimal speeds. In contrast to the Fraser River, the Shibetsu River in eastern Hokkaido, Japan is short (10.4 km from the estuary to chum salmon's spawning ground). No studies have been done in short and non-turbulent rivers to compare differences in upstream migratory behaviors of pink and chum salmon (*O. keta*).

In Japan, channel alternation for flood control and irrigation has changed general fish fauna and standing crops (Mizuno & Gose, 1981). In the Shibetsu River, the reconstruction of channel alternation to a more natural state was preliminary conducted in March 2002, and additional phases of reconstruction are planning after evaluating the results of preliminary restoration. However, there is only one study on the effect of reconstruction on upstream migratory behavior of chum salmon in the Shibetsu River (Akita et al., 2006). More information about upstream migratory behavior of chum salmon and other salmonid species are necessary for further restoration of the Shibetsu River river planned in the future.

EMG telemetry has been used for several years to investigate the factors responsible for variability in migratory energetics of sockeye salmon. EMG transmitters enable us to quantify swimming speeds of individual migrating fish (McKinley & Power, 1992; Hinch et al., 1996). Micro-data logger storaging information of ambient depths and temperature also has been used to monitor chum salmon behavior in the sea (Tanaka et al., 1998, 2000, 2001). However, there has been only one study that used EMG transmitters and depth/temperature (DT) loggers simultaneously to investigate the upstream migration of chum salmon (Akita et al., 2006). Data collected by both EMG transmitter and DT logger enable us to have coupled time-series information for swimming speed and swimming depth. Further, since we need to retrieve the DT logger after releasing a fish, we can find the fish in a river by radio signals from the EMG transmitter.

In this study, we used EMG transmitters and DT loggers to estimate swimming speed and swimming depth of the chum and pink salmon during their upstream migration in both canalized and reconstructed segment in the Shibetsu River. Our main objectives are to compare the two species' upstream migration strategies in the Shibetsu River in detail using the EMG transmitter and the DT logger.

Materials and methods

Study area

The Shibetsu River, located in eastern Hokkaido, Japan, flows to the Nemuro Straight with a total length of about 78 km and a drainage basin that occupies 671 km^2. Over the past century, the river has been altered from a meandering, complex river system, to a canalized flood-control canal. Recently a segment of the Shibetsu River located 8.4 km upstream from the river mouth was reconstructed to restore the canalized river to a more natural state. Early in 2004, artificial large woody debris (LWD) was installed in the middle of reconstructed segment. It is well known that natural LWD has many impacts on streams such as increasing channel roughness, providing high flow refuge and cover for fish, forming scour pools, and regulating the transport of gravel through the river system (Mathias 2000). Our study area covered a distance of between 7.25 km and 9.0 km from the river mouth. The canalized segment was 1.15 km in length and the reconstructed segment was 0.5 km in length (Fig. 1).

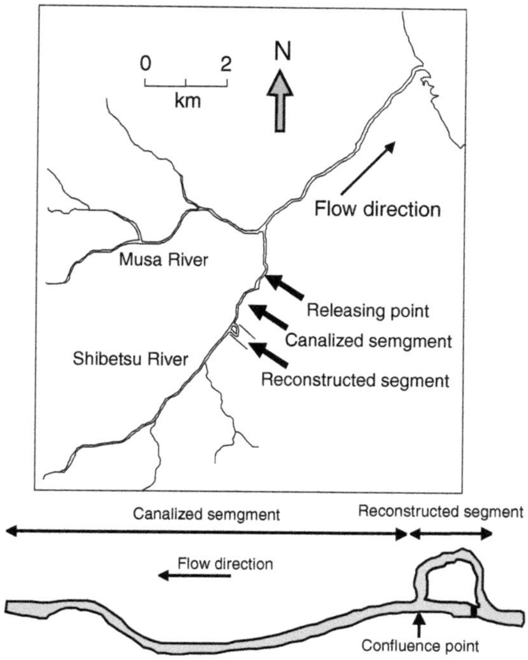

Fig. 1 Map of Shibetsu River indicating the canalized segment and the reconstructed segment

Study animals and transmitter attachment

Pink and chum salmon spawn in the Shibetsu River system, and migrate through our study area during September and October, respectively. In 2004, 5 male chum salmon (fork length, 62.5–74.9 cm; body weight 2.38–4.45 kg) and 5 male pink salmon (fork length, 51.0–59.7 cm; body weight 1.55–2.32 kg) were caught in the Shibetsu River estuary and transferred to outdoor tank at the Shibetsu Salmon Museum.

An EMG transmitter (CEMG-R11, Lotek Engineering Inc., Newmarket, Ontario: 18.3 g in air, 8.0 g in water, 16.2 mm in diameter, 53.0 mm in length) and a DT logger (M190-DT, Little Leonardo Co. Ltd, Tokyo: 18 g in air, 47.5 mm in diameter, 14.5 mm in length) were attached to all experimental fish externally. The transmitter of and logger weight did not exceed 1.67% of the body weight for both species. Fish position and swimming speed were monitored using the EMG transmitter, and swimming depth was monitored using the DT logger. Using EMG transmitters and DT loggers simultaneously allow us to monitor the position of a fish horizontally and vertically. Experimental fish were anaesthetized using FA100 (eugenol; Tanabe Seiyaku Co. Ltd, Osaka, Japan) at a concentration of 0.5 ml l^{-1} in the Shibetsu River water, and total length and body weight were measured. The fish were placed on a surgical table, ventral side down, and their gills were irrigated with water containing with FA100 to maintain sedation during the procedure. An EMG transmitters and a DT logger were externally attached to the left side of the body, below the front edge of the dorsal fin, using nylon ties. The EMG transmitter and DT logger were sutured together using nylon ties and epoxy resin before the procedure. Using a needle, the nylon ties passed through a silicon pad, the upper part of the musculature, and a silicon pad opposite the EMG transmitter and DT logger. The silicon pads were attached to minimize abrasion. The EMG transmitters functioned by detecting muscle activity via brass sensing tips (0.5 × 0.1 cm), connected to surgical steel electrode wires. The EMG electrodes were inserted subcutaneously using a hypodermic needle at about a 0.7 ratio of the body length on the left side of the fish. Two electrode tips were positioned approximately 10 mm apart, secured in the lateral red muscle toward the rear of the fish, which is primarily used in steady, non-bursting aerobic swimming activity (Geroge, 1962; Beddow & McKinley, 1999). The electrodes also appear to be sensitive to contractions of the relatively larger white muscle (McKinley & Power, 1992; Hinch et al., 1996), which is recruited primarily during burst, anaerobic swimming. EMG signals can therefore generally be related to swimming speed (Økland et al., 1997). The electrodes were sutured to avoid entangling vegetation and/or structure in the field condition. An aminoglycoside antibiotic (Akiyama Seisakujyo Co. Ltd, Tokyo, Japan) was applied to the skin around the punctures and stitches. The attachment procedures usually required approximately five minutes to complete. After recovering from anaesthesia, we observed the behavior of the chum and pink salmon with EMG transmitters and DT loggers in a hatchery pond and, found no differences in behavior manner between the experimental fish and the intact fish.

Calibration of EMG signals to swimming speeds

Following the recovery period, we analyzed an individual calibration curve for convert EMG signals from the fish into swimming speeds. For the calibration, the fish were slowly forced to swim along a track (19 m in length, 5.5 m in width, and 0.67 m in depth) with a tape measure along the edge, and their swimming distance and travel time were measured. This trial was conducted about twenty times per individual until the fish were exhausted. EMG signals data were determined by EMG radio receivers (model SRX_400; Lotek Engineering Inc.) and recorded at 3–5 s intervals. When the fish maintained a holding position, non-moving EMG signals were also gathered. After this trial, we measured a flow in this track using an electromagnetic current meter (ES7603; Yokogawa Navi-tech Co. Ltd, Tokyo, Japan.). The relation between swimming speed and EMG signal output from tagged fish was plotted and used as a standard curve to estimate speeds in free swimming fish from collected EMG signals (McFarlane et al., 2004). Fish were transferred to a recovery live-box and allowed to recover in the release point of the Shibetsu River for one.

Field study

During September and October 2004, chum and pink salmon used in the EMG calibration were individually released 7.25 km upstream from the river mouth and tracked upstream to the end of the reconstructed segment on foot using a hand-held directional Yagi antenna. SRX 400 receivers recorded EMG signals at 3–5 s intervals. Migration time through both segments was measured for each fish by subtracting time of reach exit from time of reach entry. Ground speed was then calculated by dividing the segment length by the migration time. To compare the upstream swimming patterns between chum and pink salmon, we calculated the coefficient of variation (CV) for swimming speeds for each individual in both segments. We also calculated a swimming efficiency index (SEI) by dividing each individual's ground speed by its swimming speed. These data are useful for estimating an individual's energy utilization strategy. For instance, values below 1 indicate that individuals are swimming relatively hard to achieve forward progress, values equal to 1 imply relatively easy for forward progress (analogous to moving through still water), and values greater than 1 indicate that fish are receiving some type of forward assistance. EMG signals were converted to swimming speed using equations established from the EMG calibration.

After tracking, water depth and current speed were also measured at the locations along the river, every 50 m, where fish positions were pinpointed and holding behavior were observed. Current speeds were measured at three layers (top, middle, and bottom) of the water column using an electromagnetic current meter. Daily average water temperatures were 3.9–15°C.

Statistical analyses

Regression analysis was performed by simple regression on EMG signals to swimming speed. Correlation coefficients were obtained using simple regression analysis (Excel software). Analysis of covariance (ANCOVA) was used to assess differences in condition between the specific weigh versus length regression relationships. Results were expressed as means ± standard deviation of mean. Statistical significance was assessed at the 0.05 level.

Results

Calibration of EMG signals to swimming speeds

Measurement of flow in the track revealed that there was no measurable flow, so we did not take account of flow in the track when calibrating EMG signals to swimming speeds. With simple regression analysis a significant correlation was found between EMG signals and swimming speed for both species (average of the 4 chum salmon $r^2 = 0.766$, and 3 pink salmon $r^2 = 0.809$, respectively; $P < 0.05$). We removed three salmon (one chum salmon and two pink salmon) with $r^2 < 0.6$ from analysis of swimming speed. The equations

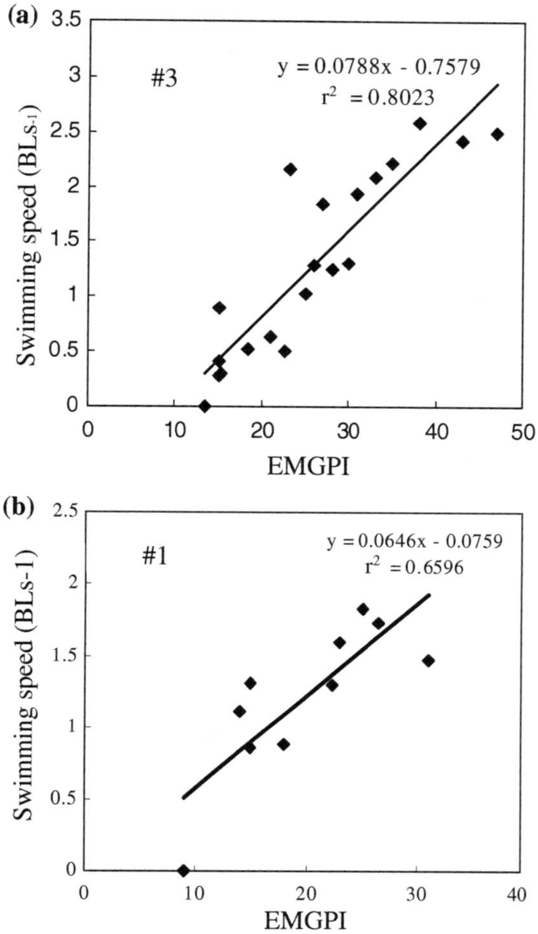

Fig. 2 Relationships between electromyogram pulse interval (EMGPI) and swimming speed (BL s^{-1}) for chum salmon #3 (**a**) and pink salmon #1 (**b**)

from the simple regression analysis were individually used to convert EMG signals of the field study to each fish's swimming speed (For example, $y = 0.044x + 0.0446$; $r^2 = 0.8556$ for chum salmon #1 Fig. 2(a), and $y = 0.06464x - 0.0759$; $r^2 = 0.6596$ for pink salmon #1 Fig 2(b)).

Field study

Upstream migration with the EMG transmitter and DT logger

All fish of both species migrated up the canalized segment, reached the confluence point with the reconstructed segment, and then entered the reconstructed meander. We retrieved 3 DT loggers for chum salmon and 2 DT loggers for pink salmon. Therefore data on swimming depths were obtained from three chum salmon and two pink salmon. Table 1 summarizes the data on the swimming depth recorded from both tagged salmon species as well as measured current speed. ANCOVA for both species revealed that chum salmon (mean fork length 72.5 ± 28.3 cm, mean mass 3328 ± 400 g) were larger than the pink salmon (mean fork length 58.4 ± 14.1 cm, mean mass 2000 ± 146.2 g ; $P < 0.05$) used in this study. An average of swimming depth and current speed of chum and pink salmon were nearly the same (Table 1). An example of the typical pattern of upstream swimming behavior of both species is demonstrated in Fig. 3. Chum and pink salmon often migrated along the riverbanks and near the bottom of the water column, sometimes displaying holding behavior (cessation of swimming during their upstream migration for 5 min). Both species sometimes displayed backtracking and crossing behavior. However, we could find no clear relationships between these behaviors and swimming depth. Typical frequency distributions of swimming depth for chum salmon #5 and pink salmon #1 are shown in Fig. 4. Although modal swimming depth of chum salmon #5 ranged 0.2–0.4 m, modal swimming depth of pink salmon #1 ranged 0.6–0.8 m. The distributions of swimming depth of chum salmon showed lower peaks than pink salmon.

The general behavior of chum and pink salmon in the canalized and reconstructed segment

Table 2 summarizes the swimming behavior data about the swimming speeds calibrated from EMG data, CV of swimming speed, ground speed and SEI from chum and pink salmon. Swimming speeds ranged from 0.0301 to 2.20 BL s^{-1} for chum salmon (mean 0.718 BL s^{-1}) and from 0.053 to 3.27 BL s^{-1} for pink salmon (mean 1.24 BL s^{-1}). Mean values of swimming speed of chum and pink salmon were not widely different. Typical frequency distributions of swimming speed for chum salmon #1 and pink salmon #1 are shown in Fig. 5. Modal swimming speeds of chum salmon #1 ranged 0.2–0.4 BL s^{-1} and that of pink salmon #1 ranged 1.2–1.4 BL s^{-1}. The modal of swim

Table 1 Sex, length, body mass, swimming depth and river current speed of chum and pink salmon

Salmon #.	Sex	Fork length (cm)	Body mass (g)	Swimming depth (m)	Current speeds (m s^{-1})
Chum Salmon					
1	M	79.6	4100	0.732 ± 0.246	0.473 ± 0.293
2	M	67.0	2380	–	–
3	M	68.3	2860	–	–
4	M	79.3	4450	0.7174 ± 0.308	0.567 ± 0.310
5	M	68.6	2850	0.342 ± 0.338	0.556 ± 0.328
Average ± SD		72.5 ± 6.32	3328 ± 894	0.597 ± 0.221	0.532 ± 0.051
Pink Salmon					
1	M	53.6	1550	0.678 ± 0.256	0.543 ± 0.309
2	M	60.4	2320	–	–
3	M	60.3	2130	–	–
4	M	61.0	2230	0.429 ± 0.244	0.590 ± 0.325
5	M	56.9	1770	–	–
Average ± SD		58.4 ± 315	2000 ± 327	0.556	0.567

Values of swimming depth and river current speed are mean ± SD

speed of the pink salmon was higher than chum salmon.

We compared mean values of swimming speed, ground speed and SEI of chum and pink salmon between the canalized and reconstructed segments. A student's *t*-test revealed that mean values of swimming speed (P = 0.478 and 0.395), ground speed (P = 0.395 and 0.275) and SEI (P = 0.848 and 0.271) of chum and pink salmon, respectively, for the canalized segment were not significantly different from the reconstructed segment.

Mean values of swimming speed between chum and pink salmon among the canalized and reconstructed segment were compared. A student's *t*-test revealed that mean values of swimming speed of both species were not significantly different in the canalized segment (P = 0.171), but these values of pink salmon were significantly higher than chum salmon in the reconstructed segment (P = 0.0037).

Mean values of ground speed between chum and pink salmon among the canalized and reconstructed segment were compared. A student's *t*-test revealed that mean values of swimming speed of pink salmon were significantly higher than chum salmon in both segments (P = 0.0027 and 0.037, respectively).

Mean values of SEI of chum and pink salmon were compared between among the canalized and reconstructed segment. A student's *t*-test revealed that these values were not significantly different in both segments (P = 0.744 and 0.155, respectively).

We compared CV of swimming speeds between chum and pink salmon among the canalized and reconstructed segments. A student's *t*-test revealed that CV of swimming speeds for chum salmon in the canalized segment were not significantly different from pink salmon (P = 0.339), but CV of swimming speeds for chum salmon in the reconstructed segment were larger than pink salmon (P = 0.025) suggesting that chum salmon were changing their swimming speed more often than pink salmon in the reconstructed segment (Table 2).

We selected two individuals (chum salmon #1 and pink salmon #4) for a detailed segment-specific behavior of temporal patterns of instantaneous swimming speeds and swimming depth from DT logger (Fig. 6). In order to clarify the differences between species that we observed in swimming patterns (Fig. 5a, b) and migration in relation to the segment characteristics. Chum and pink salmon each exhibited holding behavior (chum salmon #1: 15:25:19–15:31:15 and 16:26:15–16:31:26; pink salmon #4: 14:25:10–14:38:00 and 15:05:00–15:10:00).

Discussion

In the present study, we used EMG transmitters and DT loggers to investigate the upstream

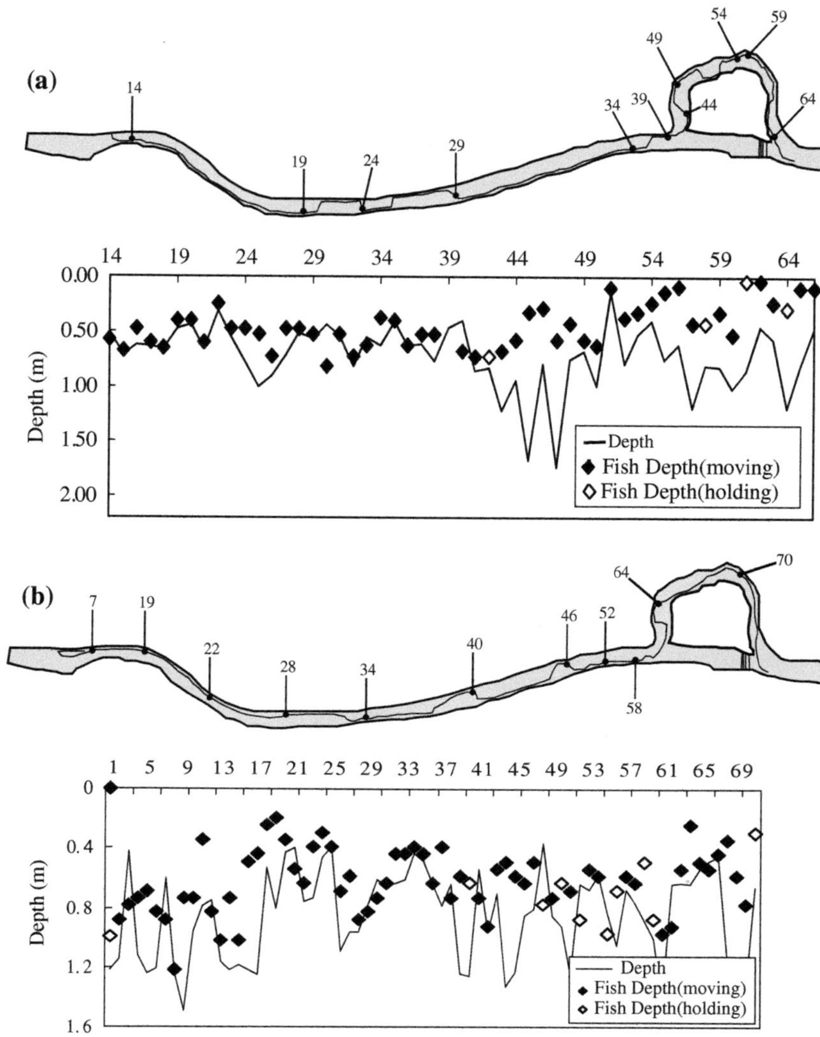

Fig. 3 Map of the study area showing the upstream migratory route (lines within river) and vertical position of chum salmon #1 (**a**) and pink salmon #1 (**b**). Numbers along the river correspond with the numbers above the diagrams below, indicating the horizontal location of the fish. The points inside the diagram show the horizontal and vertical position of the fish. Open points indicate a holding behavior, and closed points indicate a moving behavior

migratory behavior of adult male chum and pink salmon. After the experimental attachment of EMG transmitters and DT loggers, we observed the behavior of experimental fish in the swim track, and could find no clear differences in behavior from intact individuals. Further, one fish (chum salmon #3) with EMG transmitter and DT logger were caught in the upstream of fish ladder at the Kyousei bridge upstream 2.4 km from the reconstructed segment providing that the fish could pass the fish ladder. It is therefore likely that the salmon fitted with our EMG transmitters and DT loggers retain their homing motivation and maintain horizontal movement. Mellas & Haynes (1985) conducted experiments to determine the effects of various tag attachments on the swimming behavior of rainbow trout (*Salmo gairdneri*) and suggested that stomach tagging is the best method of transmitter attachment. Thorstad et al. (2000) conducted experiments to investigate the effects of external attachment and body-implanted dummy transmitters on the swimming performance of adult Atlantic salmon and suggested that there were no differences in

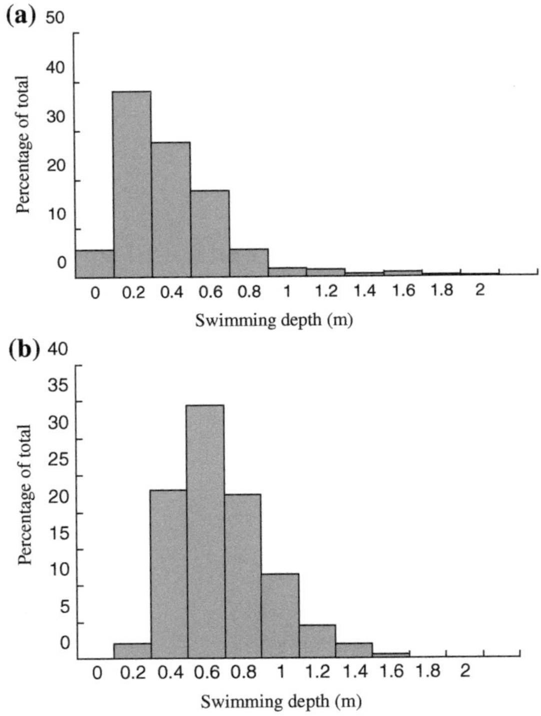

Fig. 4 Frequency distributions of swimming depth for chum salmon #5 (**a**) and pink salmon #1 (**b**) for the entire recording period

endurance among untagged fish, fish with external transmitters and body-implanted transmitters. External transmitter placement has proven useful in studying gravid fish, with surgical implantation after spawning often resulting in higher mortality rate (Winter, 1996).

We found that chum and pink salmon swam parallel to and near the riverbank. This behavior is believed to be important for upstream migration of the adult migrants (Osborne, 1961; Brett, 1983). Generally, current speed of a river increases with distance away from the riverbank and the bed of the river (Henderson, 1966). Adult pacific salmon stop feeding during their upstream migration and have to use stored energy reserves to complete their migration and spawning. So it is so important for all migrants to minimize upstream migration costs (Hughes, 2004).

We described the swimming depth and swimming speed of chum and pink salmon during upstream migration in the Shibetsu River in detail. We found that the modal swimming depth

Table 2 Summary for swimming speed, CV of swimming speed, ground speed and SEI of chum and pink salmon

Salmon #.	Swimming speeds (BL s^{-1})		CV of swimming speeds (BL s^{-1})		Ground speeds (BL s^{-1})		SEI	
	Canalized segment	Reconstructed segment	Canalized segment	Reconstructed segment	Canalized segment	Reconstructed segment	Canalized segment	Reconstructed segment
Chum Salmon								
1	0.747 ± 0.272	0.586 ± 0.317	0.463	0.68	0.23 ± 0.163	0.263 ± 0.309	0.31	0.449
2	1.44 ± 0.232	–	0.161	–	0.142 ± 0.170	–	0.09	–
3	0.478 ± 0.177	0.533 ± 0.309	0.369	0.58	0.271 ± 0.363	0.138 ± 0.141	0.567	0.259
4	0.45 ± 0.167	0.454 ± 0.211	0.445	0.464	0.364 ± 0.371	0.305 ± 0.193	0.809	0.672
5	0.749 ± 0.479	0.867 ± 0.484	–	–	0.316 ± 0.712	0.215 ± 0.116	0.421	0.248
Average ± SD	0.773 ± 0.399	0.610 ± 0.179	0.360 ± 0.138	0.575 ± 0.108	0.265 ± 0.085	0.230 ± 0.072	0.439 ± 0.270	0.407 ± 0.199
Pink Salmon								
1	0.979 ± 0.444	1.28 ± 0.421	0.453	0.329	0.49 ± 0.383	0.79 ± 0.995	0.501	0.617
2	1.26 ± 0.250	1.43 ± 0.440	0.199	0.308	0.664 ± 0.212	1.56 ± 1.47	0.527	1.09
3	–	–	–	–	–	–	–	–
4	1.23 ± 0.260	1.12 ± 0.372	0.211	0.333	0.561 ± 0.381	0.543 ± 0.411	0.456	0.485
5	–	–	–	–	–	–	–	–
Average ± SD	1.16 ± 0.154	1.28 ± 0.155	0.288 ± 0.143	0.323 ± 0.013	0.572 ± 0.087	0.964 ± 0.530	0.495 ± 0.036	0.73 ± 0.319

Values are mean ± SD

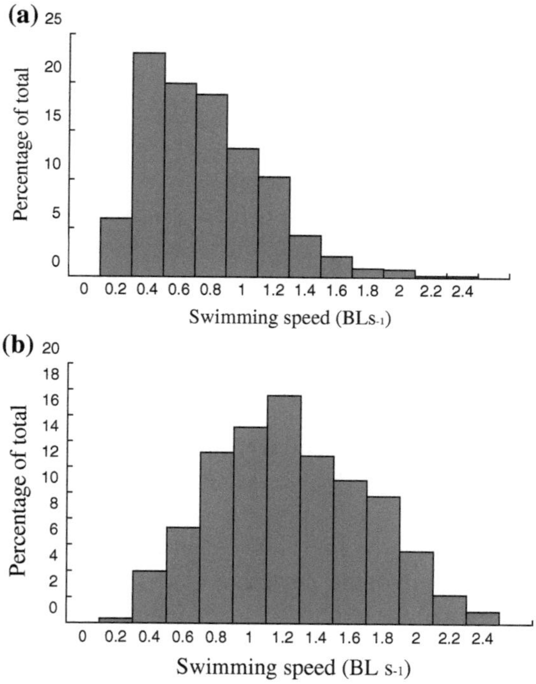

Fig. 5 Frequency distributions of swimming speed for chum salmon #1 (**a**) and pink salmon #1 (**b**) for the entire recording period

of chum salmon was lower depth (0.2–0.4 m) than pink salmon (0.6–0.8 m) suggesting that pink salmon was swimming deeper than chum salmon.

Furthermore, the modal swimming speed of chum salmon (0.4–0.6 BL s^{-1}) was lower than pink salmon (1.2–1.4 BL s^{-1}). Hertel (1966) estimated theoretically the residence that a fish experiences when swimming near the surface and found a relationship between submergence depth and the severity of wave drag. In accordance with Hertel's equation, this multiplier of energy expended will be 1 when the fish is swimming more than three body diameters beneath the surface and will peak at about 5 when the fish is swimming just a half of body diameter beneath the surface. Thus, a fish with a larger body diameter expends more energy when swimming near the surface of a river. Webb et al. (1991) investigated the effect of wave drag on the fast start performance of rainbow trout (*O. mykiss*) and found that in shallow water, wave drag accounted for 80% of the total resistance to acceleration. Mean water depth measured along the experimental section of the Shibetsu River was 0.78 m, revealing that the Shibetsu River was a relatively shallow river. In our study, the average body diameter of pink salmon (14.8 ± 5.77 cm) was larger than chum salmon (13.7 ± 7.07 cm). Thus, adult male pink salmon with a larger body diameter have to expend more energy than chum salmon. The solid lines in Fig. 6 represent a depth, maximizing the wave drag

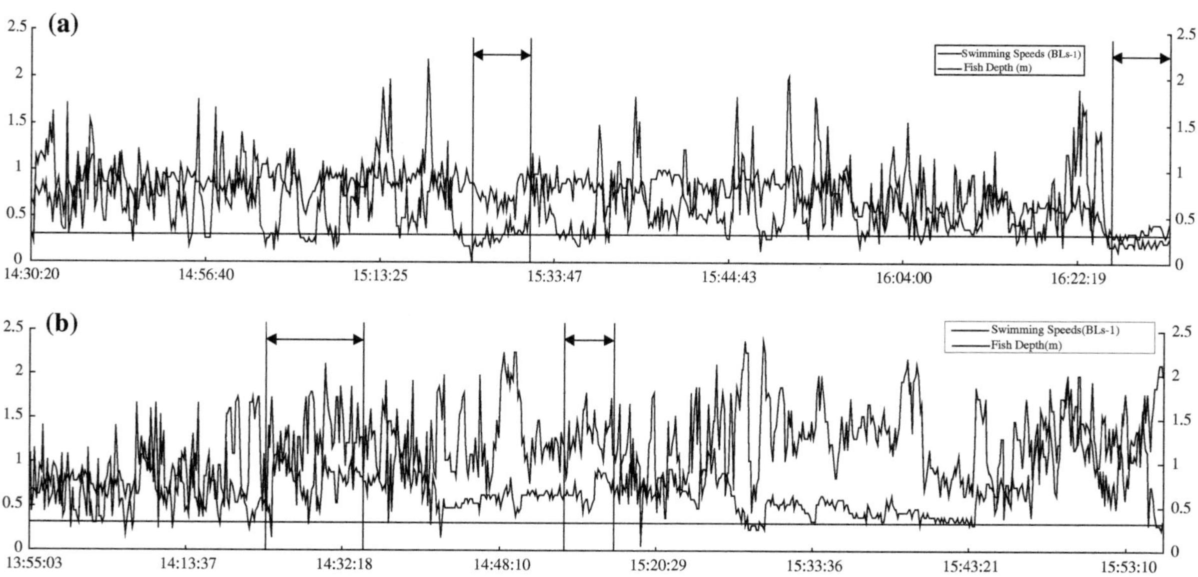

Fig. 6 Time-series plots of swimming speeds for chum salmon #1 (**a**) and pink salmon #4 (**b**). The solid lines represent a depth maximizing the wave drag calculated from individual' body diameter using Hertel's equation. The dotted lines represent holding periods of chum and pink salmon

calculated from individual body diameter using Hertel's equation. The depth maximizing the wave drag of chum salmon is shallower than pink salmon because of pink salmon's larger body diameter. Nicholas (2004) explained that large fish such as chinook salmon (*O. tshawytscha*) swim upstream further from the bank than smaller salmon such as sockeye, and found that fish optimize the trade-off between swimming in deeper water to reduce wave drag and swimming in shallower water to reduce skin friction and form drag. However, pink salmon used in our study were swimming in shallow depth in spite of expending high energy. We suppose that pink salmon could not optimize their swimming depth during upstream migration because of the shallow depth in the reconstructed segment of the Shibetsu River. We found that SEI of pink salmon was not significantly different between the canalized and reconstructed segments. However, pink salmon swam significantly faster than chum salmon in the reconstructed segment, suggesting that pink salmon used more energy than chum salmon in the reconstructed segment. It is becoming apparent that behavior is an extremely important factor affecting energy use (Rand & Hinch, 1998). In the reconstructed segment of the Shibetsu River, apparently the flow conditions are still not sufficiently complex to offer sufficient holding habitat for chum salmon in 2002 (Akita et al., 2006). We suspect that pink salmon may need to swim deeper than chum salmon to optimize energy use in the reconstructed segment. Relatively high levels of energy use may be cause pre-spawning mortality. The reconstructed segment of the Shibetsu River might have a shallower depth for upstream migration of pink salmon as a matured reconstructed segment. Thus, it is important for a river restoration to create a suitable river environment for upstream migration salmon considering species-specific migration behavior patterns.

A velocity and discharge of a river were also an important determinant of swimming patterns. Hinch et al. (1996) and Hinch & Bratty (2000) found that faster surface, turbulent and multi-directional currents in the Fraser River caused sockeye salmon to have difficulty locating appropriate migration paths, so individuals had to cross a river and they increased swimming speed variability. Our study was conducted in the Shibetsu River, which had relatively slow (0.65 m s^{-1} in September and 0.63 m s^{-1} in October) and less turbulent current. Hinch & Rand (2000) found that upstream-migrating sockeye salmon adjusted their swimming speeds to achieve ground speeds when they were at less turbulent and slow water sites that resulted in energetically minimal costs. However, when the currents encountered were relatively fast and turbulent, sockeye salmon chose to swim relatively fast and not elicit this energy saving behavior. They speculated that at fast, turbulent sites, salmon might be attempting to minimize their travel times because they only have a short while to live and significant migration impediments could seriously affect arrival times on spawning grounds and hence negatively affect spawning success. The current condition of the Shibetsu River is relatively slow and calm, and salmon in the Shibetsu River do not need to migrate a long distance. Because the Shibetsu River is a slow and short river, energy may not be determined their behavior. The holding behavior may be a way to get them into a staging area so that they can await the arrival of other salmon for optimal spawning encounters. Male salmon in the Shibetsu River may wait for the females to show up so that the timing of mate selection and reproduction is optimized. Perhaps the differences between salmon species are an effect of differences in mate and/ or nest site selection or competition among males and their dominance hierarchy. We supposed that it was important to acknowledge that in 'easier' rivers, other selection pressures rather than energy availability, may be important.

Based on the present analyze chum and pink salmon are swimming with different behavior patterns. This information would be important for a restoration of river. In the United States, many stream restoration projects are undertaken, however, the performance of these projects is rarely assessed. Post-project performance evaluations (e.g., Morris & Moses, 1999) are needed so that we can learn from our collective experiences and thereby improve future project design (Kondolf,

1995). Kondlf (2000) indicated that to be successful river restoration must be account for geomorphic influence. In order to success the restoration project, information on species-specific upstream behavior patterns may be useful to create a suitable river habitat for each upstream migration salmon respectively in the future. However, in our study we used only male experimental fish. Hinch et al. (2002) found that sockeye and pink salmon solve their migration challenges by swimming with very different strategies, although the role of large size was not clear. Sockeye females are the same size as pink males yet their respective swimming patterns were different. To further clarify the difference in upstream migratory strategy between chum and pink salmon, studies that use additional methods, for example using female experimental fish and a large swimming tunnel capable of creating water velocities, are needed for better understanding of the differences in the upstream migratory behavior of chum and pink salmon.

Acknowledgements We thank the following people for their help and co-operation: M. Mizoguchi and A. Nitta (Japan NUS Co., LTD) for technical support, Dr. J. Leonard (Northern Michigan University) for assistance with the study design, and H. Kakizaki (Nemuro Salmon Propagative Association) for providing experimental animals. We also thank the staff of Hokkaido Farming Fisheries Promotion Corporation for their support. This study was supported in part by Grant in-Aid for scientific Research. (A) (18208017) from Japanese society for the promotion of science, and from the Foundation of Riverfront Improvement and Restoration.

References

Akita, M., Y. Makiguchi, H. Nii, K. Nakao, J. F. Sandahl & H. Ueda, 2006. Upstream migration of chum salmon through a restored segment of the Shibetsu River. Ecology of Freshwater Fish 15: 125–130.

Beddow, T. A. & R. S. McKinley, 1999. Importance of electrode positioning in biotelemetry studies estimating muscle activity in fish. Journal of Fish Biology 54: 819–831.

Bernatchez, L. & J. J. Dodson, 1987. Relationship between bioenergetics and behavior in anadromous fish migrations. Canadian Journal of Fisheries and Aquatic Sciences 44: 399–407.

Brett, J. R., 1983. Life energetics of sockeye salmon, *Oncorhynchus nerka*. In Aspey, N. P. & S. I. Lustick (eds), Behavioral Energetics: The Cost of Survival in vertebrates. Ohio State University Press, Columbus, 29–63.

George, J. C., 1962. A histophysiological study of the red and white muscles of the mackerel. The American Midland Naturalist 68: 487–494.

Henderson, F. M., 1966. Open Channel Flow. Collier Macmillan Canada Ltd., New York.

Hertel, H., 1966. Structure Form and Movement. Reinhold, New York.

Hinch, S. G., E. M. Standen, M. C. Healey & A. P. Farrell, 2002. Swimming patterns and behavior of upriver-migration adult pink (*Oncorhynchus gorbuscha*) and sockeye (*O. nerka*) salmon as assessed by EMG telemetry in the Fraser River, British Columbia, Canada. Hydrobiologia 483: 147–160.

Hinch, S. G. & J. Bratty, 2000. Effects of swim speed and activity pattern on success of adult sockeye salmon migration through an area of difficult passage. Transactions of the American Fisheries Society 129: 598–606.

Hinch, S. G. & P. S. Rand, 2000. Optimal swimming speeds and forward-assisted propulsion: energy-conserving behaviors of upriver-migrating adult salmon. Canadian Journal of Fisheries and Aquatic Sciences 57: 2470–2478.

Hinch, S. G. & P. S. Rand, 1998. Swim speeds and energy use of upriver-migration sockeye salmon (*Oncorhynchus nerka*): role of local environment and fish characteristics. Canadian Journal of Fisheries and Aquatic Sciences 55: 1821–1831.

Hinch, S. G., R. E. Diewert, T. J. Lissimore, S. M. J. Prince, M. C. Healey & M. A. Henderson, 1996. Use of electromyogram telemetry to assess difficult passage areas for river-migrating adult sockeye salmon. Transactions of the American Fisheries Society 125: 253–260.

Hughes, N. F., 2004. The wave-drag hypothesis: an explanation for size-based lateral segregation during the upstream migration of salmonids. Canadian Journal of Fisheries and Aquatic Sciences 61: 103–109.

Kondolf, G. M., 1995. Learning from stream restoration projects. Proceedings of Fifth Biennial Watershed Management Conference. Ashland, Oregon, November 1994, California Water Resources Center Report No. 86: 107–110.

Kondolf, G. M., 2000. Some suggested guidelines for geomorphic aspects of anadromous salmonid habitat restoration proposals. Society for Ecological Restoration 8: 48–56.

Leonard, J. B. K. & S. D. McCormik, 1999. Effects of migration distance on whole-body and tissue-specific energy use in American shad (*Alosa sapidissima*). Canadian Journal of Fisheries and Aquatic Sciences 56: 1159–1171.

McFarlane, W. J., K. F. Cubitt, H. Williams, D. Rowsell, R. Moccia, R. Gosine & R. S. McKinley, 2004. Can feeding status and stress level be assessed by analyzing patterns of muscle activity in free swimming rainbow trout (*Oncorhynchus mykiss Walbaum*)? Aquaculture 239: 467–484.

McKinley, R. S. & G. Power, 1992. Measurement of activity and oxygen consumption for adult lake sturgen (*Acipenser fulvescens*) in the wild using radio-transmitted EMG signals. In Priede, A. G. &

S. M. Swift (eds), Wildlife Telemetry: Remote Monitoring and Tracking of Animals. Ellis Horwood, West Sussex, U.K., 307–318.

Mellas, E. J. & J. M. Haynes, 1985. Swimming performance and behavior of rainbow trout (*Salmo gairdneri*) and white perch (*Morone americana*): effects of attaching telemetry transmitters. Canadian Journal of Fisheries and Aquatic Sciences 42: 488–493.

Mizuno, N. & K. Gose, 1981. Ecology of streams. Tsukijishokan, Tokyo, 234 pp (in Japanese).

Morris, S., T. Moses, 1999. Urban stream rehabilitation: a design and construction case study. Environmental Management 23(2): 165–178.

Økland, F., B. Finstad, R. S. McKinley, E. B. Thorstad & R. K. Booth, 1997. Radio-transmitted electromyogram signals as indicators of physical activity in Atlantic salmon. Journal of Fish Biology 51: 476–488.

Osborne, M. F. M., 1961. The hydrodynamic performance of migratory salmon. Journal of experimental biology 38: 365–390.

Rand, P. S. & S. G. Hinch, 1998. Swim speeds and energy use of up-river migration sockeye salmon: simulating metabolic power and assessing risk of energy depletion. Canadian Journal of Fisheries and Aquatic Sciences 55: 1832–1841.

Tanaka, H., Y. Takagi, M. Iwata & Y. Naito, 1998. The behaviour and ambient temperature of homing chum salmon monitored by a data logger. Proceedings of the NIPR Symposium on Polar Biology 11: 62–73.

Tanaka, H., Y. Takagi & Y. Naito, 2000. Behavioural thermoregulation of chum salmon during homing migration in coastal water. The Journal of Experimental Biology 203: 1825–1833.

Tanaka, H., Y. Takagi & Y. Naito, 2001. Swimming speeds and buoyancy compensation of migrating adult chum salmon *Oncorhynchus keta* revealed by speed/depth/acceleration data logger. The Journal of Experimental Biology 204: 3895–3904.

Thorstad, E. B. F. & B. Finstad, 2000. Effects of telemetry transmitters on swimming performance of adult Atlantic salmon. Journal of Fish Biology 57: 531–535.

Webb, P. W., D. Sims & W. W. Scheltz, 1991. The effects of an air/water surface on the fast-start performance of rainbow trout (*Oncorhynchus mykiss*). The Journal of Experimental Biology 155: 219–226.

Winter, J. D., 1996. Advances in underwater biotelemetry. American Fisheries Society, Bethesda, Maryland, 555–590.

Hydrobiologia (2007) 582:55–65
DOI 10.1007/s10750-006-0541-1

FISH TELEMETRY

Riverine and early ocean migration and mortality patterns of juvenile steelhead trout (*Oncorhynchus mykiss*) from the Cheakamus River, British Columbia

Michael C. Melnychuk · David W. Welch · Carl J. Walters · Villy Christensen

Abstract High mortality (65–73%) occurred in the first month of the smolt migration in a population of wild steelhead trout. We used acoustic telemetry to monitor the downstream, estuarine, and early ocean migration of tagged smolts and estimate their mortality rates. After entering the Strait of Georgia most smolts migrated north through Johnstone and Queen Charlotte Straits rather than south through the Strait of Juan de Fuca. Of 51 smolts tagged in 2004 (49 in 2005), 36–38 (41–42) survived to leave freshwater and 14–19 (13–14) survived to leave the Strait of Georgia system. Mortality rates in separate segments of the migration were correlated with segment distances. An additional component of mobile sampling showed that few smolts died during the migration through Howe Sound. Migration rates averaged 0.7–0.9 body lengths per second (BL s^{-1}) downstream and 1.0–2.6 BL s^{-1} in ocean waters. Aggregated detection probabilities of 92–96% on lines of ocean receivers suggest that migration routes of small fishes can be quantified over several hundred kilometres, and survival rates can be estimated for even a modest number of tagged fish. Quantifying mortality patterns during the smolt migration could help to determine causes of low marine survival rates observed in recent years.

Keywords Salmonid · Marine survival · Migration · Smolt · Mortality

Guest editors: P. R. Almeida, B. R. Quintella, M. J. Costa and A. Moore
Developments in Fish Telemetry

M. C. Melnychuk (✉) · C. J. Walters · V. Christensen
Fisheries Centre, University of British Columbia, 2202 Main Mall, Vancouver, BC, Canada V6T 1Z4
e-mail: mikem@zoology.ubc.ca

D. W. Welch
Kintama Research Corp., 4737 Vista View Cres., Nanaimo, BC, Canada V9V 1N8

Introduction

Widespread declines in the abundance of Pacific salmonid populations have sparked a major conservation concern in western North America during the last decade. During the 1990s in southern British Columbia, Canada, for example, marine survival rates of steelhead trout (*Oncorhynchus mykiss*) decreased, mirroring concurrent declines in population abundances (Smith & Ward, 2000). Many steelhead populations are now at near-extinction levels. The abundance declines seem to be caused primarily by marine rather than freshwater factors (Welch et al., 2000), but these factors are not yet well understood. Descriptive data on spatial and temporal patterns of marine mortality in juvenile salmon could improve our ability to determine causes of the high mortality rates observed in recent years.

Anadromous Pacific salmon species (including steelhead) migrate as smolts from freshwater to marine habitats. Steelhead are renowned for their long-distance offshore migrations during their 1–4 years of ocean life. Some river mouths near inshore coastal areas are still hundreds of kilometres away from offshore waters where adult steelhead typically reside. Migration between riverine and offshore habitats and time spent feeding during the migratory period likely expose smolts to considerable predation risks (Walters & Juanes, 1993). Mortality during the initial migratory period is probably an important component of total marine mortality rates (Pearcy, 1992).

We conducted a broad biotelemetry study in 2004 and 2005 to measure movements and estimate mortality of salmonid smolts from several populations in British Columbia during their migratory period. The Pacific Ocean Shelf Tracking (POST) project is in the early stages of developing a continental-scale permanent telemetry array for the west coast of North America. In this paper, we present the results for one population of wild steelhead trout smolts tagged within this larger project. We used the POST array of acoustic receivers, as well as a mobile tracking platform on a smaller spatial scale, to estimate where and when mortality occurred in the first few weeks of the migration down the Cheakamus and Squamish Rivers, through Howe Sound, the Strait of Georgia, and out Juan de Fuca or Queen Charlotte Strait.

Methods

Study site

The Cheakamus River drains into the Squamish River before reaching Howe Sound, B.C. (16.7 km from the release point, Fig. 1). Howe Sound drains into the Strait of Georgia about 40 km south of the river mouth. Average annual discharge of the Squamish River is 240 $m^3\ s^{-1}$, with highest flows in June and July. Exit routes to the Pacific Ocean from the Strait of Georgia

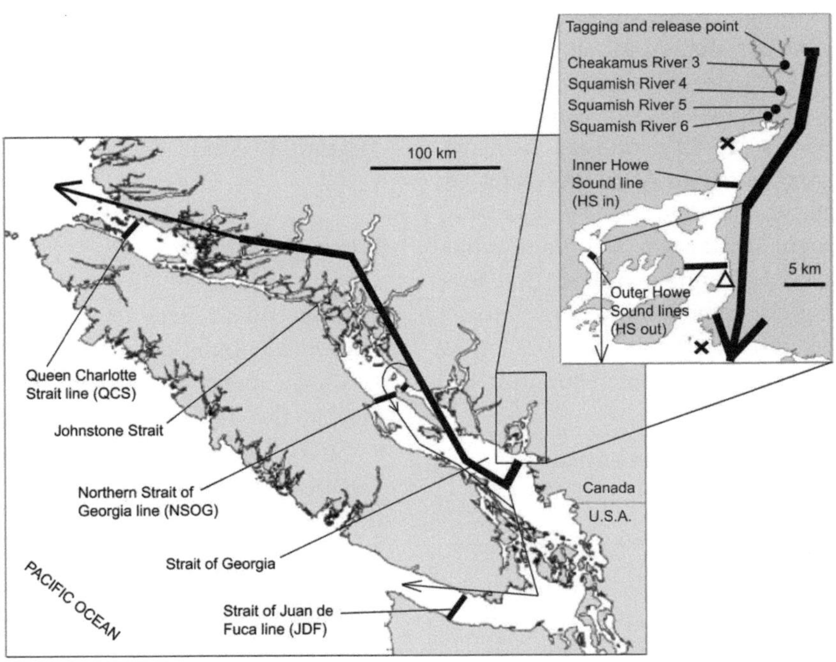

Fig. 1 Map of study area in Strait of Georgia system. Inset map shows Howe Sound. Stationary receiver locations are shown with black lines (ocean) and circles (river) and labelled. Black arrows alongside receiver locations show direction of travel across receiver lines; arrow width is proportional to the number of tags detected on that line in 2004 (similar in 2005). Locations of motionless steelhead tags identified by mobile tracking are shown by X's (2004) or an open triangle (2005)

include Johnstone Strait and Queen Charlotte Strait to the north, and the Strait of Juan de Fuca to the south. Tidal currents are relatively strong in these exit straits (Thompson, 1981).

Smolt tagging

Steelhead smolts were caught in 2004 and 2005 in side channel fence traps or rotary screw traps in the upper Paradise Valley reach of the Cheakamus River. In both years, fish were tagged and released in two groups during the migratory period. To ensure adequate sample sizes, we held the fish as they were captured daily until the time of tagging (2004: up to 21 days for first group of 42 fish and 16 days for second group of nine fish; 2005: up to 17 days for first group of 44 fish and 12 days for second group of five fish). The first group of fish in 2004 was held in the river 4–5 days after tagging to monitor fish for evidence of tag rejection, infection, or altered behaviour—none were observed. The second group in 2004 and both groups in 2005 were released 1 day after tagging. Smolts were released at dusk on May 8 and May 24 2004, and on May 6 and May 19 2005. There were no differences in survival rate estimates or travel speeds between the two release groups in either 2004 or 2005. Smolts measured 14.8–22.6 cm fork length in 2004 (FL, $\bar{x}$ = 18.4 cm) and 15.3–21.2 cm in 2005 ($\bar{x}$ = 17.8 cm). Body sizes were similar between release dates in both years.

We surgically implanted VEMCO Ltd. V8SC-6L acoustic tags into smolts (Moore et al., 1990). Tags measured 20 mm × 9 mm diameter and weighed 3.3 g in air, or 5% of average body weight and 11% of average length (Lacroix et al., 2004). Coded transmissions at 69.0 kHz allow individual tags to be identified by acoustic receivers on hydrophones. Tags emitted sound pulses at an average of 1 min^{-1} so that batteries would last >4 months. Smolts were sedated with 1 ppm metomidate HCl (Aquacalm™), and 3–4 min prior to surgery they were brought under full anaesthesia with 70 ppm MS-222 mixed with a 140 ppm $NaHCO_3$ buffer. Gills were irrigated during surgery with running water and 50 ppm MS-222 (100 ppm $NaHCO_3$) flowing into the mouth. A mucous-protectant (Vidalife™) was used in all water baths. A tag was inserted into the abdominal cavity through a mid-ventral line incision, which was closed with 2–3 polydioxanone monofilament sutures. Fish were transferred back to a trap in the river channel to recover.

Stationary acoustic array

Prior to releasing smolts, we installed VEMCO Ltd. VR-2 receivers in listening lines raised from the seabed to within 200 m depth across Howe Sound (2 lines; 4 and 10 receivers), the northern Strait of Georgia (26 receivers), Queen Charlotte Strait (22), and the Strait of Juan de Fuca (27). We also installed four receivers as single units in the Cheakamus and Squamish rivers (Fig. 1). During average river discharge and high tide, salt water reaches past Squamish River receiver 5 (>4 km). Receivers were retrieved from the rivers at least 31 days after the first smolt release and from the seabed lines at least 68 (Howe Sound) or 114 days (Straits of Juan de Fuca and Queen Charlotte) after release. Range tests in Howe Sound showed that tags were detected reliably up to 450 m away on ocean receivers (Welch et al., unpublished data): >30% of emissions were detected at tag depths of 2–20 m and >90% were detected at 40–130 m depth. Detection ranges of in-river receivers were lower (up to about 300 m) and more variable than those of ocean receivers, so detection probabilities were calculated separately for river and ocean receivers.

Mobile tracking

To complement tag detection from the stationary listening lines, we tracked smolt movements and looked for evidence of mortality within Howe Sound. Tags remaining in the same location over several days or weeks and not detected thereafter on any receiver were assumed to be mortalities, with a tag presumably lying on the seabed either in a dead smolt or previously defecated by a predator (we discuss this potential bias later). We used a VEMCO Ltd. VR-28 receiver with VH-40 directional hydrophone towed behind a boat to follow individual fish. We conducted parallel transects spaced 500–700 m apart at 10–11 km h^{-1}, and covered >90% of the area of Howe Sound. When a tag was detected, we moved the vessel toward it until the tag was decoded and its position was triangulated to within 100–200 m.

Data analyses

During the migration out of the Strait of Georgia system, tagged smolts passed up to nine detection points: four receivers in the Cheakamus and Squamish rivers, a mobile receiver, two stationary lines of receivers in Howe Sound, and one or two lines of receivers covering the southern and northern exit routes from the Strait of Georgia. Unless tag movement patterns were unusual, we assumed that tags detected sequentially along migratory routes were still in live smolts and not in predator stomachs (we discuss this potential bias later). Stationary lines of receivers were designed to provide a high but not perfect probability of detection. It was necessary to estimate detection probabilities at each river receiver or line of ocean receivers in order to estimate the extent of migration of tagged fish. We calculated the probability of detection at a given detection point i, $P(d_i)$, as the number of tagged fish detected at that point divided by the number of tagged fish known to have been there (= fish detected + fish not detected but detected at any later point). Extending this, the probability that a tagged smolt arrived at the mouth of the Squamish River undetected on all four river receivers is:

$$P(\text{no detections in rivers}) = \prod_{i=1}^{4} (1 - P(d_i)) \quad (1)$$

We estimated the survival rate to a given detection point simply from counting the number of fish detected at that point plus the number not detected but detected at any later point in their migration. Thus, we make the simplifying assumption that a smolt died between lines of receivers ‘a’ and ‘b’ if it was detected on ‘a’ but not on ‘b’ or thereafter. This estimate is uncertain because some smolts swam past one or more detection points undetected, especially past river receivers. On the three lines of ocean receivers in Howe Sound and the Strait of Georgia, where detection probabilities were higher, we estimated the number of fish that crossed each line undetected by dividing the number of fish detected on that line by the detection probability on that line. Similarly, we estimated the number of fish crossing the two outer lines of ocean receivers undetected by dividing the number of fish detected on each line by the aggregated detection probability on the three inner ocean lines. These estimates of undetected fish were used as estimates of uncertainty in mortality at each line of ocean receivers.

Pearson’s correlation coefficients were calculated to relate segment-specific mortality rates with segment distances. To determine speed of travel between stages, travel times were estimated from the last detection at one stage to the first detection at the next stage (most fish were detected multiple times at each stage). Shortest-route feasible distances were calculated between lines of receivers using the chart plotting program Memory Map 4. Travel speeds were converted to units of body length per second (BL s^{-1}).

Results

Given our assumptions of what constitutes mortality, at least 14 of 51 (27%) tagged steelhead smolts in 2004 and 13 of 49 (27%) in 2005 survived their downstream, estuarine, and inshore oceanic migrations to leave the Strait of Georgia system (Table 1). In 2004, four of these survivors left via the Strait of Juan de Fuca and 10 left to the north via Queen Charlotte Strait; in 2005 all 13 left via Queen Charlotte Strait. An additional 15 smolts in 2004 and 13 in 2005 crossed the northern Strait of Georgia line but were not detected thereafter (Fig. 1).

In total, we estimate that 38 smolts (75%) in 2004 and 42 (86%) in 2005 survived the downstream migration to the lower Squamish River. Two of these in 2004 and one in 2005 were not detected on stationary ocean receivers or in Squamish Harbour despite extensive mobile sampling, so may have died before they reached the actual river mouth 1.4 km downstream of Squamish River receiver 6 (in 2004 one of these moved past the receiver almost every night for 3 weeks,

Table 1 Steelhead smolts detected, estimate of minimum number surviving, steelhead detection probabilities, and days from release until arrival at each line of stationary receivers

Receiver line	Smolts detected		Number (%) of smolts surviving to line		% detection probability at line		Days until arrival, mean (95% conf. lim.)	
	2004	2005	2004	2005	2004	2005	2004	2005
Cheakamus River 3	2	6	38 (75)	44 (90)	5.3	13.6	3.1	0.8 (0.1–1.5)
Squamish River 4	19	19	38 (75)	42 (86)	50.0	45.2	2.1 (0.9–3.2)	2.5 (1.6–3.3)
Squamish River 5	7	27	38 (75)	42 (86)	18.4	64.3	1.4 (0.4–2.5)	3.0 (2.2–3.8)
Squamish River 6	35	23	38 (75)[b]	42 (86)[b]	92.1	54.8	2.7 (2.0–3.5)	2.6 (1.8–3.4)
Howe Sound inner	32	37	35 (69)	41 (84)	91.4	90.2	4.2 (3.3–5.2)	4.8 (3.8–5.7)
Howe Sound outer	34	39	34 (67)	40 (82)	100.0	97.5	5.6 (4.5–6.6)	5.4 (4.5–6.3)
Northern Strait of Georgia	24[a]	22	25 (49)	26 (53)	96.2	84.6	14.4 (12.8–16.1)	14.1 (12.4–15.8)
Strait of Juan de Fuca	4	0	4 (8)	0 (0)			21.8 (15.5–28.1)	
Queen Charlotte Strait	10	13	10 (20)	13 (27)			26.5 (21.2–31.8)	23.5 (22.3–24.7)
Total surviving to leave Strait of Georgia system			14 (27)	13 (27)				
Aggregated detection probability on all river receivers (%)					41.4	44.1		
Aggregated detection probability on all ocean receiver lines (%)					95.8	91.6		

[a] Plus 1 smolt that was detected on northern Strait of Georgia line before leaving via Strait of Juan de Fuca

[b] Includes two smolts in 2004 and one in 2005 that may not have reached the river mouth 1.4 km downstream of Squamish River 6

usually between dusk and dawn, so may have been in a predator). The other smolts were either never detected (12 in 2004, four in 2005), were detected downstream of the release site in the Cheakamus River only (two in 2005) or were detected upriver of the release site but not again (one in 2004, one in 2005). Of the smolts that entered Howe Sound (36–38 in 2004, 41–42 in 2005), 34 in 2004 and 40 in 2005 were detected on the outer Howe Sound line, presumably as they left the sound (Fig. 1, Table 1). Two tags in 2004 (in upper Howe Sound and just outside of Howe Sound) and one tag in 2005 (lower Howe Sound) were found when we first began mobile sampling and remained motionless for the rest of the study (1.5, 16, or 8 weeks; Fig. 1) .

Of the smolts that left Howe Sound (minimum 33 in 2004, 39 in 2005), 29 in 2004 and 26 in 2005 crossed the northern Strait of Georgia or Strait of Juan de Fuca lines, leaving several smolts unaccounted for in the Strait of Georgia (four in 2004, 13 in 2005). Some of these in 2004 may well have left undetected via the Strait of Juan de Fuca; 9 of 27 receivers were lost on that line, probably due to shrimp trawlers or strong tidal currents. If all four fish left undetected, then 18 smolts (35%) rather than 14 survived to leave the Strait of Georgia system. All receivers in 2005 were recovered. One tag in 2005 was detected in the Strait of Georgia with mobile sampling 76 days after release, but repeated sampling was not possible.

Several smolts were unaccounted for in Johnstone Strait/Queen Charlotte Strait, having crossed the northern Strait of Georgia line but not the Queen Charlotte Strait line (15 in 2004, 13 in 2005). It is unlikely that many of these smolts crossed the Queen Charlotte Strait line undetected, as detection probabilities on intact lines of ocean receivers in Howe Sound and the northern Strait of Georgia were at least 91% in 2004 and 85% in 2005 (Table 1; the uncertainties in survival to the Queen Charlotte Strait line due to probabilities of non-detection at that line are represented as error bars in Fig. 2). In contrast, detection probabilities on river receivers were generally much lower (aggregated detection probability, 41% in 2004 and 44% in 2005). It is possible that some of the smolts unaccounted for in the rivers actually bypassed one or more river receivers and made it part-way downstream, but it is unlikely they bypassed all river receivers undetected (0.031 probability for each fish in 2004 and 0.076 in 2005, Eq. 1) or made it very far through Howe Sound undetected.

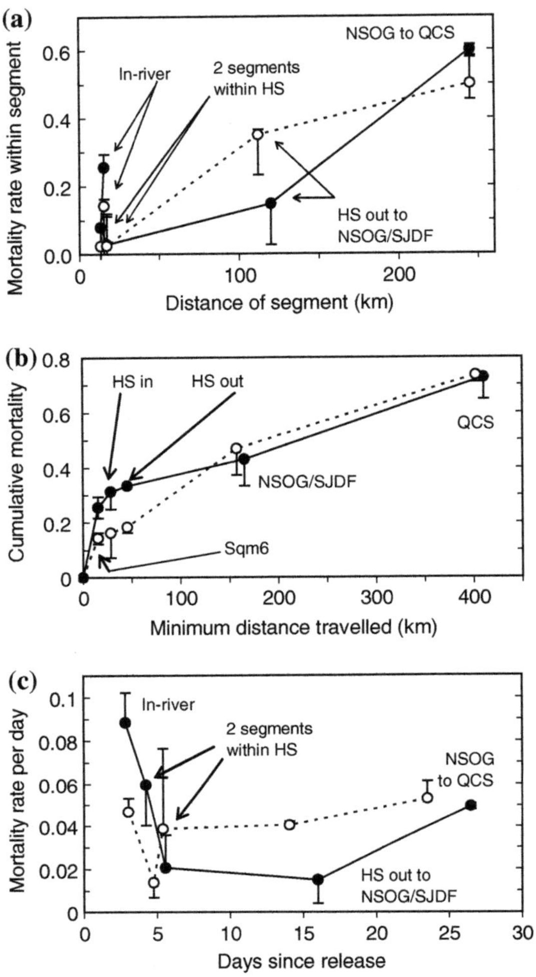

Fig. 2 (**a**) Segment-specific mortality rates between lines of receivers versus segment distances. (**b**) Cumulative mortality since release versus cumulative migration distance. (**c**) Daily segment-specific mortality rates versus cumulative migration days. Black circles and solid lines show estimates for 2004, with open circles and dashed lines for 2005. Segments and detection points are labelled, see Figure 1 for abbreviations. Error bars show added uncertainty in mortality rates due to: (i) the chance of smolts crossing a line of ocean receivers undetected either at the start or end of each segment (calculated from detection probabilities at each line); (ii) different assumptions of whether two smolts in 2004 and one in 2005 last detected in the lower river actually died in-river or between the river and HS in; (iii) up to four smolts crossing the SJDF line undetected in 2004 across unrecovered receivers. See Fig. 1 for abbreviations

Mortality rates within each segment of the migration (i.e., between lines of receivers) were correlated with segment distances (Fig. 2a; $r = 0.95$ in 2004 and 0.97 in 2005 if freshwater segments are excluded or $r = 0.85$ in 2004 and 0.95 in 2005 if freshwater segments are included; $P < 0.05$ in all cases). In both years, steelhead smolts suffered the highest segment-specific mortality (60%, 50%) between the northern Strait of Georgia and Queen Charlotte Strait lines, the longest monitored segment of their migration. The two shortest ocean segments had the lowest mortality rates (2–8%). Freshwater mortality deviated from the general trend in 2004, with higher mortality (25–29%) despite a relatively short distance of migration. Despite some differences between years in mortality rates within each segment (Fig. 2a), by the end of the migration out of the Strait of Georgia and again out of Queen Charlotte Strait, cumulative mortality rates were remarkably consistent between years (Fig.2b).

Mortality rates per day varied considerably during the smolt migration in 2004, but were relatively uniform between different segments of the migration in 2005. Daily mortality in 2004 was highest in-river (9–10% per day), decreased to 2% per day through mid-Howe Sound to the Strait of Georgia, and then increased to 5% per day between the northern Strait of Georgia and Queen Charlotte Strait lines (Fig. 2c). Excluding the segment between the Squamish River mouth and the inner Howe Sound line, daily mortality in all other portions of the 2005 migration ranged only from 4 to 5.5% per day (Fig. 2b).

Migration rates also varied over time, seen in absolute travel speeds between lines (the slopes in Fig. 3) as well as speeds standardized to body length (Fig. 4). In both years smolts took between 0.2 and 10.4 days ($\bar{x}$ = 2.7in 2004 and 2.6 days in 2005) from their time of release until arrival in the lower Squamish River, 15.3 km downstream (Fig. 3, Table 1). Migration rates between receiver lines continued to be highly variable once smolts left freshwater. In total, surviving smolts left Queen Charlotte Strait between 20–43 days after release ($\bar{x}$ = 26.5) in 2004 and between 20–28 days after release ($\bar{x}$ = 23.5) in 2005 (Table 1), having travelled a minimum distance of 406–422 km.

Body length had negligible effect on absolute travel speeds, but standardized speeds account for

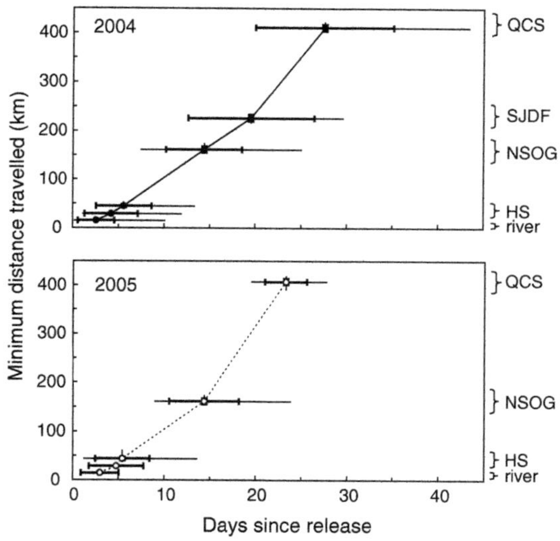

Fig. 3 Migration rate of steelhead smolts from release point in Cheakamus River downstream, through the Squamish River estuary, Howe Sound, and Strait of Georgia system. Minimum distances are shortest feasible routes between detection points. Thick error bars show 1 s.d. from the mean while whiskers show the range of times and distances across each line. See Fig. 1 for abbreviations

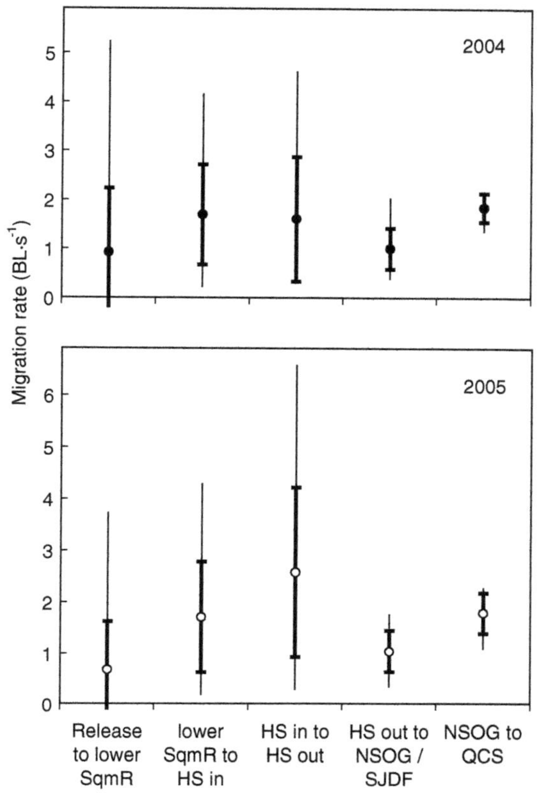

Fig. 4 Minimum travel speeds (body lengths per second) of steelhead smolts during different segments of their migration. Freshwater speeds are aggregated into a single measure from release until the first detection in the lower Squamish River. Thick error bars show 1 s.d. from the mean while whiskers show the range of travel speeds within each segment. See Fig. 1 for abbreviations

individual migration rates (Fig. 4). Migration rates downstream and through Howe Sound were highly variable in both years. The slowest average travel speeds occurred during the downstream migration ($\overline{x}$ = 0.9 BL s^{-1} in 2004 and 0.7 in 2005) despite some individuals showing rapid migrations of 3–5 BL s^{-1} (Fig. 4), probably assisted by river currents of about 1 m s^{-1}. Travel speeds were relatively fast through Howe Sound, then fell through the Strait of Georgia to 1.0 BL s^{-1}, and finally increased again in both years to 1.8 BL s^{-1} from northern Strait of Georgia to Queen Charlotte Strait. There was no consistent pattern between the estimated survival/mortality fate of individual smolts and either their absolute or standardized travel speeds; fish that died during their downstream or early ocean migration through Howe Sound, Strait of Georgia or Johnstone Strait migrated at a similar rate as surviving smolts.

Steelhead smolts showed a strong pattern of nocturnal migration in freshwater. Fish were rarely detected on the river receivers between dawn and dusk (Fig. 5). Six fish in 2004 and three in 2005 left the river before dawn after being released at 20:20–20:45 the night before, so did not migrate in-river during daylight hours. Only three fish in each year were detected on river receivers between 06:30–20:00. Two of these in 2004 were never detected on any ocean receivers or by mobile sampling, and it is possible these tags were already in predator stomachs when they were detected. The pattern of nocturnal migration broke down once smolts left freshwater. Steelhead were detected on lines of ocean receivers at all hours of the day without obvious pattern in the time they swam past.

Body size had only a small effect on the probability of survival downstream or through Howe Sound/Strait of Georgia. When the original size distribution of tagged smolts was split into three groups of freshwater mortalities, Howe

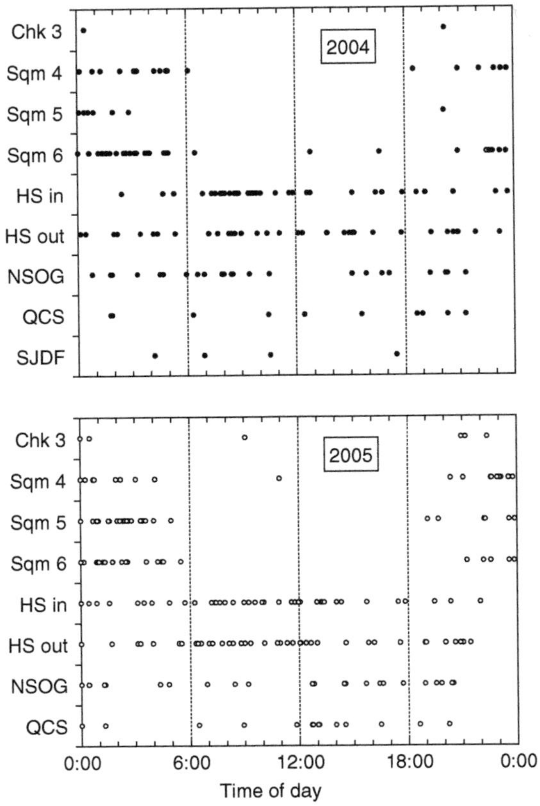

Fig. 5 Hourly downstream and ocean migration patterns of steelhead smolts. Data points show the time a fish was first detected at each detection line. Top four receiver locations are in rivers and bottom five/four are in the ocean. See Fig. 1 for abbreviations

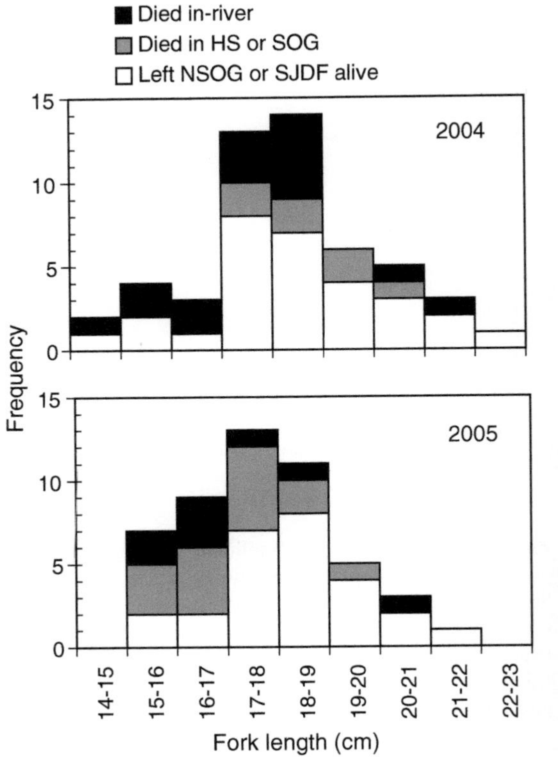

Fig. 6 Size distributions of tagged steelhead smolts. Frequencies of each size class are separated into fish presumed to have died in-river, those presumed to have died in HS or SOG, and those presumed to have left SOG alive. See Fig. 1 for abbreviations

Sound/Strait of Georgia mortalities, and survivors that crossed the northern Strait of Georgia or Strait of Juan de Fuca lines, the survivor distribution was slightly skewed towards larger fish and the mortality distributions were slightly skewed toward smaller fish, especially in 2005 (Fig. 6). Fish dying earlier in their migration tended to be among the smallest tagged, with larger fish more likely to survive the inshore migration.

Date of tagging and release between the two tag groups had no discernible effect on survival probabilities or travel speeds.

Discussion

Historical trends in total marine survival rates for this Cheakamus River steelhead population are unknown. Escapement and smolt enumeration data are now being collected, but reliable estimates of ocean mortality have not yet been made. Marine survival rates for other steelhead (e.g., Keogh River; Ward 2000) and coho salmon (*Oncorhynchus kisutch*) populations in southern British Columbia, however, declined during the 1990s and remain low. Causes for the survival declines are not well understood.

We found in both years that at least 27% of smolts tagged during their downstream migration survived to leave the Strait of Georgia system via Queen Charlotte or Juan de Fuca Straits. Survival was relatively high compared to several other steelhead, coho, and sockeye (*O. nerka*) salmon populations in southern British Columbia (Welch et al., in prep.), including a hatchery-reared coho population also from the Cheakamus/Squamish River watershed (only 24% of smolts in 2004 and

55% in 2005 left Howe Sound alive). Although early marine mortality was relatively low, the migration period was brief such that daily mortality rates were as high as 5–10% for various segments of the migration (Fig. 2c). Total marine survival rates for this population are probably <5–10% if similar to other steelhead populations, so the remaining 17–22% mortality must be distributed over the remaining 1–3 years of ocean life. These results suggest that considerable mortality occurs in both the downstream and inshore migration framed by our study area, as well as later in ocean life in offshore areas.

Daily mortality rates tended to be fairly uniform over the 2005 migration period but higher in-river than in ocean segments in 2004. Freshwater predators are prevalent: Dolly Varden char (*Salvelinus malma*) and bull trout (*Salvelinus confluentus*) inhabit the Cheakamus and Squamish Rivers and are known to prey on other salmonid smolts (Beauchamp & Van Tassell, 2001). Staghorn sculpins (*Leptocottus armatus*) inhabit the Squamish River estuary (Levy & Levings, 1978). Mergansers (*Mergus merganser*, *M. serrator*) and glaucous-winged gulls (*Larus glaucescens*) are also present in this watershed and have been important predators of coho smolts in other studies (Mace, 1983; Wood, 1987). Columbia River studies have shown that steelhead smolts are more vulnerable to avian predation than other salmonids (Collis et al., 2001). In addition, harbour seals frequent the lower Squamish River and estuary and could also capture steelhead smolts. Aside from predator density differences, the foraging efficiency of predators on smolts is conceivably greater in the shallower rivers or estuary than in deeper inshore ocean areas. Since smolts appeared to migrate downstream past receivers almost exclusively at night, they presumably held in pools or other areas of low current velocity during daylight hours, which could minimize exposure to visual predators.

The slight size-dependence in downstream and early ocean survival (Fig. 6) could represent either greater natural mortality or tag-related mortality in smaller fish. We were not able to distinguish between these possibilities. Lack of a strong body size effect on migration rate may reflect length-independent factors such as current-assisted travel downstream and through Howe Sound, or simply may not be evident given the limited sample size and a centre-heavy size distribution. Other studies have demonstrated a positive body size effect on marine survival of steelhead smolts (Ward et al., 1989).

Most smolts that survived the migration through the Strait of Georgia left via the northern route. Smolts travelled the >400 km distance at an average rate of 1.0 BL s^{-1} since release. Steelhead smolts from the Keogh River on northeastern Vancouver Island (Welch et al., 2004) and from Oregon (Pearcy et al., 1990) also showed rapid movements offshore soon after ocean entry. Exposure to predation during the early, inshore migration is probably a function of both distance travelled and residence time within inshore waters. Rapid migrations to offshore areas may reduce the temporal window of vulnerability in rivers and inshore waters where predator densities are typically higher than offshore areas (Huato, 2001).

We estimated successive mortality rates of steelhead smolts during their migration downstream and through several hundred kilometres of inshore waters on their way out of the Strait of Georgia system. In generating these estimates it was necessary to make assumptions of what we considered a surviving smolt or a mortality since fish were never recaptured. We recognise certain factors that could bias mortality estimates. We incorporated the most important of these, failure to detect a smolt as it crosses a detection point, when calculating the uncertainty in our estimates (see error bars in Fig. 2) using detection probabilities at each detection point. Other factors that could lead to overestimating natural mortality include tag loss or tag failure, tag-related mortality, and extreme delays in migration or residency away from detection points for the remainder of the study. We can not be certain of tagging effects in the wild, but laboratory studies have shown low incidences of tag-related mortality in salmonids >140 mm FL (Moore et al., 1990, Adams et al., 1998b, D. Welch et al., 2007), no or negligible effects on feeding or swimming behaviour (Moore et al., 1990, Adams et al., 1998a, b), and low incidence of tag shedding over a similar time frame to that of the present study (Moore et al., 1990, D. Welch et al., 2007). All steelhead smolts that were detected in both years showed rapid migration rates, but we can not rule out the possibilities that some fish delayed their migration several months or that some fish which were considered mortalities after repeated mobile sampling were actually residing in the same location (within 100–200 m) for several weeks. Conversely, failing to identify tags present in predator stomachs when detected could lead to underestimating mortality rates of the population. It is not unlikely that some detected tags were in predator stomachs at specific detection points. It is unlikely, however, that these predators would be detected several times at successive detection points showing similar migratory rates and routes to the conspecifics of the depredated smolt. A smolt eaten early in its migration would already be considered a mortality if it was not detected swimming across lines of ocean receivers tens to hundreds of kilometres away, so our cumulative mortality estimates are not affected. If a smolt was eaten close to one of the outer lines and carried across it, however, it is unlikely we would be able to tell it was not alive.

Detection probabilities for freshwater receivers were much lower than for lines of ocean receivers for several reasons: the noisier river environments likely reduced detection ranges; straight-line tag-to-receiver distances were often <100 m because of bends in the river; rivers were shallower with more possible obstructions between tag and receiver; and faster downstream travel speeds past receivers due to river flow meant that tags were within detection range for less time. Freshwater detection probabilities could be improved by pairing receivers in lines as they are in the ocean. The high detection probabilities evident on lines of ocean receivers in Howe Sound and the Strait of Georgia suggest that well-designed telemetry arrays can account for nearly all tags moving across lines.

A combination of tracking technologies allowed us to study the smolt migration at different spatial scales. The large-scale stationary array allowed us to estimate population survival rates over several hundred kilometres and the first month of the smolt migration. Smaller-scale mobile tracking allowed us to identify approximate times and locations of mortality for three smolts that died soon after ocean entry. The POST large-scale acoustic tracking array appears to have great potential for monitoring migrations of individual fish and estimating survival rates for salmonids and other marine fishes.

Acknowledgements We are grateful to Melinda Jacobs, Erin Rechisky, Adrian Ladouceur, Jayson Semmens, Paul Winchell, Heidi Lydersen, Scott Lincez, Orie McKenzie, Valeria Vergara, Pablo Trujillo, Dave Preikshot, Valerie Caron, Ross Thompson, Peter Rossing, David Thomson, Sherman Lai, David O'Brien, Ian Hatton, Line Christensen, Steve Martell, Dawit Tesfamichael, Pi Liu, Natalie Ban, Jonathan Anticamara, Doug Sandilands, Dal, Nathan Taylor, Brian Starzomski, Jordan Rosenfeld and Caroline Melville for their willing assistance in the field. We thank Martin Castonguay, Gilles Lacroix, and Derek Knox of Fisheries and Oceans Canada as well as Fred Goetz from the U.S. Army Corps of Engineers for generously lending acoustic telemetry equipment. We thank Sonia Batten and Isabelle Gaboury for data management help and all POST members for support. We thank Eric Taylor and two anonymous reviewers for helpful comments on the manuscript. Permission for surgical implant of tags into smolts was granted by the Department of Fisheries and Oceans Animal Care Committee. Funding was provided by the Fish America Foundation (American Sportfishing Association), the Pacific Salmon Foundation, the Gordon and Betty Moore Foundation, the Alfred P. Sloan Foundation (Census of Marine Life), and through a graduate scholarship from the National Science and Engineering Research Council of Canada to MM.

References

Adams, N. S., D. W. Rondorf, S. D. Evans & J. E. Kelly, 1998a. Effects of surgically and gastrically implanted radio transmitters on growth and feeding behavior of juvenile chinook salmon. Transactions of the American Fisheries Society 127: 128–136.

Adams, N. S., D. W. Rondorf, S. D. Evans, J. E. Kelly & R. W. Perry, 1998b. Effects of surgically and gastrically implanted radio transmitters on swimming performance and predator avoidance of juvenile chinook salmon (*Oncorhynchus tshawytscha*). Canadian Journal of Fisheries and Aquatic Sciences 55: 781–787.

Beauchamp, D. A. & J. J. Van Tassell, 2001. Modeling seasonal trophic interactions of adfluvial bull trout in Lake Billy Chinook, Oregon. Transactions of the American Fisheries Society 130: 204–216.

Collis, K., D. D. Roby, D. P. Craig, B. A. Ryan & R. D. Ledgerwood. 2001. Colonial waterbird predation on juvenile salmonids tagged with passive integrated transponders in the Columbia River estuary: Vulnerability of different salmonid species, stocks, and rearing types. Transactions of the American Fisheries Society 130: 385–396.

Huato, L. 2001. A modeling investigation of migratory behavior in fishes: A case study with sockeye salmon. Ph.D. Thesis, University of British Columbia, Vancouver, B.C.

Lacroix, G. L., D. Knox & P. McCurdy, 2004. Effects of implanted dummy acoustic transmitters on juvenile Atlantic salmon. Transactions of the American Fisheries Society 133: 211–220.

Levy, D. A. & C. D. Levings, 1978. A description of the fish community of the Squamish River estuary, British Columbia: Relative abundance, seasonal changes, and feeding habits of salmonids. Fisheries and Marine Service Manuscript Report, No. 1475.

Mace, P. M. 1983. Bird predation on juvenile salmonids in the Big Qualicum Estuary, Vancouver Island. Canadian Technical Report of Fisheries and Aquatic Sciences, No. 1176.

Moore, A., I. C. Russell & E. C. E. Potter, 1990. The effects of intraperitoneally implanted dummy acoustic transmitters on the behaviour and physiology of juvenile Atlantic salmon, *Salmo salar* L. Journal of Fish Biology 37: 713–721.

Pearcy, W. G., 1992. Ocean ecology of North Pacific salmonids. University of Washington Press, Seattle, Wash.

Pearcy, W. G., R. D. Brodeur & J. P. Fisher, 1990. Distribution and biology of juvenile cutthroat trout *Oncorhynchus clarki clarki* and steelhead *O. mykiss* in coastal waters off Oregon and Washington. Fisheries Bulletin 88: 697–711.

Smith, B. D. & B. R. Ward, 2000. Trends in wild adult steelhead (*Oncorhynchus mykiss*) abundance for coastal regions of British Columbia support the variable marine survival hypothesis. Canadian Journal of Fisheries and Aquatic Sciences 57: 271–284.

Thompson, R. E. 1981. The Oceanography of the British Columbia Coast. Canadian Special Publication of Fisheries and Aquatic Sciences. Bulletin 56, Fisheries and Oceans Canada, Ottawa, 291 pp.

Walters, C. J. & F. Juanes, 1993. Recruitment limitation as a consequence of natural selection for use of restricted feeding habitats and predation risk taking by juvenile fishes. Canadian Journal of Fisheries and Aquatic Sciences 50: 2058–2070.

Ward, B. R, 2000. Declivity in steelhead recruitment at the Keogh River over the past decade. Canadian Journal of Fisheries and Aquatic Sciences 57: 298–306.

Ward, B. R., P. A. Slaney, A. R. Facchin & R. W. Land, 1989. Size-biased survival in steelhead trout (*Oncorhynchus mykiss*): Back-calculated lengths from adults' scales compared to migrating smolts at the Keogh River, British Columbia. Canadian Journal of Fisheries and Aquatic Sciences 46: 1853–1858.

Welch, D. W., B. R. Ward, B. D. Smith & J. P. Eveson, 2000. Temporal and spatial responses of British Columbia steelhead (*Oncorhynchus mykiss*) populations to ocean climate shifts. Fisheries Oceanography 9: 17–32.

Welch, D. W., B. R. Ward & S. Batten, 2004. Early ocean survival and marine movements of hatchery and wild steelhead trout (*Oncorhynchus mykiss*) determined by an acoustic array: Queen Charlotte Strait, British Columbia. Deep-Sea Research II 51: 897–909.

Welch, D. W., S. D. Batten & B. R. Ward, 2007. Growth, survival, and tag retention of steelhead trout (*O. mykiss*) surgically implanted with dummy acoustic tags. Hydrobiologia 582: 289–299.

Wood, C. C., 1987. Predation of juvenile Pacific salmon by the common merganser (*Mergus merganser*) on eastern Vancouver Island. 1: Predation during the seaward migration. Canadian Journal of Fisheries and Aquatic Sciences 44: 941–949.

Hydrobiologia (2007) 582:67–79
DOI 10.1007/s10750-006-0558-5

FISH TELEMETRY

Conventional and EMG telemetry studies of upstream migration and tailrace attraction of adult Atlantic salmon at a hydroelectric installation on the Exploits River, Newfoundland, Canada

D. A. Scruton · R. K. Booth · C. J. Pennell · F. Cubitt · R. S. McKinley · K. D. Clarke

Abstract Conventional and electromyogram (EMG) radio telemetry studies have documented occurrence of tailrace attraction and residency, and associated energy cost, for migratory wild Atlantic at a power plant on the Exploits River, insular Newfoundland, Canada. All fish demonstrated some degree of tailrace attraction and turbine discharge was the primary factor resulting in 'false attraction'. In 2002, residency times were generally less than 1 h, although some fish demonstrated residency between 71 and 118 h. In 2004, after plant refurbishing, fish took from 6 to 11 days to reach the power plant and remained in the tailrace area from 3 to 12 days. Frequent entrances into the tailraces indicated fish were demonstrating a searching behaviour to find the upstream migration route. In 2003 and 2004, EMG data were collected from 3 fish in each year migrating to, and past, the tailraces. An Energy Index was calculated integrating EMG signal and time spent at each location to estimate potential energetic cost. Relatively high-energy expenditure was associated with tailrace attraction and residency in both years. In 2003, fish spent highest energy at the lowest tailrace (first encountered in their migration) while in 2004, there were also high-energy costs associated with a new generating unit. In both years, high EMGs both just below and above the power plant suggested these reaches were arduous and difficult to pass.

Keywords Migration · Atlantic salmon · Hydroelectric development · Electromyogram telemetry · Tailrace attraction · Energetics

Guest editors: P. R. Almeida, B. R. Quintella, M. J. Costa and A. Moore
Developments in Fish Telemetry

D. A. Scruton (✉) · C. J. Pennell · K. D. Clarke
Science Branch, Fisheries and Oceans Canada, P.O. Box 5667, St. John's, NL, Canada A1C 5X1
e-mail: scrutond@dfo-mpo.gc.ca

R. K. Booth
Lotek Wireless, 115 Pony Drive, Newmarket, ON, Canada L3Y 7B5

F. Cubitt · R. S. McKinley
Faculty of Agricultural Sciences, University of British Columbia, West Vancouver, BC, Canada V7V 1N6

Introduction

Migrating Atlantic salmon may migrate 1000s of kilometers to reach spawning and rearing habitats and, as they do not feed during upstream migrations, rely entirely on energy reserves to complete migration, gonad development, and spawning (Lucas & Baras, 2001). Hydroelectric dams are known to cause delay in the upriver migration of migratory species as fish are attracted to high water velocities and discharge, both hydrologic conditions associated with hydroelectric plant tailraces (Bernatchez & Dodson, 1987). Upstream migrants, when presented with channels of differing discharge, are attracted to the channel with highest velocity and flow resulting in "false-attraction" (Smith et al., 1997). Upstream migrants can be attracted to tailraces which result in injury and delay, increased exposure to predators, higher risk of disease, and environmental stress (Winstone et al., 1985). Migrations are energetically expensive, environmental conditions often adverse (e.g., elevated temperatures, high discharge), and distances long and difficult and significant impediments to migration could affect arrival times on spawning grounds and negatively affect spawning success (Bernatchez & Dodson, 1987). As Atlantic salmon do not feed on their upstream migration, and therefore have a fixed energy budget, any significant delays and energy expenditure related to tailrace entrainment could have costs in terms of migratory ability and gonadal production.

Little is known about the behavioural patterns, swimming strategies, and energetic demands on migrating anadromous fish in association with points of difficult passage, and in association with hydroelectric dams, power plant tailraces, and upstream fish passage systems. In this study, both conventional and physiological (electromyogram or EMG) radio telemetry was used to examine attraction and residency of anadromous Atlantic salmon to the tailraces associated with a hydroelectric plant at Grand Falls-Windsor, Newfoundland (Fig. 1). The relative energy cost used by fish at various points in their migration up the Exploits River, at the hydroelectric plant, and

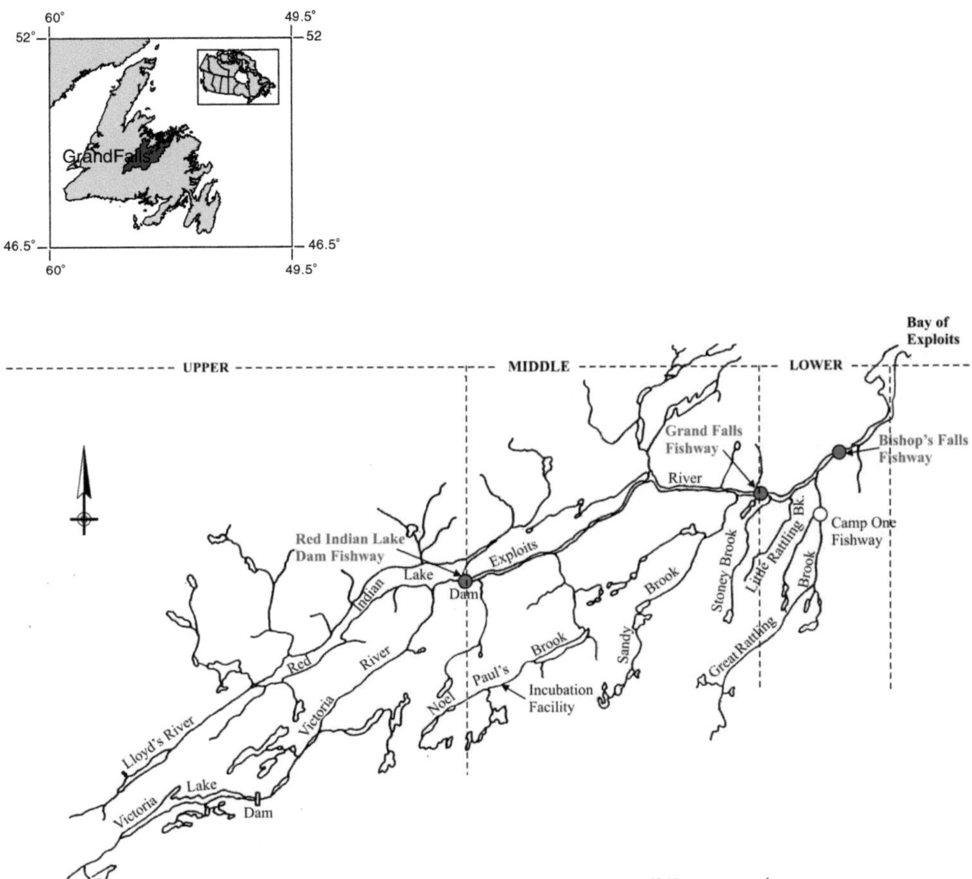

Fig. 1 Map of the Exploits River watershed, insular Newfoundland, Canada, showing the location of Grand Falls and Bishop's Falls

subsequently at an upstream fishway was determined using EMG radio telemetry. Studies were conducted before and after a major modification to the power plant which increased the generating capacity (from 45 MW to 78.5 MW), altered plant operation ('run of river' operation to 'run of river' with peaking operation), and modified the orientation of the new tailrace (more perpendicular to river flow). The paper describes the degree and nature of tailrace attraction and residency, and associated relative energy expenditure, at this facility, before and after the plant modification.

Study site

The Exploits River is the longest river (246 km in length) and largest drainage basin (11,272 km^2) on the island of Newfoundland and sustains one of the largest runs of wild adult Atlantic salmon (*Salmo salar*) in North America, developed through a major enhancement program and construction of fish passage facilities at major natural and man made barriers in the watershed (Bourgeois et al., 1994) (Fig. 1). Two large hydroelectric facilities, at Bishop's Falls and Grand Falls-Windsor, 10 km and 22 km upstream of the river mouth, respectively, have been operated by Abitibi Consolidated Company of Canada (ACCC) since the early 1900s to supply power to a pulp and paper mill. These two facilities represent important upstream and downstream migration challenges for migrating salmon. At Grand Falls, the river immediately below the power plant tailraces exits into a narrow section of the river formed by a deep gorge of some 500 m, which represents a natural difficult point of passage. The river reach above the power plant up to the fishway at Grand Falls, approximately 350 m in length, is also a difficult natural migration reach consisting of high gradient (velocity) stretches with a number of chutes and small falls.

In 2001, the ACCC proposed installing additional power generating capacity at Grand Falls-Windsor, which required a new intake and turbine (Beeton unit) and modifications to the dimensions and orientation of the tailraces. Existing generating capacity was 45 MW and addition of the new generating unit increased capacity to 78.5 MW. Prior to the refurbishing, the plant was run primarily as a 'run of river' operation, and the additional generating capacity permitted additional peaking capacity during high flow periods (primarily the spring snow melt). There was concern that operation of the larger turbine units (existing and Beeton unit; combined flow of approx. 190 m^3 s^{-1}), in combination of gorge topography, could result in higher tailrace attraction among salmon migrating past the facility, thereby delaying migration and could potentially even be a migration (velocity) barrier. Conditions for tailrace attraction would be further exacerbated during periods of minimum flows, when river flow is reduced and tailrace flow is comparatively high (90%), and these are the prevailing conditions during most of the period of adult salmon migration. Additionally, as the river is fully regulated during the migratory period, there is very little hydrological variability (e.g., 'spates') on daily or weekly scales, which are important cues for upstream migration (Jonsson, 1991).

Materials and methods

Baseline (2002/2003)

Studies were undertaken in 2002 and 2003 to establish existing conditions related to tailrace attraction at the ACCC generating facility at Grand Falls—Windsor before the construction of additional generating capacity (Beeton Unit). In 2002, studies were undertaken using conventional radio transmitters. Fish were released below the Grand Falls-Windsor facility and migration to and past the power plant was monitored by continuously recording receivers and manual tracking to document tailrace attraction and residency times. In 2003, EMG transmitters were implanted in the red (aerobic) swimming muscle of salmon and migration to and past the power plant was monitored to provide an estimate of the relative swimming effort and energetic costs (Kaseloo et al., 1992).

Conventional telemetry involved capture of salmon (n = 25) from the Grand Falls fishway (Fig. 1), surgical implantation of transmitters, release of fish below the power plant, and allowing fish to volitionally continue migration past the power plant and tailraces. Fish obtained from the fishway and moved to an incubation/ holding facility for surgery and recovery from July 15 to 19, 2002. Transmitters (Lotek Wireless, model MCFT 3GM, 8.2 by 19 mm, 1.0 g in water, 2.5 s burst rate) were surgically implanted following methods detailed in Adams et al. (1998). Surgical procedures lasted no longer than 2 min and all fish were recovered for a minimum of 10 h before release at a location 3 km downstream of the plant. The upstream movement of salmon was monitored with data logging stations at four locations consisting of a SRX 400A receiver/data logger (Lotek Wireless), 12 V power source, and four element Yagi antennae recording at 2 s intervals continuously over a 24 h period. Two stations were located downstream, one in the tailrace area (with individual antennae recording presence in separate tailraces), and a fourth at the Grand Falls fishway (Fig. 2). Data analysis included determination of presence/absence within the tailrace, length of tailrace residency (h), and determination of subsequent upstream or downstream movement. Relationships between tailrace attraction, residency time, temperature and flow were examined by linear regression (statistical significance at $P < 0.05$).

Electromyorgram (EMG) transmitters were used in 2003 to determine residency times and relative energy expenditure in relation to migration and tailrace attraction. Upstream migrating adult Atlantic salmon ($n = 8$; 500–650 mm length) were taken from a fishway to an incubation/ holding facility for implantation of EMG transmitters (Lotek Wireless model CEMG-R16-25, 16 × 59 mm, 19 g in air). Details on surgical procedures are provided in McKinley & Power (1992). EMG surgery involved implantation of electrodes, fitted with gold tips, in the axial red swimming muscle (Cooke et al., 2004). The surgical procedure lasted 4–5 min and fish were recovered for 24 h and then released, 3 km downstream from the plant. Data logging stations (as above) were set up in three distinct areas, near the bridge below the plant, at the plant, and at the Grand Falls fishway (Fig. 2). The EMG/positional data were analyzed to look at number of visits (defined as residency at a particular antenna, then departure from the antenna for a period of a minimum of 1 h), time spent (sum of all recorded signals), and EMG pulse interval (mean ± SE) for each fish for each antennae locations to provide an indication of the relative energy expenditure, and swimming effort, at each of the locations.

A Relative Energy Index (EI) was calculated to estimate the potential energy expended by each fish in association with each antennae location based on the average EMG signal and total time spent (h) at each location, as follows:

$$\text{Energy Index (EI)} = \text{Average EMG}\,(*10)_{\text{Antennae X}} \times \text{The total time spent (h)}_{\text{Antennea X}}$$

The average EMG was multiplied by ten in order to compensate for the narrow range in EMG signal as provided by the coded technology relative to the range in time spent at each antennae.

Post-Beeton Installation (2004)

Studies were again undertaken in 2004, using both conventional and EMG transmitters, after the installation of the new turbine unit (Beeton), reconfiguration of the power plant tailraces, and in consideration of the new operational regime to accommodate the additional generation capacity. The new operational scenario was based on the fulltime operation of the new Beeton turbine and an existing turbine (unit 4), at their most efficient points, followed by operation of the remaining existing vertical turbines, at a reduced capacity, as determined by flow.

As previously indicated, conventional telemetry involved capture of salmon ($n = 30$; 515–640 mm length) from the Grand Falls fishway, surgical implantation of transmitters, release of fish 3.0 km below the power plant, and allowing fish to volitionally continue migration past the power plant and tailraces. Fish were captured from July 16 to 22, 2004 and were released in four separate releases. Transmitters used in 2004 (Lotek Wireless Nanotag model NTC 6-1, 9.1 by 22.3 mm, 2.8 g in water, burst rate 2.5 s) were comparable to those used in 2002 and were

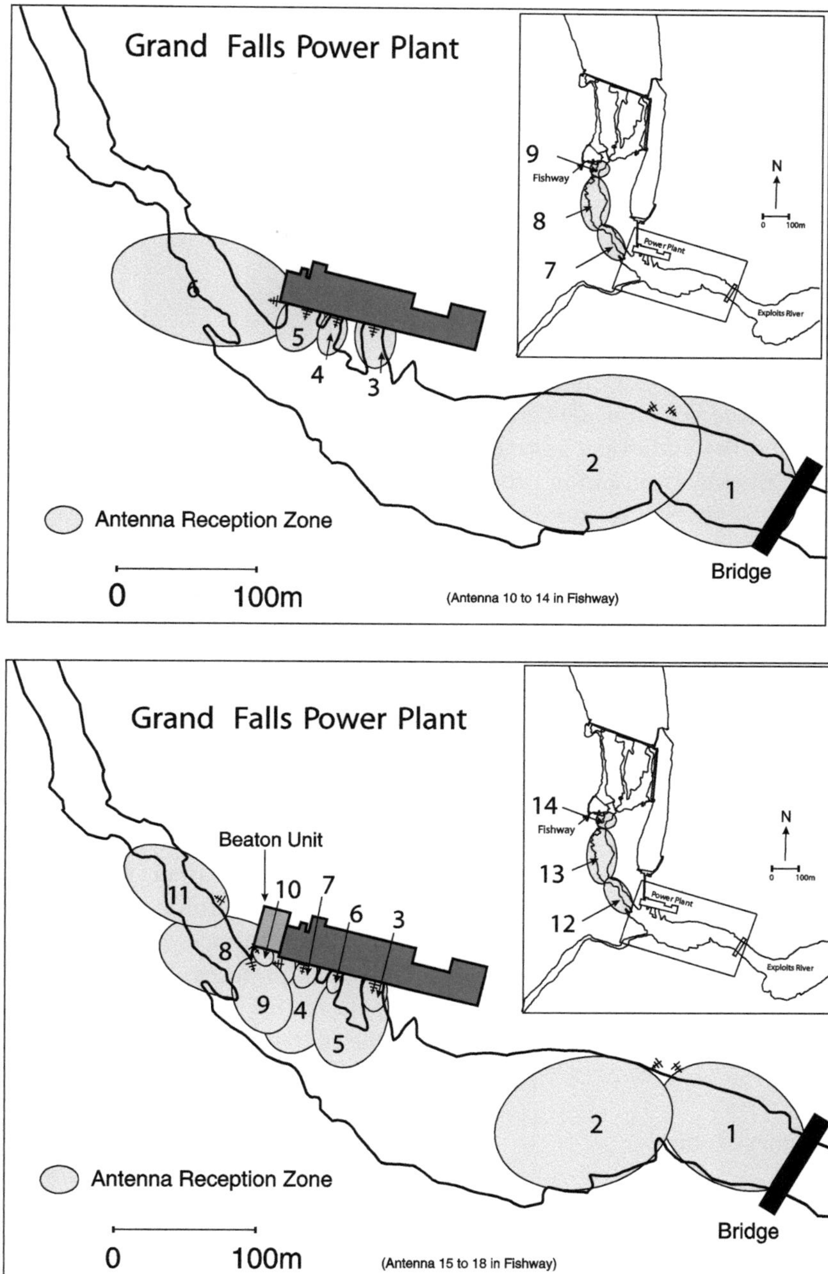

Fig. 2 Map of the Grand Falls-Windsor power plant showing the reception zones for antennae at downstream, tailrace, and upstream sites for telemetry studies in 2003 (*upper panel*) and 2004 (*lower panel*). The antennae at the upstream fishway site are not shown

surgically implanted as previously described. The upstream movement of salmon was monitored from July 16 to September 12 as described for 2002 with an additional station at the new Beeton unit tailrace (Fig. 2). Manual tracking was conducted from July 16 and 22. Analyses of data examined presence/absence in the tailrace area, time to arrival at the power station, residency time at each station (h), and upstream or downstream movement after arrival at the power plant.

Electromyogram (EMG) transmitters were also used in 2004 to determine residency times and relative energy expenditure in relation to migration and tailrace attraction. Upstream migrating adult Atlantic salmon ($n = 10$, 470–710 mm length, $\bar{x} = 569.5$, SE = 19.94) were captured and between July 19–22 ($n = 5$) and August 19–20 ($n = 5$) and handled as for 2003. EMG transmitters (Lotek Wireless model CEMG-R11-25, 11 × 61 mm, 12 g in air) were slightly smaller than in 2003 and surgery, procedures, recovery, and release procedures were as above. Data logging stations were set up in four distinct areas as in 2003, with additional antennae deployed at the new Beeton Unit and in proximity to the power plant to attempt to more clearly distinguish residency in association with individual tailraces (Fig. 2). Helicopter flights were also conducted on August 18 and October 6, 2004 to determine location of fish that had not reached the power plant. Data analyses and interpretation were as described for the 2003 data.

In 2004, attempts were made to calibrate oxygen consumption and swim speed with EMG signal in a 120 l Blaska swim speed respirometer, however, calibration trials were discontinued owing to the extremely high river water temperatures (21–24 °C) during the migration period and the associated stress to experimental animals.

Results

Conventional telemetry studies

Twenty five adult salmon were conventionally radio tagged in 2002 and 21 were subsequently detected by the data logging stations and 20 of 21 fish entered the tailrace area. River discharge did not influence tailrace attraction ($R^2 = 0.27$, $P = 0.304$, $n = 21$) however turbine flow was significantly associated with tailrace attraction ($R^2 = 0.51$, $P = 0.038$, $n = 21$). Four fish (19%) made a single entrance into the tailraces, 8 fish (38.1%) made between two and five entrances, six fish (28.6%) made between six and 10 entrances and one fish (4.8%) made 11–15 entrances into the tailrace area (Fig. 3). Two fish (9.5%)

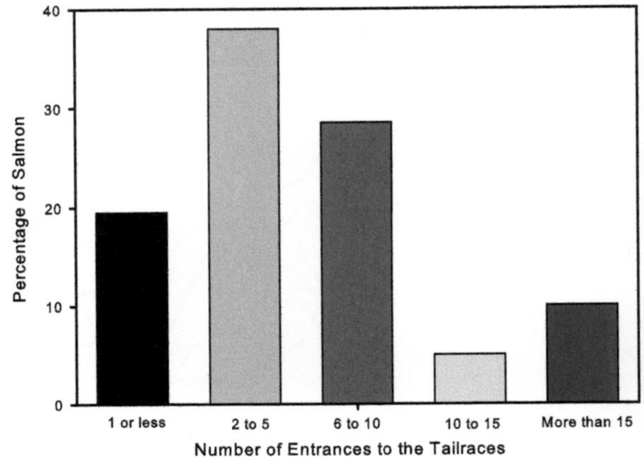

Fig. 3 Distribution of tailrace entrance attempts by conventionally radio tagged adult Atlantic salmon in 2002

made more than 15 tailrace entrances while the maximum number of individual entrances was 21. Tailrace residency varied significantly between individuals and, for most attempts, residency was generally less than 1 min. Salmon that entered the tailrace for longer periods, the average duration increased to 36.38 min. Four individuals resided in the tailrace over several days for 24, 71, 84 and 118 h. There was no relationship between tailrace residency and river flow ($R^2 = 0.17$, $P = 0.187$, $n = 21$) or turbine flow ($R^2 = 0.30$, $P = 0.089$, $n = 21$). Tailrace attraction behaviour took two distinct patterns. The first, and most prevalent pattern ($n = 18$), involved moving into and out of the tailrace over short periods, before proceeding upstream or downstream, with average cumulative tailrace residency of 0.87 h. The second pattern ($n = 4$) involved movement into the tailrace for more extended periods with average residency of 33.4 ± 27.8 h.

Thirty adult salmon were conventionally radio tagged in 2004 and 29 were subsequently detected by the data logging stations ($n = 29$) and manual tracking ($n = 21$) and all 29 fish entered the tailrace area. Most detections were associated with the unit 4 tailrace followed by the Beeton tailrace followed by the units 5–8 tailrace (Fig. 4). Occurrence of fish in the units 5–8 tailrace is important as these units were not operating during the study and therefore there was no attraction flow, suggesting residency in this area may be related to migration route (this area is

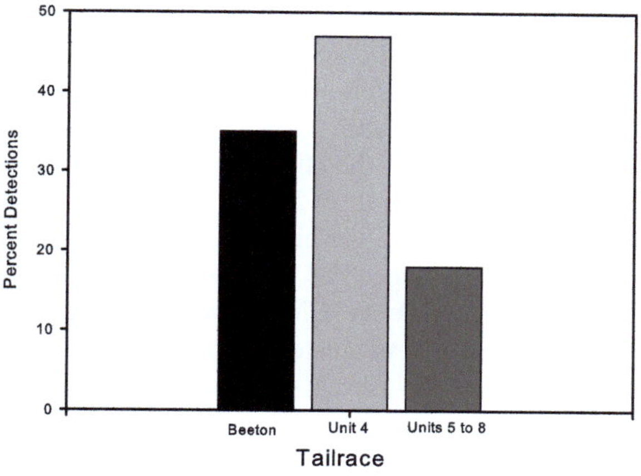

Fig. 4 Percentage of detections of conventionally tagged Atlantic salmon at each tailrace in 2004

first encountered along the migration route) or physical conditions of the tailrace (deep, minimal velocity, shaded).

In total, fish made on average 356 ± 87 (range from 45 to 2207) distinct tailrace entrances lasting on average about 4 h. Fish in the last release (#4) made 30% less entrances than other release groups. Fish in each release took on average from 6 to 11 days to reach the power plant after release, and remained in the tailrace area for, on average from 3 to 12 days (Fig. 5). Generally, with each release, the time to arrive at the tailrace increased and the duration in the tailrace area decreased, such that the times to pass the power plant were similar. For example, fish in release 1 averaged 6 days to reach the tailrace, and remained there on average for 12 days while fish from release 4 spent 11 days on average to reach the tailraces but remained there for, on average, 5 days. Three salmon were detected upstream, at the Grand Falls fishway within 5 days of release, and two were subsequently detected downstream, suggesting that at least a component of the upstream migrants undergo rapid upstream migration and then subsequent descent to holding areas when upstream migration is impeded (possible delay associated with fishway entrance). Tailrace residency (# entrances, duration) was unrelated to water temperature and total generation, which is not surprising owing to the relatively constant rate of generation over the study period (Lotek Wireless Inc., 2004).

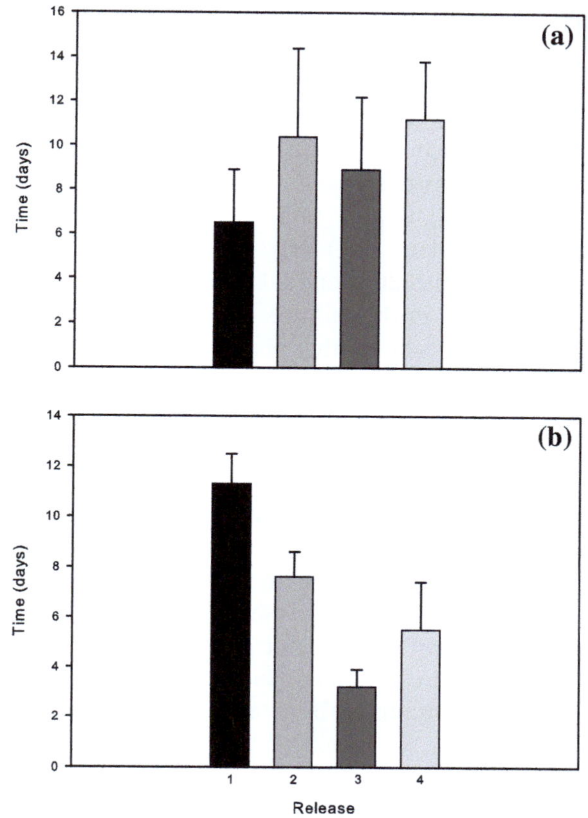

Fig. 5 The mean (±SD) time from release to arrival at the power plant (upper panel, **a**) and time spent at the power plant (lower panel, **b**) of four releases of conventionally tagged Atlantic salmon in 2004

EMG telemetry studies

In 2003, EMG tagged fish, released below the Grand Falls power plant, were subsequently recorded in the vicinity of the power plant (n = 3) and the fishway (n = 2), upstream of the plant. Five fish never returned to fishway by the completion of the monitoring period (September 25, 2003). One (1) fish was confirmed to be a mortality while helicopter monitoring on two occasions located the other four fish in downstream locations (change in location between monitoring periods suggested these fish were alive) and it is unknown why they did not attempt to migrate upstream (Scruton et al., 2004). Fish 26 made the most repeated visits to the reach below (25), the tailraces (81), and above the plant (36), and, after passing this area, moved quickly and directly to the fishway. This fish spent considerable time during two discrete periods (August 30 to September 6, September 8 to 13) in the vicinity of the power plant and tailraces. Fish 31 demonstrated lesser visits (9) to the tailraces indicating a more directed migration. Fish 32 had a large number of visits associated with residency below the plant (27), tailrace area (23), and immediately above the plant (28,) and then rapid upstream migration. Fish 26 spent 3.0 h downstream of the plant, 66.0 h in the tailrace discharge, and 79.3 h immediately above the plant and much lesser time between the plant and fishway, indicating that once false attraction from the tailraces was bypassed, upstream migration proceeded quickly. Fish 31 spent most of its time below the power plant (44.5 h) and lesser time in the tailraces (18.7 h), or above (2.8 h). Fish 32 spent 34.1 h below the plant, 47.7 h in the tailraces, 40.6 h immediately above the plant and much lesser time migrating above plant.

EMG data in 2003 (Fig. 6), suggested the relative swimming effort energy expended by Fish 26 was only slighter greater within the tailraces, than in the reaches below and above the plant, with greatest effort in the reach between the plant and fishway, with the highest EMG just below the fishway. This reach is constricted, turbulent and high gradient and, as fish spent little time in this area, suggested rapid and arduous passage. The Relative Energy Index, which integrates both EMG and time spent at each antennae, suggested most energy was expended at the lowest tailrace (antennae 3) and just above the plant (antennae 6). Fish 31 demonstrated considerable variability in EMG signal from each antenna, with EMG signals greatest in association with the upper tailrace, however fish spent very little time at this location. The Relative Energy Index suggested that the area just below the plant and the lowest tailrace (antennae 3) were most energetically taxing. Fish 32 had a similar pattern to Fish 26 in both EMG signal and Relative Energy Index and the greatest energy expenditure was associated with the area

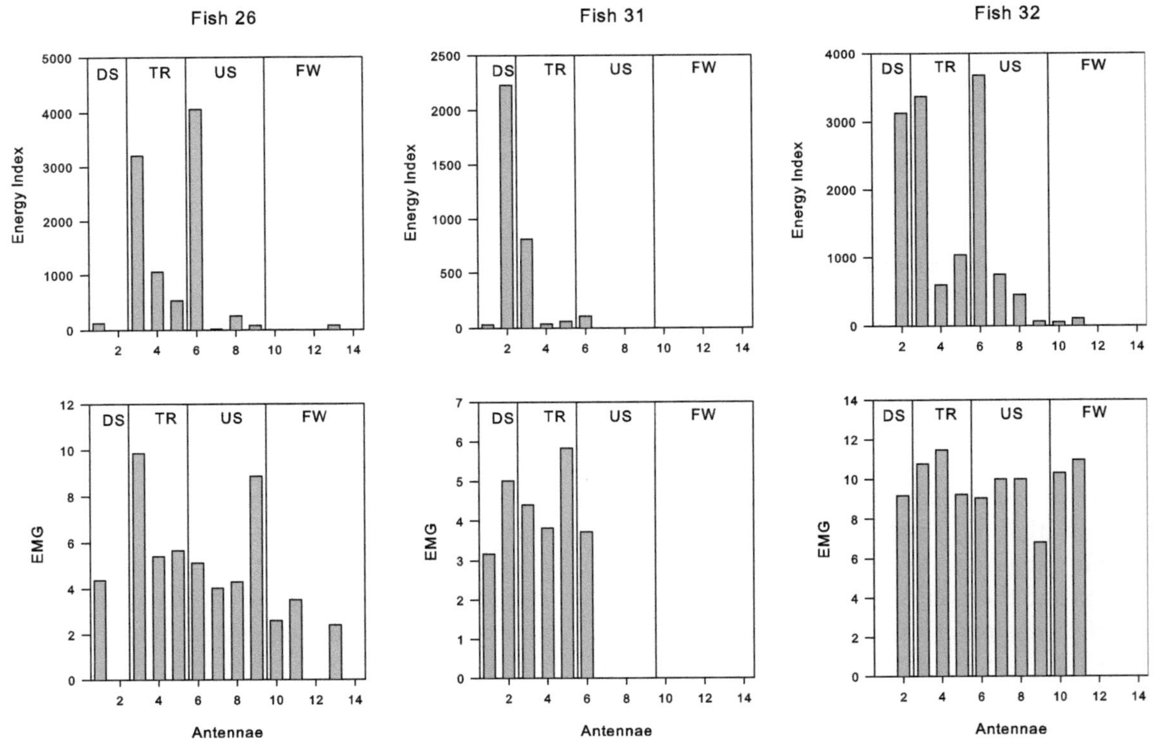

Fig. 6 The Energy Index (see text for description) and average EMG for three fish in 2003 in relation to antennae location (see Fig. 2, *upper panel*). Migration sequence (upstream) is from antenna 1 to antenna 14. DS = Downstream, TR = Tailrace, US—Between tailrace and Fishway, FW—Fishway

just below the plant, the lowest tailrace, and just above the plant. It is noteworthy that all three fish spent considerable energy in association with the lowest tailrace, the first they encounter on their upstream migration.

In 2004, three EMG tagged fish, released below the Grand Falls power plant, were subsequently recorded in the vicinity of the power plant and two of these also at the Grand Falls fishway upstream of the plant. Seven fish never returned to the fishway by the completion of the monitoring period (September 29, 2004). Two (2) fish were confirmed as mortalities while helicopter monitoring located the other four fish in downstream locations (Scruton et al., 2005). As in 2003, change in location between monitoring periods suggested these fish were alive and it is unknown why they did not attempt to migrate upstream. All three fish were initially detected, and spent considerable amount of time associated, with the tailrace for units 5–8. These units were not generating during the course of the study, and there was no discernable attraction flow. This location is the first encountered by upstream migrants and has conditions (deep, well shaded, no flow) that would be well suited for fish holding prior to resuming upstream migration. All three fish were not detected at antennae 4, 5, and 6 and were then again detected at antennae 7, in association with the inner tailrace for unit 4. Fish then spent considerable time above unit 4 (antennae 8) and in association with the Beeton unit (antennas 9 and 10), with most time in association with antennae 8, suggesting some delay before being attracted to and passing the Beeton tailrace.

EMG signals for the three fish were variable in 2004 (Fig. 7). EMGs for Fish 1 indicated EMGs associated with tailraces were similar to downstream reaches, and highest in association with the inner Beeton tailrace (antennae 10). The Relative Energy Index indicated that tailrace residency, particularly at antennae 8–10, associated with the Beeton unit, were the most energetically costly. High EMGs were also apparent

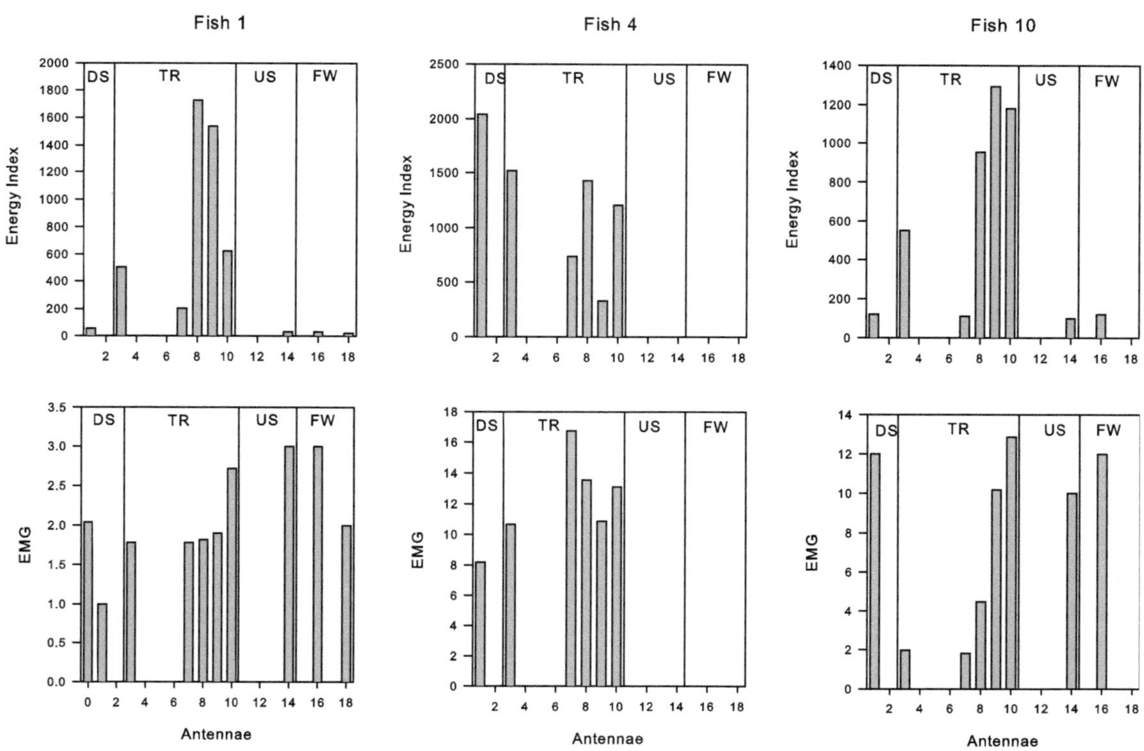

Fig. 7 The Energy Index (see text for description) and average EMG for three fish in 2004 in relation to antennae location (see Fig. 2, *lower panel*). Migration sequence (upstream) is from antenna 1 to antenna 14. DS = Downstream, TR = Tailrace, US—Between tailrace and Fishway, FW—Fishway

in the reach above the power plant and the lower fishway, similar to 2003 results, however the short amount of time spent in these areas suggested less energy expenditure than in the tailrace area. Fish 4 demonstrated highest EMGs at the inner tailrace at unit 4. The Relative Energy Index indicated highest energy expenditure below the plant, related to the large amount of time spent at this location (25 h) in comparison with the power plant (41 h distributed between all tailrace antennae). Within the tailrace area, highest energy expenditure was associated with the lowest tailrace (antennae 3) and the Beeton unit (antennae 8 and 10). Fish 10 demonstrated low tailrace EMG in association with unit 4, and progressively increasing EMGs, with the highest EMG associated with the Beeton tailrace (antennae 8–10). Highest relative energy expenditure was associated with the lowest tailrace and Beeton unit. High EMG above the plant and in the fishway did not result in high-energy expenditure as fish spent very little time in these areas. All three fish demonstrated high-energy expenditure associated with the new Beeton unit, and lesser energy use with the lowest tailrace.

Discussion

It is noteworthy that in all four studies, conventional telemetry in 2002 and 2004 and EMG telemetry in 2003 and 2004, four, one, five, and seven fish, respectively, were not detected at the power plant and upstream fixed telemetry stations. There is commonly some level of mortality associated with surgical implantation of transmitters and subsequent recovery (three known mortalities in EMG fish in our study), and mortality can be higher with EMG transmitters owing to the more invasive nature of the surgery (Adams et al., 1998; Cooke et al., 2004). The most arduous reach of the Exploits River for migration was also the study reach and fish used in studies were taken from the Grand Falls fishway, above this reach. It is possible the stress of migration, capture and transportation, handling and surgery, as well as high summer water temperatures may have played a role in these fish not resuming their migration. It is also possible these fish, after experiencing migration delay, decided to remain and possibly spawn in downstream reaches (Bernatchez and Dodson, 1987). It is assumed that the fish that did resume their migration to and past the power plant are representative of the behaviour of untagged fish. In a study from Norway, fish were collected both above and below a power plant, radio tagged and released below the plant, and there was no difference in migratory behaviour between fish that previously passed the power plant and those that had not (Thorstad et al., 2003). There are no data on the natural rate of mortality and successful migration rates in our study reach.

In 2002 (prior to refurbishing) and 2004 (after refurbishing), the power plant was responsible for tailrace entrainment among migrating Atlantic salmon and there was a relationship between attraction and total discharge from the power plant in 2002. Turbine flow generally represented in excess of 90% of the total discharge passing the facility, therefore providing the majority of attraction flow encountered by migrating salmon. Two different migratory patterns were observed, both resulting in tailrace attraction and residency, however, residency was independent of attempts to enter the tailraces and there was no relationship between residency and individual turbine flow suggesting flow may attract salmon to individual tailraces but fish continue to search for an upstream migration path. Factors within the tailrace, such as noise, vibration or dissolved gas levels create unfavourable conditions and this may explain why only four of 21 salmon in 2002 exhibited extended tailrace residency. In 2004 (after refurbishing), fish were attracted to the main three tailraces, including the new Beeton unit, but also to the tailrace associated with units 5–8, which had no attraction flow, suggesting migration route and good holding characteristics of the location may have an influence on tailrace attraction/residency. Fish generally spent more time in the tailrace area in 2004, compared to 2002, and with each release, the time to arrive at the tailrace increased and the duration in the tailrace area decreased, such that the times to pass the power plant were similar.

Fish are attracted to tailraces due to a number of hydraulic factors. Fish are attracted to tailraces as turbine discharges compete with flows in the natural river channel, and high water volume, residual energy, and turbulence of tailrace flows are conditions that attract fish seeking an upstream migratory route (FERC, 1995). Injury and/or mortality can occur in combination with tailrace attraction and will depend on physical and operating characteristics of the plant and the swimming ability of the species (FERC, 1995). Adult Atlantic salmon are strong swimmers and, while turbine velocities generally do not exceed the upper range of their swimming speeds, there have been no reported injuries associated with migration of Atlantic salmon at the Grand Falls power plant and no observations of obvious injuries at the Grand Falls fishway, above the power plant (F. Parsons, Exploits River Management Association, Grand Falls-Windsor, NL, personal communication).

River reaches with relatively complex hydraulics are associated with high swimming speed variation, indicating exploration for alternative migration routes with wandering behaviour related to searching for sustained, directional flow (Hinch et al., 2002). Complex current patterns caused by river constrictions, falls and chutes can create confusing migration cues which can impede a fish's ability to find the most appropriate pathway to continue upstream migration (Hinch et al., 2002). It is unlikely that differences in migratory route influence tailrace attraction since the different migratory patterns resulted in tailrace attraction and residency. Wandering behaviour is actually considered to be a searching behaviour for sustained, directional flow (Kynard, 1993). In our study, the number of entrances in the tailrace area in 2002–2004, including visits recorded by EMG fish, are clear examples of Atlantic salmon demonstrating this searching/wandering behaviour.

Migratory delays in relation to tailrace attraction have been well documented for salmonids in the Pacific Northwest (e.g., FERC, 1995) and delays as little as 3 days have been shown to have effects on spawning success (Fleming & Reynolds, 1991). In our studies, salmon were entrained in the tailraces for an average of 33.4 h (2002) and 3–12 days (2004), however, these delays are considerably less than those reported in the literature (e.g., citations within FERC, 1995; Thorstad et al., 2003; Gowans et al., 1999, 2003). Impact of migration delays on fish spawning will depend on a location and extent of the delay relative to spawning sites, energy consumed, the state of maturity, difficulty of the remaining migration, and environmental conditions. Thorstad et al. (2003) found migrating Atlantic salmon, attracted to a power plant tailrace, to be delayed for up to 71 days (median of 20) and several fish (7 of 17) did not successfully pass the tailrace. Upstream migration of Atlantic salmon at fish passage facilities in Scotland was delayed up to 52 days (Gowans et al., 2003). Migration delays in arctic grayling (*Thymallus articus*) caused fish to stop short of their upstream goals and displacement to downstream area resulted in use of lower quality spawning habitat (Fleming & Reynolds, 1991).

Conventional telemetry is useful in determining positions and movements of fish, however they do not account for depth, horizontal curvature swimming paths, velocity changes, etc, and therefore can provide only estimates of minimum distance swum and associated metabolic cost (Økland et al., 1997). Both EMG and conventional telemetry were useful in this study in providing insight into migratory behaviour, and relative energetic costs of migration, in association with the Grand Falls-Windsor hydrolelectric plant, associated tailrace discharge, upstream and downstream reaches, and use of a fishway. In addition to potential delays in association exclusively with the tailrace, the reach of river below the power plant, and immediately above the power plant up to below Grand Falls and the fishway entrance, are extremely constricted, turbulent, high gradient and high velocity reaches and could be considered areas of difficult passage.

Electromyogram telemetry is a tool that enables a direct measure of metabolic rates in free swimming fish in situ in relation to various environmental conditions (Cooke et al., 2004). In our study, EMG signals were not correlated with oxygen consumption, a measure of metabolic rate, owing to high water temperatures and associated stress on experimental fish. EMG data in this study were considered a direct, but relative, indicator of swimming effort (energy expended). As a true measure of energy expended includes both swimming effort (EMG) and time spent in relation to this effort, a 'Relative Energy Index' was calculated to integrate these two factors (see Materials and methods). Much of the variability in time spent (passage), EMG, and energy expenditure was related to the physical characteristics of the various locations and study reaches. There was relatively high-energy expenditure associated with tailrace attraction and residency in both years and the pattern in energy expenditure were similar in fish in each year. In 2003, fish spent highest energy at the lowest tailrace (first encountered) in their migration. In 2004, there were also high apparent energy costs at this location but highest energy costs were associated with the new Beeton unit. In both years, high EMGs above the power plant and just below the fish suggested these reaches were arduous and difficult to pass, however, these reaches were quickly passed resulting in low relative energy costs. Other studies have demonstrated reach specific variation in activity and energy expenditure and increased costs are associated with constricted reaches with higher velocities and more complex hydraulics (Standen et al., 2002). River reaches with relatively complex hydraulics and fast currents have been associated with high swimming speed variation, as determined by EMG telemetry, in upstream migrating pink and sockeye salmon (Hinch et al., 2002). It is also noteworthy that swimming in these high velocity, turbulent reaches may have been fueled by anaerobic metabolism (white swimming muscle) and the EMG signals from red swimming muscle may have underestimated the true energy expenditure (Cooke et al., 2004).

An important finding of the present study is that despite tailrace attraction, the actual consequence to the timing of migration is relatively small, and most salmon continue to migrate upstream after leaving the tailrace area. Although it is recognized that migratory delays can result in imbalances in the energy budgets of migrating salmon, the delays encountered are not thought to be of sufficient magnitude to be important in migratory or reproductive success. In fact, delays encountered as a result of the ACCC pulp and paper facility may not be as important as delays encountered by salmon reaching the Grand Falls fishway. Previous work performed at the fishway determined that salmon may spend several days milling at the fishway entrance, and making attempts to enter the fishway, before proceeding through the fishway (R. Booth, unpublished data).

Conclusions

Study results suggest that migrating adult Atlantic salmon experience some degree of tailrace attraction and migration delay at the Grand Falls-Windsor facility, however, the actual consequence to the timing of migration was considered relatively modest and most salmon resumed upstream migration upon leaving the tailrace area. Migratory delays can result in imbalances in the energy budgets of migrating salmon, however, delays encountered at the Grand Falls-Windsor facility were not considered to be of sufficient magnitude to be important in migratory or reproductive success. EMG data, and associated time spent at antennae locations, were used to provide a relative index of the energy expended by individual fish at each key points in their migration. It is clear there is energy expended in association with attraction to and delays at the power plant tailraces, but there is also considerable energy expenditure associated with reaches below and upstream of the power plants, characterized by physical conditions (gradient, flow character, constrictions) that make these reaches naturally difficult migration challenges. Migration delays and energetic costs associated with this facility must be considered in the context of the overall migratory challenges facing salmon on the Exploits River, which include the possible use of fishways on the river (Bishop's and Grand Falls, Red Indian Lake) and the physical and hydraulic conditions on the major several tributaries that salmon are utilizing for spawning and rearing.

Acknowledgements The authors would like to acknowledge the assistance of Fisheries and Oceans staff (C. Kelly, K. Smith, J. Murray, C. Bourgeois), staff of the Exploits River Management Association (F. Parsons), J. Jardour (University of British Columbia) and in particular the support and guidance of Mr. W. Eddy with the Abitibi Consolidated Company of Canada. Funding for this research was provided by the Exploits River Hydro Partnership, Fisheries and Oceans' Environmental Sciences Strategic Research Fund (ESSRF) and Natural Resources Canada's Panel for Energy Research and Development (PERD).

References

Adams, N. S., D. W. Rondonf, S. D. Evans, J. E. Kelly & R. W. Perry, 1998. Effects of surgically and gastrically implanted radio transmitters on swimming performance and predator avoidance of juvenile chinook salmon (*Oncorhynchus tshawytscha*). Canadian Journal of Fisheries and Aquatic Sciences 55: 781–787.

Bernatchez, L. & J. J Dodson, 1987. Relationship between bioenergetics and behaviour in anadromous fish migrations.). Canadian Journal of Fisheries and Aquatic Sciences 44: 399–407.

Bourgeois, C. E., J. Murray, & V. Mercer, 1994. Status of the Exploits River stock of Atlantic salmon (*Salmo salar* L.) in 1993. Canadian Department of Fisheries and Oceans, Atlantic Fisheries Research Document 94/73, 13 p.

Cooke, S. J., E. B. Thorstad & S.G. Hinch, 2004. Activity and energetics of free-swimming fish: insights from electromyogram telemetry. Fish and Fisheries 5: 21–52.

Federal Energy Regulatory Commission, 1995. Impacts of hydroelectric plant tailraces on fish passage: A report on effects of tailraces on migratory fish and use of barriers, modified project operation, and spills for reducing impacts. Federal Energy Regulatory Commission. Office of Hydropower Licensing. Washington, D.C., Paper No. DPR-9, 101 pp.

Fleming, D. F. & J. B. Reynolds, 1991. Effects of spawning-run delay on spawning migration of arctic grayling. American Fisheries Society Symposium 10: 299–305.

Gowans, A. R. D., J. D Armstrong & I. G. Priede, 1999. Movements of adult Atlantic salmon in relation to a hydroelectric dam and fish ladder. Journal of Fish Biology 54: 713–726.

Gowans, A. R. D., J. D. Armstrong, I. G. Priede & S. Mckelvey, 2003. Movement of Atlantic salmon migrating upstream through a fish-pass complex in Scotland. Ecology of Freshwater Fish 12: 177–189.

Hinch, S. G., E. M. Standen, M. C. Healy & A. P. Farrell, 2002. Swimming patterns and behaviour of upriver-migrating adult pink (*Oncorhynchus gorbuscha*) and sockeye (*O. nerka*) salmon as assessed by EMG telemetry in the Fraser River, British Columbia, Canada. Hydrobiologia 483: 147–160.

Jonnson, N., 1991. Influence of water flow, water temperature, and light on fish migration in rivers. Nordic Journal of Freshwater Research 66: 20–35.

Kaseloo, P. A., A. H. Weatherly, J. Lotimer & M. D. Farina, 1992. A biotelemetry system for recording fish activity. Journal of Fish Biology 40: 154–179.

Kynard, B.E., 1993. Anadramous fish behaviour important for fish passage. In Williams, U. P., D. A. Scruton, R. F. Goosney, C. E. Bourgeois, D. C. Orr & C. P. Ruggles (eds), Proceeding of the Workshop on Fish Passage at Hydroelectric Developments: March 26–28, 1991, St. John's, Newfoundland. Canadian Technical Report of Fisheries and Aquatic Sciences 1905, v + 153 p.

Lotek Wireless Inc., 2004. The 2004 Atlantic Salmon Monitoring Program, 102 pp + Appendices.

Lucas, M. C & E. Baras, 2001. Migration of Freshwater Fishes. Blackwell Sciences Ltd, London.

McKinley, R. S. & G. Power, 1992. Measurement of activity and oxygen consumption for adult lake sturgeon (*Acipenser fulvescens*) in the wild using radio-transmitted signals. In Preide I. G., & S. M. Swift (eds.), Wildlife Telemetry: Remote monitoring and Tracking of Animals. Ellis Horwood, New York, pp. 308–318.

Økland, F., B. Finstad, R. S. McKinley, E. B. Thorstad & R. K. Booth, 1997. Radio-transmitted electromyogram signals as indicators of physical activity in Atlantic salmon. Journal of Fish Biology 51: 476–488.

Scruton, D. A., C. J. Pennell & L. M. N. Ollerhead, 2004. Tailrace Attraction Study, Grand Falls Generating Station, 2002 and 2003. Prepared for Abitibi Consolidated Company of Canada, Grand Falls—Windsor, NL, 52 pp.

Scruton, D. A., C. J. Pennell & L. M. N. Ollerhead, 2005. Tailrace Attraction Study, Grand Falls Generating Station, 2004. Prepared for Abitibi Consolidated Company of Canada, Grand Falls—Windsor, NL, 48 pp.

Smith, I. P., A. D. F. Johnstone & G. W. Smith, 1997. Upstream migration of adult Atlantic salmon past a fish counter weir in the Aberdeenshire Dee, Scotland. Journal of Fish Biology 51: 266–274.

Standen, E. M., S. G. Hinch, M. C. Healey & A. P. Farrell, 2002. Energetics costs of migration through the Fraser River Canyon, British Columbia, in adult pink (*Oncorhynchus gorbuscha*) and sockeye (*Oncorhynchus nerka*) salmon as assessed by EMG telemetry. Canadian Journal of Fisheries and Aquatic Sciences 59: 1809–1818.

Thorstad, E. B., F. Okland, F. Kroglund & N. Jepsen, 2003. Upstream migration of Atlantic salmon at a power station on the River Nidelva, southern Norway. Fisheries Management and Ecology 10: 139–146.

Winstone, A. J., A. S. Gee & P. V. Varallo, 1985. The assessment of flow characteristics at certain weirs in relation to the upstream movements of migratory salmonids. Journal of Fish Biology 27(Suppl. A): 75–83.

Hydrobiologia (2007) 582:81–89
DOI 10.1007/s10750-006-0559-4

FISH TELEMETRY

Mid-winter activity and movement of Atlantic salmon parr during ice formation events in a Norwegian regulated river

Morten Stickler · Knut Alfredsen · David A. Scruton · Curtis Pennell · Atle Harby · Finn Økland

Abstract A telemetry study in a Norwegian regulated river was conducted through a 12-day period in mid-winter 2003. The objective was to study activity (defined as number of movement per hour) and movement (defined as distance moved per hour) during different ice formation events. Twenty-four Atlantic salmon (*Salmo salar* L.) parr were radio tagged and continuously monitored by both manually tracking ($N = 24$) and by fixed recording stations ($N = 15$). Detailed data on climate, flow and ice formation and its spatial distribution were collected and used in the analyses. Fish activity was not found to be affected by their size (L_F). There was a significant difference in activity between diel periods with highest activity during dusk (5–6 p.m.). Between high and low flow (mean ± SD, 21.1 $m^3 s^{-1}$ ± 1.7 SD and 11.1 $m^3 s^{-1}$ ± 1.7 SD, respectively) no significant difference in activity was found. During the experiment extensive anchor ice growth occurred mainly in the riffle part with thickness up to 50 cm. Juveniles tend to avoid riffle section during anchor ice formation and exploited ice covered areas, indicating critical and preferable habitats respectively. Further, a significant difference in movement was found between five selected ice events with highest mean movement during an anchor ice event and lowest mean movement during an ice break up with no anchor ice formation. No significant difference in activity or movement between parr exposed to frazil ice and parr not exposed were found.

Keywords Winter habitat · Atlantic salmon parr · *Salmo salar* L · Radio telemetry · River ice · Frazil ice

Guest editors: P. R. Almeida, B. R. Quintella, M. J. Costa and A. Moore
Developments in Fish Telemetry

M. Stickler (✉) · K. Alfredsen
Department of Hydraulic and Environmental Engineering, Norwegian University of Science and Technology (NTNU), S.P. Andersens vei 5, 7491 Trondheim, Norway
e-mail: morten.stickler@ntnu.no

D. A. Scruton · C. Pennell
Fisheries and Oceans, Box 5667, A1C 5X1 St. John's, NF, Canada

A. Harby
SINTEF Energy Research, Water Resources, 7465 Trondheim, Norway

F. Økland
Norwegian Institute for Nature Research, Tungasletta 2, 7485 Trondheim, Norway

Introduction

In sub-arctic latitudes different forms of ice are important when considering winter conditions for fish in lotic environments. Physical habitat conditions depend on several variables, and the most important are related to stream flow variables, substrate, cover and temperature (Heggenes et al., 1993; Cunjak, 1996b; Alfredsen, 1997; Tesaker, 1998; Alfredsen & Tesaker, 2002). Formation and presence of ice will influence most of these variables, and are therefore of great importance in an assessment of available winter habitat.

In the past decade winter habitat of stream fish has gained increased interest. Research implies that winter survival is a primary factor affecting fish population dynamics in cold region rivers (Power et al., 1993; Cunjak et al., 1998; Bradford & Higgins, 2001; Annear et al., 2002). As the water temperature drops below 8–10°C juvenile Atlantic salmon change from day activity to predominantly nocturnal behaviour (Fraser et al., 1993; Heggenes et al., 1993; Gries et al., 1997; Cunjak et al., 1998; Hiscock et al., 2002b). During daytime they spend most of the time sheltering within the substrate and emerging at night to feed. This behaviour has been stated as being an adaptation to avoid day active endothermic predators (Fraser et al., 1993), but has been later associated also with reduced energy expenditure (Fraser & Metcalfe, 1997). In a recent experiment conducted in a regulated by pass section by Scruton et al. (2005) fish were found to move significant less during winter than summer time. As noted by several authors (Cunjak & Power, 1986; Cunjak, 1988; Cunjak et al., 1998; Bremset, 1999; Hiscock et al., 2002b; Vehanen & Huusko, 2002) juveniles are often found in the deeper parts of the river during the cold season. In such areas hydraulic shelter is pronounced and the risk of predation is decreased. In a recent field study by Roussel et al. (2004) results implied that fish inhabiting pool during freeze up had higher emigration rates than fish in the riffle. This may indicate that the fish were exposed to some physical changes in pools that caused them to abandon the habitat. In the same study juveniles were also found beneath 10 cm anchor ice in the riffle section, suggesting suitable habitats. In another study by Heggenes et al. (1993), a hypothesis was suggested that fish moved from riffles in order to avoid anchor ice while conversely Whalen et al. (1999) observed fish to stay within riffles during an anchor ice event. Tack (1938) reported trout (*Salmo trutta*) being suffocated by frazil and Brown et al. (1994) made observations of trout shifting habitat from frazil exposed areas into habitats were they could avoid frazil and anchor ice (ice covered areas).

While several studies have been conducted on winter behaviour of juvenile salmonids very little research has focused on the interaction between the physical environment and fish behaviour (Roussel et al., 2004). This is particularly true in regulated rivers, which are often characterized by an unstable winter ice regime due to power demand. Hydropower production, particularly high-head systems with deep reservoir intakes and river releases, alters the discharge regime by releasing warm water into the river causing fluctuations of the flow. The ice regime is characterized by repeated ice break ups and increased ice production. In this paper, data obtained from a fish telemetry study conducted mid-winter 2003 in a Norwegian regulated river are analyzed and presented. The main objective of this study has been to investigate how physical events, especially different forms of ice, affect the fish behaviour. The hypothesis was that fish increased their activity during night, were more active in riffles than in pools, and that riffles were considered less suitable during anchor ice formation.

Materials and methods

The experiment was carried out in Orkla River (63°17′ N, 9°50′ E, Fig. 1). The Orkla River has a catchment's area of about 3,053 km^2 with an annual average runoff of 70 m^3 s^{-1}. It was first regulated in 1981, and today, five large hydro power plants operate in the river system. The lower 80 km of the river system is ascendable for anadromous salmonids with annual recreational catches reaching more than 30 tonnes in recent years.

The study site was located in middle portion of the river system approximately 10 km downstream of nearest hydropower discharge. The site was 300 m long and ranged from 30 m to 60 m in wetted width. The habitat is predominantly run, pool and riffle with cobbles covering the bed and

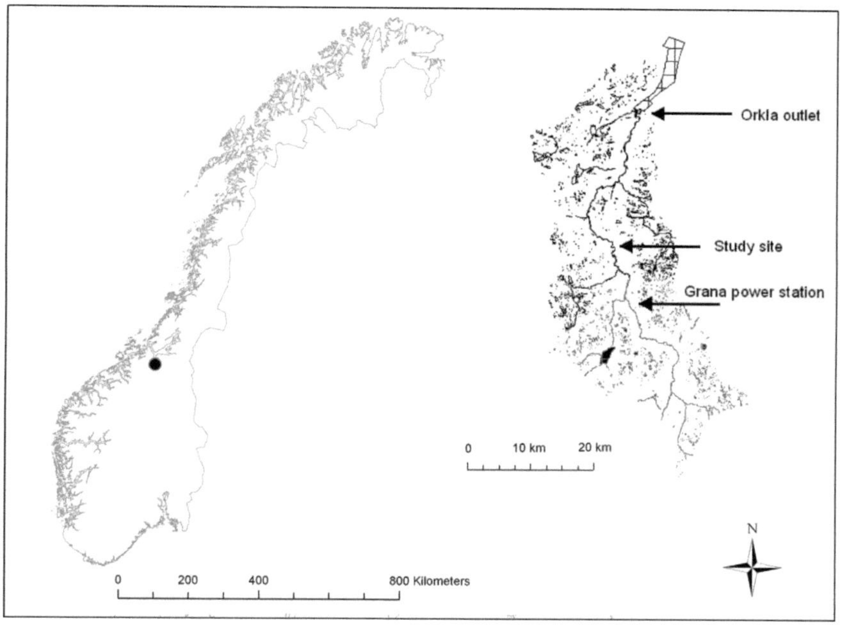

Fig. 1 Location of the study site (•), the Grana power station and the outlet. Note the short (10 km) distance between the power station and the study site. Warm production water keeps the study site free of surface ice during winter and increases the dynamic ice formation

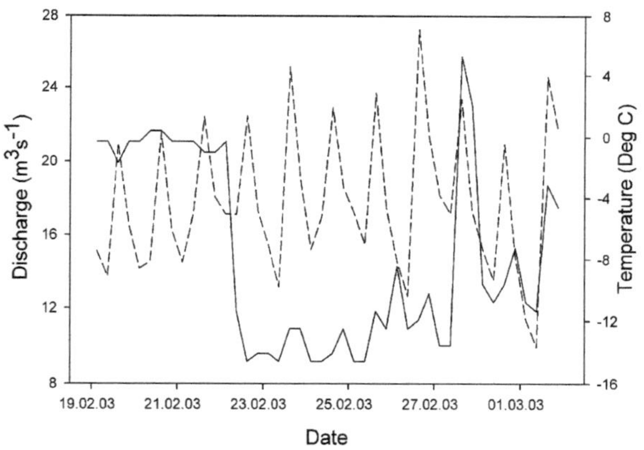

Fig. 2 Discharge (line) and temperature (dotted line) conditions during the experiment

boulders located along the margins. Minimum flow is 4 and 20 $m^3 s^{-1}$ in winter and summer, respectively. Water discharge (Fig. 2) was stable high (mean ± SD, 21.1 $m^3 s^{-1}$ ± 1.7 SD) during the first three days of the experiment, then decreased to low flow condition (11.1 $m^3 s^{-1}$ ± 1.7 SD) reducing the wetted area by 30%. The discharge level was then maintained low throughout the entire period except one day when water was released to prevent ice build up (February 28th). Maximum depth was 2 m (0.51 m ± 0.39 SD) on low flow and 2.4 m (0.75 ± 0.53 SD) on high flow. Water temperature ranged from –0.12 to 0.60°C (0.17°C ± 0.87 SD) while air temperature ranged from 2.2°C to –13.6°C (–5.3 ± 3.6 SD). Along the margins surface ice covered 27% of total wetted area. Frazil ice production and anchor ice growth (Tsang, 1982; Matousek, 1984; Prowse, 1993) had a diurnal cycle; growth during nighttime and cessation at daytime. Anchor ice formed every night in which four events had extensive ice formation (Table 1) mainly in the riffle section with thickness up to 50 cm. During the night of March 1st the air temperature dropped to –13.6°C and caused extensive growth of anchor ice, which established within the entire open area. During the night of February 22nd an artificial ice break up occurred due to a decrease in water discharge from 22 $m^3 s^{-1}$ to 11 $m^3 s^{-1}$. Here, no anchor ice formation was observed.

Radio telemetry was used to investigate the activity and movement of Atlantic salmon (*Salmo salar* L.) parr during a 12-day period mid-winter (February 19th to March 2nd). A total of 24 juveniles (mean ± SD fork length, L_F, 13.6 cm ± 0.69 SD and mass, 21.5 ± 5.0 SD) were caught using a 24 V backpack electro fisher (Smith Root Model 12). Radio transmitters (Model Lotek MBFT 7M; 7.3 × 18 m, 1.4 g in air, and 9M; 8.2 × 19 mm, 1.8 g in air, Lotec Wireless,

Table 1 Physical conditions of four selected events with significant production of frazil and anchor ice formation (event I–IV) and one event (V) with no anchor ice during the study period

Physical condition	Ice event	Ice event I Night 23rd	Ice event II Night 24th	Ice event III Night 28th	Ice event IV Night 1st[t]	Ice event V Night 22nd
Flow level		Low	Low	Low	Low	High → low
Anchor ice distrib.		Riffle	Riffle	Riffle	All open area	No anchor ice
Wetted area total: 20,048 m^2 (100%)			Surface ice area: 5,388 m^2 (27%)		Anchor ice area: 2,666 m^2 (13%)	

Wetted area is calculated on discharge 22 $m^3 s^{-1}$ and ice distribution on night 22nd to give an understanding of the physical conditions

6.7 ± 0.1% of fish body mass) were surgically implanted using similar procedure as Adams et al. (1998) and Moore et al. (1990). Fish were anesthetized by immersion in an aqueous solution of 2-phenoxy-ethanol EEC No 204 589-7 (0.5 ml per l water) and placed dorsal side down on a foam pad. Fork length (mm) and weight (0.1 g) were determined. The transmitter was inserted through a 10 mm long incision posterior to the pelvic fin. The transmitter antenna was treaded through the body wall by means of a needle (Jelco; 18 G, 38 mm) 5 mm posterior and dorsal to the incision. The incision was closed with two sutures of 4–0 Ethicon braided silk. The surgery took no longer than 4 min, and during this time, tepid water was poured over the fish's gills to prevent freezing. After surgery, tagged fish were held in a bucket for some minutes to recover before they were brought to a cage in the river. The fish were kept for 24 h before releasing them into their respective habitats.

Fish location was recorded every 6th hours (03:00, 09:00, 15:00, 21:00) using a handheld H-antenna. Position was determined by using a total station (Leica TC 307). Movement was calculated as distance moved per hour, defined as standardized movement (SM). Frequency of movement was determined by three fixed recording stations (Lotek, SRX 400) located on the river banks. Interval logging was set to every 8 min and data was downloaded every morning between 09:00 and 10:00. Data was collected from a total of 15 juveniles. Movement was determined through fluctuations in signal strength, ranging from 0 dB to 236 dB. Calibration determined that a signal change of 30 represented a movement of approximately 1 m. As field conditions (substrate, ice) influence the signal strength, the distance of fish movement was not quantified by the fixed telemetry stations. Alternatively, the approach described by David & Closs (2001) was used. Standardized activity (SA) was defined as number of movements per hour based on the time the fish was recorded at the actual station. Movement between sunrise and sunset (i.e., daylight hours) were considered daytime movement and sunset to sunrise was considered nighttime movement. Dawn and dusk was considered separately and defined as one hour between 07:00 and 08:00 and 17:00 and 18:00, respectively (The Norwegian Meteorological Institute, pers. comm.).

Detailed data on meteorological and hydrological conditions were collected during the study period. A climate station (Campbell Scientific) was located on the study site recording air temperature, radiation, humidity and wind direction. Six thermographs (Vemco model Minilog TR) were placed upstream, middle and downstream the riffle section recording water temperatures in the surface and on the bottom. In addition three Lakeland thermistors chains were placed out on same locations giving a transverse temperature profile in each of the different habitats. Discharge was monitored at the Syrstad gauging station located approximately 1 km upstream. Data on substrate and bathymetry were collected using a total station (Leica TC 307). Distribution of surface and anchor ice was mapped three times during the study period using the total station. Otherwise, visual observations were conducted every 6th hours related to 20 transects evenly spaced every 15 m along the river reach.

Statistical analysis was conducted using Minitab v. 14.1 and SPSS v. 13.0 for Windows.

Assumptions of normality and homogeneity of variance were tested and, when these conditions were violated, non parametric tests were used. Results were considered significant at the $P = 0.05$ α level. A sub period (25th February to 1st of March) with general low flow was chosen as basis for analysis of SA while four events with extensive frazil ice production and one event with ice break up were selected for analysis of SM (Table 1).

Results

Small scale movements versus diel and flow activity and response to ice

The relationship between fish length (L_F) and diel activity was not found to be significant ($r = -0.055$, $P = 0.81$) and therefore was not considered in the further analyses. A significant difference between the four diel periods was found ($S = 13.3$, $P = 0.004$) with the highest mean SA at dusk (Fig. 3). No significant difference between day and night SA ($U = 56$, $P = 0.84$) was apparent, nor between dusk and dawn ($U = 7$, $P = 1.0$). The mean SA during low and high discharge was 0.02 ± 0.004 and 0.03 ± 0.008, respectively. The fish showed no significant difference in SA between the two discharges ($U = 36$, $P = 0.10$). During periods with change in discharge no movements were observed in the monitored fishes.

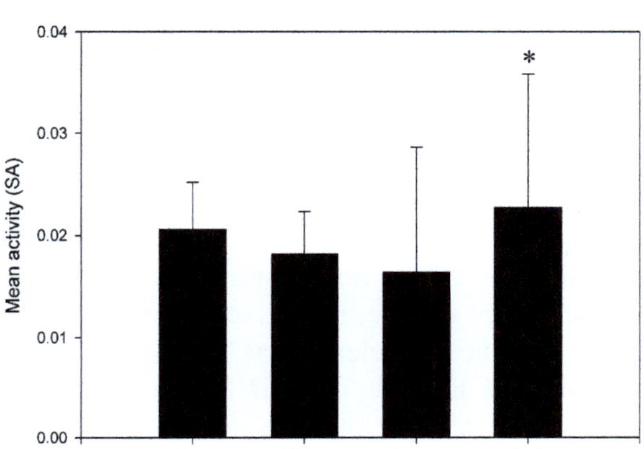

Fig. 3 Movements per hour (mean + S.E.) of diel activity of radio tagged Atlantic salmon parr

During the selected sub period (25th of February to 1st of March) twelve fish located on the antennas were consistently using surface ice covered areas while three stayed in open areas exposed to anchor ice formation. Only two fish changed position between the two areas. There were no significant difference in SA between the fish using surface ice and the fish in riffle areas ($F_{1,13}$, $P = 0.43$). Further, there was no difference in SA between fish inhabiting ice covered areas and fish in open areas in general ($U = 18.5$, $P = 0.16$).

Large scale movements versus frazil, anchor ice and ice break up

Standardized movement among the five selected ice events was significantly different ($S = 19.2$, $P = 0.001$) with highest standard movement during ice event II with anchor ice formation (4.72 ± 1.1 m/h) and lowest during event V with no anchor ice formation (0.96 ± 0.42 m/h) (Fig. 4). Of the four events with frazil ice production event IV had the lowest SM (1.34 ± 1.1 m/h). Outliers and extreme values were found in all events except event IV. Highest extreme number ($N_{\mathrm{Extreme}} = 4$) was found during event I and III. Event V was significant different from event II and III ($W = 204$, $P = 0.002$ and $W = 194$, $P = 0.001$, respectively). Flow level was stable low on event I–IV, but high on event V (Table 1). Anchor ice formed mainly in the riffle

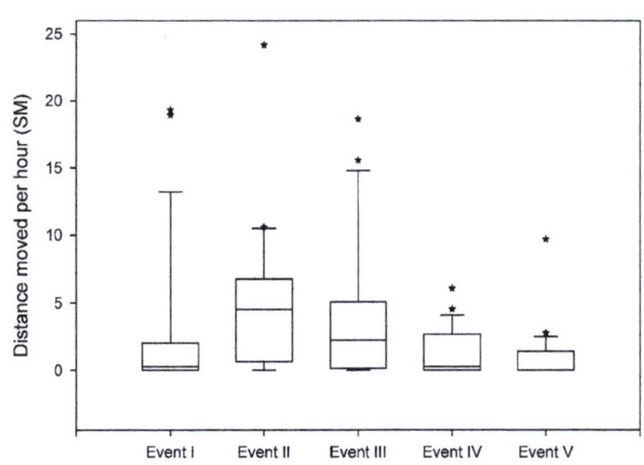

Fig. 4 Comparison of distance moved per hour of Atlantic salmon parr during four events with frazil ice production (event I–IV) and one event with ice break up (event V). *Potential outliers and extreme values

part in all events, except event IV when extensive ice formation occurred in all open areas. During event I and III none fish were observed to exploit anchor ice exposed areas (riffles). On event II four fish were located in an anchor ice exposed area before the ice formed, but during the night three fish emigrated out of the exposed area.

Only during two events (III and IV) with frazil ice production eight fish were located in open areas and exposed to drifting frazil. There was no significant difference in SM between fish exposed and fish not exposed to frazil ice ($U = 178.5$, $P = 0.19$). Further, parr ($n = 5$) inhibiting a deep pool downstream the study area were located below an established hanging dam during event IV. No significant difference in SM ($U = 18$, $P = 0.09$) was found between fish below the hanging dam and fish using ice covered areas, nor between fish below the hanging dam and fish exposed to frazil ice ($U = 6$, $P = 0.41$) located upstream in the river reach.

Discussion

Previous comparable studies, using telemetry, have been constrained by the number of fish that can be concurrently studied; for example Hiscock et al. (2002a, b) studied twelve juveniles while Robertson et al. (2004) studied eleven. Additionally, previous studies have used different approaches for analyzing observed fish behaviour including home range calculations, observed distance moved and number of movements per hour. Comparisons with previously reported studies are therefore difficult. In this study a total of 24 Atlantic salmon parr were monitored during a 12-day period mid-winter with highly variable environmental conditions using a telemetry technique as described by Hiscock et al. (2002a). The results, based on activity and movement detections, did not reveal any clear pattern in behaviour of Atlantic salmon parr despite the variable conditions. In contrast to many other studies this study indicated that juvenile activity was significantly not affected by their size, probably due to the narrow range in fish size distribution (L_F, 13.6 cm ± 0.7 SD). Heggenes et al. (2004), for example, found a positive relationship between fish size and size of home range while Hiscock et al. (2002b) found a significant decrease in activity with increasing fork length.

The hypothesis of nocturnal activity has been demonstrated in previous studies (Cunjak, 1988; Whalen & Parrish, 1999; Bremset, 2000; Hiscock et al., 2002b). In the study by Robertson et al. (2004) they found no differences in activity mid-winter among diel periods while Scruton et al. (2005) found less diel movement during winter than summer. In present study, our findings showed no significant difference in activity between day and night or dusk and dawn. There was however, a significant increase during dusk in activity pattern. Additional, the fish had contributory activity during the day, which may be attributed to the variable physical conditions we found in the regulated study river. During daytime fluctuating flow and cessation of anchor ice were observed and such conditions greatly alter the physical habitat and may force the juveniles to seek other locations.

In regulated rivers flow are frequently changed due to changes in power production. During the experiment the fish did not show a significant increase in SM between high and low flow. In the studies by Robertson et al. (2004) and Berland et al. (2004) similar results were obtained. These findings correspond well with our results although latter study was conducted during summer conditions. During event V (night 22nd, Table 1) a decrease in the discharge by 44% within 4 h triggered a break up of the border ice causing a small ice run. In some regulated rivers this is a normal operational strategy in order to prevent ice coverage from forming during low discharge and thereby avoiding larger and potentially damaging ice break ups when discharge for power production is again increased. During this event parr showed significant less movement compared to events with frazil ice production. These results may indicate that juveniles were not affected by this small ice release, most likely because of the impact of the ice release was not large enough to influence the fish in their holding positions. This does not imply that the same would be the case with a large scale ice break up. Another explanation may be that fish were positioned within coarse substrate, thereby protecting them from moving surface ice. Substrate provides important cover against predators together with hydraulic shelter and may decrease the activity and movement of juveniles (Alfredsen et al., 2004).

In winter time when the water temperature is low rapid decrease in flow at night time has been shown to cause an increase risk of pool trapped stranding, but decreased risk of beached type stranding of juveniles (Bradford et al., 1995; Bradford, 1997; Berland et al., 2004) suggesting habitats along the river margins and in shallow areas may be critical. Saltveit et al. (2001) found decreased risk of stranding during daytime at low temperatures. In the present study a rapid decrease occurred during night 22nd. Here, a group (n = 3) of juveniles was trapped in a small pool close to the river bank indicating critical habitats during flow fluctuations. Several existing winter studies, including this one, demonstrated that the habitats along the river margins are important refuges for fish during winter due to available ice cover and suitable velocity conditions. However, during fluctuating flow conditions these areas are most exposed to stranding. These findings underline the importance of a controlled discharge strategy in regulated rivers, especially during winter time.

Studies on juvenile Atlantic salmon inhibiting ice covered areas are very few, especially in natural environments. However, Whalen et al. (1999) indicated that fish avoided open areas in mid-winter. Alfredsen et al. (2004) found a positive correlation between fish located in open water and movement suggesting that fish in open water are more susceptible to move. In laboratory experiments by Finstad et al. (2004) ice cover was shown to be a positive factor in relation to energy storage in three juvenile Atlantic salmon stocks. These findings may indicate that ice covered areas during winter time are preferred by juvenile salmon owing benefits associated with energy expenditure and sheltering. During the selected sub period (25th of February to 1st of March) parr tend to use the ice covered areas (pools and runs) and avoided open riffles where anchor ice formed. The available habitat in riffle areas was highly variable between day and night because of formation of anchor ice. A test on activity between juveniles in ice covered areas and juveniles in areas exposed to anchor ice revealed no significant difference, which is somewhat not surprisingly due to the dynamic ice conditions in the riffles. In the study by Roussel et al. (2004) they observed fish dwelling beneath the anchor ice and concluded anchor ice to be a suitable habitat due to hydraulic shelter and protection from super cooled water. In the study by Cunjak & Power (1986) velocity and substrate were found to be the most important selection criteria in winter. Whalen et al. (1999) showed a similar use of anchor ice covered areas, while most other studies (Brown & MacKay, 1995; Caissie and Cunjak, 1997; Jakober et al., 1998) indicate that areas exposed to anchor ice are less suitable for fish. In this study anchor ice was observed to accumulate between the boulders thereby blocking the available substrate shelter. Test on movement between anchor ice formation events and events with no anchor ice formation showed that fish made significant more movements during ice formation. Further, during ice event I and III none individuals were found in ice exposed areas. On ice event II three of four fish emigrated from the riffle to the ice covered area in the pool. Based on our observations preferable habitats may be defined as surface ice covered and not anchor ice exposed areas (riffles).

Only a few studies have observed response of juvenile salmonids to frazil ice events. Tack (1938) observed trout (*Salmo trutta*) mortality due to tiny ice particles in the water column and Whalen et al. (1999) made only few observations of juvenile Atlantic salmon on nights with heavily frazil ice production. In the study by Roussel et al. (2004) 40% of the Atlantic salmon parr disappeared indicating long-distance-movements during the freezing period. Other radio-tracking studies (Brown & Mackay, 1995; Jakober et al., 1998; Annear et al., 2002) have related occurrence of frazil ice with more frequent winter movements of distances longer than 1 km. The data from Orkla River does not concur with these observations. Parr exposed to frazil ice showed no significant difference in movement pattern versus juveniles that were not exposed and none of the observed individuals made long emigrations. In our study less than eight fish were found in frazil exposed areas during two of the five selected ice events. Considering that very few individuals were found in these areas our results may still indicate that fish avoid areas exposed to heavy frazil transport.

Despite the general belief that winter time is critical for survival of stream fishes, the number of studies is limited compared to other seasons (Cunjak, 1996a; Roussel et al., 2004). One obvious reason for this is the difficult working environment and the methodological difficulties encountered during the winter. Bradford & Higgins (2001) showed in their study that behaviour of juvenile salmonids during winter is complex. Results from this study support that notion. More research on winter behaviour, especially in rivers with variable environmental conditions, is required in order to discover more of the factors controlling behaviour and survival of juvenile Atlantic salmon during winter time.

Acknowledgments This project is a part of the project "Effects of winter and ice conditions on the behaviour and habitat selection of juvenile Atlantic salmon" which is a part of the Wild salmon research program funded by the Norwegian Research Council. Funding for Canadian participation in this research was provided by Fisheries and Oceans Canada's Environmental Sciences Strategic Research Fund (ESSRF) and Natural Resources Canada's Panel for Energy Research and Development (PERD). The authors wish to thank all persons who participated in the field work and Tommi Linnansaari for reviewing a draft of this paper.

References

Adams, N. S., D. W. Rondorf, S. D. Evans, J. E. Kelly & R. W. Perry, 1998. Effects of surgically and gastrically implanted radio transmitters on swimming performance and predator avoidance of juvenile chinook salmon (*Onchorhynchus tshawytscha*). Canadian Journal of Fisheries and Aquatic Sciences 55: 781–787.

Alfredsen, K., 1997. A Modelling System for Estimation of Impacts on Fish Habitat. In Proceedings of the XXVIIth IAHR Congress. Environmental and Coastal Hydraulics: Protecting the Aquatic Habitat, Wang SSY, ASCE Publications.

Alfredsen, K., M. Stickler, D. A. Scruton, C. J. Pennel, C. Kelley, J. H. Halleraker, A. Harby, H. P. Fjeldstad, J. Heggenes & F. Økland, 2004. A Telemetry Study of Winter Behaviour of Juvenile Atlantic Salmon (Salmo salar) in a River with a Dynamic Ice Regime. Fifth Symposioum on Ecohydraulics. Aquatic habitats: Analysis & Restoration, Madrid, Spain, IAHR.

Alfredsen, K. & E. Tesaker, 2002. Winter habitat assessment strategies and incorporatin of winter habitat in the Norwegian habitat assessment tools. Hydrological Processes 16: 927–936.

Annear, T. C., W. Hubert, D. Simpkins & L. Hebdon, 2002. Behavioural and phsyciological response of trout to winter habitat in tailwaters in Wyoming, USA. Hydrological Processes 16: 915–925.

Berland, G., T. Nickelsen, J. Heggenes, F. Økland, E. B. Thorstad & J. Halleraker, 2004. Movements of wild Atlantic salmon parr in relation to peaking flows below a hydropower station. River Research and Applications 20: 957–966.

Bradford, M. J., 1997. An experimental study of stranding of juvenile salmonids on gravel bars and side channels during rapid flow decreases under winter conditions. Regulated Rivers 13: 395–401.

Bradford, M. J. & P. S. Higgins, 2001. Habitat, season- and size-specific variation in diel activity patterns of juvenile Chinok salmon (*Oncorhynchus tshawytscha*) and steel head trout (*Oncorhynchus mykiss*). Can J Aquat. Sci 58: 365–374.

Bradford, M. J., G. C. Taylor, J. A. Allan & P. A. Higgins, 1995. An experimental study of the juvenile coho salmon and rainbow trout during rapid flow decreases under winter conditions. North American Journal of Fisheries and Management 15: 473–479.

Bremset, G., 1999. Young Atlantic salmon (*Salmo salar* L.) and brown trout (*Salmo trutta* L.) inhabiting the deep pool habitat, with special reference to their habitat use, habitat preferences and competitive interactions. Department of Biology. Trondheim, Norway, NTNU, dep. of zoology: 1–10.

Bremset, G., 2000. Seasonal and diel changes in behaviour, microhabitat use and preferences by young pool-dwelling Atlantic salmon, *Salmo salar*, and brown trout, *salmo trutta*. Environmental Biology of Fishes 59: 163–179.

Brown, R. S. & W. C. Mackay, 1995. Fall and winter movements of and habitat use by cutthroat trout in the Ram River, Alberta. Transactions of the American Fisheries Society 124: 873–885.

Brown, R. S., S. S. Stanislawski & W. C. Mackay, 1994. Effects of frazil ice on fish. Workshop on environ.aspects of river ice, Saskatoon, Canada, Enir. Canada, National Hydrology Research Institute.

Cunjak, R. A., 1988. Behavior and microhabitat of young Atlantic Salmo (*Salmo salar*) during winter. Canadian Journal of Fisheries and Aquatic Sciences 45: 2156–2160.

Cunjak, R. A., 1996a. Winter habitat of selected stream fishes and potential impacts from land-use activity. Canadian Journal of Fisheries and Aquatic Sciences 53: 267–295.

Cunjak, R. A., 1996b. Winter habitat of selected stream fishes and potential impacts from land-use activity. Canadian Journal of Fisheries and Aquatic Sciences 53: 267–282.

Cunjak, R. A. & G. Power, 1986. Winter habitat utilization by stream resident brook trout (*Salvelinus fontinalis*) and brown trout (*Salmo trutta*). Canadian Journal of Fisheries and Aquatic Sciences 43: 1970–1981.

Cunjak, R. A., T. D. Prowse & D. L. Parrish, 1998. Atlantic salmon (*Salmo salar*) in winter: "the season of parr discontent"? Canadian Journal of Fisheries and Aquatic Sciences 55: 161–180.

David, B. O. & G. P. Closs, 2001. Contionus remote monitoring of fish activity with restricted home ranges using radiotelemetry. Journal of Fish Biology 59: 705–715.

Finstad, A., T. Forseth, T. F. Næsje & O. Ugedal, 2004. The importance of Atlantic salmon (*Salmo salar*). Journal of Animal Ecology 73: 959–966.

Fraser, N. H. C. & N. B. Metcalfe, 1997. The costs of becoming nocturnal; feeding efficiency in relation to light intensity in juvenile Atlantic salmon. Functional Ecology 11: 385–391.

Fraser, N. H. C., N. B. Metcalfe & J. E. Thorpe, 1993. Tempereture dependent swithch between diurnal and nocturnal foraging salmon. Proceedings of the Royal Society of London B. 252: 135–139.

Gries, G., K. G. Whalen, F. Juanes, & D. L. Parrish, 1997. Atlantic salmon (*Salmo salar*) in later summer; evidence for diel activity partitioning. Canadian Journal of Fisheries and Aquatic Sciences 54: 1408–1413.

Heggenes, J., O. M. W. Krog, O. R. Lindås, J. G. Dokk & T. Bremnes, 1993. Homeostatic behavioural responses in a changing environment: brown trout (*Salmo trutta*) become nocturnal during winter. Journal of Animal Ecology 62: 295–308.

Hiscock, M. J., D. A. Scruton, J. A. Brown & D. C. Clarke, 2002a. Winter movement of radio tagged juvenile Atlantic salmon on Northeast Brook, Newfoundland. Transactions of the American Fisheries Society 131: 577–581.

Hiscock, M. J., D. A. Scruton, J. A. Brown & C. J. Pennell, 2002b. Diel activity pattern of juvenile Atlantic salmon (*Salmo salar*) in early and late winter. Hydrobiologia 483: 161–165.

Jakober, M. J., T. E. Mcmahon, R. F. Thurow & C. G. Clancy, 1998. Role of stream ice on fall and winter movements and habitat use by bull trout and cutthroat trout in Montana headwater streams. Transactions of the American Fisheries Society 127: 223–235.

Matousek, V., 1984. Types of Ice Run and Conditions for their Formation. IAHR International Symposium on Ice, Hamburg, Germany.

Moore, A., I. C. Russel & E. C. E. Potter, 1990. The effects of intraperiotoneally implanted dummy acoustic transmitters on the behaviour and physiology of juvenile Atlantic salmon, *Salmo salar*. Journal of Fish Biology 37: 713–721.

Power, G., R. A. Cunjak, J. Flannagan & C. Katopodis, 1993. Chapter 4: Biological Effects of River Ice in Encironmental Aspects of River Ice. National Hydrology Research Institute, Environment Canada.

Robertson M. J., C. J. Pennel, D. A. Scruton, G. J. Robertson & J. A. Brown, 2004. Effect of increased flow on the behaviour of Atlantic salmon parr in winter. Journal of Fish Biology 65: 1070–1079.

Roussel J. M., R. A. Cunjak, R. Newbury, D. Caissie & A. Haro, 2004. Movements and habitat use by PIT-tagged Atlantic salmon in early winer: the influence of anchor ice. Freshwater Biology 49: 1026–1035.

Saltveit, S. J., J. Halleraker, J. V. Arnekleiv & A. Harby, 2001. Field experiments on stranding in juvenile Atlantic salmon (*Salmo salar*) and brown trout (*Salmo trutta*) from rapid flow decreases caused by hydropeaking. Journal of Regulated Rivers 17: 609–622.

Scruton, D. A., C. Pennell, L. M. N. Ollerhead, K. Alfredsen, M. Stickler, A. Harby, M. Robertson, K. D. Clarke & L. J. Ledrew, 2005. Movement of juvenile salmonids on diel and seasonal scales in response to experimental hydropeaking in Newfoundland, Canada. COST 626 European Aquatic Modelling Network, Silkeborg, Denmark. 19–21 May.

Tack, E., 1938. Trout mortality from the formation of suspended ice crystals. Fisherei-Zeitung 41: 42.

Tesaker, E., 1998. Mitigation of ice effects of the habitat in regulated and natural rivers. Proceedings of the 14th international symposium on ice, Potsdam/New york/USA.

Tsang, G., 1982. Frazil and anchor ice, a monograph. Ottawa, National research counscil of Canada, Subcommitee on hydraulics of ice-covered rivers.

Vehanen, T. & A. Huusko, 2002. Behaviour and habitat use of young-of-the-year Atlantic salmon (*Salmo salar*) at the onset of winter in artificial streams. Hydrobiologia 154: 133–150.

Whalen, K. G. & D. L. Parrish, 1999. Nocturnal habitat use of Atlantic salmon parr in winter. Canadian Journal of Fisheries and Aquatic Sciences 56: 1543–1550.

Whalen, K. G., D. L. Parrish & M. E. Mather, 1999. Effect of ice formation on selection of habitats and winter distribution of post-young-of-the-year Atlantic salmon parr in winter. Canadian Journal of Fisheries and Aquatic Sciences 56: 1543–1550.

Hydrobiologia (2007) 582:91–97
DOI 10.1007/s10750-006-0552-y

FISH TELEMETRY

Results of satellite tagging of Atlantic bluefin tuna, *Thunnus thynnus*, off the coast of Ireland

Michael J. W. Stokesbury · Ronan Cosgrove · Andre Boustany · Daragh Browne · Steven L. H. Teo · Ronald K. O'Dor · Barbara A. Block

Abstract Pop-up satellite archival tags were attached to six Atlantic bluefin tuna (*Thunnus thynnus*) off the west coast of Ireland in autumn 2003 and 2004. The satellite tags measured pressure, ambient temperature and light for the term of deployment. Radio pop-up satellite end-point positions, light and sea surface temperature estimations of geolocation indicate that two fish tagged minutes apart off the coast of County Donegal, migrated to the eastern and western Atlantic Ocean over the following 8 months. The two fish were 5218 km apart at the termination of the experiment. After tagging in September and popping up the following March and April, one fish had traveled to the western Atlantic while the other was located in the waters off the southwest coast of Portugal. A third fish tagged off the coast of County Donegal in October 2004 moved into the Mediterranean Sea and was caught by a fishing vessel southeast of Malta on 11 June 2005. The results link bluefin tuna feeding on European foraging grounds with known eastern breeding regions and western Atlantic waters.

Keywords Bluefin tuna · Ireland · Satellite tags · Movement

Introduction

The study of fish migration and stock identification are closely related (Mather et al., 1995). Tagging studies initiated in the western Atlantic Ocean and using Pop-up Satellite Archival Transmitting Tags (PATs) and implantable archival tags have investigated the movement patterns of Atlantic bluefin tuna (*Thunnus thynnus* L.; Block et al., 2001; Stokesbury et al., 2004; Block et al., 2005; Wilson et al., 2005). Bluefin tuna are commercially important, highly migratory members of the scombridae. Their migration patterns and stock structure is complex (Mather, 1995, Block et al., 2001; 2005) and understanding their movement patterns is crucial for their proper management and conservation (NRC 1994).

Guest editors: P. R. Almeida, B. R. Quintella, M. J. Costa and A. Moore
Developments in Fish Telemetry

M. J. W. Stokesbury (✉) · R. K. O'Dor
Biology Department, Dalhousie University, 1355 Oxford St., Halifax, Nova Scotia, Canada B3H 4J1
e-mail: mstokesb@dal.ca

M. J. W. Stokesbury · A. Boustany · S. L. H. Teo · B. A. Block
Tuna Research and Conservation Center, Stanford University, Hopkins Marine Station, Oceanview Boulevard, Pacific Grove, CA 93950, USA

R. Cosgrove · D. Browne
An Bord Iascaigh Mhara (BIM), The Irish Sea Fisheries Board, New Docks, Co. Galway, Ireland

Fisheries for Atlantic bluefin tuna are managed by the International Commission for the Conservation of Atlantic Tunas (ICCAT) under a two stock hypothesis (NRC, 1994). One eastern Atlantic Ocean and Mediterranean Sea stock that spawns in the Mediterranean Sea (Richards, 1976) and another western Atlantic Ocean stock that spawns in the Gulf of Mexico (McGowan and Richards, 1989) and the Florida Straits (Rivas, 1954). The stock boundary line used by ICCAT to divide the stocks is at 45 W Longitude. Bluefin tuna are managed as if there is a low mixing rate (~2–4% year^{-1} NRC, 1994) between the two stocks. However, a recent electronic tagging study indicated that the trans-Atlantic movement rate of bluefin tuna was likely to be higher because 37.2% of the recaptures of Atlantic bluefin that were tagged in the western Atlantic Ocean, occurred in the eastern Atlantic Ocean and Mediterranean Sea (Block et al., 2005). The electronic tagging in western Atlantic locations suggests that a significant component from the eastern spawned stock must migrate into west Atlantic waters for foraging. Similarly, some large giants from the Gulf of Mexico spawned stock, migrate into central and East Atlantic waters to forage. Discerning the extent of this mixing, and the affect this has on the assessment of abundance of each stock, is critical to the future management of Atlantic bluefin tuna.

In 2000, a small sport fishery for Atlantic bluefin tuna began off the coast of County Donegal and more recently one has also begun off County Mayo, on the mid-west coast of Ireland. Little is known about the life history of the giant bluefin tuna now being targeted by sports fishers in these fisheries. Bluefin tuna that enter Norwegian waters most likely traveled through the waters off the northwest coast of Ireland (Mather et al., 1995). Also, fish may pass western Ireland on route from waters in the western Atlantic, to the spawning grounds in the Mediterranean Sea (Mather et al., 1995). Interestingly, bluefin tuna present off Norway in the autumn have also been linked to the Western Atlantic Ocean as nine large bluefin tuna tagged in the Bahamas were captured off Norway (Mason et al., 1977). Also, recent electronic tagging experiments indicated that some bluefin tuna tagged off the coast of North Carolina in winter moved in spring and summer months to the Eastern Atlantic Ocean and to the waters offshore of Ireland (Block et al., 2005). Some of these fish continued on into the Mediterranean Sea to spawn indicating that eastern stock fish were foraging in the west and east Atlantic.

The migration pattern and stock of origin of the Atlantic bluefin tuna currently being exploited off the west coast of Ireland is unknown. In this study we employed the PAT tags and external tagging methodologies used in the western Atlantic Ocean (Block et al., 2001), to tag giant bluefin tuna captured in the eastern Atlantic Ocean. The study enabled the examination of movements and environmental preferences of several individuals of giant bluefin tuna released off the west coast of Ireland.

Materials and methods

Atlantic bluefin tuna were fished in two different locations off the west coast of Ireland. Fish were captured by trolling squid spreader bars with rod and reel off County Donegal and County Mayo. The fish were caught and brought on board the tagging vessel, measured, and tagged with PATs using procedures described by Stokesbury et al., (2004). The PATs (hardware PAT4, software 4.01e) provided an end point location based on the Doppler shift of the tags radio transmission to the Argos satellites (root mean square error of <350 m; Taillade, 1992). PATs also measured and archived light, ambient temperature and pressure at 60 s intervals. Depth and temperature data were summarized into 12 h bins prior to transmission. Temperature vs. depth profiles were generated with minimum and maximum temperature at each of the eight equally spaced depths. The tags were equipped with a pre-release program and an auto-depth correction. So, if the tag prematurely detached from the fish and floated to the surface it reported to the satellite 4 days after detachment.

The daily movements of Atlantic bluefin tuna were estimated from light-based calculation of longitude using Wildlife Computers WC-GPE 1.01.0005 (Hill and Braun, 2001), and latitude estimates based on satellite derived sea surface temperatures (SST) at the derived longitude compared with SST's recorded by the tag (Teo et al., 2004). Geolocation estimates for PATs deployed on Atlantic bluefin tuna had a root mean square error of 1.30 for light-based longitude and 1.89 for SST-based latitude (Teo et al., 2004).

Results

A total of six Atlantic bluefin tuna were tagged and released with second generation PATs off the west coast of Ireland in 2003 and 2004. In 2003 two fish (221 and 225 cm curved fork length (CFL)) were tagged off County Donegal (55.36 N, 7.52 W) on 20 September, each with a single PAT tag (#03–267 and 03–191, respectively) and an externally placed conventional tag. These two fish were hooked by the same boat at approximately the same time and were tagged and released within 15 min of each other. A third fish (264 cm CFL) was tagged off County Mayo on 15 October 2003 with PSAT #03–275. The tags deployed on the two fish released off County Donegal reported 177 (PSAT #03–267) and 205 (PSAT #03–191) days later from different sides of the Atlantic Ocean. Tag #03–275 failed to report. In 2004 two fish (226 and 230 cm curved fork length (CFL)) were tagged off County Donegal (55.43 N, 7.38 W) on 9 October, each with a single PAT tag (#04–267 and 04–268, respectively) and an externally placed conventional tag. These two fish were hooked by the same boat at approximately the same time and were tagged and released within 10 min of each other. A third fish (230 cm CFL) was tagged off County Donegal (55.41 N, 7.46 W) on 8 October 2004 with PSAT #04–269. The tag deployed on the fish released off County Donegal on 8 October was later returned by fisherman after they caught the fish in the Mediterranean Sea on 11 June 2005. Tags #04–267 and 04–268 failed to report.

Tag 03–267 reported to Argos satellites on 15 March 2004 from the Bahamas at 25.67 N, 67.39 W, a straight-line distance of 5805 km from the tagging site (Fig. 1). Geolocation positions indicate that this fish resided in waters over the Irish shelf from tagging until 26 October. It then moved to off shelf waters for a period of 24 days, corroborated by multiple dives to depths deeper than 400 m (Fig. 2a). The fish then returned to waters over the Irish shelf on 19 November 2004, where it remained until 18 December, 2004. It then again moved off shelf until the tag reported from the Bahamas in March, 2004 (Fig. 1).

Fish 03–267 spent 79.9% of its time in the top 100 m of the water column including 42.8% in the top 5 m (Fig. 3a). The fish spent 55.5% of its time in water with temperatures between 12 and 16°C

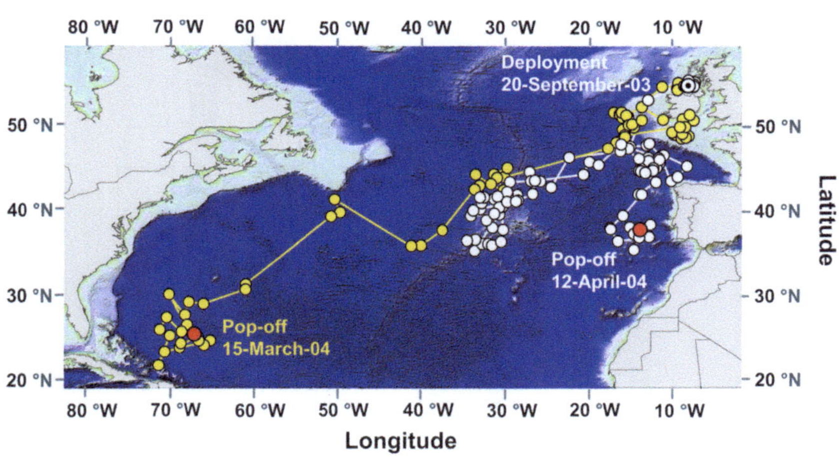

Fig. 1 Location data from pop-up satellite tags deployed on two Atlantic bluefin tuna (yellow circles = fish 03–267, white circles = fish 03–191) off Ireland (circle with dot = release location) in autumn 2003, obtained from radio transmission end points to the Argos satellite system (red circles), and geolocation estimates based on light level longitude and sea surface temperature latitude estimates (yellow and white circles; $N = 184$)

Fig. 2 Maximum depth (◇), maximum (□) and minimum (Δ) temperatures reported by PATs deployed on Atlantic bluefin tuna: (**a**) 03–267 and (**b**) 03–191, both tagged off the west coast of Ireland in autumn 2003

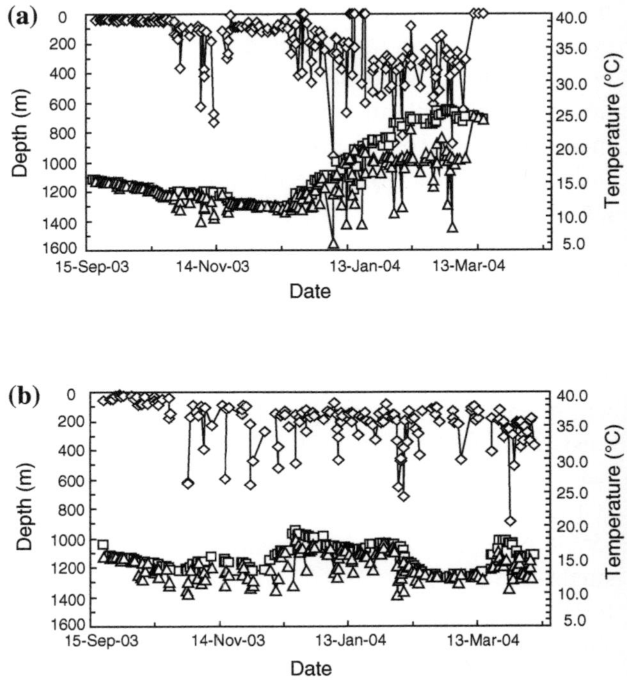

(Fig. 3a). Sea surface temperatures from 12 to 26°C were recorded by the tag while the fish crossed the mid-Atlantic and entered the waters of the Bahamas (Fig. 4a).

Tag 03–191 reported on 12 April 2004 from waters off southwest Portugal, at 38.17 N, 13.33 W, a straight-line distance of 1962 km from the tagging site (Fig. 1). Geolocation positions (Fig. 1) and its maximum depth profile (Fig. 2b) indicate that this fish resided in waters over the Irish shelf from tagging until 21 October, and then moved off shelf. It then moved off shelf to waters over the mid-Atlantic Ridge and back to the Eastern Atlantic Ocean. Finally the fish resided in

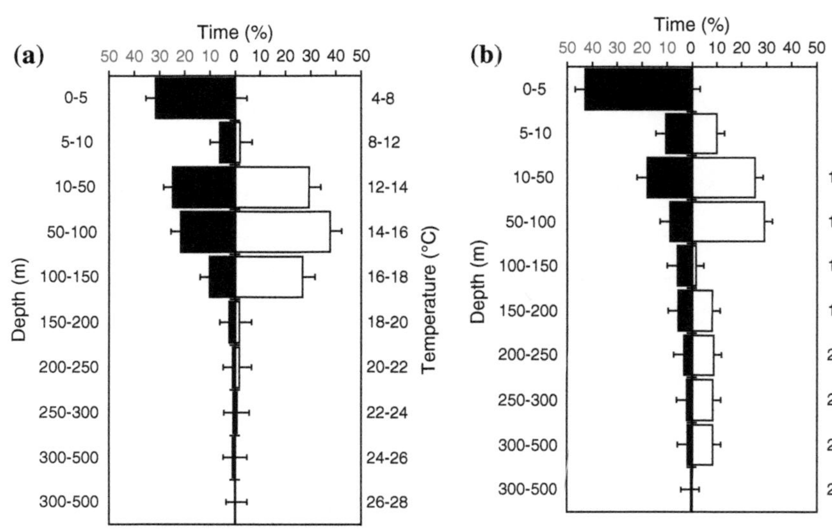

Fig. 3 Bluefin tuna (**a**) 03–267's and (**b**) 03–191's time at depth distribution (black) and time at temperature distribution (white)

Fig. 4 Water column temperature profile generated by (a) tag 03–267 and (b) tag 03–191

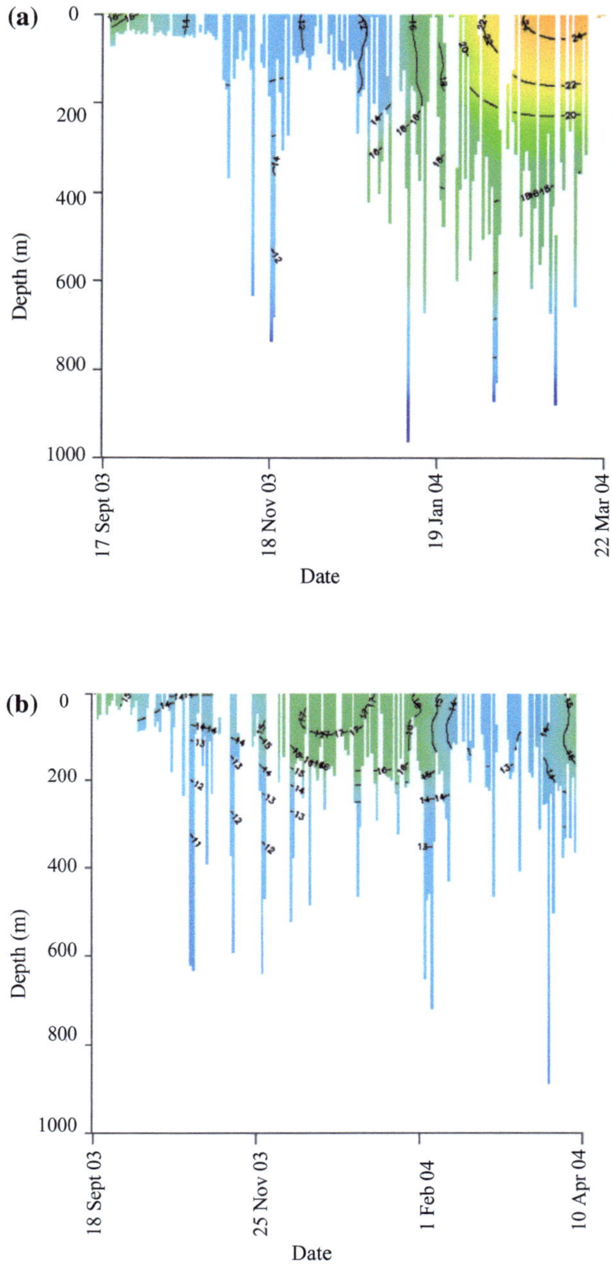

shelf waters, southwest of Portugal until the tag reported (Fig. 1).

Fish 03–191 spent 84.25% of it's time in the top 100 m of the water column (Fig. 3b). The fish spent 93.84% it's time in water with temperatures between 12 and 18°C (Fig. 3b). The temperature depth profile indicated that the water column was mixed and cool during the period on the Irish shelf. However, when the fish moved to the mid-Atlantic Ocean, the water column became more stratified and the thermocline was deeper. Finally, off Portugal the water column again became mixed and cool (Fig. 4b).

Tag 04–269 was captured in a purse seine on 11 June 2005 in waters southwest of Malta, at 31.57 N, 17.42 W, a straight-line distance of

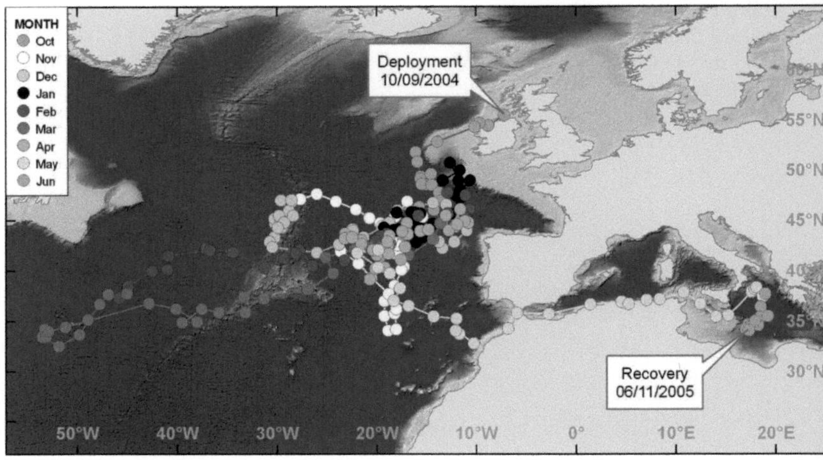

Fig. 5 Location data from a pop-up satellite tag deployed on an Atlantic bluefin tuna (04–269) off Ireland in autumn 2004, obtained from radio transmission end points to the Argos satellite system, and geolocation estimates based on light level longitude and sea surface temperature latitude estimates (circles are color coded for month)

2766 km from the tagging site (Fig. 5). Geolocation positions (Fig. 5) indicate that this fish moved southwest from waters off Ireland in October and remained in the North Atlantic until May 2005. The fish then crossed into the Mediterranean Sea in June 2005 and was captured by a purse seine fishing vessel 11 July 2005.

Discussion

Atlantic bluefin tuna caught and tagged on feeding grounds in waters off the northwest coast of Ireland in autumn 2003 and 2004 subsequently moved to waters in the West Atlantic Ocean and the Mediterranean Sea. These results link bluefin tuna feeding on European foraging grounds with known eastern breeding regions and western management zone waters

Two fish simultaneously caught and tagged at approximately the same time in September 2003 off the west coast of Ireland, moved to opposite sides of the Atlantic Ocean by March and April, 2004. Bluefin tuna 03–267 traveled to the waters of the Bahamas, close to the Florida Straits, a proposed spawning ground for bluefin tuna in the northwest Atlantic Ocean (Rivas, 1954). The water in this area was warm enough for a bluefin tuna to spawn (>24°C; Block et al., 2001). Electronically tagged bluefin tuna from previous studies have moved to these waters in winter and spring and encountered ambient water temperatures warm enough to be indicative of spawning (Block et al., 2001). This fish was tagged 2342 km east of the ICCAT stock boundary line, and moved to a Western Atlantic Ocean location, during the spawning season. Fish 03–191, traveled to waters off southwestern Portugal near the Ibero-Moroccan Bay. However, the water temperatures at the time the tag reported were well below the temperatures in which bluefin tuna are hypothesized to spawn (Block et al., 2001; 2005). The tag reported approximately 755 km from the Straits of Gibraltar. Possibly this fish was on its way to the Mediterranean spawning ground when the tag popped up.

Both fish spent a large portion of their time in the top 5 m of the water column. This may be due to the overall thermal regime that the fish are exposed to in the offshore waters off Ireland. Overall the temperatures were cool and the occupancy of surface waters may be associated with the thermal tolerances of both prey and giant bluefin. The tuna spent the majority of their time in a fairly narrow temperature range during the period they were in east Atlantic waters. However, the two temperature ranges differed significantly with fish 03–267 experiencing warmer and cooler time-at-temperature profiles than 03–191. This was caused by bluefin tuna 03–267 moving into the west Atlantic in winter where it encountered the warmer waters of the Bahamas while bluefin tuna 03–191 occupied the cooler waters of the mid and eastern Atlantic Ocean in winter.

Fish 04–269 traveled to waters southwest of Malta in the Mediterranean Sea in the months of

May and June. This fish moved from the feeding grounds off northwestern Ireland to the spawning grounds in the Mediterranean Sea precisely during the time that other electronically tagged fish move into these regions from the west Atlantic releases (Block et al., 2001; 2005). This result suggests that at least a component of the Irish offshore fishery is composed of Mediterranean spawned breeders.

The results of this study are in agreement with previous studies showing linkages between Atlantic bluefin tuna off the coast of North America and those found off the coast of Ireland (Block et al., 2005), and also support a potential link between Atlantic bluefin tuna on the feeding grounds off Ireland and the eastern Atlantic Ocean spawning grounds off Malta in the Mediterranean Sea. Additionally linkage between the foraging grounds off Ireland and the west Atlantic spawning grounds is tentatively shown here however technically the position of the pop-up satellite tag was just to the North of the known ICCAT areas where bluefin are putatively thought to spawn in the west Atlantic outside the Gulf of Mexico. Given the small sample size, caution must be exercised in over interpreting these tracks. However they are similar to several tracks noted in Block et al. (2005), indicating potentially that independent breeding stocks are mixing on feeding grounds and then sorting to breeding grounds. It is important to increase the tagging effort in the eastern Atlantic Ocean to gain further knowledge of the movement patterns and stock structure of the autumn aggregation of bluefin tuna that is present off the west coast of Ireland.

Acknowledgements This study was supported by BIM, a grant from the David and Lucile Packard Foundation and the Monterey Bay Aquarium Foundation. MJWS was supported by a Dalhousie scholarship, and NSERC IPGS and an NSERC grant to RKO. We thank A. Malloy and J. Brittain, for vessel access that made this study possible, H. Tan for logistical support, and A. Walli and K. Weng for help with figure preparation.

References

Block, B. A., H. Dewar, S. B. Blackwell, T. D. Williams, E. D. Prince, C. J. Farwell, A. Boustany, S. L. H. Teo, A. Seitz, A. Walli & D. Fudge, 2001. Migratory movements, depth preferences, and thermal biology of Atlantic bluefin tuna. Science 293: 1310–1314.

Block, B. A., S. L. H. Teo, A. Walli, A. Boustany, M. J. W. Stokesbury, C. Farwell, T. Williams, K. Weng & H. Dewar, 2005. Electronic tagging and population structure of Atlantic bluefin tuna. Nature 434: 1121–1127.

Hill, R. D. & M. J. Braun, 2001. Geolocation by light-level. The next step: latitude. In Sibert, J. & J. Neilson (eds), Methods and Technologies in Fish Biology and Fisheries. Vol. 1. Kluwer Academic, Dordrecht, The Netherlands, 443–456.

Mason, J. M. Jr., F. J. III Mather & R. E. Baglin Jr., 1977. Results of United States cooperative tagging of Atlantic bluefin tuna, October 1974 through October 1976. ICCAT, Collective Volume of Scientific Papers 6: 259–263.

Mather, F. J., J. M. Mason & A. C. Jones, 1995. Historical document: life history and fisheries of Atlantic bluefin tuna. NOAA Technical Memorandum NMFS-SE-FSC- 370, p. 165.

McGowan, M. F. & W. J. Richards, 1989. Bluefin tuna, *Thunnus thynnus*, larvae in the Gulf Stream off the Southeastern United States: satellite and shipboard observations of their environment. Fisheries Bulletin 87: 613–631.

National Research Council, 1994. An assessment of Atlantic bluefin tuna. National Academy Press, Washington D.C, 143 pp.

Richards, W. J, 1976. Spawning of bluefin tuna (*Thunnus thynnus*) in the Atlantic Ocean and adjacent seas. ICCAT Collective Volume of Scientific Papers 2: 03–267–278.

Rivas, L. R., 1954. A preliminary report on the spawning of the western north Atlantic bluefin tuna (*Thunnus thynnus*) in the Straits of Florida. Bulletin of Marine Science Gulf and Caribbean 4: 302–322.

Stokesbury, M. J. W., S. L. H. Teo, A. Seitz, R. K. O'Dor & B. A. Block, 2004. Movement of Atlantic bluefin tuna (*Thunnus thynnus*) as determined by satellite tagging experiments initiated off New England. Canadian Journal of Fisheries and Aquatic Sciences 61: 1976–1987.

Taillade M., 1992. Animal tracking by satellite. In Priede, I. M. & S. M. Swift (eds), Wildlife Telemetry Remote Monitoring and Tracking of Animals. Ellis Horwood, New York, NY. pp. 149–160.

Teo, S. L. H., A. Boustany, S. Blackwell, A. Walli, K. C. Weng & B. A. Block, 2004. Validation of geolocation estimates based on light level and sea surface temperature from electronic tags. Marine Ecology Progress Series 283: 81–98.

Wilson, S. G., M. E. Lutcavage, R. W. Brill, M. P. Genovese, A. B. Cooper & A. W. Everly, 2005. Movements of bluefin tuna (*Thunnus thynnus*) in the northwestern Atlantic Ocean recorded by pop-up satellite archival tags. Marine biology 146: 409–423.

Hydrobiologia (2007) 582:99–107
DOI 10.1007/s10750-006-0548-7

FISH TELEMETRY

Fjord migration and survival of wild and hatchery-reared Atlantic salmon and wild brown trout post-smolts

Eva B. Thorstad · Finn Økland · Bengt Finstad · Rolf Sivertsgård · Núria Plantalech · Pål Arne Bjørn · R. Scott McKinley

Abstract The behaviour of wild (n = 43, mean L_T = 152 mm) and hatchery-reared (n = 71, mean L_T = 198 mm) Atlantic salmon and wild anadromous brown trout (n = 34, mean L_T = 171 mm) post-smolts with acoustic transmitters was compared in a Norwegian fjord system. There was no difference in survival between wild and hatchery reared salmon from release in the river mouth to passing receiver sites 9.5 km and 37.0 km from the release site. Mortality approached 65% during the first 37 km of the marine migration for both groups. There was no difference between wild and hatchery-reared salmon either in time from release to first recording at 9.5 km (mean 135 and 80 h), or in the rate of movement through the fjord (mean 0.53 and 0.56 bl s^{-1}). Hatchery-reared salmon reached the 37 km site sooner after release than the wild salmon (mean 168 and 450 h), but rate of movement in terms of body lengths per second did not differ (mean 0.56 and 0.77 bl s^{-1}). The brown trout remained a longer period in the inner part of the fjord system, with much slower rates of movement during the first 9.5 km (mean 0.06 bl s^{-1}).

Guest editors: P. R. Almeida, B. R. Quintella, M. J. Costa and A. Moore
Developments in Fish Telemetry

E. B. Thorstad (✉) · F. Økland · B. Finstad · N. Plantalech
Norwegian Institute for Nature Research (NINA), Tungasletta 2, 7485 Trondheim, Norway
e-mail: eva.thorstad@nina.no

R. Sivertsgård
Norwegian College of Fishery Science, University of Tromsø, 9037 Tromsø, Norway

P. A. Bjørn
Norwegian Institute of Fisheries and Aquaculture Research, Breivika, 9291 Tromsø, Norway

R. S. McKinley
West Vancouver Laboratory, The University of British Columbia, VTV 1N6 British Columbia, Canada

Keywords Acoustic telemetry · Fjord migration · Swimming speed · *Salmo trutta* · *Salmo salar*

Introduction

Survival and behaviour during the early stages after the critical transition from freshwater to seawater are important factors for anadromous fish. At the end of the freshwater stage, Atlantic salmon *Salmo salar* L. and anadromous brown trout *Salmo trutta* L. juveniles transform physiologically and morphologically and migrate to sea as smolts (Wedemeyer et al., 1980; Hoar, 1988; Finstad & Jonsson, 2001). Atlantic salmon migrate to the open ocean for feeding, whereas the brown trout remain feeding in the fjord and coastal areas (Jonsson, 1985; Hansen et al., 2003; Klemetsen et al., 2003; Rikardsen, 2004). The understanding of the early marine phase of the Atlantic salmon and the environmental factors that may influence their behaviour and distribution in the sea is limited (Moore et al., 2000), and even less is known about the brown trout. This lack of information is particularly critical because the heaviest mortality of salmon in the sea apparently takes place during the first months after the smolts leave fresh water (Hansen et al., 2003).

The first stages of the marine migration of post-smolts can be monitored using telemetry techniques (Moore et al., 2000; Lacroix & Voegeli, 2000). For Atlantic salmon, the method has been limited by the relatively large size of acoustic transmitters, thus most studies have investigated behaviour of larger hatchery-reared posts-molts (e.g. LaBar et al., 1978; Lacroix & McCurdy, 1996; Finstad et al., 2005; Thorstad et al., 2004). However, the behaviour of hatchery-reared post-smolts may differ from that of wild post-smolts (e.g. Jonsson et al., 1991), and as pointed out by Lacroix & McCurdy (1996), there is a general need to obtain information on the movements of normally sized wild Atlantic salmon post-smolts.

Smaller acoustic transmitters are now available, allowing tagging of wild Atlantic salmon post-smolts. The objective of this study was to compare the rate of movement and survival of wild Atlantic salmon post-smolts with hatchery-reared Atlantic salmon and wild brown trout post-smolts in a Norwegian fjord system. It was hypothesised that the rate of movement of Atlantic salmon post-smolts through the fjord system was not dependent on their origin (as suggested by Lacroix & McCurdy, 1996; Voegeli et al., 1998; Lacroix et al., 2004). Furthermore, it was hypothesised that survival did not differ between wild and hatchery-reared Atlantic salmon (as suggested by Hvidsten & Lund, 1988). It was also hypothesised that the sea trout post-smolts remained in the inner part of the fjord system to a larger extent than the wild Atlantic salmon post-smolts (as suggested by Finstad et al., 2005).

Materials and methods

Study area

The study was carried out in the Romsdalsfjord system on the West Coast of middle Norway, 62°40′ N 8°10′ E (Fig. 1, described in Finstad et al., 2005).

Handling, tagging and release of fish

Wild (n = 43) and hatchery-reared (n = 71) Atlantic salmon and wild brown trout (n = 34) smolts were tagged with coded acoustic transmitters (Table 1). The smallest 27 wild salmon were tagged with Thelma (Trondheim, Norway) transmitters, model THS-04 (7 × 19 mm, mass in air/water of 1.9/1.2 g, guaranteed life time 28 days), whereas the remaining smolts were tagged with VEMCO Ltd. (Nova Scotia, Canada) transmitters, model V8SC-6L-R256 (9 × 20 mm, mass in air/water of 3.3/2.0 g, guaranteed life time 70 days). The tag to body mass ratio was larger in wild than in hatchery salmon (Mann–Whitney U test, $P < 0.001$, Table 1). All transmitters were on frequency 69 kHz and had a power output of 139 db re 1uPa@1 m.

The wild Atlantic salmon and sea trout were caught in a trap close to the river mouth during their downstream migration in the River Eira, transported to the Statkraft hatchery in Eresfjord in a tank with oxygenated water and tagged zero to five days later (this variation was due to practical reasons). The hatchery salmon were 2-year-old smolts from the Statkraft hatchery with wild salmon from the River Eira as parents. The fish were tagged during 6 May–4 June 2004, depending on capture rates of wild smolts, using methods described in Finstad et al. (2005). Water temperature in the hatchery was 2.7–7.0°C. A seawater tolerance test (Blackburn & Clarke, 1987) performed on the hatchery salmon on 6 May revealed plasma chloride levels at 147 mM, indicative that the smolts were ready to be released into seawater (Sigholt & Finstad, 1990).

Groups of tagged smolts were transported in water-filled plastic bags to a cage (volume 0.5 m^3)

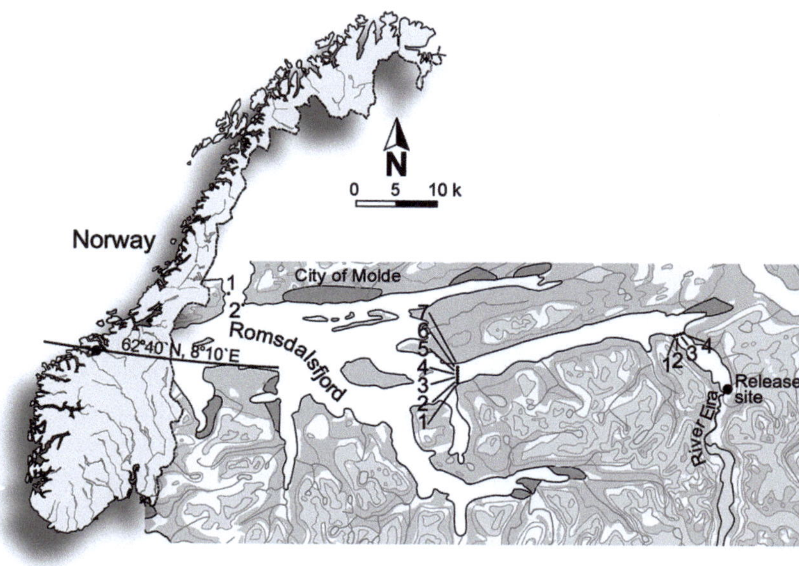

Fig. 1 The Romsdalsfjord system in Middle Norway with the release site for acoustically tagged Atlantic salmon and anadromous brown trout in 2004 and automatic receiver sites 9.5 km (4 receiver sites), 37.0 km (7 receiver sites) and 65.0 km (2 receiver sites) from the release site

in the fjord at the mouth of the River Eira 0–2 days after tagging. The fish were kept in the cage for 2–21 h prior to release (Fig. 1, Finstad et al., 2005). Fish during the first release were held in the cage from one day to the next, but appeared stressed by being held in the cage. The procedure with keeping the fish in the cage was maintained to adjust the fish to the marine environment, but the time was reduced. All fish appeared to be in good condition at release. Schools of wild and Carlin-tagged Atlantic salmon and brown trout post-smolts were frequently seen at the release site during the study period. All fish were released during high tide in daylight (from 3 h before to 1 h after the highest tide).

Recording of fish after release

Fish were recorded when passing 19 individual receiver/data loggers (VEMCO model VR2) moored at sites 9.5, 37.0 and 65.0 km from the fish release site at the estuary of the River Eira (Fig. 1). At 9.5 km from the release site, where the fjord is 1.48 km wide, four moorings were evenly distributed across the fjord. Receivers were attached to each mooring at 3 m depth. An additional receiver was attached to two of the moorings at 10 m depth (site 1 and 3, Fig. 1). At 37.0 km from the release site, where the fjord is 2.60 km wide, seven moorings were evenly distributed across the fjord. Receivers were attached to each mooring at 5 m depth. An additional

Table 1 Atlantic salmon (wild and hatchery) and anadromous brown trout (wild) post-smolts tagged with acoustic transmitters and released at the mouth of the River Eira in the Romsdalsfjord system in 2004

Group	Number tagged	Mean body length (L_T), mm (range, SE)	Mean body mass, g (range, SE)	Mean transmitter mass in water to fish mass ratio (%) (range, SE)	Release dates
Wild salmon	43	152 (136–173, 1.2)	25 (18–38, 0.66)	6.0 (4.0–8.0, 0.17)	7, 8, 9, 12, 20 and 22 May
Hatchery salmon	71	198 (160–245, 2.6)	70 (37–124, 2.7)	3.1 (1.6–5.4, 0.12)	7, 8, 9, 12, 20 and 31 May
Wild trout	34	171 (143–242, 4.6)	42 (23–114, 3.8)	5.6 (1.8–8.7, 0.31)	7, 8, 9, 12, 20, 22, 26 and 31 May, 6 June

receiver was attached to four of the moorings at 10 m depth (site 1, 3, 5 and 7, Fig. 1). At 65.0 km from the release site, where the fjord is 2.46 km wide, one receiver was attached on each side of the fjord at 5 m depth. The receiver at site 1, 65 km distant (Fig. 1), disappeared before data were downloaded.

The sea depth at receiver moorings was 24–288 m. Receiver range for detecting signals from transmitters was typically a radius of 200–260 m for transmitters at 0.5–3.0 m depth, but varied considerably (radius of 45–620 m during range tests) with factors such as wave action and salinity. The receivers collected data until 24 or 25 July 2004. Manual tracking from a boat using a VEMCO VR60 receiver was carried out in the area around the mouth of the River Eira on 27 July 2004.

Data analysis

Data collected did not meet the assumptions for parametric tests, and therefore, non-parametric tests were employed. Rates of movement in terms of body lengths per second are based on time from release to first recording at the different receiver sites. This must not be regarded as actual swimming speeds, but as minimum speeds, since the fish most likely did not follow the shortest migration routes (Thorstad et al., 2004). For analysis of where in the fjord the fish passed, only recordings by receivers at 3 m depth were used 9.5 km from the release site, and at 5 m depth 37.0 km from the release site, for standardisation.

Results

Water temperature and salinity

Water temperature and salinity were measured at eight sites in the middle of the fjord 1–60 km from the river mouth on 4, 14 and 30 May and 11 June at 1 m depth. Water temperature varied between 7.5 and 13.8°C (mean 10.8). Salinity varied between 7.7 and 31.9 ppt (mean 25.5).

Proportions of fish recorded after release

There was no difference in the proportion of wild and hatchery salmon reaching either the 9.5 km or 37.0 km receiver lines (Fisher's exact test, 9.5 km: $P = 0.85$, 37.0 km: $P = 1.00$, Table 2). However, the proportion of trout recorded was smaller than the proportion of salmon at both places (Fisher's exact test, 9.5 km: $P = 0.001$, 37.0 km: $P = 0.002$, Table 2). There were no differences in body length or transmitter/body mass ratio between individuals recorded and those not recorded either 9.5 or 37.0 km from the receiver site for any of the groups (wild salmon, hatchery salmon and trout, Mann–Whitney U tests, $P = 0.17$–0.92), except that wild salmon recorded 37.0 km from the release site were larger than those not recorded ($P = 0.039$).

Only one fish recorded 37.0 km from the release site was not recorded when passing the receivers 9.5 km from the release site (a wild salmon). The three hatchery salmon recorded 65.0 km from the release site (Table 2) were all recorded when passing receivers both 9.5 and 37.0 km from the receiver site.

On 27 July, transmitters from 6 wild salmon (14%), 6 hatchery salmon (8%) and 7 trout (21%) were located in the area close to the river mouth by manual tracking. One transmitter that had belonged to a hatchery salmon (body size 171 mm, 43 g) was found in a saithe *Pollachius virens* L. that was caught by an angler in this area.

Table 2 Number and proportion of Atlantic salmon and anadromous brown trout post smolts recorded at receiver sites 9.5 km, 37.0 and 65.0 km from the release site in the Romsdalsfjord system in 2004

Group	Number (%) recorded 9.5 km from the release site	Number (%) recorded 37.0 km from the release site	Number (%) recorded 65.0 km from the release site
Wild salmon	25 (58%)	15 (35%)	0
Hatchery salmon	39 (55%)	25 (35%)	3 (4%)
Wild trout	8 (24%)	3 (9%)	0

Rates of movement

There was no difference between wild and hatchery salmon either in time from release to first recording 9.5 km from the release site (Mann–Whitney *U*-test, $P = 0.11$), or in the rate of movement in terms of body lengths per second (Mann–Whitney *U*-test, $P = 0.43$) (Table 3). At 37.0 km from the release site, hatchery salmon were recorded sooner after release than the wild salmon (Mann–Whitney *U*-test, $P = 0.033$), but the rate of movement in terms of body lengths per second did not differ (Mann–Whitney *U*-test, $P = 0.18$) (Table 3). Hatchery salmon recorded at 37.0 km from the release site ($n = 25$) had a faster migration to 9.5 km from the release site than those not recorded at 37.0 km ($n = 14$) (Mann–Whitney *U*-test, $P = 0.038$), indicating a higher loss of the slowest migrating fish. Such a difference was not found for wild salmon post-smolts ($n = 14$ recorded and 11 not recorded, $P = 0.27$).

The trout took a longer time from release to first recording at 9.5 km than the salmon (wild and hatchery salmon combined, Mann–Whitney *U*-test, $P < 0.001$, Table 3). Also the rate of movement in terms of body lengths per second was slower for the trout than for the salmon (Mann–Whitney *U*-test, $P < 0.001$, Table 3). At 37.0 km from the release site, only three trout were recorded (Table 3).

Migration routes and patterns

Nearly all fish were recorded by more than one of the receivers when passing 9.5 and 37.0 km from the release site. For each fish, the receiver with the highest number of recordings was used to indicate where in the fjord the fish had passed (Figs. 2, 3), although exact distance to shore cannot be determined. At 9.5 km from the release site, the wild and hatchery salmon were recorded in the middle of the fjord as well as closer to the shore, whereas the trout were recorded closer to shore (Fig. 2). Combining data for all groups indicated that more post-smolts passed on the south-western side of the fjord than on the north-eastern side (Fig. 2). Also 37.0 km from the release site, the salmon were recorded both in the middle of the fjord and along the shore,

Table 3 Time and rate of movement for tagged Atlantic salmon and anadromous brown trout post-smolts from release to first recording at receiver sites 9.5 km, 37.0 and 65.0 km from the release site in the Romsdalsfjord system in 2004

Group	Mean time (h) from release to first recording 9.5 km from release site (range, SE)	Mean rate of movement (bl s^{-1}) from release to first recording 9.5 km from release site (range, SE)	Mean time (h) from release to first recording 37.0 km from release site (range, SE)	Mean rate of movement (bl s^{-1}) from release to first recording 37.0 km from release site (range, SE)	Mean time (h) from release to first recording 65.0 km from release site (h) (range, SE)	Mean rate of movement (bl s^{-1}) from release to first recording 65.0 km from release site (range, SE)
Wild salmon	135 (9–667, 41)	0.53 (0.03–1.88, 0.10)	450 (35–1493, 133)	0.56 (0.04–1.89, 0.14)	–	–
Hatchery salmon	80 (7–949, 27)	0.56 (0.01–1.62, 0.07)	168 (26–670, 40)	0.77 (0.08–1.85, 0.12)	154 (82–201, 37)	0.64 (0.40–1.00, 0.18)
Wild trout	350 (134–646, 70)	0.06 (0.02–0.12, 0.01)	1309 (1199–1528, 109)	0.05 (0.04–0.06, 0.01)	–	–

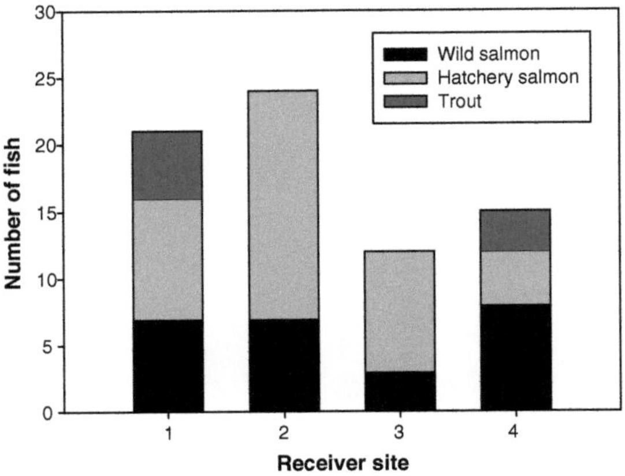

Fig. 2 Number of tagged Atlantic salmon and anadromous brown trout post-smolts detected at the four receiver sites 9.5 km from the release site in the Romsdalfjord system in 2004. Individuals are counted at the site with the highest total count of recordings. Receiver site numbers correspond to those in Fig. 1

whereas the trout stayed closer to shore (Fig. 3). There was no indication that the fish passed more often on one side of the fjord than on the other side, except that there were few recordings on the receiver closest to shore on the northern side of the fjord (Fig. 3).

Most of the salmon followed a one-way route out of the fjord, with little evidence of animals reversing their migration route. The exceptions were 1 wild salmon and 5 hatchery salmon (no difference in the proportion between groups, Fisher's exact test, $P = 0.38$) subsequently recorded 9.5 km from the release site after they had been recorded 37.0 km from the release site. None of the three trout recorded at 37.0 km returned later to inner site.

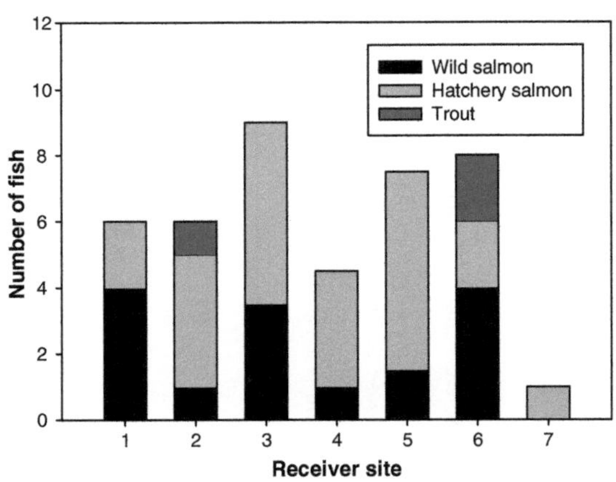

Fig. 3 Number of tagged Atlantic salmon and anadromous brown trout post-smolts detected at the seven receiver sites 37 km from the release site in the Romsdalfjord system in 2004. Individuals are counted at the site with the highest total count of recordings. Receiver site numbers correspond to those in Fig. 1

System performance

Based on range tests it is likely that nearly all fish were recorded when they passed receiver sites at 9.5 and 37.0 km, since 185 m was the longest distance from a receiver possible to pass when passing the receiver lines. This is supported by the fact that only one of 43 fish (2%) that was recorded in the outer areas had not been recorded when passing receiver sites further in. Furthermore, the fish passing were in nearly all cases recorded by several of the receivers and with a high number of recordings per fish, indicating a good coverage by the receivers.

Discussion

This study is consistent with the results and conclusions from previous studies of behaviour of hatchery-reared Atlantic salmon post-smolts (e.g. LaBar et al., 1978; Lacroix & McCurdy, 1996; Thorstad et al., 2004; Finstad et al., 2005), and suggests that those findings may also be representative of wild Atlantic salmon. The proportions of fish recorded 9.5 and 37.0 km from the river mouth (survival) did not differ between wild and hatchery-reared post-smolts. Similarly, time from release to first recording and rate of movement in terms of body lengths per second from release to 9.5 km from the river mouth did not differ. The only difference found was that time from release to first recording 37.0 km from the river mouth was shorter for hatchery-reared than for wild fish. However, this difference can be attributed to the larger body size of the hatchery-reared salmon, since the rate of movement in terms of body lengths per second was similar. It is therefore concluded that no behavioural differences between wild and hatchery-reared salmon post-smolts were detected, but that time from release to reaching 37.0 km from the release site was dependent on body size.

The rates of movement for the salmon (0.53–0.77 body lengths per second) were similar to those previously recorded for larger hatchery-reared salmon in the same fjord system (mean body length 29.5 cm, rates of movement 0.54–0.69 body lengths per second, Finstad et al., 2005). Higher ground speeds have been recorded for Atlantic salmon post-smolts in estuaries (e.g. LaBar et al., 1978; Fried et al., 1978; Moore et al., 1992, 1995) and in Passamaquoddy Bay in Canada (Lacroix & McCurdy, 1996), probably at least partly due to higher water velocities in these localities. Ground speed depends on water velocities and by which methods the rates of movement are recorded, and may therefore be difficult to compare among studies.

Swimming speeds have been recorded in the same fjord system as the present study for hatchery-reared Atlantic salmon post-smolts by continuous manual tracking, with post-smolts swimming at a mean ground speed of 1.27 body lengths per second over 10-min periods (Thorstad et al., 2004). If these ground speeds are representative, it is likely that the salmon post-smolts did not follow the straightest route out of the fjord in the present study, but had actually swum approximately twice this distance. However, it must be pointed out that the individual variation in the rate of movement, and thereby likely in the migration pattern, was large. Also, a few individuals returned to 9.5 km from the release site after they had been recorded 37.0 km away, indicating that some salmon post-smolts may move around in the fjord system for some time. Thorstad et al. (2004) found that post-smolts were actively swimming (at a speed of 1.32 body lengths per second when corrected for the speed and direction of the water current) and not simply drifting passively with the current.

Trout had a slower rate of outward movement than the salmon, and a smaller proportion of the brown trout post-smolts was recorded reaching both the 9.5 and 37.0 km receiver lines. Thus, the trout remained longer in the inner fjord than the wild and hatchery-reared Atlantic salmon, although a few individuals reached the outer fjord areas (37.0 km from the river mouth) within the 2 month study period. The results agree with a previous study on anadromous brown trout (Finstad et al., 2005). The longer time spent in the inner part of the fjord system could potentially expose the trout to a higher predation pressure than the Atlantic salmon (see below).

Both the wild and hatchery-reared Atlantic salmon post-smolts migrated in the middle of the fjord as well as closer to the shore, with a large individual variation. However, the results indicated that there may be some patterns in fjord migration routes, with more fish being recorded on one side of the fjord than on the other side. It is not known whether this corresponds with environmental patterns (such as currents) in the fjord system.

Assuming all passing fish were recorded at the receiver sites, the mortality of the salmon post-smolts was 40–45% during the 9.5 first km (the individual not recorded at 9.5 km but at 37.0 is included), and 36–42% during the next 27.5 km, with a total mortality during the first 37.0 km of the sea migration of 65%. These proportions must be regarded as maximum mortality compared to non-tagged post-smolts, because a few fish may theoretically have passed receivers without being recorded, and because the mortality may be increased due to handling and tagging (see below). However, the results correspond with reports of high predation rates in other Norwegian estuaries by cod *Gadus morhua* L. and saithe on post-smolts. In the estuary of the River Orkla, cod predation was estimated to be 20% (Hvidsten & Lund, 1988), and in a small area in the estuary of the River Surna to be 25% (Hvidsten & Møkkelgjerd, 1987). A large number of saithe with wild untagged post-smolts in their stomach were caught close to the river mouth during the present study, suggesting a high loss of post-smolts due to predation (Jepsen et al., 2006). The 12 salmon transmitters located in this area in late July were probably from fish being eaten by saithe or cod, and the transmitter later being ejected from the predator. The possibility of some tags being detected while in the stomachs of large predators passing the receiver lines cannot be ruled out. The movement behaviour of saithe and cod in the fjord system is not known, and therefore not the likelihood of these fish to move outwards in the fjord and being detected by the receivers. However, the fast gastric evacuation rates of gadoid predators (Andersen, 1999; 2001) indicate that this is not a large methodological problem in this study.

The present study confirms that the first phase of the marine migration seems to imply high mortality on the Atlantic salmon post-smolts, as suggested by Hansen et al. (2003). However, when comparing these results with the total mortality during the marine phase for Atlantic salmon from the nearby River Orkla (80–98% for one-sea-winter salmon, 89–99% for multi-sea-winter salmon, Hvidsten et al., 2004), it seems that there must also be a high mortality during later stages of the marine migration. The same conclusion was indicated by a study of acoustically tagged steelhead trout *Oncorhynchus mykiss*, with survival rates of ≥55% in the early ocean phase (Welch et al., 2004).

Hvidsten & Lund (1988) found no difference in predation rates between wild and hatchery-reared salmon smolts and no evidence of selective predation on the smallest wild and hatchery-reared smolts. This corresponds with the results from the present study, except that there was an indication of a higher survival to 37 km from the release site for larger than for smaller wild salmon smolts. Hatchery-reared post-smolts recorded at 37.0 km from the release site had a faster migration to 9.5 km from the release site than those not recorded at 37.0 km, indicating a higher loss among the slowest migrating fish. Thus, the risk of being predated may increase with time spent in the fjord system. An alternative explanation may be that the fish with the lowest migration speeds were the weakest fish, and therefore more likely to be predated.

Handling and tagging may influence the behaviour of the fish. For Atlantic salmon post-smolt studies in the sea, Moore et al. (2000) recommended tags to be less than 5% of fish mass to minimise effects on behaviour and survival. Some individuals had larger tags in the present study, and the wild post-smolts had relatively larger transmitters than the hatchery-reared post-smolts. According to Peake et al. (1997), wild salmon smolts may also be generally more affected by tagging than hatchery-reared smolts. The most likely potential negative effects of transmitters in the present study are higher mortality and lower swimming speeds for the smallest fish (e.g. McCleave & Stred, 1975), although transmitters weighing up to 12% of the body mass did not affect the swimming performance of juvenile rainbow trout (*Oncorhynchus mykiss*) (Brown et al., 1999). No difference in tag to body mass ratio between fish recorded or not recorded after release was found in the present study, indicating that potential negative effects of the tagging did not affect the conclusions of the study. However, if conclusions were affected it was most likely by overestimating mortality and underestimating swimming speeds for the wild compared to hatchery-reared post-smolts.

Acknowledgements We would like to thank the staff at the Statkraft hatchery in Eresfjord, Bjørg Anne Vike, Petter Sira and Torbjørn Utigard, for extensive help and co-operation during the project. Stian R. Almestad, Espen Holthe, Jan Gunnar Jensås, Niels Jepsen and Stig Sandring are also thanked for help during the project The study was partially financed by the European Commission contract no: Q5RS-2002–00730, AquaNet Canada (RSM), the Norwegian Institute for Nature Research and Statkraft Energy AS. The experimental procedures used conformed to Norwegian ethical requirements.

References

Andersen, N. G., 1999. The effects of predator size, temperature, and prey characteristics on gastric evacuation in whiting. Journal of Fish Biology 54: 287–301.

Andersen, N. G., 2001. A gastric evacuation model for three predatory gadoids and implications of using pooled field data of stomach contents to estimate food rations. Journal of Fish Biology 59: 1198–1217.

Blackburn, J. & W. C. Clarke, 1987. Revised procedure for the 24 hour seawater challenge test to measure seawater adaptability of juvenile salmonides. Canadian Technical Report of Fisheries and Aquatic Sciences 1515: 1–39.

Brown, R. S., S. J. Cooke, W. G. Anderson & R. S. McKinley, 1999. Evidence to challenge the "2% rule" for biotelemetry. North American Journal of Fisheries Management 19: 867–871.

Finstad, B. & N. Jonsson, 2001. Factors influencing the yield of smolt releases in Norway. Nordic Journal of Freshwater Research 75: 37–55.

Finstad, B., F. Økland, E. B. Thorstad, P. A. Bjørn & R. S. McKinley, 2005. Migration of Atlantic salmon and sea

trout post-smolts in a Norwegian fjord system. Journal of Fish Biology 66: 86–96.
Fried, S. M., J. D. McCleave & G. W. LaBar, 1978. Seaward migration of hatchery-reared Atlantic salmon, *Salmo salar*, smolts in the Penoboscot River estuary, Maine: riverine movements. Journal of the Fisheries Research Board of Canada 35: 76–87.
Hansen, L. P., M. Holm, J. C. Holst & J. A. Jakobsen, 2003. The ecology of post-smolts of Atlantic salmon. In Mills, D. (ed.), Salmon at the Edge. Blackwell Science, Oxford, 25–39.
Hoar, W. S., 1988. The physiology of smolting salmonids. In Hoar, W. S. & D. J. Randall (eds), Fish Physiology: The Physiology of Developing Fish. Viviparity and Posthatching Juveniles, Volume XIB. Academic Press, New York, NY, 275–343.
Hvidsten, N. A. & R. A. Lund, 1988. Predation on hatchery-reared and wild smolts of Atlantic salmon, *Salmo salar* L., in the estuary of River Orkla. Journal of Fish Biology 33: 121–126.
Hvidsten, N. A. & P. I. Møkkelgjerd, 1987. Predation on salmon smolts, *Salmo salar* L., in the estuary of the River Surna. Journal of Fish Biology 30: 273–280.
Hvidsten, N. A., B. O. Johnsen, A. J. Jensen, P. Fiske, O. Ugedal, E. B. Thorstad, J. G. Jensås, Ø. Bakke & T. Forseth, 2004. Orkla – et nasjonalt referansevassdrag for studier av bestandsregulerende faktorer hos laks. NINA Fagrapport 079: 1–96. Norwegian Institute for Nature Research, Trondheim (In Norwegian with English abstract).
Jepsen, N., E. Holthe & F. Økland, 2006. Observations of predation on salmon and trout smolts in a river mouth. Fisheries Management and Ecology 13: 341–343.
Jonsson, B., 1985. Life history patterns of freshwater resident and sea-run migrant brown trout in Norway. Transactions of the American Fisheries Society 114: 182–194.
Jonsson, B., N. Jonsson & L. P. Hansen, 1991. Differences in life history and migratory behaviour between wild and hatchery-reared Atlantic salmon in nature. Aquaculture 98: 69–78.
Klemetsen, A., P.-A. Amundsen, J. B. Dempson, B. Jonsson, N. Jonsson, M. F. O'Connell & E. Mortensen, 2003. Atlantic salmon *Salmo salar* L., brown trout *Salmo trutta* L. and Arctic charr *Salvelinus alpinus* (L.): a review of aspects of their life histories. Ecology of Freshwater Fish 12: 1–59.
LaBar, G. W., J. D. McCleave & S. M. Fried, 1978. Seaward migration of hatchery-reared Atlantic salmon (*Salmo salar*) smolts in the Penobscot River estuary, Maine: open water movements. Journal du Conseil international pour l'Exploration de la Mer 38: 257–269.
Lacroix, G. L. & P. McCurdy, 1996. Migratory behaviour of post-smolt Atlantic salmon during initial stages of seaward migration. Journal of Fish Biology 49: 1086–1101.
Lacroix, G. L. & F. A. Voegeli, 2000. Development of automated monitoring systems for ultrasonic transmitters. In Moore, A. & I. Russell (eds), Advances in Fish Telemetry. Proceedings of the third Conference on Fish Telemetry in Europe. CEFAS, Lowestoft, 37–50.
Lacroix, G. L., P. McCurdy & D. Knox, 2004. Migration of Atlantic salmon post-smolts in relation to habitat use in a coastal system. Transactions of the American Fisheries Society 133: 1455–1471.
McCleave, J. D. & K. A. Stred, 1975. Effect of dummy telemetry transmitters on stamina of Atlantic salmon (*Salmo salar*) smolts. Journal of the Fisheries Research Board of Canada 32: 559–563.
Moore, A., G. L. Lacroix & J. Sturlaugsson, 2000. Tracking Atlantic salmon post-smolts in the sea. In Mills, D. (ed.), The Ocean Life of Atlantic Salmon – Environmental and Biological Factors Influencing Survival. Fishing News Books, Oxford, 49–64.
Moore, A., E. C. E. Potter & A. A. Buckley, 1992. Estuarine behaviour of migrating Atlantic salmon (*Salmo salar* L.) smolts. In Priede, I. M. & S. M. Swift (eds), Wildlife Telemetry. Ellis Horwood Limited, Chichester, 390–399.
Moore, A., E. C. E. Potter, N. J. Milner & S. Bamber, 1995. The migratory behaviour of wild Atlantic salmon (*Salmo salar*) smolts in the estuary of the River Conwy, North Wales. Canadian Journal of Fisheries and Aquatic Sciences 52: 1923–1935.
Peake, S., R. S. McKinley, D. A. Scruton & R. Moccia, 1997. Influence of transmitter attachment procedures on swimming performance of wild and hatchery-reared Atlantic salmon smolts. Transactions of the American Fisheries Society 126: 707–714.
Rikardsen, A. H., 2004. Seasonal occurrence of sea lice *Lepeophtheirus salmonis* on sea trout in two north Norwegian fjords. Journal of Fish Biology 65: 711–722.
Sigholt, T. & B. Finstad, 1990. Effect of low temperature on sea-water tolerance in Atlantic salmon (*Salmo salar*) smolts. Aquaculture 84: 167–172.
Thorstad, E. B., F. Økland, B. Finstad, R. Sivertsgård, P. A. Bjørn & R. S. McKinley, 2004. Migration speed and orientation of Atlantic salmon and sea trout post-smolts in a Norwegian fjord system. Environmental Biology of Fishes 71: 305–311.
Voegeli, F. A., G. L. Lacroix & J. M. Anderson, 1998. Development of miniature pingers for tracking Atlantic salmon smolts at sea. Hydrobiologia 371/372: 35–46.
Wedemeyer, G. A., R. L. Saunders & W. C. Clarke, 1980. Environmental factors affecting smoltification and early marine survival of anadromous salmonids. Marine Fisheries Review 42: 1–14.
Welch, D. W., B. R. Ward & S. D. Batten, 2004. Early ocean survival and marine movements of hatchery and wild steelhead trout (*Onconhynchus mykiss*) determined by an acoustic array: Queen Charlotte Strait, British Columbia. Deep-Sea Research II 51: 897–909.

Hydrobiologia (2007) 582:109–120
DOI 10.1007/s10750-006-0544-y

FISH TELEMETRY

Hatchery-reared fish have less consistent behavioral pattern compared to wild individuals, exemplified by red tilefish studied using video observation and acoustic telemetry tracking

Takashi Yokota · Reiji Masuda · Nobuaki Arai · Hiromichi Mitamura · Yasushi Mitsunaga · Hiroyuki Takeuchi · Tatsuo Tsuzaki

Abstract The behavior of wild and hatchery-reared red tilefish *Branchiostegus japonicus* was analyzed using two different methods: video observation and acoustic telemetry tracking. In the laboratory, digging and swimming activity of seven wild and five hatchery-reared fish were recorded for 2–4 days in an experimental aquarium and related to changes in light intensity. The activity of wild individuals increased with light intensity, while hatchery-reared individuals were active during both day and night. In the field, 18 wild and 9 hatchery-reared fish were released and tracked using an acoustic telemetry system during the winter and the summer in Maizuru Bay, Kyoto (Japan). Seven wild and three hatchery-reared fish settled within 2 km of the release point for 21–200 days whilst 11 wild and 6 hatchery-reared fish were detected in the central part of the bay within 13 days. The settled fish demonstrated a diel pattern of behavior; most of them probably stayed outside their burrows in the daytime but inside their burrows during the nighttime. During the winter wild fish tended to stay near the release site for longer periods than the hatchery-reared fish. However, in summer the hatchery-reared fish tended to remain close to the release site for longer than the wild fish. Based on these two experiments, we suggest that the activity of wild red tilefish respond to changes in low levels of light intensity (0–0.1 μmol m^{-2} s^{-1}) at dawn and display a diel behavior. Hatchery-reared fish may differ from wild fish with regard to their behavior, response to water temperature and light conditions.

Guest editors: P. R. Almeida, B. R. Quintella, M. J. Costa and A. Moore
Developments in Fish Telemetry

Electronic supplementary material The online version of this article (doi: 10.1007/978-1-4020-6237-7_12) contains supplementary material, which is available to authorized users.

T. Yokota (✉) · N. Arai · H. Mitamura
Graduate School of Informatics, Kyoto University, Sakyo, Kyoto 606-8501, Japan
e-mail: tyokota@bre.soc.i.kyoto-u.ac.jp

R. Masuda
Maizuru Fisheries Research Station, Kyoto University, Maizuru, Kyoto 625-0086, Japan

Y. Mitsunaga
Faculty of Agriculture, Kinki University, Nara 631-8505, Japan

H. Takeuchi
Miyazu Station, National Center for Stock Enhancement, Fisheries Research Agency, Miyazu, Kyoto 626-0052, Japan

T. Tsuzaki
Tamano Station, National Center for Stock Enhancement, Fisheries Research Agency, Tamano, Okayama 706-0002, Japan

Keywords Red tilefish · *Branchiostegus japonicus* · Hatchery-reared fish · Rearing and releasing strategy · Site fidelity · Diel behavioral activity

Introduction

Red tilefish (*Branchiostegus japonicus*) is an important fishery resource found at depths of 30–200 m in the coastal waters of the Japanese Archipelago and in the East China Sea (Okumura, 2000). During the 1980's increased exploitation by the commercial fishery resulted in a rapid depletion of the stock. Because Kiyono et al. (1977) suggested that the red tilefish are not highly migratory, sea-farming was considers the most efficient strategy for stock enhancement.

In the 1990s, yearly production of more than one million fertilized eggs and more than one hundred thousand juveniles was achieved in the Miyazu Station of the Japan Sea-Farming Association, which later became the Miyazu Station, National Center for Stock Enhancement (NCSE), Fisheries Research Agency (FRA) (Hondoh et al., 2001, 2002). Since 1998 the Miyazu Station, NCSE, FRA and the Kyoto Institute of Oceanic and Fishery Science have conducted tag and release experiments using hatchery-reared fingerlings in order to develop a protocol for the most efficient release of the juvenile red tilefish. However, although as many as 40,000 fingerlings had been released in Wakasa Bay from May 1998 to June 2004, only 24 fish had been captured by December 2004 and the rate of recovery is still extremely low.

Previous studies on the development of sea-farming have suggested that the behavioral characteristics of released fish are considered to affect post-stocking mortality (Tsukamoto et al., 1999). However, few studies have attempted to describe the behavioral characteristic of red tilefish, and there is only a single observation of a fish in the wild entering it's burrow at a depth of 117 m (Toriyama, 1975). In rearing tanks, wild red tilefish display territorial behavior (Okumura, 2000), and hatchery-reared fish >75 mm total length (TL) construct burrows within a muddy substrate (Hondoh et al., 2002). Although a previous acoustic telemetry tracking study of red tilefish has been conducted in Wakasa Bay at depths of 30–43.6 m, the released fish dispersed so rapidly that the fish could not be tracked for more than 7 days (Yokota et al., 2004; Mitamura et al., 2005).

For the development of a rearing and releasing strategy, it is essential to understand the behavioral ecology of wild and hatchery-reared red tilefish and to compare them, since in order to survive in the natural environments released hatchery-reared fingerlings should behave in a similar manner to wild fish. Therefore, in order to obtain long-term information about the behavioral ecology of red tilefish in the natural environment, an acoustic telemetry tracking study commenced in January 2003 in Maizuru Bay. The study area, which forms the western part of Wakasa Bay, is a typical shallow semi-closed body of water. Yokota et al. (2004) and Mitamura et al. (2005) have already reported the results of a tracking study examining the behavior of red tilefish in this area. Hatchery and wild fish were both shown to settle and demonstrate a diel pattern of activity.

In order to provide a novel insight into the behavior of the red tilefish and compare the behavioral characteristics of wild and hatchery reared fish a further study was implemented utilizing acoustic telemetry and 24 h videotaping techniques to directly observe the diurnal activity of the fish.

In this paper the results of two experiments are described. Firstly, a laboratory based study examining the diurnal activity of both wild and hatchery reared fish. Secondly, a field based telemetry study describing the diurnal patterns of movement of both wild and hatchery reared fish. The results of these studies are discussed in relation to differences in behavior of wild and hatchery reared fish and the possible causes for the low recovery of the released fingerlings in terms of their behavioral characteristics.

Materials and methods

Sample fish

Twenty wild red tilefish (W1-20, 319 ± 27 mm TL) were captured by fishermen in the western part of Wakasa Bay and maintained in indoor tanks (5 kl) in the Miyazu Station of NCSE, FRA (see Electronic Supplementary Material). The wild fish experienced water temperatures in the wild and in rearing tanks between 10–23°C; the open sea temperatures were determined from on-line data published on the web site of the Japan Oceanographic Data Center (http://www.jodc.go.jp/service.htm). Ten hatchery-reared fish (HR1-10, 255 ± 53 mm TL) reared in indoor tanks (25 kl) in the Miyazu Station of NCSE, FRA at temperatures between 10°C and 27°C for 2–4 years were also used in the study (see Electronic Supplementary Material). The light condition during rearing followed the natural ambient levels.

Video observation in the laboratory

The experiment was carried out in the Maizuru Fishery Research Station (MFRS), Kyoto University. One part of the experimental room, equipped with a light manager (NK28691, National, Japan) to regulate light intensity according to a time schedule, was isolated with a blackout curtain. An acrylic aquarium (90 cm × 45 cm × 45 cm) was set just under the light. A video camera (AI18CIR-AFM, HOGA, Japan) was placed 1 m from the aquarium to record the entire aquarium area. An infrared illuminator (C4-170, HOGA, Japan) was also placed beside the aquarium to record data during the period of darkness. The bottom of the aquarium was covered with gravel (grain diameter <10 mm) to a depth of 8 cm. Sea water and air were supplied continuously to the aquarium. Light intensity was set to vary according to a time schedule, so that the day was divided into four phases based on light conditions. In the 'dawn-phase (6:00–7:30)', the light intensity gradually increased from 0 to 2.25 μmol m^{-2} s^{-1} (Compact-LW, ALEC ELECTRONICS, Japan) at the surface of the aquarium. In the 'day-phase (7:30–18:00)', the light intensity was constant (2.25 μmol m^{-2} s^{-1}). In the 'dusk-phase (18:00–19:30)', the light intensity gradually decreased from 2.25 to 0 μmol m^{-2} s^{-1}. In the 'night-phase (19:30–6:00)', the light was completely turned off (0 μmol m^{-2} s^{-1}). Seven wild fish (W14-20, 316 ± 25 mm TL) and five hatchery-reared fish (HR5, 6, 8–10, 217 ± 11 mm TL) were used in this experiment (see Electronic Supplementary Material). One fish was introduced into the aquarium during the day-phase and recorded for 2–4 days continuously. The day-phase in which the fish was introduced was defined as day0-phase, and the subsequent phases were also defined as dusk0-phase, night0-, dawn1-, day1-, dusk1-, and so on. The fish demonstrated two particular types of behavior during the study; 'turn' and 'digging'. The former was defined as the behavior by which the fish changed its orientation by more than 180°. The latter was defined as the behavior in which the fish scooped the gravel from the substrate with its mouth and moved it to an adjacent position. The number of turns and diggings were counted every 10 min. The water temperature varied between 15–17°C throughout this experiment.

The turn frequency was assumed to reflect the fish activity. In evaluating the fluctuation of their activity under different light conditions, the data from the same five phases, from the dawn1-phase to the dawn2-phase, were used for each individual, in order to eliminate any initial handling stress. Since the total numbers of turns in the five phases were considerably different among the individuals (927–3117 for wild; 1161–3779 for hatchery-reared), the fluctuation of fish activity under different light conditions was analyzed as follows. The turn frequencies in each phase were calculated, and then the ratio of the five frequencies was calculated for each individual. Because the ratio was assumed to reflect the fluctuation of fish activity, the ratio was converted into percentages of activity shared among each of the five phases, where total activity in the five phases was defined as 100%. The activities among the five phases were calculated for each individual. Comparisons were then conducted between the activities during each of the five phases using the Kruskal–Wallis test, for both wild and hatchery-reared individuals. If there was statistically significant differences among the five phases, we performed non-parametric multiple comparisons (Nemenyi test) (Zar, 1999).

For wild individuals, every two successive number of turns per 10 min was compared using the Wilcoxon test from 5:30 to 8:00. The fluctuations of the numbers were also compared to the variation of the light intensity in the aquarium during the same period. Digging was assumed to be part of a behavior to construct a burrow in the substrate.

Acoustic telemetry tracking in the field

This experiment was conducted in Maizuru Bay (Fig. 1). The majority of the bay has a water depth of less than 20 m with a muddy silt substrate. Eighteen wild fish (W1-18, 321 ± 25 mm TL) and nine hatchery-reared fish (HR1-9, 261 ± 53 mm TL) were used in this experiment (see Electronic Supplementary Material). Two types of acoustic transmitters were used, a V8SC-6L (dimensions 9.0 mm × 20.0 mm, 2.0 g in water) and a V7-4L (dimensions 7.0 mm × 20 mm, 0.8 g in water) both from VEMCO Ltd., Canada. The tags transmit a single coded signal at a random interval of between 20 and 60 s (average 40 s) and at a constant frequency of 69 kHz. Acoustic transmitters were surgically inserted into the peritoneal cavity of the fish, in a manner similar to that described in Mitamura et al. (2002). After the surgical procedure, the fish were left to recover in black polycarbonate tanks (500 l) for a period of 2–7 days. The fish were then released from the surface in the central part of Maizuru Bay in winter and summer periods (Fig. 1, see Electronic Supplementary Material). After the release, the fish were monitored using 4–7 fixed acoustic receivers (VR2, VEMCO Ltd., Canada) placed to cover the release point and adjacent area. The receiver automatically recorded the ID number and the time of signal detection when a tagged fish traveled within the receiving range (350 m for V8SC-6L, and 150 m for V7-4L transmitters).

In analyzing VR2 records, we counted the number of signal detections every 60 min and observed the fluctuations of their frequency for each individual. If a diel fluctuation along with the circadian rhythm was recognized for more than 3 days, comparisons were conducted between the detection frequencies in the daytime and that during nighttime using the Wilcoxon test. The time of changes in detection frequency were also observed and compared with the time of sunrise and sunset. Data of sunrise and sunset hours was available on the web site of National Astronomical Observatory of Japan (http://www.nao.ac.jp/reki/).

The period starting from the release until the last tracking record was defined as the 'track term (days)'. Track terms were compared among the four groups (wild individuals released in winter, W1-4, 14-18; hatchery-reared in winter, HR5-9; wild in summer, W5-13; hatchery-reared in summer, HR1-4) using the Kruskal–Wallis test fol-

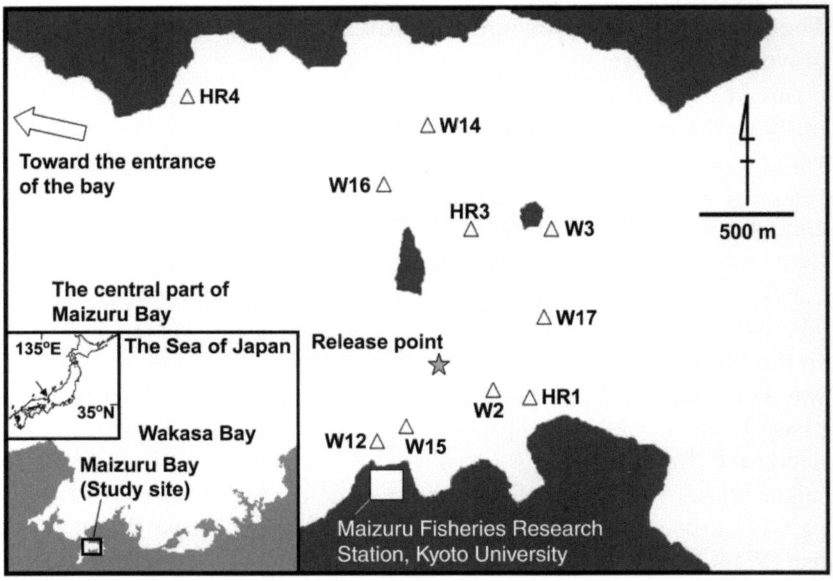

Fig. 1 Map of the study site, the central part of Maizuru Bay: a typical semi-closed water bay that forms the western part of Wakasa Bay, which is open to the Sea of Japan

lowed by the non-parametric multiple comparisons (Dunn's procedure) (Zar, 1999).

Water temperatures at a depth of 7 m, measured by a conductivity/temperature recorder (Compact-CT, ALEC ELECTRONICS, Japan) along the MFRS, were used for the study. Light intensity was measured at the same point using a miniature light-intensity data logger (Compact-LW, ALEC ELECTRONICS, Japan) from 29 January to 15 February 2005. Light intensity was also measured around noon (at 11:00–13:00) on 16 March 2005 at a depth of 50, 70, 80 and 110 m in the western part of Wakasa Bay where wild red tilefish are naturally distributed.

Results

Video observation in the laboratory

The turn frequency was different among individuals, as indicated by the total numbers of turns from dawn1-phase to dawn2-phase among the individuals (see Fig. 2 as an example). The number of turns per 10 min fluctuated even under the constant light conditions (2.25 μmol m^{-2} s^{-1}) during day-phases (Fig. 2). In the case of wild fish, the percentages of activity shared among the five phases were statistically different among the five phases (Kruskal–Wallis test, $H = 10.92$, $N = 35$, $df = 4$, $P < 0.05$); the fish activity was statistically higher in the dawn1-phase compared to the day1-phase (Nemenyi test, $P < 0.05$) (Fig. 3, Table 1). Although the other pairs were not statistically different, the fish activity in the dawn2-phase was highest next to that in dawn 1-phase. In the case of hatchery-reared fish, in contrast, the percentages were not statistically different among the five phases (Kruskal–Wallis test, $H = 7.24$, $N = 25$, $df = 4$, $P = 0.12$), even though the fluctuation of the average percentages was apparently similar to that of wild fish (Fig. 3). Two hatchery-reared fish (HR9, 10) were active in the dawn-phases, one fish (HR8) was active in the dusk1-phase, and

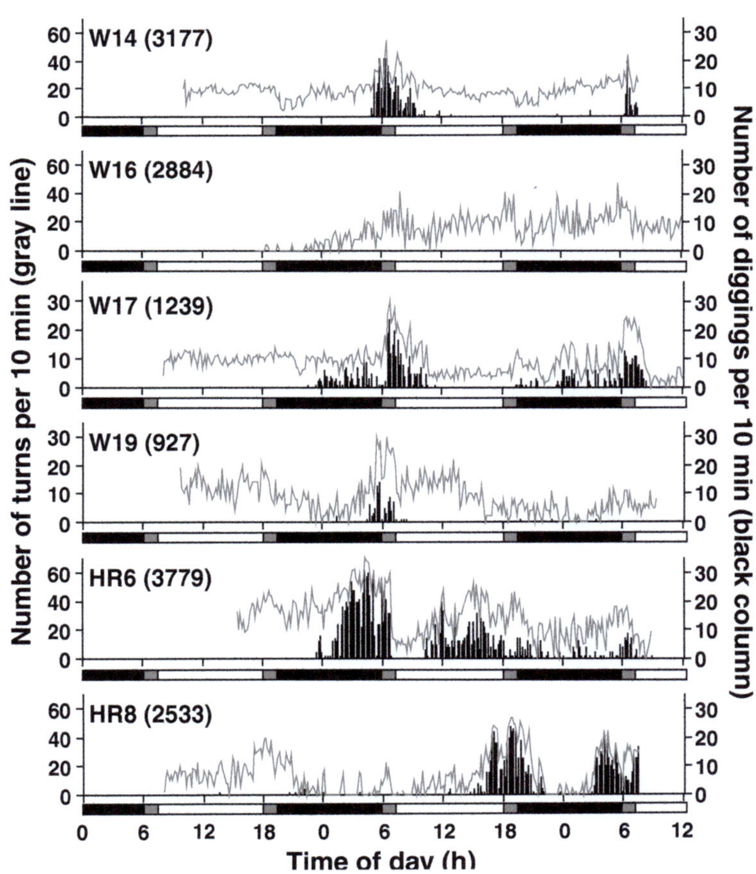

Fig. 2 Examples of turns and diggings. The gray lines show the number of turns per 10 min, and the black columns show the number of diggings per 10 min. The numbers in the brackets are the total number of turns displayed from the dawn1-phase to the dawn2-phase. The bottom bar of each graph shows phases defined by light conditions

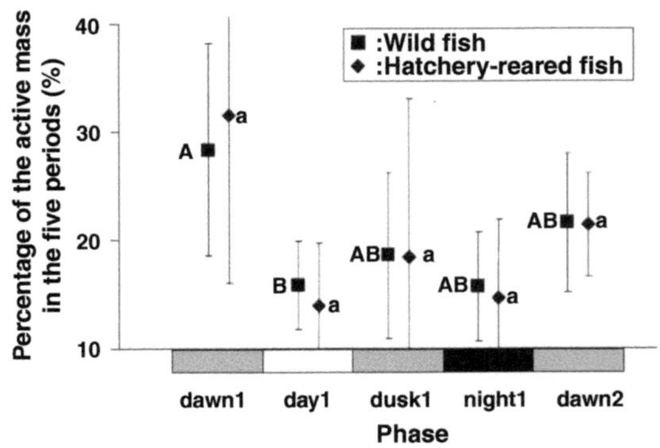

Fig. 3 The percentages of activity for each of the five phases shown for each fish origin (Mean ± SD). The capital letters indicate the significant difference among the four phases for the wild individuals, and small letters indicate that for only the hatchery-reared individuals. See Table 1 for the detailed statistical results

Table 1 Detail of the non-parametric multiple comparisons (Nemenyi test) of the activity for wild fish in the five phases

Phase	Percentage of activity (%) (Mean ± SD)	q-value			
		Day1	Dusk1	Night1	Dawn2
Dawn1	28.3 ± 9.8	3.91*	2.77	3.79	1.51
Day1	15.8 ± 4.1	–	1.14	0.11	2.40
Dusk1	18.6 ± 7.7	–	–	1.03	1.25
Night1	15.7 ± 5.0	–	–	–	2.29
Dawn2	21.6 ± 6.4	–	–	–	–

For the hatchery-reared fish, the activities were not significantly different among the five phases in the Kruskal-Wallis test, so multiple comparisons were not conducted for them. There is a statistical difference between two phases, if q-value is larger than 3.858 (= $q_{0.05, 4, \infty}$)

$^{*}P < 0.05$

activities of the other two (HR5, 6) were relatively constant through the five phases. One of the most noteworthy observations was that the variation among the hatchery-reared fish was larger than among the wild ones, as indicated by the larger standard deviations of the hatchery-reared, exception for the dawn2-phase (Fig. 3).

For the wild fish, the number of turns was not correlated with an increase in light intensity (from 0 to 2.25 μmol m^{-2} s^{-1}) throughout the dawn-phase (Fig. 4b). However, the number of turns at 6:00–6:10 (the first 10 min in dawn-phase) was statistically greater than that at 5:50–6:00 (the last 10 min in night-phases) (Wilcoxon test, $P < 0.01$), even though light intensities at 5:50–6:10 were nearly 0 μmol m^{-2} s^{-1} (Fig. 4a). From 6:10 to 7:30, there was no 10 min period when the number of turns showed statistically larger value compared to the adjacent 10 min (Wilcoxon test, $P > 0.05$).

Digging behavior was displayed by three wild fish (W14, 16, 17) and two hatchery-reared fish (HR6, 8) (Fig. 2). All of the wild fish displayed digging behaviors in or around the dawn-phase. In contrast, the two hatchery-reared fish dug throughout the study period with the exception of the dawn-phases (Fig. 2).

Acoustic telemetry tracking in the field

Soon after the release, signals of each fish were recorded by various VR2s (see Fig. 5 for an example). Sixty-seven percent of wild individuals released in the winter (6/9: W2, 3, 14, 15, 16, 17), 0% of hatchery-reared ones released in the winter (0/5), 11% of wild ones released in the summer (1/9: W12) and 75% of hatchery-reared ones released in the summer (3/4: HR1, 3, 4) were gradually detected at a specific VR2 within 2 km of the release point (Figs. 1, 5), and were successfully monitored for periods of between 21 and 200 days (see Electronic Supplementary Material). The seven wild and three hatchery-reared individuals which remained in Maizuru Bay were defined as 'settled-fish'. The remaining fish were located after a period of 13 days in the central part of Maizuru Bay.

In the VR2 records of settled-fish, we detected a 'diel-period' of signal frequency; fish were mainly detected in the daytime but rarely during nighttime (see Fig. 6 for an example). However, only one hatchery-reared fish was detected with higher frequency during nighttime compared to in the daytime (Wilcoxon test, $P < 0.01$) (see Electronic Supplementary Material). The signal frequency was different among the settled-fish except fish HR1 (9.2–99.5 signals per hour in the daytime, 0.7–22.9 during nighttime), owing to the types of acoustic transmitter and the number of VR2s monitoring the fish. However, the frequencies in the daytime were significantly larger than those during nighttime for each individual (Wilcoxon test, $P < 0.01$).

In the VR2 records of settled-fish except for fish HR1 during their diel-periods, changes in

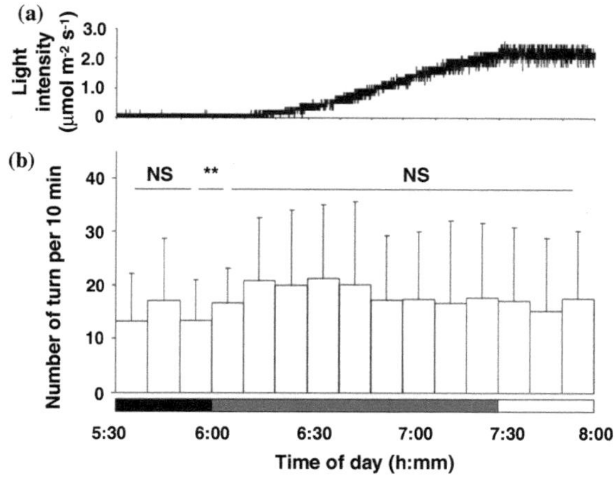

Fig. 4 (**a**) Light intensity from 5:30 to 8:00. (**b**) The average number of turns per 10 min shown by the wild individuals (Mean ± SD). 'NS' and '**' indicate statistical differences between every two successive 10 min periods; NS: $P > 0.05$, $^{**}P < 0.01$

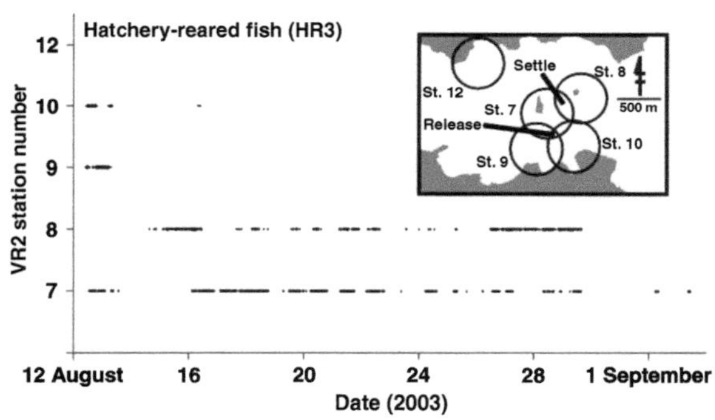

Fig. 5 An example of signal detections by the VR2s placed to monitor the released fish

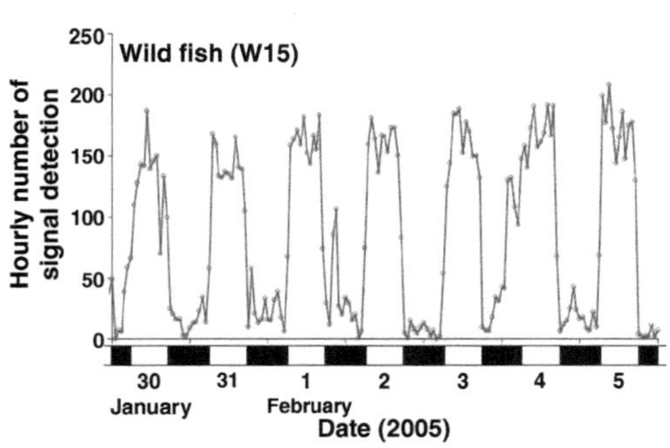

Fig. 6 An example of the regular diel fluctuation of signal detection frequency during diel-period. Day-□, Night -■

the frequency of signal detection were equally synchronized with the variations of the time at sunrise and sunset, occurring at <30 min before sunrise and <30 min after sunset (see Fig. 7 for an example). In the records of the fish settled within the receiving ranges of more than two VR2s, the fluctuations in the frequencies of signal detection by each VR2 were considerably similar to one another (see Electronic Supplementary Material for an example). In the three VR2 records of wild fish W2, changes of frequency occurred <30 min before sunrise and <30 min after sunset, and the mean numbers of signal detections in the daytime were fairly constant especially at St. 4 (28.5–32.2 signals: about two thirds of the total signals transmitted every 30 min theoretically). During diel-period, wild fish W15 settled at 7 m deep, the same depth where light intensity was measured; signals from fish W15 and its ambient light condition were monitored at the same time, from 30 January to 6 February 2005. The record of the light-intensity data logger revealed that light intensity started to increase 30–20 min before sunrise, reaching to 0.5 μmol m^{-2} s^{-1} at sunrise, and decreasing to 0 μmol m^{-2} s^{-1} 20–30 min after sunset (Fig. 8). Therefore, the change in the pattern of signal detection was synchronized with the onset of increase and the end of decrease in light intensity (Figs. 7 and 8).

During the diel-period of settled-fish, the hourly numbers of signal detection fluctuated in the daytime (see Fig. 6 for an example). For fish W15, settled at the same depth of the light-intensity data logger, the light intensity every 10 min was compared to detection frequency per 10 min from 10:00 to 14:00, but there was no relationship between them ($N = 204$, $R^2 = 0.0076$, $P = 0.22$). Light intensity at depths of 50–110 m in the western part of Wakasa Bay ranged 0–0.2 μmol m^{-2} s^{-1} around noon (at 11:00–13:00) on 16 March 2005.

Track terms were statistically different among the four groups (wild individuals released in winter, hatchery-reared in winter, wild in summer, hatchery-reared in summer) (Kruskal–Wallis test, $H = 11.62$, $N = 27$, $df = 3$, $P < 0.01$). Although there was no significant difference between each of the pairs (Dunn's procedure, $P > 0.05$), the wild individuals released in winter (W1-4, 14-18, 67% settled, mean 58 days, 1–131 days) and the hatchery-reared individuals

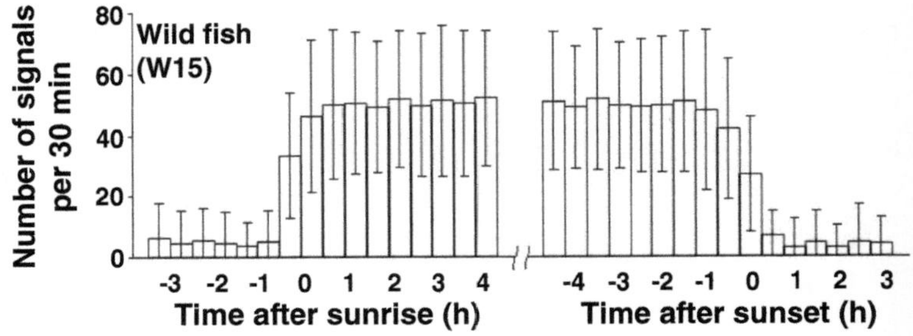

Fig. 7 An example of the number of signal detections per 30 min after sunrise and sunset during diel-period (Mean ± SD)

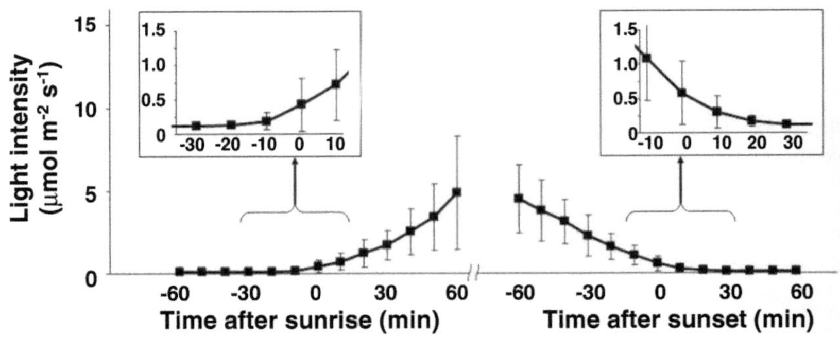

Fig. 8 Fluctuation of light intensity at 7 m deep during sunrise and sunset (Mean ± SD). The upper two graphs show an enlarged view of the light intensity during the exact time of sunrise and sunset

released in the summer (HR1-4, 75% settled, mean 80 days, 5–200 days) tended to stay longer compared to the others: the wild individuals released in summer (W5-13, 11% settled, mean 6 days, 0–32 days) and the hatchery-reared individuals released in the winter (HR5-9, 0% settled, mean 3 days, 0–7 days). The water temperatures experienced by the wild fish released in the winter (11–16°C, W1-4, in January 2003; 11–19°C, W14-18, in November 2004) were within the range the fish had experienced before the release (10–23°C) (see Electronic Supplementary Material). In contrast, the water temperatures experienced by the wild fish released in the summer (24–25°C, W5-10, in August 2003; 20–24°C, W11-13, in June 2004) were near or beyond the upper limit the fish had experienced before the release (10–23°C) (see Electronic Supplementary Material). In the case of hatchery-reared fish, the water temperatures experienced by the fish released in both seasons (11–27°C, HR1-4, in August 2003; about 18°C, HR5-9, in November 2004) were within the range of temperatures experienced by the fish before their release (i.e., 10–27°C) (see Electronic Supplementary Material).

Discussion

Site fidelity and diel behavioral activity

Since wild fish in the aquarium tended to be more active during the dawn-phases (Figs. 2, 3), red tilefish might be active when the light intensity increases. In the field, the monitoring records of settled-fish indicated that each fish moved for a period of time over the substrate after release before settling (Fig. 5). The circadian rhythm records might indicate a diel behavioral pattern of fish, where they moved out of the burrow during daytime and retreated into the burrow at night (Figs. 6, 7). This observation is supported by the presence of muddy silt substrate in Maizuru Bay, the habit of constructing burrows (Toriyama, 1975; Hondoh et al., 2002) and the attenuation of acoustic signals by obstructions such as bottom substrate existing in their path (Starr et al., 2002; Mitamura et al., 2005). Although we can not completely rule out that the fish stayed beyond the receiving ranges of VR2s horizontally during the absence of signal detections, the location of the settled-fish might not have changed horizontally throughout a single day for the following reasons. The similar fluctuation of signal detections by the three VR2s (see Electronic Supplementary Material) probably demonstrates that the horizontal position of fish W2 did not differ more than a few 100 m between the daytime and during nighttime, and also indicates that the individual did not move out of the receiving ranges of the three VR2s for more than 30 min continuously at a certain period of the daytime. Because two thirds of the total signals were detected at St. 4 each 30 min in the daytime, the average fish absence within the receiving range of VR2 at St. 4 each 30 min, lasted 10 min at the most. Therefore, considering that red tilefish display territorial behavior (Okumura, 2000), fish W2 may not have made large horizontal movements throughout a day but changed its position within limited areas, inside or outside burrows, for example.

Based on two experiments, it can be said that wild red tilefish might leave their burrow to forage, or to construct or maintain burrows intensively along with the onset of increase in light intensity at dawn. After that period wild red tilefish may become relatively inactive and stay outside near their burrows, and might retreat into their burrows as light intensity decreases to almost 0 μmol m^{-2} s^{-1}. Based on two reports of observation from a submersible, Hondoh et al. (2002) suggested that burrows play a role of refuge from predators; the two species of Branchiostegidae, blueline tilefish *Caulolatilus microps* and golden tilefish *Lopholatilus chamaeleonticeps*, were observed to hide in their burrows immediately before the appearance of a hammerhead shark (Able et al., 1987), and one red tilefish was observed entering its burrow immediately before the approach of the submersible 'Shinkai' (Toriyama, 1975). Diel behaviors and activities of fish have been reported in many fish species (Inoue, 1978; Tabata, 1988; Tabata et al., 1989), and these phenomena are probably critical for survival, such as forage for prey and avoidance of predators (Sakamoto, 1988; Sudo, 1988). Sudo (1988) also suggested that circadian rhythms of fish are closely related to those of their prey and space competitors. Previous studies have shown that red tilefish primarily forage on benthic invertebrates such as polychaetes and crustaceans (Funada, 1963; Kiyono & Hayashi, 1977; Ozaki et al., 2004). Although information about the predators and competitors of this species is limited, we consider some possible advantages for the site fidelity and the dawn activity of red tilefish. If available prey is abundant surrounding their burrows, staying near their burrow might reduce their encounter with predators and competitors. It is probably less risky for red tilefish to be active at dawn owing to the low frequency of encounter with predators and competitors for prey or space, if most of them are nocturnal or active around noon. However, the exact explanation for those phenomena is yet to be determined.

Light intensity and behavioral activity

In the aquarium, wild individuals might have responded to a slight increase in low levels of light intensity (0–0.1 μmol m^{-2} s^{-1}) and therefore, became more active (Fig. 4a, b). In the field, the fish W15 might have changed its behavioral pattern with the onset of increase and the end of decrease in light intensity (Figs. 7, 8), supporting the laboratorial results. Previous reports have shown that some pelagic fishes can respond to no more than 1.0×10^{-3} μmol m^{-2} s^{-1} and form schools in experimental tanks (Hunter, 1968; Hunter & Nicholl, 1985; Glass et al., 1986; Miyazaki et al., 2000). Although we did not measure the light intensity in the natural habitat of wild red tilefish (at depths of 30–200 m) continuously for 24 h, they might spend a major part of a day under the light intensity of nearly 0 μmol m^{-2} s^{-1} based on our measurement around noon. Therefore, wild red tilefish probably respond to low levels of light intensity which are too low to be measured by the light-intensity data logger.

On the contrary, are their activities influenced by high levels of light intensity, at no less than 0.1 μmol m^{-2} s^{-1}, for example? In the aquarium, instead of the increase of light intensity, the activity of wild fish did not significantly increase from 6:10 to 7:30 (Fig. 4b), and tended to be lower in the day-phase than the dawn-phase (Fig. 3). Even though the light intensity was constant during the day-phase, the number of turns per 10 min fluctuated for every fish (Fig. 2).

In the field, signal detections fluctuated during daytime (Fig. 6). In the VR2 records, the frequency of signal detections is assumed to reflect the ratio of time that released fish were outside their burrows, since they may not habitually move more than a few 100 m. However, there was no relationship between the ambient light intensity and the signal frequency of fish W15 during the diel-period ($R^2 = 0.0076$, $P = 0.22$). From the two experiments, once being active along with the increase of light intensity at dawn, high levels of light intensity does not influence the activity and behavioral pattern of red tilefish in the subsequent periods (>0.1 μmol m^{-2} s^{-1}). The factors causing the behavioral activity under high levels of light intensity remain unknown. Previous studies have shown that some fish species have circadian rhythms of locomotor activities which are controlled by some kinds of endogenous factors as well as exogenous factors, such as light conditions (Nelson & Johnson, 1970; Ooka-Souda et al., 1985; Tabata, 1988). We can not determine whether or not the behavioral activity of red tilefish is controlled by endogenous factors. However, it is considered possible that these factors explain the fluctuation of activity and the relative inactive state of red tilefish in the daytime after becoming active at dawn. Further studies on circadian rhythms should be carried out.

Differences between wild and hatchery-reared fish

Among the wild red tilefish, the diel behavioral activities may be similar to one another, since the activities and the digging frequencies of wild fish in the aquarium tended to be higher at the dawn-phase (Figs. 2, 3), and all the wild fish settled in Maizuru Bay were detected in the daytime but rarely during nighttime (Fig. 6, see Electronic Supplementary Material). In contrast, among the hatchery-reared red tilefish, the diel behavioral activities may differ, since the activities and the digging frequencies in the aquarium experiment increased during different phases (Figs. 2, 3), and one of the three settled hatchery-reared fish was detected exceptionally with high frequency during nighttime but rarely in the daytime (see Electronic Supplementary Material). If the pattern of the wild fish is assumed to be the regular pattern of red tilefish, the hatchery-reared fish might have included many irregular individuals whose diel behavioral patterns and/or responses to the fluctuations of light conditions were different from the regular pattern. The exact cause for the inconsistency is to be determined, but there are at least three possible explanations. (1) Since the diel behavioral activity of fish may be a critical factor for their survival (Sakamoto, 1988), irregular individuals might be less likely to survive in the natural environment than regular ones. Therefore, the wild individuals having unfavorable behaviors for survival might have gone naturally extinct before these experiments, whereas hatchery-reared individuals might not. (2) Tabata (1988) and Tabata et al. (1989) reported that rapid changes in environmental conditions might cause the change in the circadian rhythm of fish. Since the light condition during the rearing had not been strictly controlled, another possible explanation of the inconsistency is that many hatchery-reared individuals showing irregular patterns might have been influenced by light illumination during dark periods before these experiments. (3) The size of hatchery-reared fish observed in the aquarium (217 ± 11 mm TL) was smaller than that of wild ones (316 ± 25 mm TL). A previous study has shown that black bullhead *Ictalurus melas* at sizes of <120 mm standard length become more active in the daytime than during nighttime even though fish >120 mm in length are nocturnal (Darnell & Meierotto, 1965). Therefore, we cannot completely rule out that a change from an irregular to a regular pattern of behavior might have occurred between the size of 200 mm TL and 300 mm TL.

In the field, the ratios of settlement and track terms were different between the periods of release and wild or hatchery reared fish (see Electronic Supplementary Material). In the case of wild fish, one possible explanation for the settlement is that they settled in Maizuru Bay where the temperatures were within the range of what they had experienced before the release (10–23°C). When the water temperatures in the study area were near or beyond the upper limit of what the fish had experienced before the release (10–23°C), wild fish did not settle because the temperatures were probably too high to survive. The wild fish released in the summer (especially in August 2003) might have moved to look for a cooler environment soon after the release. In the case of hatchery-reared fish, maintained at a temperature of 10–27°C, the fish released in the winter (in November 2004) did not settle but moved away soon after release even though the water temperature was about 18°C. It is impossible to explain the settlement of hatchery-reared fish only from the perspective of water temperature. Previous studies have shown that environmental factors other than water temperature and light condition may affect fish behavior and migration (Greer Walker et al., 1978; Metcalfe et al., 1992; Wannamaker & Rice, 2000). Therefore, further studies to reveal the other environmental cues causing the settlement of red tilefish are still required.

Quality of hatchery-reared fingerlings for sea-farming

Since the hatchery-reared fish tended to differ from the wild fish in their diel behavioral activity and response to water temperature in the two experiments, it is possible that the fingerlings released in Wakasa Bay included individuals whose behavior was different from that of wild ones in the natural habitats. Therefore, the released fingerlings might have become active in foraging and constructing burrows when the wild fish were inactive or inside burrows, and moved to areas where water temperatures were higher compared to that in the natural habitats. If this is the case, the fingerlings may easily have been preyed upon by the larger carnivorous fish that are nocturnal or active around noon and are found in the warmer area. The fingerlings may also have the disadvantage in foraging for prey, if the distributions and/or activities of the released fingerings do not overlap with those of preys. Although the sizes of hatchery-reared fish used in this study (204–332 mm TL) were larger than those of the released fingerlings (about 100 mm TL), comparative study of behavioral characteristic among the smaller different sizes have not been conducted except for a study on the development of burrow construction ability (Hondoh et al., 2002). Therefore, we cannot deny the possibility that behaviors of released fingerlings are different from those of the fish in the present study, and the exact explanation for the low recovery rate of released fingerlings is yet to be determined. However, a low survivability derived from the different diel behavioral activity and response to water temperature could be one of the critical causes for the low recovery rate.

In the present study, novel information about the behavior of red tilefish and the difference between fish of both origins has been acquired. In order to reveal the exact factors regulating the survival of released fingerlings, further studies on the behavioral characteristics and comparisons between the wild and hatchery reared fish should be conducted. In addition, it is also necessary to compare the behavior of hatchery-reared fish of different sizes in order to understand the high mortality of released fish.

Acknowledgements We would like to acknowledge Captain K. Sato who operated the research boat, and thank Dr K. Uchida, Mr. T. Ide, Mr. K. Matsuda, Mr. U. Fukunishi, Mr. S. Nakatsuma and Ms. H. Makino who provided much assistance. We also want to thank ALEC ELECTRONICS, Chateau Marine Survey Co., Ltd., and NIPPON KAIYO CO., LTD. for letting us use many apparatus for our experiments. We are grateful to the staff and students at the laboratory of Biosphere Informatics, the Graduate School of Informatics, Kyoto University, and the two anonymous referees for detailed suggestion on improving this manuscript.

References

Able, K. W., D. C. Twichell, C. B. Grimes & R. S. Jones, 1987. Tilefishes of the genus *Caulolatilus* construct burrows in the sea floor. Bulletin of Marine Science 40: 1–10.

Darnell, R. M. & R. R. Meierotto, 1965. Diurnal periodicity in the black bullhead, *Ictalurus melas* (Rafinesque). Transaction of the American Fisheries Society 94: 1–8.

Funada, H., 1963. Research on the stock of red tilefish in Wakasa Bay—Biological study on red tilefish. Research Performance of Kyoto Fisheries Experiment Station 15: 1–24.

Glass, C. W., C. S. Wardle & W. R. Mojsiewicz, 1986. A light intensity threshold for schooling in the Atlantic mackerel *Scomber scombrus*. Journal of Fish Biology 29(Suppl A): 71–81.

Greer Walker, M., F. R. Harden Jones & G. P. Arnold, 1978. The movements of plaice (*Pleuronectes platessa* L.) tracked in the open sea. Journal du Conseil 38: 58–86.

Hondoh, Y., N. Murakami, M. Watanabe, H. Takeuchi, Y. Fujinami & T. Tsuzaki, 2001. Seed production of red tilefish *Branchiostegus japonicus* using an artificial fertilized eggs. Saibai Gyogyou Gijutsu Kaihatu Kenkyu 28: 73–79.

Hondoh, Y., R. Masuda & T. Tsuzaki, 2002. Development of burrowing in hatchery-reared red tilefish *Branchiostegus japonicus*. Saibai Gyogyou Gijutsu Kaihatu Kenkyu 29: 85–89.

Hunter, J. R., 1968. Effects of light on schooling and feeding of jack mackerel *Trachurus symmetricus*. Journal of the Fisheries Research Board of Canada 25: 393–407.

Hunter, J. & R. Nicholl, 1985. Visual threshold for schooling in northern anchovy *Engraulis mordax*. Fishery Bulletin 83: 235–242.

Inoue, M., 1978. Fish Behavior and Fish-catching Methods. Koseisya Koseikaku, Tokyo.

Kiyono, S. & B. Hayashi, 1977. Ecological studies on the red tilefish in Wakasa Bay-2: Distributions of immatures and adults. Bulletin of the Kyoto Institute of Oceanic and Fishery Science 1: 15–28.

Kiyono, S., B. Hayashi & T. Komiyama, 1977. Ecological studies on the red tilefish in Wakasa Bay-1: Distributions of immatures and adults. Bulletin of the Kyoto Institute of Oceanic and Fishery Science 1: 1–14.

Metcalfe, J. D., M. Fulcher & T. J. Storeton-West, 1992. Progress and developments in telemetry for monitoring the migratory behavior of plaice in the North Sea. In Priede, I. G. & S. M. Swift (eds), Wildlife Telemetry: Remote Monitoring and Tracking of Animals. Ellis Horwood, New York, London, Toronto, Sydney, Tokyo, Singapore, 359–366.

Mitamura, H., N. Arai, W. Sakamoto, Y. Mitsunaga, T. Maruo, Y. Mukai, K. Nakamura, M. Sasaki & Y. Yoneda, 2002. Evidence of homing of black rockfish *Sebastes inermis* using biotelemetry. Fisheries Science 68: 1189–1196.

Mitamura, H., N. Arai, H. Mitsunaga, T. Yokota, H. Takeuchi, T. Tsuzaki & M. Itani, 2005. Directed movements and diel burrow fidelity patterns of red tilefish, *Branchiostegus japonicus*, determined using ultrasonic telemetry. Fisheries Science 71: 491–498.

Miyazaki, T., S. Shiozawa, T. Kogane, R. Masuda, K. Maruyama & K. Tsukamoto, 2000. Developmental changes of the light intensity threshold for school formation in the striped jack *Pseudocaranx dentex*. Marine Ecology Progress Series 192: 267–275.

Nelson, D. R. & R. H. Johnson, 1970. Diel activity rhythms in the nocturnal, bottom-dwelling sharks, *Heterodontus francisci* and *Cephaloscyllium ventriosum*. COPEIA 4: 732–739.

Okumura, S., 2000. Studies on broodstock management and seed production of the red tilefish, *Branchiostegus japonicus*. Japan Sea-Farming Association Report of Special Research 16: 1–43.

Ooka-Souda, S., H. Kabasawa & S. Kinoshita, 1985. Circadian rhythms in locomotor activity in the hagfish, *Eptatretus burgeri*, and the effect of reversal of light-dark cycle. Zoological Science 2: 749–754.

Ozaki, H., Y. Hamanaka, M. Itani & K. Takeno, 2004. Feeding habit of immature red tilefish in the western part of Wakasa Bay. Bulletin of the Kyoto Institute of Oceanic and Fishery Science 26: 55–57.

Sakamoto, W., 1988. Diel activity and survival in larval fish. In Hanyu, I. & M. Tabata (eds), Daily Rhythmic Activities in Aquatic Animals. Koseisya Koseikaku, Tokyo: 101–116.

Starr, R. M., J. N. Heine, J. M. Felton & G. M. Cailliet, 2002. Movements of bocaccio (*Sebastes paucispinis*) and green spotted (*S. chlorostictus*) rock fishes in a Monterey submarine canyon: implications for the design of marine reserves. Fishery Bulletin 100: 324–337.

Sudo, H., 1988. Diel predator-prey interactions between fishes and gammarids. In Hanyu, I. & M. Tabata (eds.), Daily Rhythmic Activities in Aquatic Animals. Koseisya Koseikaku, Tokyo, 117–133.

Tabata, M., 1988. Diel and circadian locomotor activity in fish. In Hanyu, I. & M. Tabata (eds), Daily Rhythmic Activities in Aquatic Animals. Koseisya Koseikaku, Tokyo, 79–100.

Tabata, M., M. Minh-Nyo, H. Niwa & M. Oguri, 1989. Circadian rhythm of locomotor activity in a teleost, *Silurus asotus*. Zoological Science 6: 367–375.

Toriyama, M., 1975. Tilefish and some other species observed from the submersible, 'Shinkai'. Nansei Suiken News 13: 12.

Tsukamoto, K., H. Kuwada, K. Uchida, R. Masuda & Y. Sakakura, 1999. Fish quality and stocking effectiveness: behavioral approach. In Howell, B. R., E. Moksness & T. Svåsand (eds), Stock Enhancement and Sea Ranching. Fishing News Books, Blackwell Science, London, 205–218.

Yokota, T., H. Mitamura, N. Arai,Y. Mitsunaga, H. Takeuchi, T. Tsuzaki & M. Itani, 2004. A method of studying fish behavior using underwater acoustic biotelemetry—Tracking of the red tilefish released in the Wakasa Bay and the Maizuru Bay. Journal of Advanced Marine Science and Technology Society 10(1): 29–40.

Wannamaker, C. M. & J. A. Rice, 2000. Effects of hypoxia on movements and behavior of selected estuarine organisms from the southeastern United States. Journal of Experimental Marine Biology and Ecology 249: 145–163.

Zar, J. H., 1999. Biostatistical Analysis, 4th edn. Prentice-Hall, New Jersey.

Hydrobiologia (2007) 582:121–132
DOI 10.1007/s10750-006-0540-2

FISH TELEMETRY

Sea lamprey (*Petromyzon marinus* L.) spawning migration in the Vouga river basin (Portugal): poaching impact, preferential resting sites and spawning grounds

N. O. Andrade · B. R. Quintella · J. Ferreira · S. Pinela · I. Póvoa · S. Pedro · P. R. Almeida

Abstract Historical spawning grounds for sea lampreys (*Petromyzon marinus* L.) in most Portuguese river basins are becoming inaccessible due to the construction of impassable dams and/or weirs. Studies like the one described in this paper are particularly important in areas like the Vouga river basin, where there is a considerable fishing effort from both professional fishermen and poachers. In fact, for management and conservation purposes, it is important to clarify several aspects of the sea lamprey spawning run in this particular watershed. Therefore, a total of 30 radio tagged, migrating sea lampreys were released in the River Vouga and some of its main tributaries during 2004 and 2005. Results from the tracking sessions were used to determine the effect of poaching on the spawners' population and the characteristics of the resting sites used during the upstream movement. The rivers' stretches were also characterized according to the type of substrate present in the riverbed and flow type, in order to determine its aptitude to constitute appropriate spawning habitats for sea lampreys. We have identified in the upstream stretches of River Vouga, and in the tributary River Caima, characteristics that are particularly suitable for the construction of nests by the spawners. Presence of larvae provided evidence that spawners migrated into the River Vouga's upper reaches. However, abundance and age class diversity appeared to be higher downstream of Sernada and Carvoeiro weirs, corroborating the telemetry data which suggested difficulty in passing these obstacles during low precipitation years. In the River Caima, migration was most predictable, which might be related to daily water releases from a small hydropower dam at dusk, that stimulated the lampreys to resume migration. Poaching has had a considerable negative effect on the success of the tagged lampreys' migration: 76% of the animals released during 2005 were captured.

Keywords *Petromyzon marinus* · Radio telemetry · Resting sites · Spawning migration · Poaching

Guest editors: P. R. Almeida, B. R. Quintella, M. J. Costa and A. Moore
Developments in Fish Telemetry

Electronic supplementary material The online version of this article (doi: 10.1007/978-1-4020-6237-7_13) contains supplementary material, which is available to authorized users.

N. O. Andrade · B. R. Quintella · J. Ferreira · S. Pinela · I. Póvoa · S. Pedro · P. R. Almeida (✉)
Institute of Oceanography, Faculty of Sciences of the University of Lisbon, Campo Grande 1749-016 Lisboa, Portugal
e-mail: pmraposo@mail.telepac.pt

P. R. Almeida
Department of Biology, University of Évora, 7002-554 Evora, Portugal

Introduction

The sea lamprey (*Petromyzon marinus* L.) is an anadromous species that spawns in the main Portuguese River basins. Sea lamprey spawners initiate their migration in the Portuguese rivers in the middle of December, with the migration peaking between February and April and spawning usually taking place between May and June, depending on the meteorological conditions (Almeida et al., 2000). Studies carried out with holobiotic specimens showed no evidence of homing behaviour (Bergstedt & Seelye, 1995). Other authors suggest that some environmental factors such as the water temperature, the type of substrate and river flow (Morman et al., 1980; Young et al., 1990) or the release of a migratory pheromone by stream-dwelling larval lamprey (Teeter, 1980; Li et al., 1995; Vrieze & Sorensen, 2001; Sorensen & Vrieze, 2003; Fine et al., 2004; Sorensen et al., 2005) may act as a fundamental stimuli to the adult sea lamprey, guiding them to spawning streams. Mature, spermiated, male sea lamprey also release a potent sex pheromone that induces preference and searching behaviour in ovulated female lampreys (Li et al., 2002). During this period, sea lampreys exhibit a prominent pattern on the migratory activity, being active invariably through the night period and seeking rocky substrates for shelter at dawn (Hardisty & Potter, 1971; Stier & Kynard, 1986; Kelso & Gardner, 2000; Almeida et al., 2000; Almeida et al. 2002a; Quintella et al., 2004).

The gastronomic importance of sea lampreys is reflected by their high commercial value, which can easily reach € 45 per animal during the peak of the season. The high economic value of the sea lamprey in Portugal, and particularly in the Vouga River basin, makes them preferred target of both professional fishermen and poachers, creating a major threat to the sustainability and conservation of Vouga's sea lamprey population. Gravel extraction, dredging, pollution and habitat destruction resulting from the river regulation create additional threats to the survival of this species in the Portuguese rivers (Assis, 1990; Almeida et al., 2000; Almeida et al., 2002a; Rogado et al., 2005). In particular, the interruption of the Vouga's longitudinal continuity has clearly contributed to significant reduction in the habitat available to both spawners and ammocoetes. There are already several physical obstacles that interrupt the migratory path of the sea lampreys and a dam is also projected to be built soon in the River Vouga, an enterprise that will promote additional alterations in the aquatic community equilibrium, particularly for diadromous species present in this basin (i.e. sea lamprey; river lamprey *Lampetra fluviatilis*, L.; European eel *Anguilla anguilla*, L.; allis shad *Alosa alosa*, L.; twaite shad *Alosa fallax*, Lacépède, 1803).

Since the available spawning grounds for this species in most of the Portuguese river basins is quite limited due to the construction of impassable dams, it seems extremely relevant to clarify some aspects concerning the sea lamprey spawning run in the Vouga river basin. Therefore, the goal of this study was to determine the effect of poaching on the migrating sea lamprey in the Vouga river basin, to identify the main obstacles to the migration of adult sea lamprey spawners in the River Vouga and its main tributaries, and to identify and characterize the sites where the sea lampreys rest during their spawning migration. Additionally, a characterization and evaluation of these rivers' aptitudes according to the river bed flow type and the existing substrate type as appropriate spawning grounds for this species was also performed. This study assumes particular importance in rivers like Vouga, where there is a considerable fishing effort from both professional fishermen and poachers. During the 2004/05 spawning season, this situation was aggravated by an abnormally dry winter season, with extremely low precipitation.

Study area

The River Vouga is a medium size Portuguese river (148 km) whose basin covers an area of 3,635 km^2 in the northern part of the country (Fig. 1), being limited by the Douro and the Mondego river basins to the North and to the South, respectively (Loureiro & Macedo, 1986). It drains into the Atlantic Ocean in the Ria de Aveiro (45 km^2), an extremely productive estuary with an important fishing industrial activity.

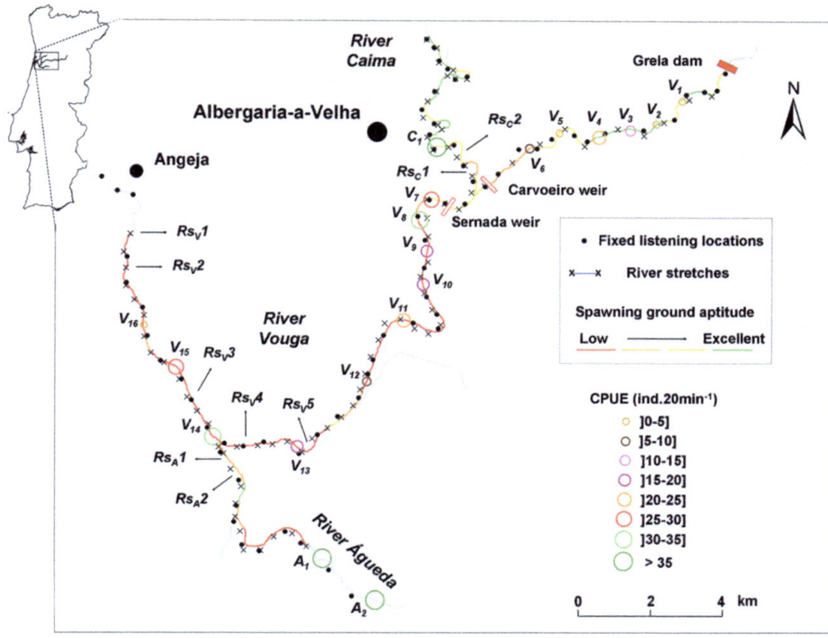

Fig. 1 Map of the Vouga river basin and results from the 2004 electrofishing campaign represented in captures per unit of effort (CPUE–individuals captured in 20 min fishing) in selected sampling sites along the studied watercourses. River stretches limits and their aptitude as spawning grounds are also represented. RS_Vn, Release sites in River Vouga; RS_Cn, Release sites in River Caima; RS_An, Release sites in River Águeda; V_n, Sites electrofished in River Vouga; C_n, Sites electrofished in River Caima; A_n, Sites electrofished in River Águeda

Sea lampreys are captured in the estuary with trammel nets, and in the river with a type of fyke net that is locally called a "gamboa." Besides hundreds of professional fishers, poaching also has great relevancy in the River Vouga and its tributaries, the Caima (196 km^2) on the northside of the basin and the River Águeda (972 km^2) with its important tributaries, the rivers Cértima (541 km^2) and Alfusqueiro (205 km^2) in the south (Loureiro & Macedo, 1986). The River Antuã (149 km^2) belongs to Vouga's river basin but it drains directly into the northern part of the Ria de Aveiro estuary.

The Vouga's flow is controlled by the discharge from four dams, the Ribafeita and the Drizes in the River Vouga and the Padrastos and the Burgães in the upper part of the River Caima. Other enterprises planned in the River Vouga will probably also constrain anadromous species: the Ribeiradio hydroelectric power dam, to be located approximately 15 km upstream of the Grela dam (53 km from the river mouth), a small hydroelectric power dam built in the 1930's that became the upper limit of the sea lamprey migration on this river (Costa et al., 2001); and the new Carvoeiro weir, 34 km from the river mouth. Several weirs were built in this river basin, most of them to power mills, but are already abandoned and partially destroyed, like the one in River Caima (8.5 km from the River Vouga and 51 km from the River Vouga's mouth) that in normal conditions is impassable for sea lampreys. There are also two large blockstone weirs in River Vouga, the Sernada (42 km), and the Carvoeiro (43.5 km), that can equally reveal themselves quite difficult to be successfully overcomed by fish.

Materials and methods

In the autumn of 2004, an electrofishing (electrofisher Hans Grassl EL62, 3.0 KW, direct current 300/600 V) campaign was carried out in the River Vouga and its main tributaries, namely, the rivers Antuã, Caima and Águeda. The main purpose of this survey was to identify both ammocoete beds and areas of the river that were used by the sea lampreys to spawn in former years in order to avoid releasing tagged sea lampreys in rivers with weak attractive chemical stimuli from larval population to spawners (Sorensen et al., 2005). Captured ammocoete age classes were defined according to Quintella et al. (2003).

Radio tracking telemetry procedures

In the spawning season of 2004, five sea lampreys were purchased from local fishermen in the area near the village of Angeja (c. 16 km from the mouth of River Vouga). These animals were tagged and released between February and April in the River Vouga and used as a preliminary approach to the study of sea lamprey up river spawning migration in this watershed. During 2005, 25 sea lampreys caught in the fishing area near Angeja were purchased from local fishermen and then tagged and released in several locations of the rivers Vouga (n = 13), Caima (n = 8) and Águeda (n = 4) from mid January to mid April (Fig. 1).

In both seasons, the tagging procedure was carried out in situ using externally mounted radio transmitters from ATS (model type F2020). These tags were cylindrical, with dimensions of 12 × 39 mm, weighing 8.6 g in air (transmitter plus wires, discs and sleeves) and were powered by 3.5 V, 0.09 mA lithium batteries which guaranteed an 80-day longevity (max. 160). They operated between 150.000 and 150.990 MHz, with a pulse rate and width of 40 ppm and 20 ms, respectively. The complete procedure was adapted from Almeida et al. (2000) in order to become faster and less stressful to the animals and so it would take, roughly, 10 min for induction in a anaesthetic solution of 0.3 ml 2-phenoxyethanol per litre of water, and c. 3 min to weigh, measure and tag the animal. Before being released in the wild, lampreys were left to recover in the river margin inside a hoop net for a period of 1–2 h. In order to determine the poaching impact on the released animals a reward was offered for each recovered transmitter.

The animals were located using the R2000 receiver and a Yagi antenna (ATS) and then monitored continuously, either by car or on foot, from release at dusk until dawn or early morning. Position of tagged animals and time were recorded in previously selected, listening locations in a total of 45 sites in River Vouga, 14 in River Caima and 10 in River Águeda. The sex of the animals used could not be determined due to their early stages of gonad maturation during which spawners do not present any external sexual dimorphism.

Characterization of the resting sites and the rivers' appetence as suitable spawning grounds

During the 2005 migration season, the radio tracking telemetry was used to study not only the upstream movement, but also some aspects related to the selection of the resting sites during the migration. The variables used to characterize resting sites are shown in Table 1. On each day following a tracking session, the precise location of the tracked animal would be identified within an area of approximately 1 m^2, and its diurnal resting site (within 2 m diameter range) would be characterized according to those variables. After the continuous monitoring of any given animal, the resting sites characterization would still be done regularly even, if the lamprey changed refuge. To characterize the rivers' abilities as suitable spawning grounds the rivers were divided in 500 m long stretches, from the lower part of the study area to the upstream limit to migrants. Then, according to both the riverbed flow type and substrate type the stretches were given a value in an aptitude scale (Table 2). So, respectively, in the rivers Vouga, Caima and Águeda, 51, 18 and 10 stretches were considered and further characterized (Fig. 1).

Data analysis

The average ground speed (GS), expressed in body lengths per minute (BL min^{-1}), was determined as:

$$\mathrm{GS} = \frac{s}{\Delta t \times L_t},$$

where s is the total distance the animal travelled during continuous monitoring and between t and

Table 1 Variables used for the characterization of the sea lamprey refuges and the river stretches aptitude for spawning grounds

Variables	Types/Definition	Classes
Rbt	Sinuate; braided; regular (natural); regular (artificial)	<30%; 30–60%; >60%
Ft	Run; riffle; pool; slack	<30%; 30–60%; >60%
Cs	Artificial; rock; stone; gravel; sand; silt	<30%; 30–60%; >60%
Cvc	Mean cover	0%; <50%; >50%; 100%
S	Mean cover	0%; <50%; >50%; 100%
Astr	Present/absent	Easy; medium; hard; impassable
Pobs	Present/absent	Easy; medium; hard; impassable
C_w	Mean width	[0–15]m; [15–30]m; [30–45]m; [45–60]m; [60–75]m
Deb	Present/absent	–
W_{depth}	Mean water depth	≤1 m; [1–3]m; [3–5]m; ≥5 m
SGdep	Present/absent	–
Temp	Mean water temperature observed during continuous monitoring sessions (°C)	

Rbt, River bed type; Ft, Flow type; Cs, Channel substrate; Cvc, Channel vegetation cover; S, Shading of channel/resting site; Astr, Artificial structures; Pobs, Physical obstacles; C_w, Channel width; Deb, Debris; W_{depth}, Water depth; SGdep, Sand/Gravel deposits; Temp, Temperature

t^{-1} (Δt in min), and L_t is the animal total length (in cm). The total distance moved (TDM) and the total time tracked (TTT) are both only associated with the continuous monitoring and not obligatory related to the final position of the animal in the end of the season.

To calculate the halting probability of the released sea lampreys when related with the TDM, a logistic expression was adjusted. The animals that behave atypically, like the lampreys that stopped shortly after being released, and those that stopped in areas with difficult obstacles to transpose were not used in this analysis.

Results

Electrofishing in 2004 revealed a clear distinction between the abundance of sea lamprey larvae downstream and upstream of the Sernada and Carvoeiro weirs (Fig. 1). Presence of larvae upstream from these obstacles provided evidence that spawners migrated into the River Vouga's upper reaches and established larvae populations. However, abundance and age class diversity appeared to be higher downstream from these obstacles, between the sampling sites V_7 and V_{15} with peaks in sites V_8, V_{11} and V_{14} (Fig. 2). The higher ammocoete abundance downstream corroborates the telemetry data which suggested difficulty in passing Sernada and Carvoeiro weirs. Downstream from V_{15}, the sediment gradually becomes siltier, and also not suitable as ammocoetes' habitat. In the lower part of the study area (i.e. sampling site V16), only sea lamprey juveniles were captured, probably during their downstream trophic migration. Upstream from the last surveyed site (V16) we found favourable envi-

Table 2 Ecological characteristics considered for the classification of the aptitude of the stretches' to sea lamprey spawners reproduction

	Aptitude			
Variables	Low	Medium	Good	Excellent
River bed flow type	Absence of riffles	Run + riffle >60%	Run + riffle >60% and presence of riffle	Run + riffle >60% and riffle >30%
Substrate type	Absence of Stone and Gravel <30%	Gravel + Stone <30% and presence of Stone	30% <Gravel + Stone <60% and presence of Stone	Gravel + Stone >60% and Stone >30% or Gravel >60%

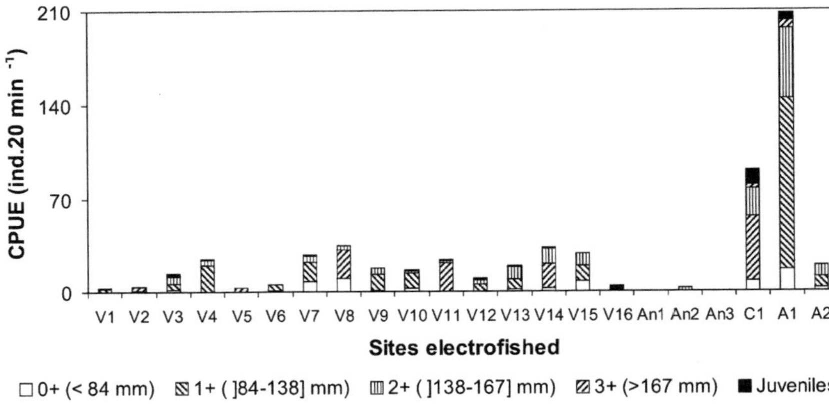

Fig. 2 CPUE of sea lamprey ammocoetes and juveniles in each electrofished site in the rivers Vouga, Antuã Caima and Águeda. V_n, Sites electrofished in River Vouga; A_n, sites electrofished in River Antuã; C_n, Sites electrofished in River Caima; A_n, Sites electrofished in River Águeda. Ammocoete age classes were defined according to Quintella et al. (2003)

ronmental conditions for these populations until, approximately, 2 km downstream from the Grela dam (V_1) constituting, roughly, 30 km of adequate habitat in River Vouga.

In tributaries Caima and Águeda, ammocoete beds were much easier to find and with much higher densities of animals. In fact, there were very good natural conditions for these animals throughout the watercourses. In River Antuã, larvae density was very low, being extremely hard to find any lamprey (i.e. *P. marinus* and *Lampetra* sp.) ammocoete whatsoever and, therefore, it was decided to exclude this particular river from the monitoring sessions in the following year, due to the fact that there would be a strong possibility that this river would not have enough pheromonal stimuli from larvae to attract spawners. Larvae and juveniles of both species were found in all rivers surveyed, a total of 741 *P. marinus* and 82 *Lampetra* sp. were collected.

The tagged lampreys released in River Vouga during 2004 spawning season had a total length (L_t) between 82 and 92 cm (mean ± SD—880 ± 40 cm) and a total weight (W_t) between 1.1 and 1.4 kg (1.3 ± 0.14 kg). The total time tracked (TTT) ranged from 10 h 05 min to 62 h 00 min and the total distance moved (TDM) in the upstream direction varied from 4.4 to 20 km. Only one of the animals released during this season showed the typical behaviour, moving upstream after release, but the rest of the lampreys either moved downstream or halted right after being released. One of them, released upstream from the Grela dam in order to assess their behaviour when migrating on a river stretch without stimuli from pheromones released neither by ammocoetes nor by other spawners, showed a peculiar behaviour moving up and downstream several times like it was somehow disoriented till it passed over this dam, remaining a few hundred meters downstream from it for several weeks.

Only one of the animals managed to overcome the Sernada and Carvoeiro weirs (at the end of the spawning season, only 33.3% of the tagged animals that reached these obstacles managed to overcome them) during continuous monitoring, taking 6 h 10 min and 4 h 20 min, respectively, to do so. Besides this animal in 2004, just two more in the 2005 spawning season successfully passed the Sernada weir, eventually residing between the rocks of the Carvoeiro weir, not being able to pass that structure. In the first year of study, two tagged sea lampreys were captured by poachers and another one that moved downstream after release was recaptured by professional fishermen.

In the 2005 spawning season, the animals used had a L_t ranging from 78 to 98 cm (mean ± SD—90 ± 5.3 cm) and the W_t varied between 0.8 and 1.9 kg (1.5 ± 0.3 kg) (see Electronic supplementary material). The TDM by the tagged sea lampreys during continuous monitoring, considering either movements done up or downstream, ranged from just 0.1 km to c. 13 km

(6.4 ± 4.8 km) and the TTT ranged between 00 h 28 min interval in the River Vouga and 11 h 52 min in the River Caima. The water temperature (*T*) at release ranged from 5.38°C to 15.28°C (9.88 ± 3.82°C). Both extreme values were registered in the River Caima. In the River Vouga, tagged animals were released in five locations and in the rivers Caima and Águeda in two locations (Fig. 1) in order to avoid lampreys being captured by professional fishing nets, since these entrapment gears were not consistently set in the same places throughout the spawning season.

Monitored lampreys exhibited strong diel patterns of activity in both years, moving only at night. In the night after release, most sea lampreys ($n = 6$) resumed their migration from 1 h 12 min to 3 h 15 min (2 h 01 min ± 56 min) after sunset. After this period if the lampreys had not yet commenced the upstream movement, it would not be initiated until the following night.

During continuous monitoring sessions, three tagged sea lampreys were poached in the River Vouga and one in the River Caima. In these situations, it took the poachers between 28 min and 11 h 52 min to recapture them.

In 2005, 69% ($n = 9$) of the animals released in River Vouga were poached and of those, four were captured downstream from the Sernada and Carvoeiro weirs. The signal of one of the transmitters was lost during the season, which means that it either malfunctioned or that the lamprey was caught by poachers. The latter is more probable based on returns of other tags for the reward. From the River Caima, 62.5% ($n = 5$) of tagged sea lampreys were poached and 25% ($n = 2$) were confirmed to be predated, probably by otters due to the marks of the recovered specimens. From the River Águeda, 25% ($n = 1$) of tagged lampreys were captured by poachers and the signal was lost from 75% ($n = 3$) of the animals.

Overall, 60% ($n = 15$) of the total tagged sea lampreys in the Vouga River basin were captured by poachers and the signal was lost from 16% ($n = 4$). We believe that the sum of the animals, i.e. 76% ($n = 19$) of the lampreys, represents the poaching activity during 2005. Predation occurred for 8% ($n = 2$) of total tagged lampreys and 16% ($n = 4$) were still there at the beginning of May.

The migration in River Caima was more predictable. The ground speeds (GS) observed for the animals released varied significantly between rivers (Kruskal-Wallis, $P < 0.05$). In rivers Vouga, Caima and Águeda the mean GS were respectively 25.75 ± 3.77, 15.91 ± 3.56 and 12.01 ± 7.12 BL min^{-1} (Fig. 3). In River Águeda, tagged individuals showed an atypical behaviour according to what has been previously described (Almeida et al., 2000, 2002b, 2005; Quintella et al., 2004), moving just a few hundred meters upstream and then halting for several weeks.

The sites sea lampreys selected to rest or shelter during daytime ($n = 39$) averaged 0.84 ± 0.43 m in depth, with slow current speeds (0.03 ± 0.04 ms^{-1}), and with high oxygen concentration (99.69 ± 9.28%). They also selected refuges completely shaded or darkened either by the existence of debris, rocks or vegetation (aquatic or riverbank) on 58% of occasions. Debris was found in 49% of resting sites and may be important for these animals by creating concealment and darkness. The existence of thick channel vegetation coverage (Cvc) was not an important feature for these animals; 41% of selected sites, had less than 50% Cvc. Also, 38% of the tagged lampreys chose resting sites without this type of cover. The dominant channel substrate (Cs) found in the selected resting sites was sand (39%) and the subdominant, which in this case may be a more determinant characteristic, consisted mainly of gravel (41%) and stone (27%).

Suitable spawning grounds for adult sea lampreys are just found in the upper reaches of the River Vouga (Fig. 1). Unfortunately, this area is not always reachable throughout the entire migration season, being extremely dependent on precipitation and river discharge. Punctually, some fairly suitable stretches can be found downstream and lampreys may, in alternative, spawn there (Fig. 1). This observation was strengthened by informations gathered from interviews to local professional fishermen and poachers. Bypassing the Sernada weir, River Caima also appears to be an important alternative, although in dry winter seasons such as occurred in 2004 and 2005, the river flow may be insufficient. Most of the area comprehending

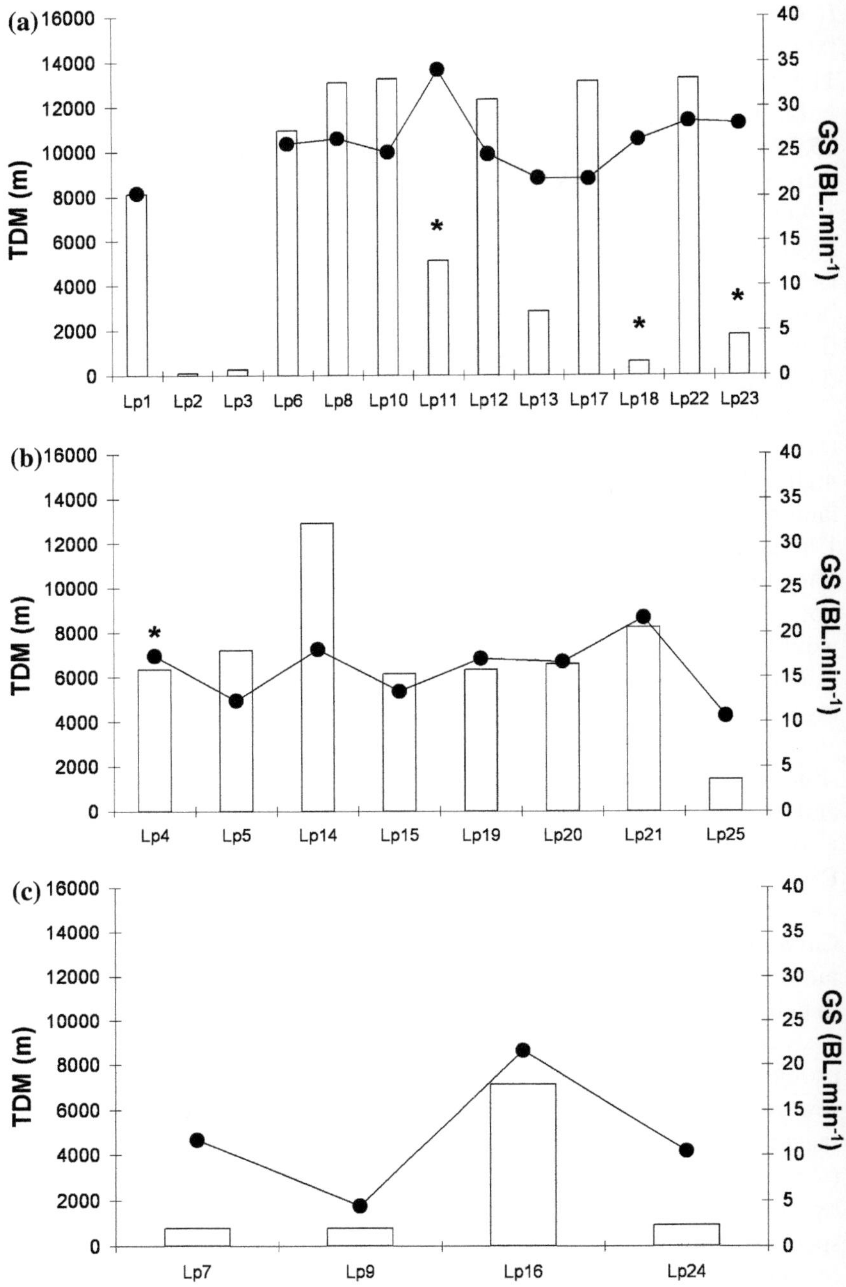

Fig. 3 Total distance moved (TDM - bars) and the mean ground speed (GS - lines) for the released tagged sea lampreys in the rivers Vouga (**a**), Caima (**b**) and Águeda (**c**) during continuous monitoring. The asterisks identify the animals captured by poachers during these sessions

the lower reaches of River Águeda is unsuited to be used by lampreys as spawning grounds since the substrate is mostly composed of sand and silt.

Excluding the animals released in River Águeda due to their atypical behaviour and those from the other studied watercourses which were in situations where they had obstacles like the existence of a difficult weir to overcome, 10 individuals were used to determine the probability of halting (CH) (Fig. 4), at the following equation:

$$CH = \frac{1}{\frac{1}{101} + 0.2427 \times 0.9995^{\mathrm{TDM}}}$$

(R^2 – 0.96; df – 6; F – 149.32; $P<0.001$)

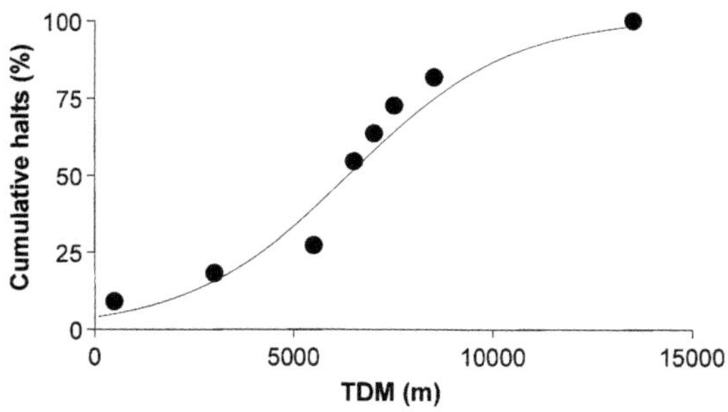

Fig. 4 Adjusted logistic expression to calculate the halting probability for sea lampreys in the rivers Vouga and Caima. TDM, total distance moved

where CH represents the probability of halting (cumulative) and TDM the total distance moved (m). According to the equation, the probability of an animal halting and looking for shelter reaches 50% when TDM reaches 6.4 km.

Discussion

The distinction in the levels of both abundance and age class diversity of lamprey ammocoetes downstream and upstream the Sernada and Carvoeiro weirs is clear. This may diminish the chemical stimuli to adults produced by resident larval lamprey that in a situation of no disturbance or when impassable obstacles are further away from the estuary would be felt from the rivers' upper reaches. However, the distribution of these animals showed a generalized occupation of the entire riverbed, unlike what has been described for the River Mondego where colonization is confined to a relatively small area downstream from the Açude-Ponte dam in Coimbra, which has been sea lamprey's natural distribution upper limit since it was built (Quintella et al., 2003).

In the River Vouga, larger ammocoete populations can be more easily found downstream from the Carvoeiro an Sernada weirs, particularly, from the sampling sites V_7–V_{15} which are, probably, preferential habitats for the establishment of ammocoetes beds since some abundance peaks were observed in these areas, mainly associated with the existence of younger individuals, belonging to the 0^+ age class. Moreover, above the Carvoeiro weir no animal belonging to this age class was captured during the electrofishing campaign. These results corroborate those obtained from the present radio telemetry studies that revealed the extreme difficulty these animals have to overcome these obstacles, which are increased in periods of low precipitation like the 2004 and 2005 spawning seasons. In the River Vouga's tributaries, rivers Caima and Águeda, ammocoete beds were easier to identify and the higher values of CPUE were also obtained in these rivers. The reasons are probably also related to the fact that these are narrower watercourses with lower depths and, therefore, being also easier to survey. They also suffer from less human interventions and both the professional fishing activity and poaching have a more reduced expression and so, the adult spawners that are able to enter into one of these rivers will probably be less likely to get caught and, consequently, will have more chance to successfully spawn.

The adequate natural conditions for the establishment of ammocoetes beds are gathered in the whole extension of the River Águeda, in the first 8 km of the River Caima and in the 30 km downstream of the Grela dam in the River Vouga.

In the spawning season of 2004, the behaviour demonstrated by these animals was atypical, at least in relation to what has been previously described for this species, particularly, in the River Mondego (Almeida et al., 2000, 2002a, b; Quintella et al., 2004; Almeida et al., 2005). Except for one of the animals, the rest of them either moved downstream or halted right after being released. The hesitation and disorientation of the animal released above the Grela dam may be explained by the more than probable lack of stimuli caused by the inexistence of both ammocoetes and other spawners upstream.

The management of the river flow downstream from the Grela dam is for now a determinant factor for the success of sea lamprey spawning migration in the River Vouga. The present data supports previous findings in which the sea lamprey spawners' cryptic pattern of behaviour is described (Hardisty & Potter, 1971; Stier & Kynard, 1986; Almeida et al., 2000; Kelso &

Gardner, 2000; Almeida et al., 2002a, b; Quintella et al., 2004), which may be considered to be a strategy developed by this species mainly to avoid predation during this sensitive period. Almeida et al. (2002b) have shown that heightened night-time dam discharge seemed to act as a positive stimulus for the sea lamprey to migrate by promoting the increase of the animals' ground speed, but daytime releases did not influence the lampreys to move. In River Caima, an identical tendency could also be observed and the migration in this river always appeared to be much more predictable regardless of the individuals. The median GS values observed in the River Vouga (Median = 25.87 BL min^{-1}) are within those formerly obtained by Almeida et al. (2002) in the River Mondego during low (Median = 33.3 BL min^{-1}) and high (Median = 28.9 BL min^{-1}) river discharge periods. On the other hand, the mean GS observed between the lampreys released in River Caima was more similar than when considering the ones released in the other rivers and the migration was also reinitiated more often but in this case the median GS (Median = 16.88 BL min^{-1}) was much lower. The slower velocities generally observed in the River Caima are probably related to this river being shallower, much narrower and sinuate, with lower caudal and having more riffle areas and debris than the River Vouga, creating areas of difficult passage where higher ground speeds are harder to obtain.

The increase of the current velocity and level at dusk seems to have a determinant role on the physical stimulation of these animals and thus to positively influence their upstream migration (Almeida et al., 2002b). Therefore, it would be beneficial for the sea lamprey if the discharges originated by structures such as hydropower dams were coincident with its nocturnal period of activity or at least between 1 and 3 h after sunset, from the middle of December to the end of May or early June. In these circumstances, one should bare in mind the importance of the period between dusk and dawn or early morning as a fundamental aspect for the success of these animals' spawning migration. Nevertheless, more studies applied to this matter are still needed.

The fishing activity and, particularly, poaching, has an important negative effect on the number of sea lampreys that effectively spawn in the Vouga river basin. Furthermore, in such dry winter seasons with extremely low precipitation levels like 2004 and, specially, 2005, this effect is certainly more worrying and in the present situation it had a preponderant effect on the number of animals caught since it was much easier to capture these animals in a situation of low river discharge and for the lampreys was much harder to find any suitable places to shelter and spawn. In fact, in this spawning season there were 15 (60%) confirmed captures done by poachers in the three rivers and for four of the tagged animals the respective radio signal was lost, meaning that probably these have also been captured but either for being scared or just uninformed, poachers did not reclaim the transmitter recovery reward. If this hypothesis is correct, 76% (n = 19) of the released animals in 2005 in this river basin were poached. This result is quite relevant since it may have serious implications in this year's recruitment and that, associated with the progressive limitation of suitable available habitats for both ammocoetes and adults to spawn, will certainly be prejudicial for this species conservation in this river basin. It urges for more effective control of the fishing activity, both professional and illegal, by the competent authorities in this river basin along with a no fishing period during the peak migration (i.e. from February to April).

In the River Vouga, the Sernada and Carvoeiro weirs are extremely difficult obstacles to overcome and in 2005, four of the animals released in this river were captured downstream from it. Due to the extreme difficulty in bypassing this kind of obstacle, sea lampreys are usually obliged to remain downstream of these structures but, nevertheless, they still keep on trying to pass over them. In fact, all the animals that managed to overcome those weirs during the two seasons did it sometimes a month after being released. During this period of time they are quite vulnerable to poachers, who use several methods (i.e. trammel nets, lamprey-spears; underwater fishing) to capture them in these assembling areas. Additionally, suitable spawning grounds in the River Vouga are mainly found upstream the latter weir and, therefore, whenever the river discharge is low it turns quite difficult for the animals to reach them. River Caima appears to have quite good ecological conditions for the lampreys to spawn, assuming that they manage to bypass the Sernada weir. For this reason competent authorities should consider the hypothesis of removing the Sernada and Carvoeiro weir, or at least to build effective fish passes in these structures for the target species. The design of the fishway installed in the Grela weir is clearly not appropriate to be use by migratory and potamodromous fish species present in the Vouga basin. The redesigning and transformation of the Grela fish pass would considerably increase the available habitat for sea lamprey spawners and for the growing of ammocoetes.

Sea lamprey spawners select not very deep places with very slow current velocity and high concentrations of oxygen as refuges for the daytime, dark, either due to dense aquatic vegetation, rock formations or other factors. The existence of debris is also an important feature in the riverbed creating this kind of preferential shelters. According to the present results, the probability of an animal halting and looking for shelter reaches 50% when TDM reaches 6.4 km. This means that even if there were suitable conditions for these animals to halt before that distance, it is probable that they will not do so.

Further studies regarding the sea lamprey's preferential resting sites and migratory movement should be based on the characterization of shorter river stretches considering the halting probability for these animals. The present results reinforce the idea that the river discharge, either from natural or human origin, is a factor that acts as positive stimulus in the migration of these animals. Its application is quite important to the conservation of sea lamprey populations and must be taken in consideration by the proper institutions that manage river flow related structures such as dams and barrages and when considering the construction of fish passages.

Acknowledgements This research was financially supported by two Portuguese Institutions, the Water Institute (INAG) of the Ministry of Environment and by a Ph.D. grant from the Foundation for Science and Technology (FCT) to B.R. Quintella with the following reference SFRH/BD/9125/2002. The authors would also like to thank Gonçalo Tomaz for the useful assistance on the fieldwork.

References

Almeida, P. R., H. T. Silva & B. Quintella, 2000. The migratory behaviour of the sea lamprey *Petromyzon marinus* L., observed by acoustic telemetry in River Mondego (Portugal). In Moore, A. & I. Russel (eds), Advances in fish telemetry. CEFAS: Lowestoft, Suffolk, 99–108.

Almeida, P. R., H. T. Silva & B. R. Quintella, 2002a. The spawning migration of the sea lamprey (*Petromyzon marinus* L.), in the River Mondego (Portugal). In Pardal, M. A., J. C. Marques & M. A. Graça (eds), Aquatic ecology of the Mondego river basin. Global importance of local experience. Imprensa da Universidade de Coimbra, Coimbra, 381–386.

Almeida, P. R., B. R. Quintella & N. M. Dias, 2002b. Movement of radio-tagged anadromous sea lamprey during the spawning migration in the River Mondego (Portugal). Hydrobiologia 483: 1–8.

Almeida, P. R., B. R. Quintella, A. Koed & N. O. Andrade, 2005. Using electromyogram telemetry to study the spawning migration of sea lamprey (*Petromyzon marinus* L.). In Spedicato, M. T., G. Lembo & G. Marmulla (eds), Aquatic Telemetry, advances and applications. FAO-COISPA, Rome, 3–11.

Assis, C. A, 1990. Threats to the survival of anadromous fishes in the River Tagus, Portugal. Journal of Fish Biology 37: 225–226.

Bergstedt, R. A. & J. G. Seelye, 1995. Evidence for lack of homing by sea lampreys. Transactions of the American Fisheries Society 124: 235–239.

Costa, M. J., P. R. Almeida, I. M. Domingos, J. L. Costa, M. J. Correia, M. L. Chaves & C. M. Teixeira, 2001. Present status of the main shads' populations in Portugal. Bulletin Français de la Pêche et de la Pisciculture 362/363: 1109–1116.

Fine, J. M., L. A. Vrieze & P. W. Sorensen, 2004. Evidence that Petromyzontid lampreys employ a common migratory pheromone that is partially comprised of bile acids. Journal of Chemical Ecology 30: 2091–2110.

Hardisty, M. W. & I. C. Potter, 1971. The general biology of adult lampreys. In Hardisty, M. W. & I. C. Potter (eds), The biology of lampreys, Vol. I. Academic Press, London, 127–206.

Kelso, J. R. M. & W. M. Gardner, 2000. Emigration, upstream movement, and habitat use by sterile and fertile sea lampreys in three Lake Superior tributaries. North American Journal of Fisheries Management 20: 144–153.

Li, W, P. W., Sorensen & D. D. Gallaher, 1995. The olfactory system of migratory adult sea lamprey (*Petromyzon marinus*) is specifically and acutely sensitive to unique bile acids released by conspecific larvae. The Journal of General Physiology 105: 569–587.

Li, W., A. P. Scott, M. J. Siefkes, H. Yan, Q. Liu, S. Yun & D. A. Gage, 2002. Bile acid secreted by male sea lamprey that acts as a sex pheromone. Science 296: 138–141.

Loureiro, J. & M. E. Macedo, 1986. Bacia hidrográfica do Rio Vouga. In Direcção-Geral dos recursos e aproveitamentos hidráulicos (eds). Monografias hidrológicas dos principais cursos de água de Portugal continental. Divisão de Hidrometria, Lisboa, 207–240.

Morman, R. H., D. W. Cuddy & P. C. Rugen, 1980. Factors influencing the distribution of sea lamprey (*Petromyzon marinus*) in the Great Lakes. Canadian Journal of Fisheries and Aquatic Sciences 37: 1811–1826.

Quintella, B. R., N. O. Andrade & P. R. Almeida, 2003. Distribution, larval stage duration and growth of the sea lamprey ammocoetes, *Petromyzon marinus* L., in a highly modified river basin. Ecology of Freshwater Fish 12: 1–8.

Quintella, B. R., N. O. Andrade, A. Koed & P. R. Almeida, 2004. Behavioural patterns of sea lamprey spawning migration during difficult passage areas studied by electromyogram telemetry. Journal of Fish Biology 65: 961–972.

Rogado, L. (coord.), P. Alexandrino, P. R. Almeida, J. Alves, J. Bochechas, R. Cortes, I. Domingos, F. Filipe, J. Madeira & F. Magalhães, 2005. *Petromyzon marinus* Lampreia-marinha. In Cabral, M.J., J. Almeida, P.R. Almeida, T. Delinger, N. Ferrand de Almeida, M.E. Oliveira, J.M. Palmeirim, A.L. Queiroz, L. Rogado & M. Santos-Reis (eds), Livro vermelho dos Vertebrados de Portugal. Instituto de Conservação da Natureza, Lisboa, 65–66.

Stier, K. & B. Kynard, 1986. Movement of sea-run sea lampreys, *Petromyzon marinus*, during the spawning migration in the Connecticut river. Fishery Bulletin 84: 749–753.

Sorensen, P. W. & L. A. Vrieze, 2003. The chemical ecology and potential application of the sea lamprey migratory pheromone. Journal of the Great Lakes Research 29: 66–84.

Sorensen, P. W., J. M. Fine, V. Dvornikovs, C. S. Jeffrey, F. Shao, J. Wang, L. A. Vrieze, K. R. Anderson & T. R. Hoye, 2005. Mixture of new sulfated steroids functions as a migratory pheromone in the sea lamprey. Nature Chemical Biology 1: 324–328.

Teeter, J., 1980. Pheromone communication in sea lamprey (*Petromyzon marinus*): implications for population management. Canadian Journal of Fisheries and Aquatic Sciences 37: 2123–2132.

Vrieze, L. A. & P. W. Sorensen, 2001. Laboratory assessment of the role of a larval pheromone and natural stream odor in spawning stream localization by migratory sea lamprey (*Petromyzon marinus*). Canadian Journal of Fisheries and Aquatic Sciences 58: 2374–2385.

Young, R. J., K. A. Houston, J. G. Weise & J. R. M. Kelso, 1990. The effect of environmental variables on the population dynamics of sea lamprey, *Petromyzon marinus*. Canadian Technical Report of Fisheries and Aquatic Sciences 1736, 33 pp.

Hydrobiologia (2007) 582:133–142
DOI 10.1007/s10750-006-0550-0

FISH TELEMETRY

Habitat use and fish activity of landlocked Atlantic salmon and brook charr in a newly developed habitat compensation facility

Eva C. Enders · Karen E. Smokorowski · Curtis J. Pennell · Keith D. Clarke · Brent Sellars · David A. Scruton

Abstract Degradation and destruction of valuable spawning and rearing habitat due to anthropogenic changes (e.g., flow modification and channelisation) is known to have dramatic impacts on fish populations. To compensate for habitat losses due to hydropower development, an artificial fluvial habitat channel ('Compensation Creek') was constructed in south-central Newfoundland, Canada. The creek was designed to include appropriate habitat features for the two dominant salmonid fish species, landlocked Atlantic salmon (*Salmo salar* L.) and brook charr (*Salvenius fontinalis* Mitchell). The study examines the habitat use of landlocked Atlantic salmon and brook charr in the Compensation Creek using electromyogram (EMG) radio telemetry. Ten landlocked Atlantic salmon and eight brook charr were captured and tagged with EMG transmitters. In laboratory swimming experiments, the EMG values were calibrated against swimming speed. Fish were then released in the Compensation Creek and tracked on a daily basis. The results show that (1) during residence in the creek, both species used preferentially the habitat features designed to match their rearing habitat preferences, and (2) swimming speed did not vary among habitat types for either species.

Guest editors: P. R. Almeida, B. R. Quintella, M. J. Costa and A. Moore
Developments in Fish Telemetry

E. C. Enders (✉) · C. J. Pennell · K. D. Clarke · D. A. Scruton
Ecological Sciences Section, Fisheries and Oceans Canada, 80 East White Hills Road, P. O. Box 5667, St. John's, NL, Canada A1C 5X1
e-mail: enderse@dfo-mpo.gc.ca

K. E. Smokorowski
Great Lakes Laboratory for Fisheries and Aquatic Sciences, Fisheries and Oceans Canada, 1 Canal Drive, Sault Ste. Marie, ON, Canada P6A 6W4

B. Sellars
Newfoundland and Labrador Hydro, Columbus Drive, P. O. Box 12400, St. John's, NL Canada A1B 4K7

Keywords Habitat restoration · Fish activity · Spawning · EMG radio telemetry · Salmonid

Introduction

Hydropower is an important source of clean and reliable electrical energy and further development is therefore of increasing importance to assure supply for future energy demands. Hydropower brings clear advantages as a renewable energy resource, however, possible unfavourable effects due to the anthropogenic flow changes on fish habitat and energy expenditure must be considered (Poff et al., 1997; Saltveit et al., 2001). Key impacts of hydropower development on river systems are physical obstructions and water diversions causing separation of habitats. Denying the access and free migration between valuable habitats may lead to decline or extinction of affected fish populations (Bain et al., 1988; Cowx & Gould, 1989). To prevent potential negative impacts from alteration, disruption or destruction of fish habitat, environmental policies [e.g., European Water Framework Directive (European Parliament, 2000) and "No net loss guiding principle" of the Canadian fish habitat management policy (DFO, 1986)] require the compensation of habitat losses by either restoring or constructing habitat elsewhere, or for bypassing the obstruction. Substantial attention has been paid to assess and to minimise potential negative effects of proposed hydropower developments. However, less attention has been focussed on the actual impact on fish populations after the development (Minns, 1997). Post construction assessments focus in general on fish production in the form of biological indices such as biomass or catch per unit effort using conventional sampling methods such as passive netting and electrofishing (DesLandes et al., 1995). Ecologically critical time periods such as juvenile rearing, migration and spawning, that would provide valuable information for the evaluation of important habitats, are rarely examined (Langton et al., 1996). In this respect, electromyogram (EMG) radio-telemetry may provide a valuable tool to analyse habitat utilisation, fish movement and associated energy expenditure in situ and in real time (Cooke et al., 2004). EMG radio tags transmit a value proportional to muscle activity, allowing the continuous recording of fish location and activity by a remote receiver. The technology is well suited to study habitat utilisation and migratory behaviour of fish, in relation to prevailing environmental conditions on both spatial and temporal scales, and consequently has been applied to study fish in natural and impacted environments (Briggs & Post, 1997; Hinch & Bratty, 2000). EMG values have to be calibrated using swimming performance tests to obtain estimates on the swimming speed. EMG values are correlated to swimming speed; however, relationships may differ for individual fish and between species, which necessitates individual calibration of EMG values for each fish to obtain quantitative estimates of energy expenditure (Økland et al., 1997; Thorstad et al., 2000).

In south-central Newfoundland, Canada, an artificial fluvial habitat channel ('Compensation Creek') was constructed to compensate for loss of valuable rearing and spawning habitat due to hydropower development and the resulting diversion of water. Previous studies determined that the lost habitat was likely the main spawning area for the native salmonid fish species, and was critical for maintaining populations in the existing reservoir system (Scruton & LeDrew, 1997). Therefore, Compensation Creek was designed to include appropriate rearing and spawning habitat features for the two dominant salmonid fish species, landlocked Atlantic salmon (*Salmo salar* L.) and brook charr (*Salvenius fontinalis* Mitchell) (AMEC 2001). Using EMG radio-telemetry, habitat utilisation and movement patterns of adult landlocked Atlantic salmon and brook charr were analysed during operation of the Granite Canal Hydropower Generating Facility to determine the effectiveness of the constructed fluvial habitat.

The objectives of this study were to (1) calibrate the EMG values of landlocked Atlantic salmon and brook charr to swimming speed, (2) analyse the habitat utilisation of landlocked Atlantic salmon and brook charr in the artificial fluvial habitat channel, and (3) estimate fish activity during residence in the channel using EMG radio-telemetry.

Materials and methods

Study site

The study was conducted at Granite Canal, a hydropower generating facility (Granite Canal Generating Station, Newfoundland and Labrador Hydro) in south-central Newfoundland, Canada (48°12′ N; 56°48′ W). Water flows from Granite Lake through the hydro plant into the Southwest Arm of Meelpaeg Lake (Fig. 1). Due to the diversion of water from Granite Lake to the power plant, valuable spawning habitat was lost and Newfoundland and Labrador Hydro constructed a highly engineered Compensation Creek to provide rearing and spawning habitat for the two dominant fish species of the ecosystem (Scruton & LeDrew, 1997). Before the spawning season, from 1 September to 15 October, the hydropower plant was shut down during night (1900–0600) to attract migrating fish to the Compensation Creek for spawning. Discharge in Compensation Creek was maintained at 2.5 $m^3 s^{-1}$. The creek has a length of 1.5 km and is divided in two distinct channel types that provide habitat for landlocked Atlantic salmon and brook charr as determined from literature-derived preferences. The main channel, designed to provide habitat for landlocked Atlantic salmon, is dominated by riffle, pocket water and pool sequences with cobble-boulder substratum and has a width ranging from 15 m to 20 m (Fig. 1). Two side channels, designed to provide habitat for brook charr, are dominated by pools, runs, undercut banks and woody debris having a width of 5–8 m. During the study period, water temperature ranged from 7.4–11.4°C.

Fish surgery

From 25 September to 18 October 2004, 10 landlocked Atlantic salmon and 8 brook charr were caught with Fyke nets in the Southwest Arm of Meelpaeg Lake (1 km east of the hydropower generating facility; Fig. 1). After capture, fish were brought to the mobile field station at the Granite Canal Generating Station where fish were immediately anaesthetized (clove oil; 0.6 mg l^{-1}), weighed (g wet) and measured (fork length; cm). Fish were tagged with coded EMG transmitters. The cylindrical transmitters (model CEMG-R11-12 Lotek Wireless Inc.; 48 mm in length and 11 mm in diameter, mass in air of 8.8 g, guaranteed battery life of 19 days and model cEMG-R11-18 Lotek Wireless Inc.; 54 mm in length and 11 mm in diameter, mass in air of 10.0 g, guaranteed battery life of 50 days) were equipped with an antenna and two Teflon-coated electrodes. The end of each electrode was attached to a gold tip sensor (7 × 1 mm) that serves as electrical and mechanical contact with muscle tissue and secures the electrode in the musculature. Coded EMG transmitters sample the rectified and amplified bioelectrical potential changes between the two electrodes due to muscle contractions at a rate of 6000 Hz. An

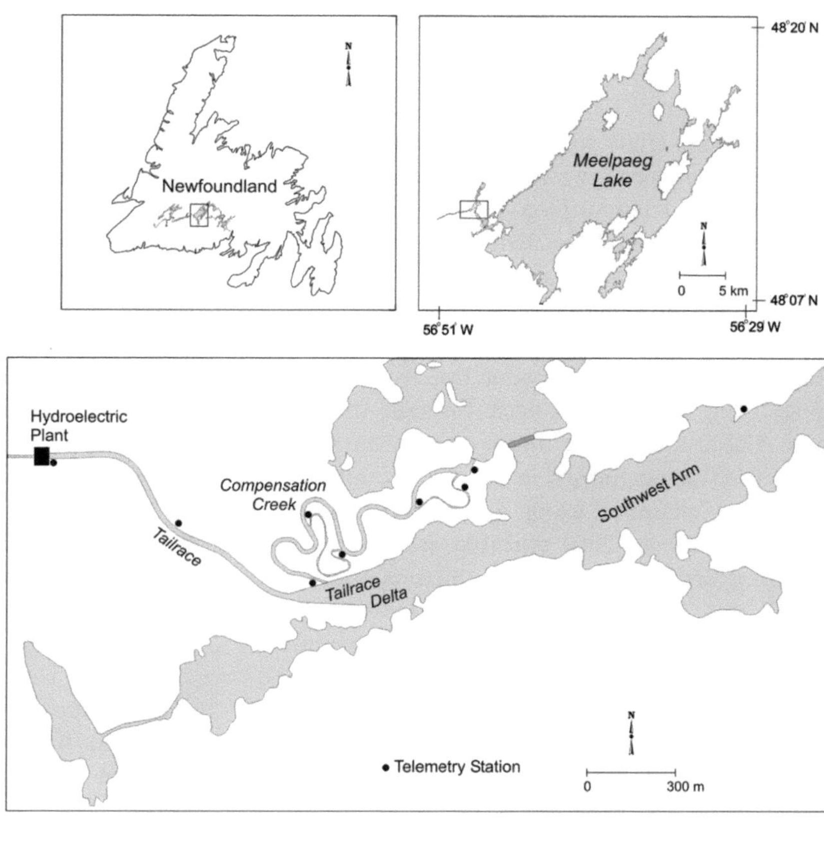

Fig. 1 Map of Compensation Creek with positions of the fixed automatic tracking stations

average voltage is calculated every second. The mean over five 1-s averages is subsequently converted in a relative activity level that ranges between 0 (no activity) and 50 (extreme activity). Due to noise and other variables, primarily the placement of the electrodes, the EMG output may vary significantly between individuals but also within individual fish.

Electromyogram transmitters were surgically implanted via a 3 cm incision on the ventral surface posterior to the pelvic girdle. The transmitter was inserted through the incision and smoothly pushed forward into the body cavity. The antenna was threaded through the body wall using a hollow needle. The two gold electrodes were inserted 10 mm apart in the narrow band of red oxidative musculature just underneath the skin using a surgical plunger device similar to that described by Bunt (1999). The incision was closed using three independent silk sutures (2-0 Sof-Silk™). Fish were allowed to recovery over night in an 800 l holding tank at the mobile field station and resting EMG values are obtained from fish observed at complete rest, without any swimming motions, at the bottom of the holding tank.

EMG calibration

Forced swim trials were used to establish relationships between swimming speed and EMG values. Swim trials were conducted using a 112 l Blazka-type swimming chamber (124 cm in length, 33 cm in internal diameter). Water flow within the swimming chamber was generated by an impeller connected to an electrical motor. The impeller drew water passing the fish from the inner tube and pushed back through the outer tube. Calibration of the motor revolutions to flow velocities was performed in the mid-section of the swimming chamber using a flowmeter (Marsh-McBirney, model 2000 portable flowmeter). For further description of the swimming chamber, see Booth et al. (1997). Fish represented less than 4% of the cross-section surface of the swimming chamber; consequently blocking effect could be ignored (Bell & Terhune, 1970; Beamish, 1978). The EMG values were recorded every 5 s using a radio telemetry receiver (Lotek Wireless Inc., model SRX_400_W32CT).

After surgery, fish were placed in the swim chamber and allowed for 2 h adaptation time under flow through conditions at a swimming speed of 10 cm s^{-1}. During each swim trial, flow velocity was increased by 20 cm s^{-1} increments every 2 min until the fish exhibited obvious fatigue and fell back to the downstream grid. Simultaneously, fish were filmed and their EMG values recorded. Data were synchronised using the time stamps on both the video recording and the radio telemetry receiver. Water temperature for the swim trials ranged between 10.0°C and 14.5°C. The video tapes were analysed frame-by-frame (30 frames s^{-1}). For each individual fish, linear regression analyses were used to obtain relationships between swimming speed and EMG values. The significance level for the statistical tests was $P \leq 0.05$.

Fish tracking

After the swimming experiments, fish were allowed to recover over night in the holding tank and were released the next morning in the Compensation Creek. Tagged fish were tracked during 27 September to 9 November 2004 using 10 fixed automatic data logging receivers (SRX_400 receiver; Lotek Wireless Inc.). The receivers were set up at strategic locations along the creek, at the tailrace and in the Southwest Arm (Fig. 1). Each station was equipped with a 3-element Yagi antenna and three underwater cable antenna (stripped coaxial cable, model RG-58/U, Belden). EMG values were recorded continuously every 5 s. Additionally, fish were manually tracked once a day using an SRX_400 receiver and a handheld H-antenna (AF Antronics Inc.). Fish position was recorded on a detailed site map (1:1000). To exclude potential effects from the surgery and the relocation, telemetry data from the first 24 h after release were not used in further data analysis.

Data analysis

Fish positions were georeferenced and transferred to ArcView 3.2 to analyse utilisation of different habitat types within Compensation Creek (Table 1.). The percentage of time fish

Table 1 Availability (%) and characterisation of different habitat categories in Compensation Creek

Habitat category	Availability (%)	Characterisation	Mean flow velocity (cm s^{-1})	Water depth (cm)
Riffle	38.1	Shallow, moderate flow velocities, moderate turbulence, boulder-cobble substrate	21.3	24.2
Pocket water	20.0	Shallow, swiftly flowing water, considerable turbulence, substrate largely composed of boulders	25.1	22.8
Pool	37.6	Deep, low flow velocities, very low turbulence, gravel substrate	10.7	94.6
Run	2.8	Moderately shallow, slow flowing water, low turbulence, cobble-gravel substrate	9.1	50.3
Undercut bank	0.9	Artificial overhang in the stream	2.0	100.0
Woody debris	0.5	Logs artificially placed in the stream to provide fish habitat	2.0	50.0

Mean flow velocity (cm s^{-1}) and water depth (cm) in specific habitat (AMEC 2005)

spent at a given habitat type was estimated and the percentage of surface area of each habitat type within the total surface area of Compensation Creek was calculated. Ivlev's electivity index was used to analyse if fish selected specific habitat types in comparison to their availability in the environment (Ivlev, 1961).

$$E_i = \frac{r_i - n_i}{r_i + n_i} \qquad (1)$$

where E_i is the Ivlev's electivity index, r_i the percentage of habitat i used by the fish and n_i the percentage of habitat i available in the environment. The electivity index varies from –1 to +1. Values between 0 and +1 indicate habitat preference whereas values between –1 and 0 indicate habitat avoidance. To compare habitat preference for the two species, all available habitat types were included in the analysis, regardless if species were located in that habitat type or not.

Results

EMG calibration in laboratory

Resting EMG values varied by 9-fold (range 1–9) between individual landlocked Atlantic salmon and by 8-fold (range 1–8) between individual brook charr. EMG values increased significantly with an increase in swimming speed (Fig. 2). With an increase of flow velocity from 10 cm s^{-1} to 90 cm s^{-1}, EMG values increased significantly by on average 4.0-fold for landlocked Atlantic salmon and buy 2.8-fold for brook charr (Table 2).

Habitat utilisation and associated swimming speed

Landlocked Atlantic salmon were observed on average over a 21 days ± 13 days time period. The corresponding time period for brook charr was 9 days ± 10 days. During that time period, on average 9737 ± 15337 EMG values, associated with fish locations, were recorded (Table 2). Nine out of ten landlocked Atlantic salmon and six out of eight brook charr were observed in more than one habitat type. During their residence in the creek, fish of each species preferentially used habitat types designed to match their habitat preferences (Table 3). Landlocked Atlantic salmon were found predominately in the main channel. On average, landlocked Atlantic salmon was observed 36% of the time in riffles, 21% in pocket waters and 43% in pools, however no significant differences in habitat use were observed between these three habitats (Kruskal Wallis test, $P = 0.46$). Nine out of 10 landlocked Atlantic salmon moved out of the Compensation Creek during the observation period. Four fish explored the tailrace leading to the power plant and two fish migrated into Meelpaeg Lake. Brook charr were observed in a wider range of habitat and significant differences were observed in habitat use (Kruskal-Wallis test, $P = 0.005$).

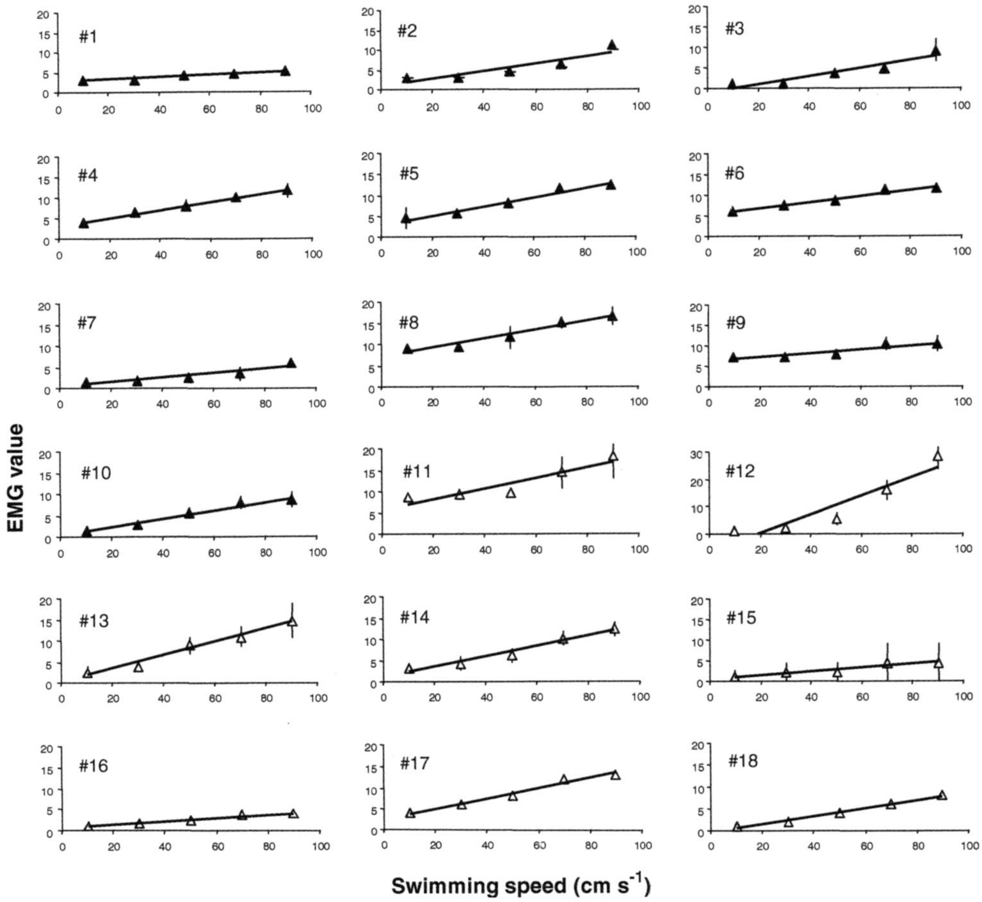

Fig. 2 Relationship between swimming speed (cm s^{-1}) and mean EMG values (+SD) for 10 landlocked Atlantic salmon (in solid triangles) and for eight brook charr (in open triangles)

Brook charr predominately used habitat features such as pools (on average 17%), undercut banks (on average 52%) and woody debris (on average 25%) in the side channels. However, significant differences were only observed in the use of under cut banks (52%) in comparison to both riffle (7%) and pocket water (5%) (Tuckey test, $P < 0.05$). During the observation period, seven out of eight brook charr moved out of the Compensation Creek.

Ivlev's electivity index showed that landlocked Atlantic salmon used riffle, pocket water and pool habitat in equal proportions to their availability in Compensation Creek (Fig. 3). Landlocked Atlantic salmon avoided habitat types found in the side channels (run, undercut bank and woody debris; $E_i < -0.95$). In contrast, brook charr avoided riffles and runs ($E_i < -0.95$) and had strong preference for undercut banks and woody debris ($E_i > 0.95$).

Mean swimming speed observed in the field for individual fish varied by 1.1- to 3.4- fold between different habitats ranging from 8.8 cm s^{-1} to 60.0 cm s^{-1} (Table 4). However, mean swimming speed did not differ between habitat types for landlocked Atlantic salmon and brook charr (Kruskal Wallis test, $P = 0.16$ and $P = 0.48$, respectively) and swimming speed ranges were highly overlapping between different habitat types (Fig. 4).

Discussion

Electromyogram telemetry has been suggested to provide valuable insights into habitat utilisation

Table 2 Data of individual landlocked Atlantic salmon (AS) and brook charr (BC), mass, fork length (L_F), experimental water temperature (T), date of surgery and swim trail, linear regression statistics of the calibration of swim speed (U in cm s^{-1}) to EMG values, time period fish were observed in the field and number of EMG values obtained in the field

Fish					Date of		Regression equation	r^2	P	Field observation		
ID	Species	Mass (g)	L_F (cm)	T (°C)	Surgery	Swim trail				df	Time period	N of EMGs
1	AS	1068	47.0	14.0	24 Sep	25 Sep	U = 30.71 EMG – 73.43	0.93	0.005	3	18 d 9 h	10160
2	AS	646	38.5	14.5	25 Sep	26 Sep	U = 8.70 EMG – 1.68	0.77	0.03	3	28 d 4 h	3027
3	AS	909	44.0	14.0	26 Sep	27 Sep	U = 9.01 EMG + 14.87	0.85	0.02	3	1 d 11 h	998
4	AS	615	39.0	13.0	27 Sep	28 Sep	U = 10.14 EMG – 30.91	0.99	<0.001	3	42 d 3 h	2194
5	AS	821	44.0	13.0	28 Sep	29 Sep	U = 8.89 EMG – 24.09	0.95	0.003	3	36 d 10 h	3533
6	AS	963	42.0	14.0	30 Sep	1 Oct	U = 13.18 EMG – 66.61	0.96	0.003	3	18 d 14 h	252
7	AS	692	41.0	14.0	30 Sep	1 Oct	U = 16.58 EMG + 1.07	0.82	0.02	3	23 d 8 h	13700
8	AS	457	35.5	12.5	1 Oct	2 Oct	U = 9.17 EMG – 62.85	0.93	0.005	3	1 d 20h	1129
9	AS	1363	50.0	12.0	6 Oct	7 Oct	U = 18.40 EMG – 107.31	0.83	0.02	3	11 d 19 h	1311
10	AS	588	38.0	12.5	29 Sep	30 Sep	U = 10.04 EMG – 2.63	0.96	0.003	3	33 d 2 h	29297
11	BC	306	30.0	14.0	23 Sep	24 Sep	U = 7.03 EMG – 33.97	0.82	0.02	3	3 d 23 h	1410
12	BC	255	29.0	13.0	26 Sep	27 Sep	U = 2.54 EMG + 23.02	0.83	0.02	3	2 d 4 h	3487
13	BC	231	29.0	12.5	26 Sep	27 Sep	U = 6.28 EMG – 1.07	0.96	0.002	3	21 d 22 h	2410
14	BC	330	30.0	13.0	27 Sep	28 Sep	U = 8.01 EMG – 7.72	0.95	0.003	3	4 d 22 h	12882
15	BC	278	30.0	13.0	28 Sep	29 Sep	U = 20.30 EMG – 7.86	0.86	0.02	3	5 d 0 h	16238
16	BC	539	38.0	12.5	29 Sep	30 Sep	U = 24.18 EMG – 12.11	0.95	0.003	3	31 d 0 h	63410
17	BC	380	33.5	12.7	17 Oct	18 Oct	U = 8.11 EMG – 19.79	0.96	0.002	3	3 d 3 h	6812
18	BC	432	36.0	11.3	18 Oct	19 Oct	U = 10.98 EMG + 3.90	0.98	<0.001	3	4 d 19 h	3025

Table 3 Mean and SE of time (in %) landlocked Atlantic salmon (AS) and brook charr (BC) were observed in specific habitat types in Compensation Creek during the tracking period

Fish ID	Species	Riffle	Pocket water	Pool	Run	Undercut bank	Woody debris
Mean	AS	36.0	21.1	42.8	0.1	0.0	0.0
SE	AS	10.1	8.9	10.3	0.1	0.0	0.0
Mean	BC	2.3	1.1	16.2	3.5	52.6	22.2
SE	BC	1.5	0.9	9.2	2.4	15.1	10.9

and associated energy expenditures of fish (Cooke et al., 2004). In the present study, EMG telemetry was used to identify the utilisation of specific habitat types of landlocked Atlantic salmon and brook charr in a newly developed fluvial habitat. Landlocked Atlantic salmon

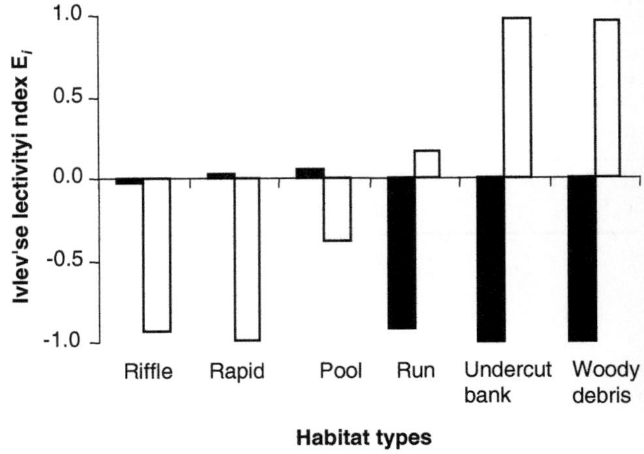

Fig. 3 Selection of different habitat types expressed as Ivlev's electivity index for landlocked Atlantic salmon (solid bars) and brook charr (open bars)

preferred faster flow and deeper pools in the main channel, whereas brook charr was mainly found in undercut banks and woody debris in the side channels of Compensation Creek. These results suggest that the constructed fluvial habitat features, based on knowledge of the appropriate rearing and spawning habitats for the two dominant salmonid fish species, are being preferentially utilized, as designed. No difference in mean swimming speed were observed in relation to different habitat types, which is most likely due to the small differences in flow velocities among different habitat types.

In the present study, coded EMG tags (cEMG-R11-12 and cEMG-R11-18, Lotek Wireless Inc.) have been used. These tags measure the bioelectrical voltage due to muscle activity over a 5 s period and then calculate a mean value over this period. The mean value is subsequently converted to an EMG value that ranges from 0 to 50. This EMG value is transmitted to the radio-telemetry receiver. Although the theoretical range of EMG values ranges from 0 to 50, EMGs observed in Compensation Creek rarely differed by more than 3-fold from the resting EMG of the corresponding fish. The small ranges of EMG values observed in this study may suggest that fish effectively chose microhabitats that allowed them to reduce energy expenditures associated with movements. Unfortunately, the algorithm of the conversion from the bioelectrical voltage to the EMG values is unknown. Therefore, the coded EMG tag appears to be a 'black box', where it is impossible to evaluate what the voltage and/or the realistic range of EMG values are for any given fish. Consequently, the conventional EMG transmitter may provide more biological meaningful data than the coded EMG transmitter. The integration of bioelectrical voltage over 5 s may further attenuate the effects of short term burst swimming motions, which may last only one second. The analyses of 1 s averages of EMG values may therefore lead to an increased range of EMG values.

In conclusion, EMG telemetry provided insights into habitat utilisation of landlocked Atlantic salmon and brook charr in a newly developed artificial stream. The technique provided valuable activity estimates. Differences were observed in the habitat preferences of both species. However,

Table 4 Mean and SE of swimming speed (cm s^{-1}) in specific habitat types for landlocked Atlantic salmon (AS) and brook charr (BC)

Fish ID	Species	Riffle	Pocket water	Pool	Run	Undercut bank	Woody debris
Mean	AS	27.9	37.4	25.2	–	–	–
SE	AS	4.6	4.2	2.3	–	–	–
Mean	BC	34.9	25.2	23.1	23.2	28.6	23.2
SE	BC	8.6	0.0	2.0	0.2	2.3	1.5

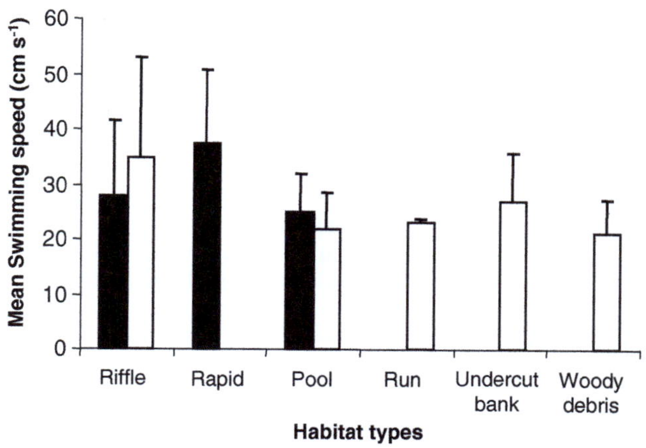

Fig. 4 Mean (±) swimming speed of individual means in different habitat types for landlocked Atlantic salmon (solid bars) and brook charr (open bars)

mean swimming speeds did not differ significantly between the habitat types.

Acknowledgement We thank C. Kelly (Fisheries and Oceans Canada) for assistance in the field, J. McCarthy (AMEC) for information on spawning activity and rearing habitat, S. Finucan (Ontario Ministry of Natural Resources) for providing telemetry equipment and the Environmental Services and Properties Department of Newfoundland and Labrador Hydro for in-kind and financial support at the Granite Canal Hydro Development. This research was funded by Natural Resources Canada's (NRCan) Panel for Energy Research and Development (PERD), Northeast Science and Information Section of the Ontario Ministry of Natural Resources and Brascan Power Limited. E.C. Enders was supported by a Visiting Fellowship in Canadian Government Laboratories from the Natural Sciences and Engineering Research Council of Canada (NSERC).

References

AMEC, 2001. Granite Canal Hydroelectric Development Fish Habitat Compensation Plan. Prepared for Newfoundland and Labrador Hydro, St. John's, NL.

Bain, M. B., J. T. Finn & H. E. Booke, 1988. Streamflow regulation and fish community structure. Ecology 69: 382–392.

Beamish, F. W. H., 1978. Swimming Capacity. In Hoar W. S. & J. R. Randall (eds), Fish Physiology—Locomotion, vol. VII. Academic Press, New York, 101–187.

Bell, W. H. & L. D. B. Terhune, 1970. Water tunnel design for fisheries research. Fisheries Research Board of Canada Technical Report 195: 1–69.

Booth, R. K., R. S. McKinley, F. Økland & M. M. Sisak, 1997. In situ measurement of swimming performance of wild Atlantic salmon (*Salmo salar*) using radio transmitted electromyogram signals. Aquatic Living Resources 10: 213–219.

Briggs, C. T. & J. R. Post, 1997. In situ activity metabolism of rainbow trout (*Oncorhynchus mykiss*): Estimates obtained from telemetry of axial muscle electromyograms. Canadian Journal of Fisheries and Aquatic Sciences 54: 859–866.

Bunt, C. M., 1999. A tool to facilitate implantation of electrodes for electromyographic telemetry experiments. Journal of Fish Biology 55: 1123–1128.

Cooke, S. J., E. B. Thorstad & S. G. Hinch, 2004. Activity and energetics of free-swimming fish: insights from electromyogram telemetry. Fish and Fisheries 5: 21–52.

Cowx, I. G. & R. A. Gould, 1989. Effects of stream regulation on Atlantic salmon, *Salmo salar* L., and brown trout, *Salmo trutta* L., in the upper Severn catchment, UK. Regulated Rivers: Research & Management 3: 235–245.

DesLandes, J. C., S. Guenette, Y. Prairie, D. Roy, R. Verdon & R. Fortin, 1995. Changes in fish populations affected by the construction of the La Grande complex (Phase I), James Bay region, Quebec. Canadian Journal of Zoology 73: 1860–1877.

DFO, 1986. The Department of Fisheries and Oceans policy for the management of fish habitat. DFO, Ottawa.

European Parliament & Council of the European Union, 2000. Directive 2000/60/EC of the European Parliament and of the Council of 23 October 2000 establishing a framework for Community action in the field of Water Policy. Official Journal of the European Communities, L 327/1–72.

Hinch, S. G. & J. Bratty, 2000. Effects of swim speed and activity pattern on success of adult sockeye salmon migration through an area of difficult passage. Transactions of the American Fisheries Society 129: 598–606.

Ivlev, V. S., 1961. Experimental ecology of the feeding in fishes. Yale University Press, NewHaven.

Langton, R. W., R. S. Steneck, V. Gotceitas, F. Juanes & P. Lawton, 1996. The interface between fisheries research and habitat management. North American Journal of Fisheries Management 16: 1–7.

Minns, C. K., 1997. Quantifying 'no net loss' of productivity of fish habitats. Canadian Journal of Fisheries and Aquatic Sciences 54: 2463–2473.

Økland, F., B. Finstad, R. S. McKinley, E. B. Thorstad & R. K. Booth, 1997. Radio-transmitted electromyogram signals as indicators of physical activity in Atlantic salmon. Journal of Fish Biology 51: 476–488.

Poff, N. L., J. D. Allan, M. B. Bain, J. R. Karr, K. L. Prestegaard, B. D. Richter, R. E. Sparks & J. E. Stromberg, 1997. The natural flow regime: a paradigm for river conservation and restoration. Bioscience 47: 769–784.

Saltveit, S. J., J. H. Halleraker, J. V. Arnekleiv & A. Harby, 2001. Field experiments on stranding in juvenile Atlantic salmon (*Salmo salar*) and brown trout (*Salmo trutta*) during rapid flow decreases caused by hydropeaking. Regulated Rivers: Research & Management 17: 4–5.

Scruton, D. A. & L. J. LeDrew, 1997. A retrospective assessment of the flow regulation of the West Salmon River, Newfoundland, Canada. Fisheries Management & Ecology 4: 467–480.

Thorstad, E. B., F. Økland & B. Finstad, 2000. Effects of telemetry transmitters on swimming performance of adult Atlantic salmon. Journal of Fish Biology 57, 531–535.

Hydrobiologia (2007) 582:143–153
DOI 10.1007/s10750-006-0561-x

FISH TELEMETRY

Mobility of individual roach *Rutilus rutilus* (L.) in three weir-fragmented Belgian rivers

Caroline Geeraerts · Michaël Ovidio · Hilde Verbiest · David Buysse · Johan Coeck · Claude Belpaire · Jean-Claude Philippart

Abstract Adult roach *Rutilus rutilus* (L.) ($N = 24$; 19.9–36.1 cm FL) from three highly fragmented Belgian rivers were tagged with surgically implanted radio transmitters. Their seasonal movements were observed from March to August 2004 (circum reproduction period) in river stretches delimited by two physical barriers. In the three rivers, roach displayed similar patterns of movements which were mainly influenced by the date of observation (movements increased in late April–May) and water temperature (travel distances were more important when water temperature ranged between 10°C and 14°C). Roach sometimes cleared physical obstacles. The mean distances travelled in each river were relatively short (max. 2.5 km) and mainly influenced by the length of the study reach, which was delimited by physical barriers.

Guest editors: P. R. Almeida, B. R. Quintella, M. J. Costa and A. Moore
Developments in Fish Telemetry

C. Geeraerts (✉) · H. Verbiest · D. Buysse · J. Coeck · C. Belpaire
Research Institute for Nature and Forest, Duboislaan 14, 1560 Groenendaal (Hoeilaart), Belgium
e-mail: caroline.geeraerts@inbo.be

M. Ovidio · J.-C. Philippart
Biology of Behaviour Unit, Laboratory of Fish Demography and Hydro ecology, University of Liège, 10 Chemin de la Justice, 4500 Tihange, Belgium

Keywords Roach · Migration · Telemetry · Barrier · Fragmentation · Seasonal movements

Introduction

Roach *Rutilus rutilus* (L.) is a dominant cyprinid fish in many rivers throughout the Eurasian mainland. It is common in rivers, lakes, canals and reservoirs, in streaming as well as in standing waters. It is a benthopelagic, potamodromous species that can survive in poor quality and fragmented rivers. It is a non-rogue fish, which lives in schools and forages in various environments such as the neighbourhood of aquatic vegetation but also in open water. Regardless of its strong tolerance to water disturbance, roach populations have rarely been protected and minimally studied (Vandelannoote et al., 1998).

Information on the behaviour and mobility of individual roach is necessary to understand its ecology and adaptive behaviour to survive in highly disturbed environments. Earlier studies during the 1960s in lakes and rivers suggested that roach move over short distances (Williams, 1965; Stott, 1967) but more recent studies demonstrated that roach sometimes migrate over several kilometers (Baade & Fredrich, 1998) and display homing behaviour to spawning areas (Goldspink, 1977; L'Abée-Lund & Vøllestad, 1985). The objective of this study is to analyse the seasonal movements of individual roach in three highly fragmented rivers in Belgium and to observe how roach adapt their use of space in river stretches delimited by up- and downstream physical obstacles.

Materials and methods

Study area

The study was performed simultaneously in three rivers in northern and southern Belgium (Fig. 1). The River Kleine Nete is a small lowland river in the northern part of Belgium, which belongs to the bream zone (Huet, 1949) of the River Scheldt basin (Fig. 1(1)). It is a 50 km long river with a drainage area of 815 km^2. The River Kleine Nete has been canalised and straightened to improve drainage of the valley. Actually five weirs and two siphons fragment the main course of the Kleine Nete. The study site was situated in the slow flowing middle reach of the river in between weirs A and B (Fig. 1(1)). Weir B is equipped with a fish pass. The distance between these weirs is 7 km. One tributary, the River Aa, enters into the Kleine Nete just upstream of the weir A. About 30 fish species are found in the River Kleine Nete of which roach is the most abundant. Annual discharge in the study reach ranges between 2.1 and 26.3 m^3 s^{-1} and water temperature ranges between 0°C and 23°C (data from 1997 to 1999).

The River Grote Nete is a 60 km long small lowland river in the northern part of Belgium with a drainage area of 730 km^2. The river also belongs to the River Scheldt basin (Fig. 1(2)) and is characterised as a bream zone (Huet, 1949). The upper part of the River Grote Nete has a strong meandering course while the middle and lower reach of the river is canalised and straightened. Six physical obstacles and four siphons

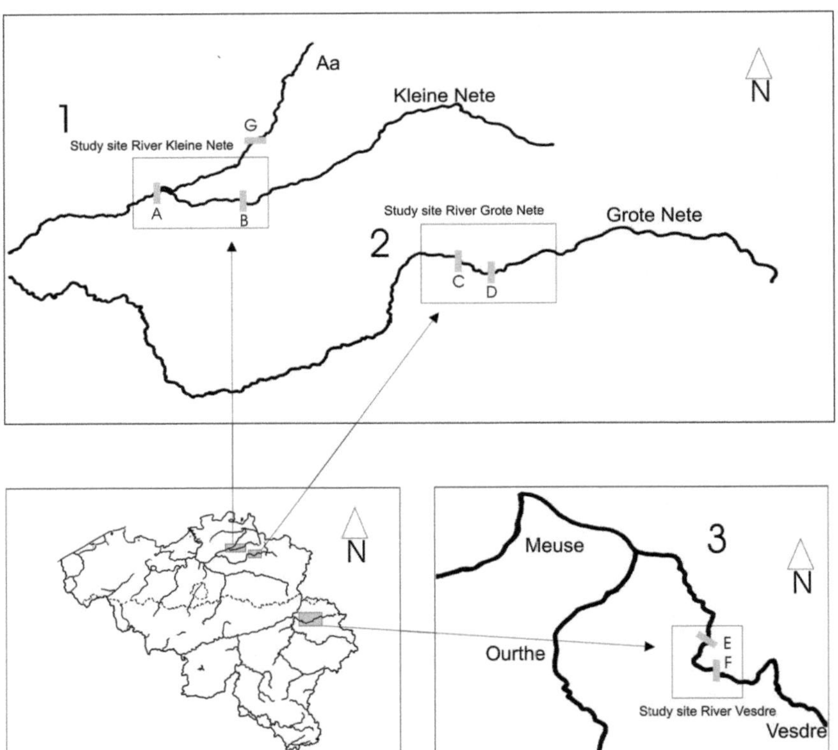

Fig. 1 Location of the three study areas in Belgium: (**1**) the River Kleine Nete, (**2**) the River Grote Nete and (**3**) the River Vesdre. Transverse bars represent the barriers that may interfere with the free circulation of fish. (A) weir Grobbendonk, (B) weir Herentals equipped with a fish pass, (C) weir Meerhout, (D) water mill Meerhout, (E) weir Chaudfontaine, (F) weir Hauster and (G) weir on the River Aa, a tributary of the River Kleine Nete

fragment the main course of the Grote Nete. The study site was situated in the slow flowing middle reach of the river in between weir C and water mill D (Fig. 1(2)). The distance between these obstacles is 3 km. The fish community consists of 25 species. Gudgeon *Gobio gobio* (L.) is the most abundant species, followed by roach. Annual discharge, ranges between 0.3 and 5.4 $m^3 s^{-1}$ and water temperature ranges between 0°C and 23°C (data from 1997 to 1999).

The River Vesdre is a 72 km long upland river in the southern part of Belgium and is part of the Meuse basin (sub-basin: 702 km^2) (Fig. 1(3)). Twenty-nine physical obstacles fragment the main course of the River Vesdre. From 1850 to 1998 the Vesdre was highly polluted. Since 1998 water quality has improved due to the construction of water purification stations. The study site was situated in the lower reach of the river in between obstacles E and F (Fig. 1(3)). The distance between these obstacles is 1.2 km. In the lower part of the Vesdre, the fish assemblage is a mix of the barbel/grayling zone (Huet, 1949) and consists of 20 species. Annual discharge ranges between 4 and 120 $m^3 s^{-1}$ (data from Water Division, D.G.R.N.E.) and water temperature ranges between 0°C and 23°C.

Capture, tagging and tracking

In each river system $N = 8$ roach (>150 mm FL and >150 g) were sampled by DC electric fishing from 12 March to 8 April 2004 (Table 1). In the rivers Kleine and Grote Nete, roach were captured at different sites in the study reach. In the River Vesdre, roach were captured at the same place. Roach were anaesthetised in a solution of 2-phenoxy-ethanol (0.2 mg L^{-1}), and 40–42 MHz internal implant radio transmitters (ATS Inc.) were inserted into the body cavity of the fish through a midventral incision (Ovidio & Philippart, 2002). In order to avoid any adverse effect of long-term post-operation care on their behaviour, the fish were released at their exact capture site as soon as they had recovered and showed spontaneous swimming activity (about 20–30 min after surgery).

Tracking started the day after tagging. Locations were recorded during daylight, by triangulation using an ATS R2000 or Fieldmaster receiver with a loop antenna (ATS Inc.) from labelled marks lining the banks of the river. Roach were located daily in the rivers Grote Nete and Kleine Nete and located 2–5 times a week in the River Vesdre. Fish were tracked during a maximum of 160 days in the Grote Nete and Kleine Nete and during 114 days in the Vesdre (from March to July or August) in order to include the pre-spawning, spawning and post-spawning period of the species (Vøllestad & L'Abée-Lund, 1987; Baade & Fredrich, 1998). Water temperature was monitored hourly by data loggers (TidBit, Onset Computer Corp.) and water flow was recorded hourly (data from the Water Division) in each river. To take into account the differences in intervals between fish locations in the three rivers, the movements were standardised to 'weekly mean distance travelled'

Table 1 Characteristics of tracked roach and specifications of implanted miniature radio-transmitters

	River Kleine Nete	River Grote Nete	River Vesdre
Fish number	1–8	9–16	17–24
Mean length, (min–max) (mm)	239 (199–361)	234 (205–315)	232 (215–252)
Mean weight (min–max) (g)	204 (156–308)	225 (196–559)	238 (172–295)
Sex ratio	4 females/4 males	3 females/5 males	4 females/4males
Capture dates	22/03–08/04/2004	12–17/03/2004	23/03/2004
Tracking duration (days)	34–140	17–160	11–114
Transmitter model	ATS-F1170	ATS-F1470	ATS-F1580
Transmitter weight in air (g)	4	4	3.6
(% of fish body mass) (min–max)	2.0 (2.6–1.3)	1.8 (2.0–1.0)	1.5 (2.1–1.2)
Transmitter length (mm)	31	31	25
Antenna	Internal coiled	Internal coiled	External (15 mm)
Battery life (days)	200	390	390

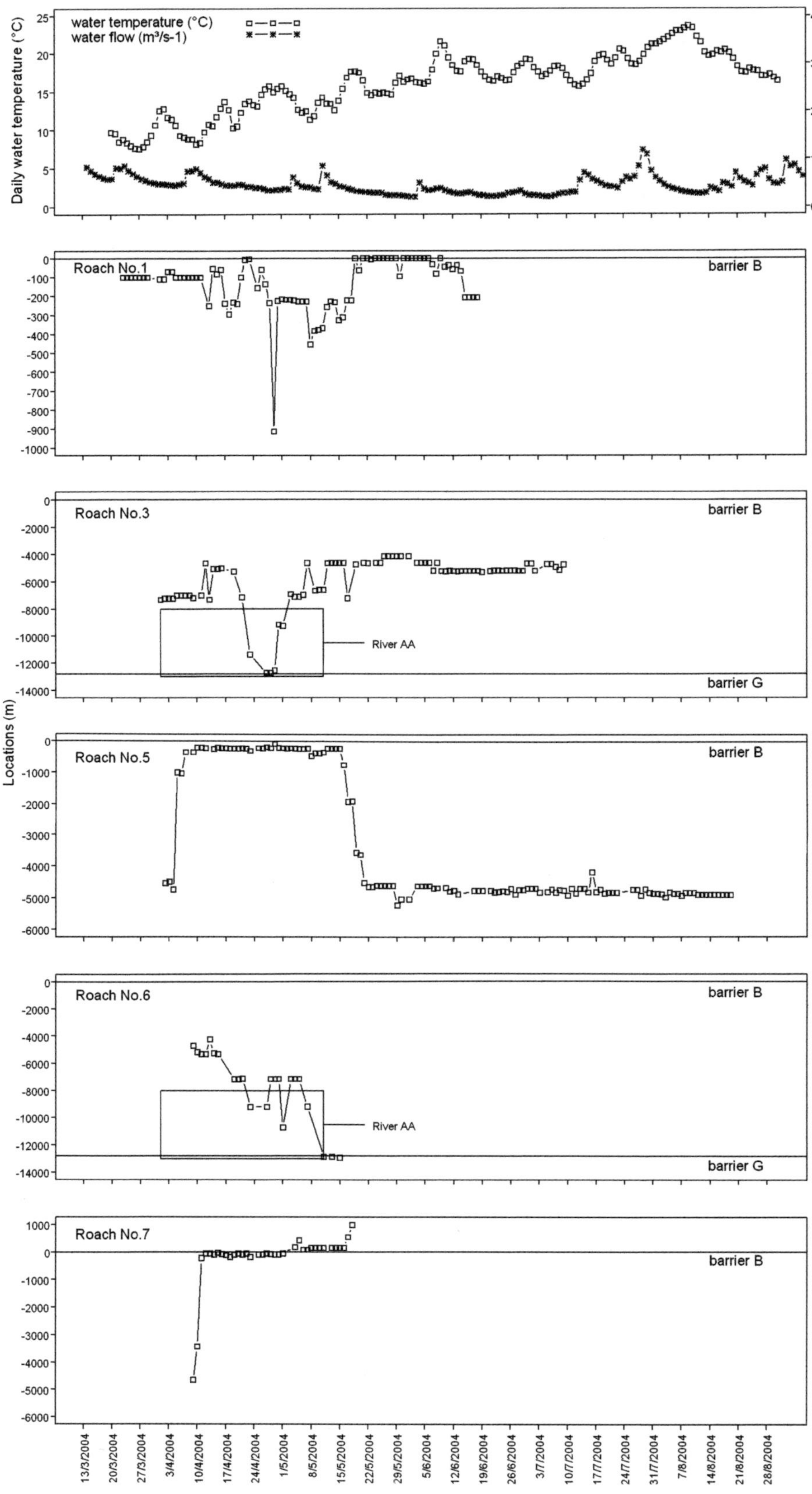
water temperature (°C)
water flow (m³/s-1)
Daily water temperature (°C)
Daily water flow (m³/s)
Roach No.1
barrier B
Roach No.3
River AA
barrier G
Locations (m)
Roach No.5
Roach No.6
Roach No.7
13/3/2004
20/3/2004
27/3/2004
3/4/2004
10/4/2004
17/4/2004
24/4/2004
1/5/2004
8/5/2004
15/5/2004
22/5/2004
29/5/2004
5/6/2004
12/6/2004
19/6/2004
26/6/2004
3/7/2004
10/7/2004
17/7/2004
24/7/2004
31/7/2004
7/8/2004
14/8/2004
21/8/2004
28/8/2004

◀ **Fig. 2** Daily variations of water temperature (°C) and water flow (m^3 s^{-1}) in the River Kleine Nete from 13 March to 31 August 2004. Positions (m) of 5 radio-tagged roach in the River Kleine Nete. The point 0 corresponds to the most upstream barrier of the study reach. R2, R4 and R8 are not represented. R4 was lost and R2 and R8 show similar behaviour as R3

to perform statistical analysis. It corresponds to the mean distance travelled by roach during a one-week interval.

Results

The individual movements of representative tagged roach in the three rivers are illustrated in Fig. 2 (Kleine Nete), Fig. 3 (Grote Nete), and Fig. 4 (Vesdre).

At the beginning of April, when the water temperature approached 12°C, roach in the River Kleine Nete demonstrated two distinct behavioural patterns (Fig. 2). (i) Three out of eight roach (R1, R5, R7) showed increased mobility and were frequently located in the upstream part of the study reach, in a faster flowing zone 200 m downstream of the weir and fish pass (obstacle B). They stayed there until the beginning of May, after which they homed to their capture site. Roach R7 made an upstream movement and cleared the fish pass (of obstacle B) after which it was lost. (ii) The other individuals (R2, R3, R6, R8) stayed close to their capture site until water temperature rose above 14°C. They then made a downstream migration (about 3–6 km) into a tributary, the River Aa (Fig. 1(1)). Roach R3 and R8 made several up- and downstream movements before returning at the beginning of May to a location nearby their capture site in the River Kleine Nete. Roach R6 continued ascending the River Aa and cleared a temporarily flat lying weir (obstacle G) after which it was lost. After this period when the water temperature rose to 23°C, roach showed lower mobility and the net length of the daily journeys generally decreased. Roach R2 and R8 are not represented in Fig. 2 but had similar movement patterns as described in (ii). Roach R4 was lost for unknown reasons and its movements are not described.

From late March to mid-April, when the water temperature increased from 7°C to 13°C, the roach (R9, R10, R11, R13 and R14) in the River Grote Nete exhibited frequent up- and downstream movements (max. 860 m between two locations) alternated with periods of lower mobility (Fig. 3). From 14 April to 29 May, when the water temperature varied from 11°C to 18°C, roach showed increased mobility and were frequently located in the upstream part of the study reach, in a faster flowing zone 500 m downstream the water mill (obstacle D). After this period, when the water temperature rose to 22°C, roach frequently moved between two locations but the net length of the daily journeys generally decreased. Roach R10 and R13, contrary to the other roach, stayed in an upstream part of the river. Roach R12, R15, R16 are not represented in Fig. 3. Roach R12 was lost after a month. Roach R15 and R16 showed similar movement patterns as described for roach R9 and R10, respectively.

In the River Vesdre, roach R18, R20 and R23 were lost for unknown reasons. Until late April, in the River Vesdre, roach demonstrated low mobility and were mainly concentrated near their capture site (Fig. 4). However, some individuals displayed up- and downstream movements during increased flow conditions. When water temperature rose above 12°C, in late April, roach demonstrated two main behavioural patterns. (i) Some individuals (R17 and R24) moved downstream. Roach R17 made a fast downstream movement of about 1,700 m on 22 April 2004 and passed the Chaudfontaine weir (obstacle E). Roach R24 moved upstream over 300 m on 26 April and immediately started a progressive downstream migration of about 2,000 m that finished on 10 May. Then it moved upstream over 1 km to a faster flowing part of the river situated downstream the obstacle E. (ii) The other individuals stayed in the 1.2 km long stretch between the

Fig. 3 Daily variations of water temperature (°C) and water flow (m^3 s^{-1}) in the River Grote Nete from 13 March to 31 August 2004. Positions (m) of 5 radio-tagged roach in the River Grote Nete. The point 0 corresponds to the most upstream barrier of the study reach. R12, R15 and R16 are not represented. R12 was lost and R15 and R16 show similar behaviour as R9 and R10 respectively ▶

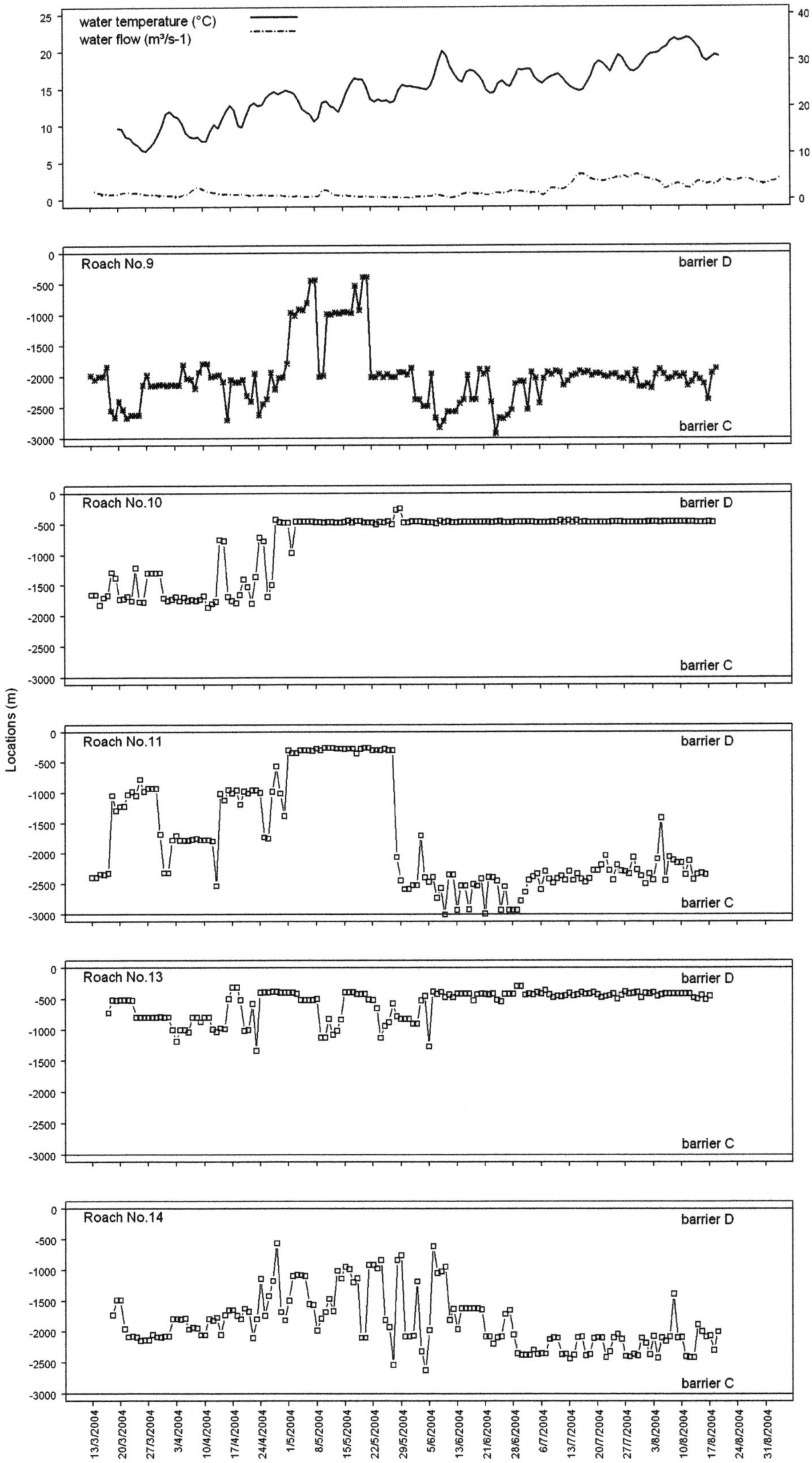
water temperature (°C)
water flow (m³/s-1)
Roach No.9
Roach No.10
Roach No.11
Roach No.13
Roach No.14
barrier D
barrier C
Locations (m)
13/3/2004
20/3/2004
27/3/2004
3/4/2004
10/4/2004
17/4/2004
24/4/2004
1/5/2004
8/5/2004
15/5/2004
22/5/2004
29/5/2004
5/6/2004
13/6/2004
21/6/2004
28/6/2004
6/7/2004
13/7/2004
20/7/2004
27/7/2004
3/8/2004
10/8/2004
17/8/2004
24/8/2004
31/8/2004

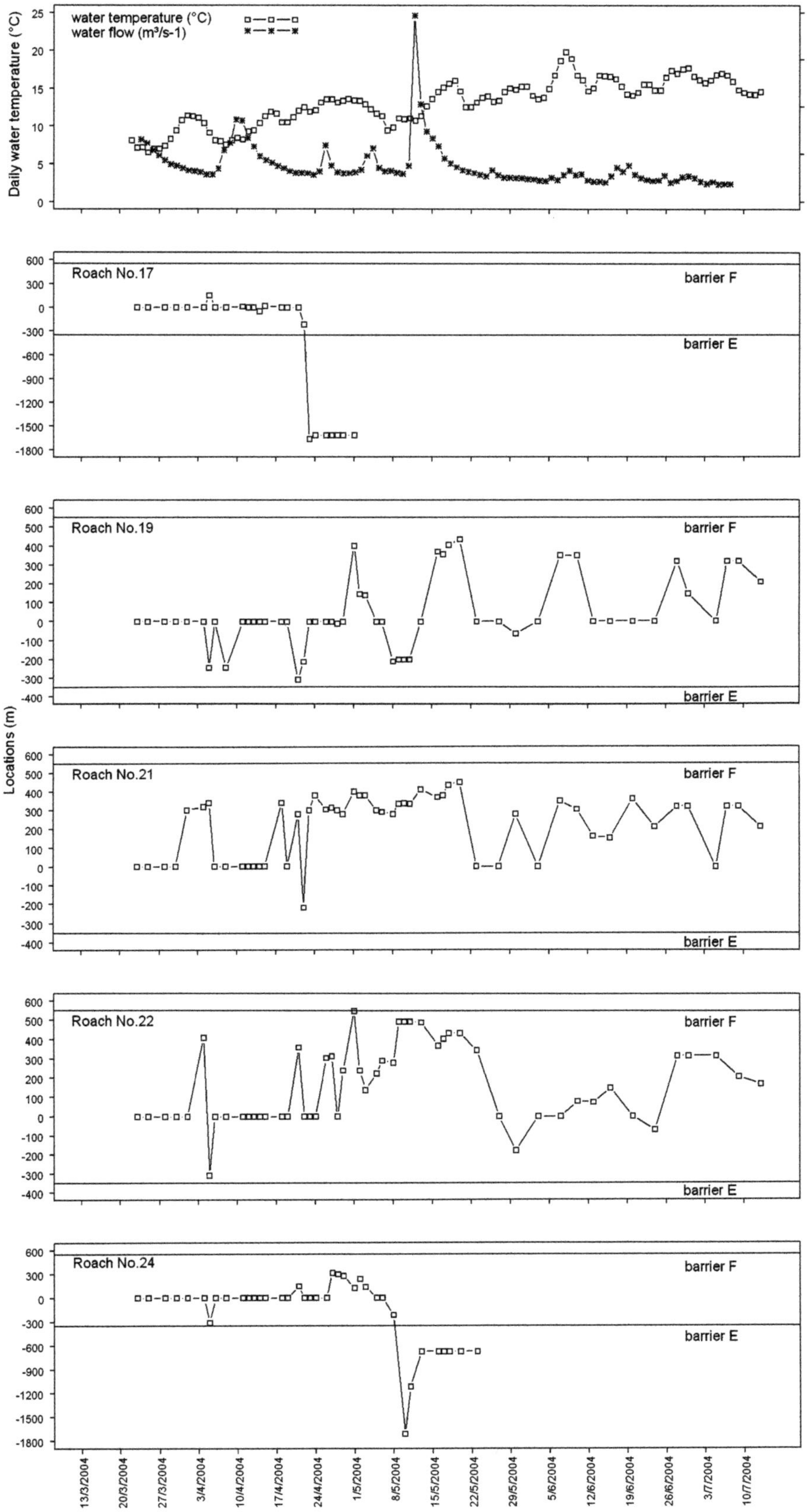
water temperature (°C)
water flow (m³/s-1)
Daily water temperature (°C)
Daily water flow (m³/s)
Roach No.17
Roach No.19
Roach No.21
Roach No.22
Roach No.24
barrier F
barrier E
Locations (m)
13/3/2004
20/3/2004
27/3/2004
3/4/2004
10/4/2004
17/4/2004
24/4/2004
1/5/2004
8/5/2004
15/5/2004
22/5/2004
29/5/2004
5/6/2004
12/6/2004
19/6/2004
26/6/2004
3/7/2004
10/7/2004

◀ **Fig. 4** Daily variations of water temperature (°C) and water flow (m^3 s^{-1}) in the River Vesdre from 13 March to 10 July 2004. (2) Positions (m) of 5 radio-tagged roach in the River Vesdre. The point 0 corresponds to the capture site of the roach. Roach R18, R20 and R23 were lost for unknown reasons and were not represented

obstacles E and F. In mid-May, when the water temperature increased and reached 14°C for the first time, roach R19, R21 and R22 showed increased mobility and were frequently located in the upstream part of the study reach in a faster flowing zone of 300 m downstream the obstacle F. After this period, when the water temperature rose to 20°C, roach continued to move between the downstream and upstream parts of the study area but the net length of the daily journeys generally decreased and they never cleared the obstacles in upstream direction.

When grouping the weekly mean distance travelled by roach in the three rivers (Fig. 5), it appeared that movements increased from the beginning of April until mid-May when water temperature fluctuated between 10°C and 14°C. From mid-May, when water temperature rose above 14°C and the water flow decreased, the weekly mean distance travelled by the fish decreased and roach were less active. On 25 July and 15 August, two striking peaks (in weekly mean distance) corresponded to increased upstream distances travelled by three fish in the rivers Kleine Nete and Grote Nete (Fig. 5).

Roach movements were most important in the 10–14°C water temperature intervals that mainly occurred during April and May (potential spawning period) except for the two peaks in July and August. A significant difference in the weekly mean distance travelled was observed when grouping the roach movements (of the three water courses) into three categories of water temperature (<10°C; 10–14°C; >14°C; Kruskal–Wallis, $p < 0.05$). Similar analyses were performed to test the influence of water flow on roach movements, but no statistical relationship was found.

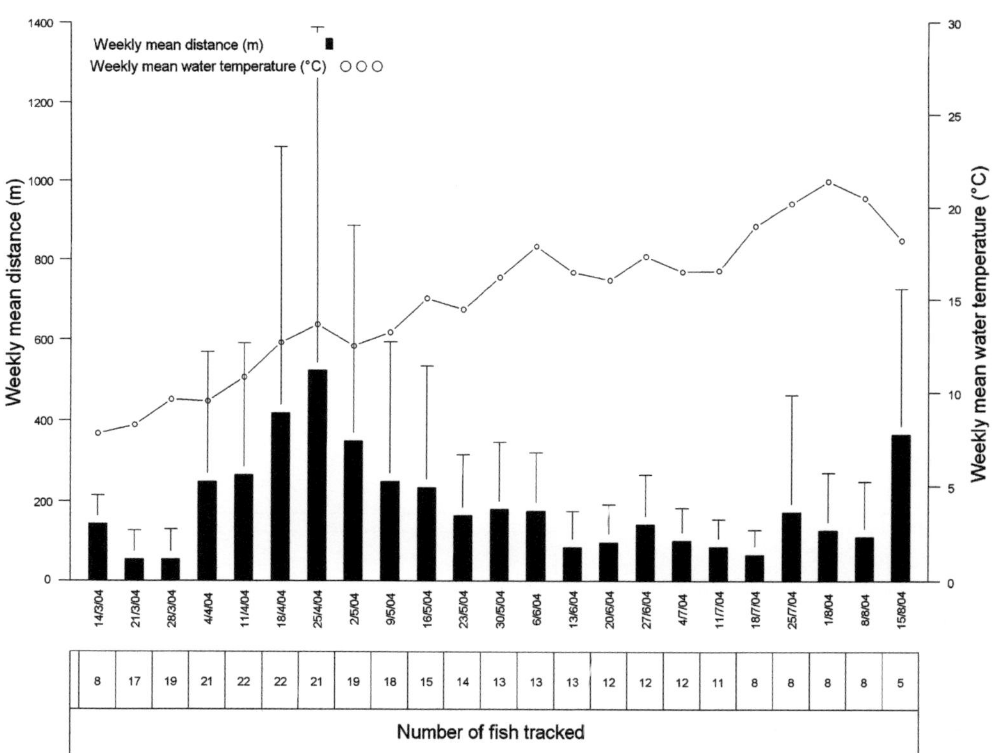

Fig. 5 Weekly mean distance (m) travelled by roach, in the three rivers, in relation to the mean weekly water temperature (°C) of the three rivers. Error bars represent standard deviation

The weekly mean distances travelled by roach during the overall study period were different between rivers (Fig. 6). In the River Kleine Nete, where fish migration is unobstructed for the first 14 km of the reach (first barrier in the River Aa), the mean distance travelled was about 475 m (Fig. 6). In the rivers Grote Nete and Vesdre, the free migration reach is limited to respectively 3 and 1 km, by which the weekly mean distance travelled diminished to 145 and 86 m, respectively. Weekly mean distances travelled were significantly different between the rivers Kleine Nete and Grote Nete ($p < 0.0001$; Scheffe *f*-test) and between the rivers Kleine Nete and Vesdre ($p < 0.0001$; Scheffe *f*-test). No differences were observed between the rivers Grote Nete and Vesdre ($p = 0.48$; Scheffe *f*-test).

Discussion

Transmitter weight, as a percentage of the fish's body mass, is the primary constraint to limit the effect of the tagging procedure on fish behaviour (Mulcahy, 2003). A commonly cited recommendation (Winter, 1983) is to limit the weight of internal transmitters to ≤2% of the fish's body mass. However, recent studies demonstrated that, at least for some species, this value might be largely increased, sometimes up to 12% (Jepsen et al., 2002). In this study, in order to minimise the potential effect of the tag on roach behaviour, the transmitters represented a maximum of 1.5% of the fish body mass.

The duration and dynamics of the roach movements were generally quite similar between individuals in the tracked upland and lowland rivers. Roach showed maximum activity from the beginning of April until the end of May and they were frequently located in the faster flowing parts of their study reaches during that time period. Even though spawning activity could not be observed, it can be assumed that these movements were related to spawning activity as this period corresponds to the reproduction time of the species in similar environments (Vøllestad & L'Abée-Lund, 1987; Baade & Fredrich, 1998, Lucas et al., 1998; Poncin, 1994). Fast flowing zones in rivers were already described as potential spawning areas for roach (Holcik & Hruska, 1966; Diamond, 1985). Outside of this period (during the pre- and post-spawning period) roach frequently moved between different locations but the net length of the daily journeys were generally smaller. Distances travelled by roach increased significantly when water temperature varied between 10°C and 14°C, which also corresponds to the late April-May period. Baade & Fredrich

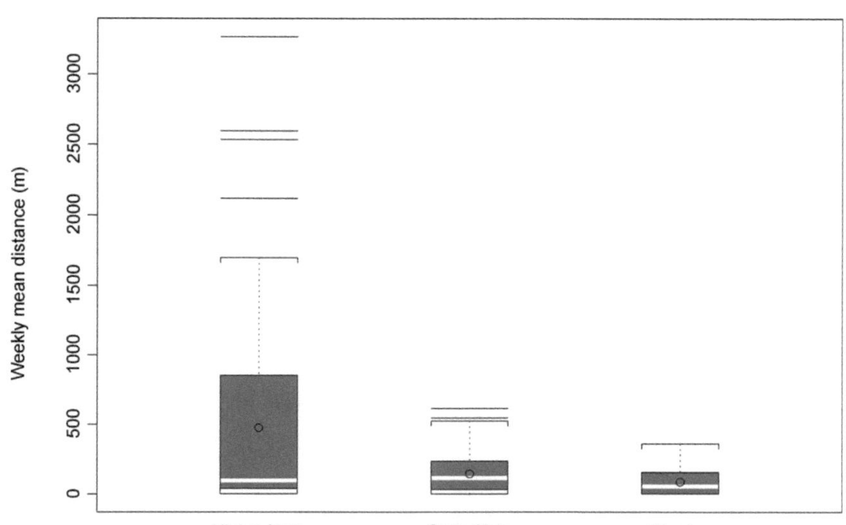

Fig. 6 Weekly mean distance travelled by tagged roach in the three rivers during the overall study period. Values are median, percentiles 5, 25, 75 and 95. Bars indicate outlier values and circles indicate weekly mean distances

(1998) noticed that there is a highly significant distinction in mobility rates and activity levels between April and May (when fish are most active) and other times. The longer movements were observed during spawning season.

No statistical relationship was observed between water flow and distance travelled although some rare winter movements were observed during very high flow events. In lake Årungen (south-eastern Norway), Vøllestad & L'Abée-Lund (1987) suggested that spawning activity is regulated by water flow and water temperature and suggested that roach spawned synchronously in years with rapid increases in temperature, whereas they had a prolonged spawning period in years with low or with slow increases in water temperature. It has previously been demonstrated that photoperiod is the principal factor synchronizing the start of the spawning in roach (Worthington et al., 1982). In unstable environments, the use of predictable cues such as day length will be positively selected. Temperature however is important in regulating the intensity and duration of the spawning (Worthington et al., 1982).

Although the dynamic of the roach movements and the activity levels were similar in the three rivers, our results demonstrate that the extent of the movements observed was mainly related to the distance between physical barriers in the study areas. Distances travelled were more pronounced in the River Kleine Nete where the distance between the physical barriers is 14 km (first barrier on the River Aa tributary). Upstream from the downstream weir, free entrance in the River Aa tributary is possible, and the upstream weir (in the River Kleine Nete) is equipped with a fish pass (Fig. 1(1)). On the other hand, in the River Vesdre, where the distance between the physical barriers is only 1.2 km, the distances travelled were much shorter and the proportion of roach moving downstream during the reproduction period was more pronounced.

Our results suggest that roach were not frequent obstacle leapers (at least in the upstream direction) and they were able to complete all of their biological activities in limited stretches of rivers in highly fragmented environments. As well, they sometimes demonstrated up- and downstream movements during the reproduction period in search of available spawning habitat. In a 32 km unfragmented stretch of the River Spree (Germany), Baade & Fredrich (1998) observed that roach migrated up to 10 km upstream to spawn. Some studies however showed that roach could sometimes clear small physical obstacles. Lucas et al. (1998) demonstrated that radio-tracked roach ascended the Skip Bridge weir on the Nidd (United Kingdom) and moved further upstream to spawning areas. Svärdson (1951) observed a roach that was trapped in a wire-netting, jumping at least 15 cm above the water level.

The roach tracked in three fragmented rivers showed similar patterns of movements that were mainly influenced by the date of observation and the water temperature. The distances travelled were relatively short (max. 12.7 km) and mainly influenced by the length of the study reach, delimited by physical barriers. Nevertheless roach were able to adapt their behaviour to the habitat fragmentation by completing their life cycles in reaches sometimes shorter than 2 km. This might partially explain their subsistence in highly disturbed environments, however further studies are needed for a better understanding of the long-term effects of the habitat fragmentation on roach populations.

Acknowledgements This study was part of the FISHGUARD project 'Impact assessment and remediation of anthropogenic interventions on fish populations' financed by the Belgian Federal Office for Scientific, Technical and Cultural Affairs. The authors wish to thank D. Hennebel, R. Baeyens, S. Martens, G. Rimbaud and Y. Neus for field support. We are grateful to P. Verschelde and M. Melsen for their assistance with the draft of the graphs, F. Coopman for drawing the maps and two anonymous referees for their critical comments. We also would like to thank J. Carr for reviewing and providing constructive comments.

References

Baade, U. & F. Fredrich, 1998. Movement and pattern of activity of the roach in the River Spree, Germany. Journal of Fish Biology 52: 1165–1174.

Diamond, M., 1985. Some observations of spawning by roach, *Rutilus rutilus* L., and bream, *Abramis brama* L., and their implications for management. Aquaculture and Fisheries Management 16: 359–367.

Goldspink, C. R., 1977. The return of marked roach (*Rutilus rutilus* L.) to spawning grounds in Tjeukemeer, The Netherlands. Journal of Fish Biology 11: 599–603.

Huet, M., 1949. Aperçu de la relation entre la pente et les populations piscicoles des eaux courantes. Schweizerische Zeitschrift für Hydrologie 11: 332–351.

Holcik, J. & V. Hruska, 1966. On the spawning substrate of roach, *Rutilus rutilus* L., and bream, *Abramis brama* L. and notes on the ecological characteristics of some European fishes. Vestnik Ceskoslvenske Spolecnosti Zoologicke 30: 22–29.

Jepsen, N., A. Koed, E. B. Thorstad & E. Baras, 2002. Surgical implantation of telemetry transmitters in fish: how much have we learned. Hydrobiologia 483: 239–248.

L'Abée-Lund, J. H. & L. A. Vøllestad, 1985. Homing precision of roach *Rutilus rutilus* in Lake Aarungen, Norway. Environmental Biology of Fishes 13: 235–239.

Lucas, M. C., T. J. Thom, A. Duncan & O. Slavik, 1998. Coarse Fish Migration Occurrence, Causes and Implications. Technical Report W152, Environment Agency, 161.

Mulcahy, D. M., 2003. Surgical implantation of transmitters into fish. ILAR Journal 44(4): 295–306.

Ovidio, M. & J. C. Philippart, 2002. The impact of small physical obstacles on upstream movement of six species of fish (Synthesis of a—year telemetry study in the River Meuse basin). Hydrobiologia 483: 55–69.

Poncin, P., 1994. La reproduction des poissons de nos rivières. Cahiers d'Ethologie 13: 317–342.

Stott, B., 1967. The movements and population densities of roach (*Rutilus rutilus* L.) and gudgeon (*Gobio gobio* L.) in the River Mole. Journal of Animal Ecology 36: 407–423.

Svärdson, G., 1951. Spawning behaviour of *Leuciscus leuciscus* (L.). Reports of the Institute of Freshwater Research, Drottingholm 33: 199–203.

Vandelannoote, A., R. Yseboodt, B. Bruylants, R. Verheyen, J. Coeck, J. Maes, C. Belpaire, G. Van Thuyne, B. Denayer, J. Beyens, D. De Charleroy & P. Vandenabeele, 1998. Atlas van de Vlaamse Beek- en Riviervissen. Water-Energik-Vlario (WEL), Wijnegem, 1–303.

Vøllestad, L. A. & J. H. L'Abée-Lund, 1987. Reproductive biology of stream-spawning roach, *Rutilus rutilus*. Environmental Biology of Fishes 18: 219–227.

Williams, W. P., 1965. The population density of four species of freshwater fish, roach (*Rutilus rutilus* L.), bleak (*Alburnus alburnus* L.), dace (*Leuciscus leuciscus* L.) and perch (*Perca fluviatilis* L.) in the River Thames at Reading. Journal of Animal Ecology 36: 407–423.

Winter, J. D., 1983. Underwater Biotelemetry. In Nielsen, L. A. & D. L. Johnson (eds), Fisheries Techniques. American Fisheries Society, Bethesda, Maryland, 371–396.

Worthington, A. D., N. A. A. Macfarlane & K. W. Easton, 1982. Controlled reproduction in the roach (*Rutilus rutilus* L). In Richter, C. J. J. & H. J. H. Goos (eds), Reproductive Physiology of Fish. Proceedings of an International Symposium, Wageningen, 220–223.

Hydrobiologia (2007) 582:155–169
DOI 10.1007/s10750-006-0557-6

FISH TELEMETRY

Assessment of a retrofitted downstream fish bypass system for wild Atlantic salmon (*Salmo salar*) smolts and kelts at a hydroelectric facility on the Exploits River, Newfoundland, Canada

D. A. Scruton · C. J. Pennell · C. E. Bourgeois · R. F. Goosney · T. R. Porter · K. D. Clarke

Abstract In 2002 and 2003, the Bishops Falls hydroelectric generating facility on the Exploits River, insular Newfoundland, Canada, underwent extensive refurbishing including replacement of turbines and installation of a 'retrofitted' bypass and fish handling system. The effectiveness of this new bypass system has been assessed during the annual downstream run of wild Atlantic salmon smolt and kelt in 2003 and 2004. In 2003, 195 smolt were radio tagged and released between June 9 and July 2, in the forebay of the hydro plant (19 releases) and one upstream (in-river) release. Fish guidance efficiency (FGE) of the system overall was 63% (123 of 195 fish) with 36 fish passing through the turbines, and six known mortalities. In 2004, between June 9 and July 2, a total of 358 smolt and 103 kelt were released in the forebay in 45 and 13 releases ($n = 8$ per release), respectively. The FGE of the system for smolt was 71.7% (257 of 358 fish) and for kelt was 92.3% (95 of 103 fish). In 2004, 96 tagged smolt passed through the turbines and 43 (44.8%) were detected at a downstream station confirming they had survived turbine passage, suggesting an overall survival of smolt passage of the Bishops Falls hydro facility in the order of 85%. A total of seven kelts (6.8%) passed through the turbines and were not detected 1.5 km downstream suggesting they did not survive turbine passage. Smolt spent on average 39.8 h in the forebay before exiting in 2003 and forebay residency averaged 26 h in 2004. In both years, most smolt selected their passage route, actively or passively, within the first 10 h with secondary peaks at 25–30 and 50–55 h, corresponding to evening passage in the second and third night, after release. Few smolt were bypassed or entrained into turbines during daylight hours. In both years turbine passed smolt spent more time in the forebay suggesting the longer fish reside in the forebay the greater the likelihood of turbine entrainment. Kelt were either bypassed or turbine entrained relatively quickly, within 2 h of release, and virtually all kelts were bypassed/turbine entrained during the hours of 18:00 and 01:00. These data on fish behavior and residency in the forebay will assist further refinement of operations of the bypass facility to optimize survival.

Keywords Fish telemetry · Hydroelectric developments · Fish passage · Atlantic salmon · Smolt · Downstream migration · Bypasses

Guest editors: P. R. Almeida, B. R. Quintella, M. J. Costa and A. Moore
Developments in Fish Telemetry

D. A. Scruton (✉) · C. J. Pennell · C. E. Bourgeois · R. F. Goosney · T. R. Porter · K. D. Clarke
Fisheries and Oceans Canada, Science Branch, 80 East White Hills Road, P.O. Box 5667, John's, NL, Canada A1C 5X1
e-mail: scrutond@dfo-mpo.gc.ca

Introduction

Migratory anadromous salmonids must traverse dams and other barriers to maintain populations in upstream habitats. The mortality of Atlantic salmon smolt at dams, during downstream migration, represents a potential population reduction that does not allow for normal biological compensatory mechanisms and, as such, could adversely affect adult recruitment and yield from the stock originating above the barrier (Ruggles et al., 1993). Smolt descent in rivers is a hazardous period in the life history of Atlantic salmon and passage over falls and dams, desmoltification, and predation are major sources of mortality for migrating juveniles (Hvidsten & Johnsen, 1997). Significant delays associated with passage may also be important in smolt survival as they may desmoltify or reach the sea under environmental conditions that may reduce survival (Ruggles, 1980). In many anadromous Atlantic salmon stocks, adults over-winter in the river system after spawning and return to the sea as kelts. These kelts often migrate during the same period as smolts, and, as repeat spawners, can be an important component of the returning adult salmon stock (Porter, 1975). Effective passage of downstream migrating life stages of anadromous salmonids has been a particularly difficult aspect of hydroelectric development to manage and efforts to divert smolts and kelts from turbines have met with varying degrees of success (U.S. Congress, 1995).

Downstream migrating fish have several potential passage routes as they approach a hydroelectric facility. They can enter the forebay, where they may enter a bypass or pass through the turbines, or they can pass over the dam or through a spillway, depending on the design and operation of the power plant. Design of fish passage systems is highly influenced by fish behavior and site conditions. A fundamental behavior pattern of smolt approaching dams is that of surface orientation, following maximum flows. Physiological and behavioral changes for adaptation to salt water affect swimming proficiency and buoyancy, determine position and performance of fish in the water column, and hence influence susceptibility to fish protection systems and turbine entrainment. Kelt, on the other hand, are extremely emaciated after spawning and over wintering and their physiological condition and swimming ability is compromised (Booth et al., 1997). Peak smolt and kelt migration occurs during high flow conditions in the spring and the majority pass in spillage during this time, depending on the water storage and operational scheme of the hydroelectric system. Conversely, as peak flow conditions abate, migrating smolt and kelt must actively avoid entrainment, as more of the bulk flow is directed to the power plant (Coutant & Whitney, 2000). Design and operation of fish bypasses at hydroelectric installations requires knowledge of swimming behavior, orientation of movement, environmental cues for migration, and sustained and burst swimming ability of target fish. Mechanical, structural, and operational features of hydropower installations are the primary features influencing the potential for successful guidance and downstream passage. Site specific studies are needed to relate fish behavior and movements to hydraulic (flow direction and velocity) and other environmental cues encountered by fish as they approach a dam and enter forebays, bypasses, or turbine intakes in order to optimize the design and operation, and hence the effectiveness, of fish protection schemes (Coutant, 1999).

In 2002, Abitibi Consolidated Company of Canada (ACCC) upgraded a hydroelectric generating facility at Bishop's Falls, on the Exploits River, Newfoundland, Canada and, as part of this retrofit, constructed a new downstream fish bypass facility. ACCC and the Department of Fisheries and Oceans (DFO) have since undertaken a multi-year project to evaluate the efficiency of the new fish bypass including: (i) determining the fish guidance efficiency (FGE) of the bypass for smolt and kelt; (ii); residency time for fish in the forebay relative to FGE; (iii) impact of bypass on condition of migrants, and (iv) subsequent survival downstream of the facility. This paper focuses on the results of the radio telemetry studies including fish guidance efficiencies, residency time in the forebay, time of day of passage and entrainment, and post turbine passage survival of migrating fish. These studies are intended to assist in improvement of design and operation and function of the bypass to optimization fish passage and survival.

Study site and bypass system

The Exploits River is the longest river (246 km) and has the largest drainage basin (11,272 km^2) on the island of Newfoundland (Fig. 1). It has a mean annual discharge of approximately 290 $m^3 s^{-1}$ and sustains one of the largest populations of anadromous Atlantic salmon (*Salmo salar*) in North America. During the early 1900s, a pulp and paper mill was constructed at Grand Falls along with three dams, one at Red Indian Lake (112 km upstream from the estuary), one at Grand Falls (22 km upstream), and one at Bishop's Falls (10 km upstream) (Fig. 1). Two large hydroelectric facilities, at Bishop's Falls and Grand Falls, have been constructed in order to supply power to the pulp and paper mill at Grand Falls.

Historically, anadromous Atlantic salmon were restricted to 10% of the watershed due to natural obstructions at Grand Falls and on several tributaries, and fishways were constructed to improve access at Bishop's Falls (1959), Grand Falls (1974), and Red Indian Lake (1990) on the river mainstem. Enhancement of the Atlantic salmon stock began with an adult transfer program from 1957 to 1965 and continued from 1967 to 1992, with stocking of unfed fry in various sections throughout the drainage (Taylor & Bauld, 1973; O'Connell & Bourgeois, 1987; Bourgeois et al., 1994; Mullins et al., 2003). The returning adult salmon population increased as a result of the enhancement activities with adult returns peaking at about 32,000 Atlantic salmon 1996 and an average of about 22,000 over the last decade (O'Connell et al., 2003).

In 2002, Abitibi Consolidated Company of Canada (ACCC) upgraded the hydroelectric

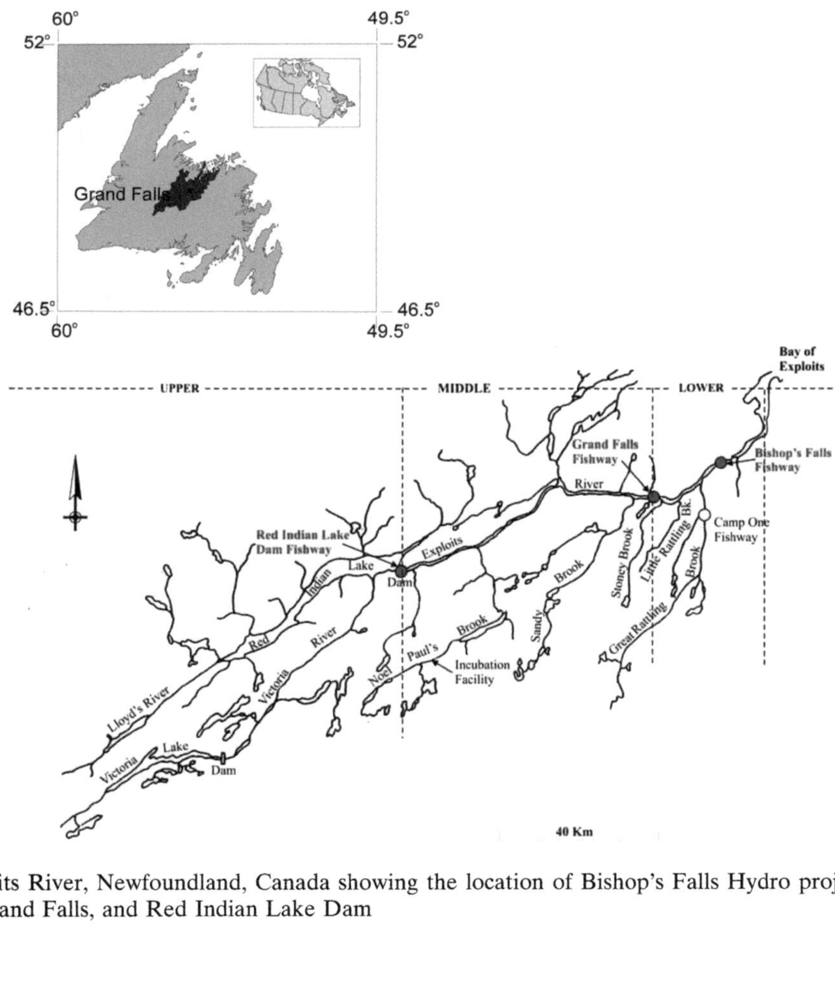

Fig. 1 The Exploits River, Newfoundland, Canada showing the location of Bishop's Falls Hydro project and fishways at Bishop's Falls, Grand Falls, and Red Indian Lake Dam

generating capacity at Bishop's Falls and the new operating regime, modified the water utilization pattern through the turbines with potential to increase entrainment of fish into the forebay and turbines. Prior to the retrofit, passage routes for downstream migrating fish were: (i) over the dam, (ii) through a 'smolt' opening in the dam, (iii) through spillway gates in the dam (only during the early part of the run), (iv) into the forebay and bypassed back into the river through an existing bypass (reverse fishway), and (v) and through the turbines. As part of this retrofit, in 2003 ACCC constructed a downstream fish bypass to reduce entrainment of downstream migrating smolt and kelt into the turbines (Fig. 2). The fish bypass was constructed in an existing spillway gate exiting the forebay. The bypass was surface spill type designed to extract water, with Atlantic salmon smolt and kelt, directly from the forebay into a plunge pool and then return fish to the Exploits River, downstream of the dam. The plunge pool was designed to be drained to a level where fish were confined to a holding compartment for enumeration and evaluation, and subsequent release.

Materials and methods

2003 Studies

Bypass efficiency was assessed by releasing experimental groups of smolt in the forebay between June 8 and July 2, 2003 (Scruton et al., 2004a). All smolt entering the by-pass were enumerated daily and test smolt were taken from the plunge pool (08:00 h to 10:00 h) and moved to a 1,000-l insulated container. All fish selected for telemetry study were measured for length (mm) and weight (g), anesthetized, and transmitters (Lotek Wireless Model NTC-4-2S 17.0 mm × 8.0 mm 1.65 g in air, pulse rate of 1.5 s, battery life of approx. 20 days) surgically implanted following methods described in Adams et al. (1998) and Moore et al. (1990). The surgical procedure lasted approximately 3–4 min, after which fish were recovered, transferred to a release-cage (0.9 m × 0.6 m × 0.6 m) in the forebay (at approximately 12:00 h), acclimated, and released at 18:00 h. Tagged fish were allowed to recover (approx. 6 h) and then released by opening the floor of the cage and allowing smolt to volitionally exit. The release site was in quiet water at the farthest

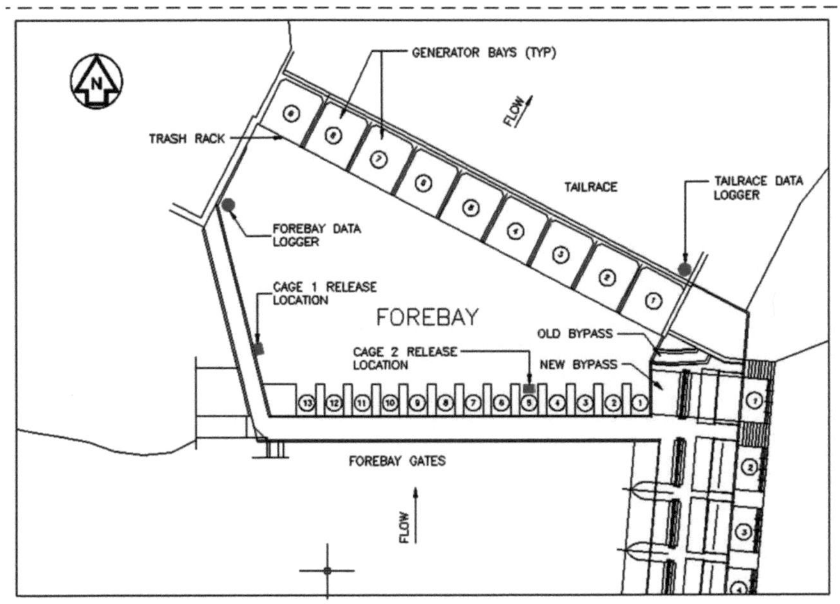

Fig. 2 Bishop's Falls forebay showing location of intake gates, new bypass, generator bays, and release cages for smolt and kelt studies

distance from the bypass entrance (Fig. 2), and could be considered a 'worst case' scenario as fish have to traverse the entire forebay, past the entrance gallery to turbines, to locate and use the bypass.

Fish used in the study were 're-cycled' from fish that had entered the forebay and bypass. Capturing fish from the river above the power plant would have been extremely difficult and would have introduced considerable capture and handling stress. We felt the capture and handling stress would impose a greater bias than 're-cycling' fish from the bypass for experiments.

In 2003, data logging stations (Lotek Wireless SRX_400 telemetry receivers attached to a series of underwater and/or Yagi antennae) were strategically placed throughout the forebay to cover all possible passage routes; (i) entrance into to the bypass; (ii) tailrace monitoring turbine passage; (iii) end of the dam monitoring passage over the dam and/or moving back upstream. Residency times and bypass efficiency were determined from data logged on receivers covering various exit locations. Antennae were deployed in the forebay to determine relative location within the forebay (comparing relative signal strength from various antennas, after calibration). Another receiver was deployed approximately 1.5 km downstream from Bishop's Falls Dam, to monitor radio tagged smolt having passing the dam and/or through the turbines. We are confident that fish detected 1.5 km below the plant were indeed alive as we did not observe fish drifting downstream at this location and injured or dead fish in the vicinity of the plant are quickly predated upon by herring gulls. Additionally, there is an area in the river below the power plant where dead fish (both smolts and kelts) naturally accumulate due to river hydraulics and, although dead fish were observed in this location, no radio signals were detected.

Telemetry trials were conducted from June 9 to July 2, 2003 and included a total of 195 smolt in 19 forebay releases (1 with $n = 5$, 1 with $n = 9$, 1 with $n = 13$, 16 with $n = 10$) and 1 in-river release ($n = 8$). The in-river release was done to examine the possible effect of entry into the forebay on fish guidance efficiencies. FGE was determined from the ratio (%) of released fish that were subsequently detected in the bypass, by release. The residency time of radio tagged smolt in the forebay was evaluated between each group of fish by exit route (bypassed, turbine passed, over the dam, returned upstream) and compared statistically (T-test or Mann–Whitney rank sum test, at $P < 0.05$). The time of day that smolt exited through the bypass or turbines was evaluated with respect to day/night, time of release, and in relation to operation of the bypass.

2004 Studies

Radio telemetry was used to investigate passage route, bypass efficiency (fish guidance efficiency or FGE), forebay residency, and movements of Atlantic salmon smolt (May 29 to July 6) and kelt (May 28 to June 18) in 2004. Smolt and kelt were removed from the plunge pool (08:00 h to 10:00 h), moved to a 300-l insulated container, examined as to condition (only fish in good condition were used in trails) and measured for length (mm) and weight (g). Radio transmitters (Lotek Wireless Model NTC-4-2S 17.0 mm × 8.0 mm 1.65 g in air, pulse rate of 1.5 s, battery life of approx. 20 days) were surgically implanted as above.

Fish were then transferred to the release cages in the forebay at approximately 12:00 h, and held there until release at 18:00 h. In 2004, two release sites were used for all forebay releases; Site 1 was the same location as in 2003 while Site 2 was located at intake gate # 5, and was selected to represent a location where fish would normally enter the forebay from the headpond through the submerged gates, and their residency and ability to locate and use the bypass would be more representative of naturally migrating smolt and kelt (Fig. 2). Smolt and kelt were released as described above. Forebay trials included a total of 358 smolt in 45 releases (43 with $n = 8$ and 2 with $n = 7$) and 103 kelt in 13 releases (12 with $n = 8$ and 1 with $n = 7$).

Tagged smolt and kelt were monitored and data logged using a series of telemetry receivers (Lotek Wireless SRX_400) attached to a series of underwater and/or Yagi antennae as described above. In 2004, more attention was paid to monitoring all possible passage routes from the forebay including: entrance to the plunge pool and in the bypass, both sides of the tailrace (turbine exit), upstream end of the dam, at the fishway (Fig. 3). Another receiver was deployed approximately 1.5 km downstream to monitor smolt and kelt moving downstream after passing over the dam or through the turbines. Constraints at this location prevented full coverage of the river width. Tagged fish were also manually tracked daily in the forebay, and both upstream and downstream of the hydro facility.

Data analyses

FGE was determined from the ratio (%) of released fish that were subsequently detected in the bypass, by release. The fate of radio tagged fish with respect to fish passage was also determined, by release. The residency time (h) of radio tagged fish in the forebay was evaluated statistically (*T*-test or Mann–Whitney rank sum test, $P < 0.05$), by passage route (bypassed, turbine passed, over the dam, returned upstream). The amount of total spent in the forebay by bypassed and turbine passed fish was examined. The time of day that fish were either bypassed or turbine entrained was evaluated with respect to day/night, time of release, and in relation to operation of the bypass.

Results

A summary of relevant information during the study period for smolt and kelt trials in 2003 and 2004 is provided in Table 1 and more detail can be found in Bourgeois et al. (2004, 2005). Data on river discharge, water temperature, plant discharge, and bypass flows during the smolt and kelt study period is provided. Generally temperatures were lower in 2004 than in 2003, largely because of the earlier onset of the study (May 28 2004 versus June 8 2003). This earlier onset in 2004 also permitted more availability of kelts for study as kelts generally migrate at the tail end of the snow melt discharge (Porter, 1975). River discharge was greater in 2003 than in 2004, due to natural environmental conditions, however discharge through the power plant was identical and these are the conditions smolt/kelt would experience upon release in the forebay. There was some variability in flows into the bypass in 2004, as this was a parameter being examined, but this was quickly adjusted to 0.9 $m^3\ s^{-1}$ (as in 2003), owing to problems experienced with fish handling. An examination of daily power generation (MW) and daily bypass flow ($m^3\ s^{-1}$) in relation FGEs for individual smolt trials in 2003 revealed that these

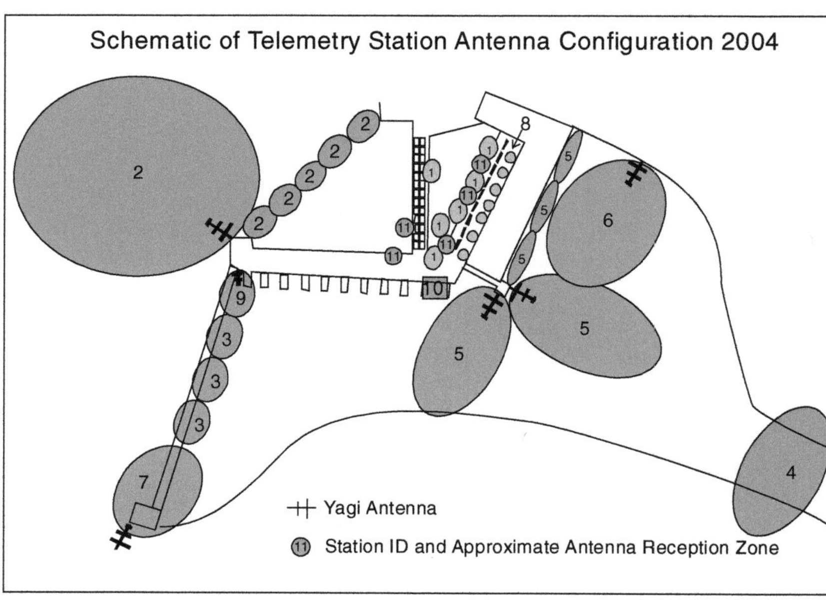

Fig. 3 Schematic location of the location of the fixed station telemetry antennae, including the approximate field of detection (as determined through calibration) in relation to the Bishop's Falls hydroelectric facility in 2003 and 2004

Table 1 A summary of details for the study period for smolt and kelt trials in 2003 and 2004

Parameter	2003	2004
Study period (smolt)	June 8–July 2	May 29–July 6
Study period (kelt)	N/A	May 28–June 18
River discharge	194–211 $m^3 s^{-1}$ (= 201 ± 0.92 SE)	155-214 $m^3 s^{-1}$ (= 173 ± 4.84 SE)
Plant discharge	255 $m^3 s^{-1}$ (20 MW)	255 $m^3 s^{-1}$ (20 MW)
Bypass flows	0.9 $m^3 s^{-1}$	0.6–1.3 $m^3 s^{-1}$
Water temperature	9.9–20.2°C (= 14.5 ± 0.78 SE)	6.0–18.5°C (= 11.9 ± 0.69 SE)
Total smolts passed	109,038	159,592
Total kelts passed	500	5,890

These include river discharge, water temperature, plant discharge, and bypass flows during the study period. The total number of smolt and kelt bypassed for the entire year is also provided

hydraulic factors had no influence on FGE (Bourgeois et al., 2004). Larger numbers of both kelts and smolts were bypassed in 2004 and this is also likely related to the earlier start-up of the bypass operation.

2003

The fate (passage route) of all radio tagged fish, by release, is shown for smolt in Fig. 4 (upper). Thirty-six (36) smolt passed through the turbines, six smolt exited upstream out of the forebay and passed over the dam, one smolt exited upstream out of the forebay and was not subsequently detected, while the fate of 29 smolt were not determined (either undetected [n = 14] or were detected at (at least) one antennae but insufficient data to conclusively determine passage direction [n = 15]). The pattern in FGEs for the 19 releases is shown in Fig. 4 (lower). The overall FGE of the system for smolt, from 19 forebay releases, was 62.3% SE ± 4.0 (117 of 187 fish), varying from 30% to 100% . Generally, five releases had FGEs 80% or higher and these higher FGEs were in two consecutive releases, June 23–25 and June 29–July 1, while there were six releases with FGEs of less than 60%. A total of 36 radio tagged smolt (19%) were determined to have entered the turbines, with entrainment ranging from 0% to 40%. Sixteen of these fish (44%) were detected in downstream reaches with six (17%) determined to be mortalities while 10 (28%) survived turbine passage. Fish were verified as mortalities as the radio signals were detected in manual tracking and fish did not move over repeated trackings. Live fish were those that passed the fixed monitoring station 1.5 km below the plant. The FGE for the one in-river release was 75% (6 of 8), with the other two smolt passing over the dam.

The time spent by smolt in the forebay, after release, summarized by passage route (Fig. 5 (upper)) indicates the mean time spent (±SE) for

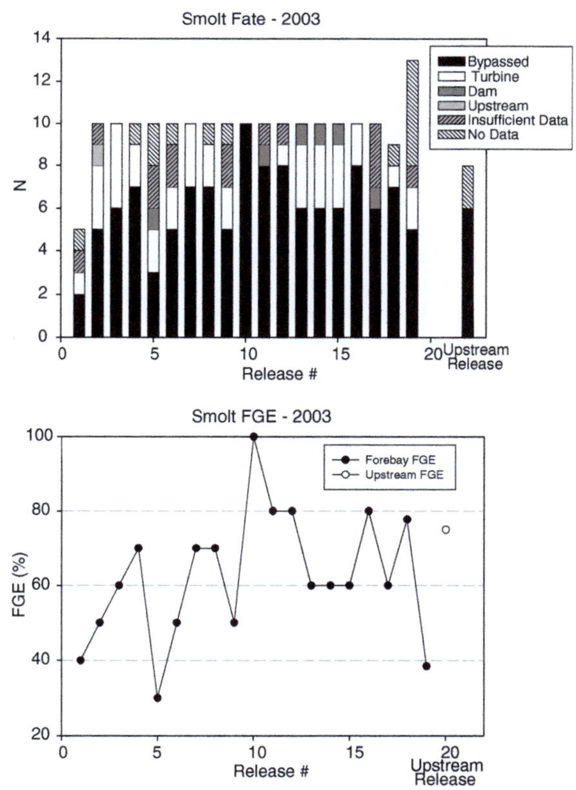

Fig. 4 The fate of radio tagged smolt in relation to passage of the Bishop's Falls hydro facility in 2003 (*upper panel*) and the fish guidance efficiency (FGE) of the Bishop's Falls bypass for smolt, by release, in 2003 (*lower panel*)

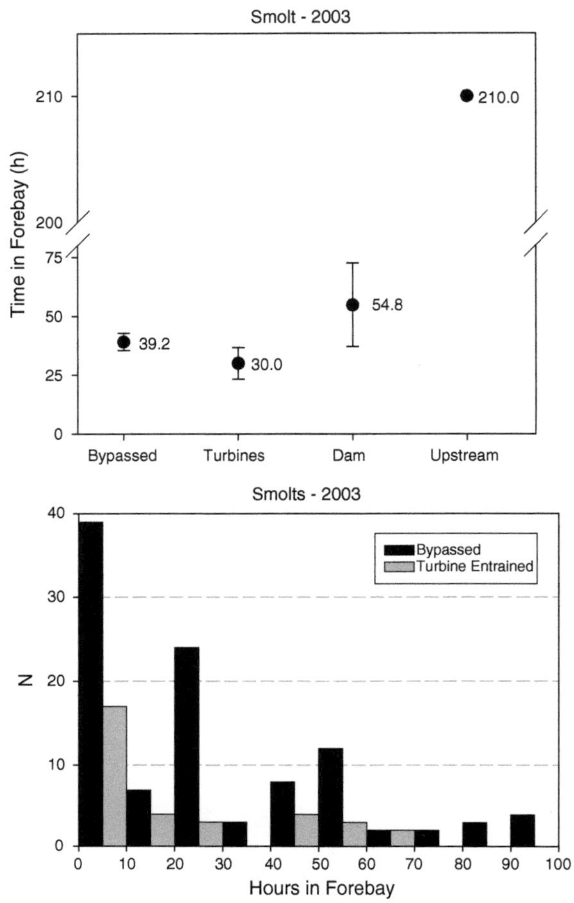

Fig. 5 The mean (±SE) amount of time spent in the Bishop's falls forebay by (i) bypassed smolt, (ii) turbine entrained smolt, (iii) smolt that passed over the dam, or (iv) smolt that returned upstream in 2003 (*upper panel*). The distribution of time spent in the forbay by bypassed and turbine entrained fish in 2003 (*lower panel*)

bypassed fish, turbine entrained fish, fish that passed over the dam, and for the one fish that returned upstream was 39.2 h (±3.7), 30.0 h (±6.7), 54.8 h (±17.7), and 210.0 h, respectively. The mean time spent by bypassed smolt ranged from 14.8 to 79.9 h, while the mean time spent by turbine entrained smolt ranged from 1.2 to 86.3 h. The difference in time spent between smolt that were bypassed and turbine entrained was marginally statistically significant (Mann–Whitney rank sum test, $P = 0.041$). Time spent in the forebay by smolt that passed over the dam was greater than for bypassed or turbine entrained smolt, however, there was no significant difference (Mann–Whitney rank sum test, $P = 0.22$ and $P = 0.084$, respectively). The distribution of time spent in the forebay by bypassed and turbine entrained (Fig. 5 (lower)) indicated a tendency for smolt to select passage route, either actively or passively, fairly quickly, within the first 10 h. For bypassed smolt, there are secondary peaks at 25–30 h and 50–55 h, corresponding to evening passage in the second and third night, after release.

Most fish were bypassed or turbine entrained between 18:00 and 02:00 (100 fish [81%] and 23 fish [64%], respectively) (Fig. 6). Peak smolt turbine entrainment (6 fish, 17%) occurred within 1 h of release. A peak in proportion of bypassed smolt was evident at 21:00, coincident with onset of dusk, while few smolt (26) entered the bypass between dawn (05:00 h) and the release time (18:00 h), as only 26% (44 fish) were bypassed during daylight hours. The bypass system was closed during 09:00 h to 12:00 h, and smolt could only exit via the turbines or forebay intake gates, a period when 8 of 36 (25%) were turbine passed.

2004

The mean overall FGE for smolt, from 45 forebay releases, was 71.7% SE ± 3.09. There were two release locations in the forebay with 26 releases from cage 1 (May 29 and July 6, $n = 8$ for 25 releases, $n = 7$ for 1 release) and 19 releases from cage 2 (between June 14 to July 6, $n = 8$ for 18

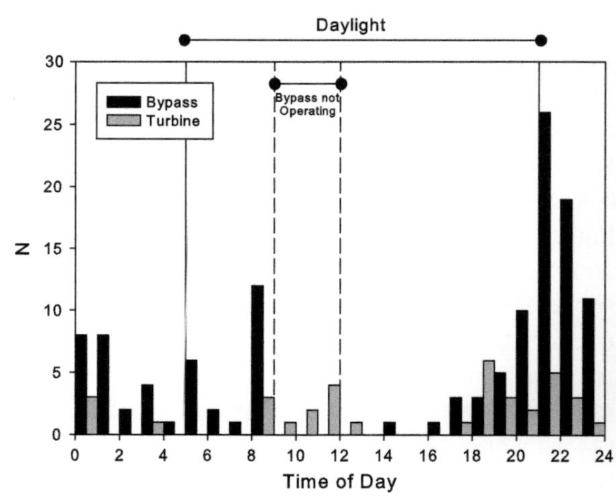

Fig. 6 Distribution of radio tagged smolt by time of day (hour) that exited via bypass and turbines. The release time, daylight hours, and the time the bypass was not operating (closed to count fish) are indicated

releases, $n = 7$ for 1 release). FGEs from cage 1 averaged 70.0 ± (SE) 3.89%, and ranged from 25% (two trials) to 100% (three trials) while FGEs from cage 2 averaged 74.2 ± (SE) 5.10%, and ranged from 25% (one trial) to 100% (five trials) and the difference in FGE from the two release locations was not significant (*t*-test, $P = 0.512$). The pattern in smolt FGE (26 from cage 1, 19 from cage 2) is shown in Fig. 7 (lower). Twenty-four (24) of 26 releases were 50% or greater and 15 of 26 releases were 75% or higher from cage 1, while 17 of 19 releases were 50% or greater and 11 of 19 releases were 75% or higher in cage 2. There was no apparent trend to the timing of higher or lower FGEs, however the occurrence of low FGEs was consistent between cages, suggesting there were aspects of the experimental conditions which resulted in the lower FGEs.

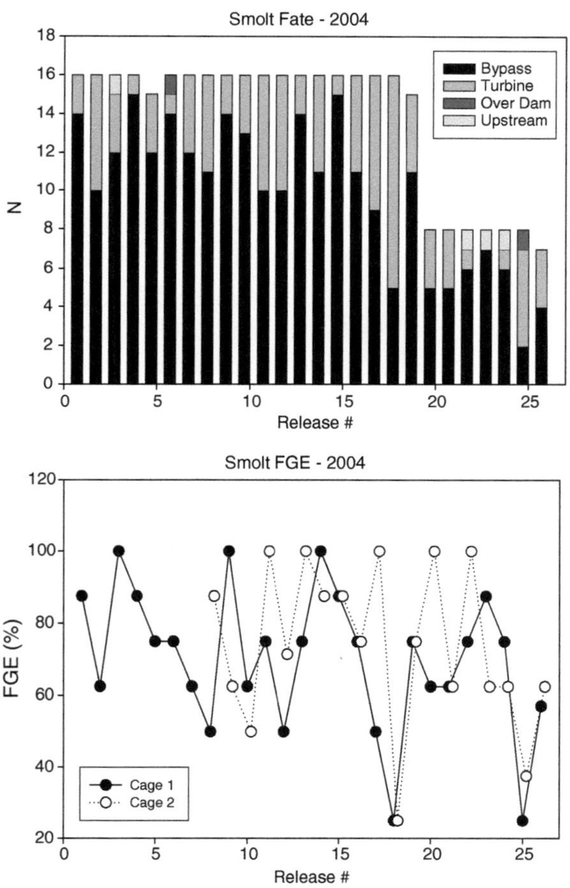

Fig. 7 The fate of radio tagged smolt in relation to passage of the Bishop's Falls hydro facility in 2004 (*upper panel*). Note results for releases 1 through 19 are the average for 2 releases from each of the release cages on the same day. The lower panel presents the fish guidance efficiency (FGE) of the Bishop's Falls bypass for smolt, by release in 2004

Two hundred and fifty-seven (257) smolt were bypassed, 96 smolt passed through the turbines, two smolt exited upstream out of the forebay and passed over the dam, and three smolt exited upstream out of the forebay and subsequently were not detected passing the facility, while one smolt remained in the forebay (likely mortality) (Fig. 7 (upper)). Forty-three (43) of 96 turbine passed fish (45%) were subsequently detected at a downstream reception station (Sir Robert Bond Bridge) and were considered to have survived turbine passage. The combined survival of smolt passing the Bishops Falls facility, combining both bypassed fish and those that survived turbine passage, is in the order of 83.8% (300 of 358 fish released) and this is a conservative estimate as we could not assume 100% coverage at the downstream location.

One hundred and three (103) kelt were tagged and released from both cages (cage 1, $n = 11$, cage 2, $n = 2$) between May 28 and June 18, 2004 in 13 forebay releases ($n = 8$ for 12 trails, $n = 7$ for one trial). The mean overall FGE for kelt was 92.3 ± (SE) 1.76% and ranged from 87.5% (eight trials) to 100% (five trials) (Fig. 8 (lower)). Similarly, the pattern in FGE for the 13 kelt releases indicated all releases had an FGE of 87.5% or greater and five releases (38%) had 100% FGE (Fig. 8 (bottom)). Ninety five (95) kelt were bypassed, seven passed through the turbines, while the remaining two kelt exited upstream out of the forebay and subsequently were not detected passing the facility (Fig. 8 (upper)). None of the seven turbine passed fish were detected at the downstream reception station (Sir Robert Bond Bridge), therefore none were considered to have survived turbine passage, therefore the combined survival of all kelt passing through the Bishops Falls forebay, is the same as the FGE of the bypass, at 92.3%.

The amount of time that smolt spent in the forebay, after release, is summarized by passage route in Fig. 9 (upper). The mean time spent (±SE) for bypassed smolt, turbine entrained smolt, smolt that passed over the dam, and smolt that returned upstream was 24.3 h (SE ± 2.7),

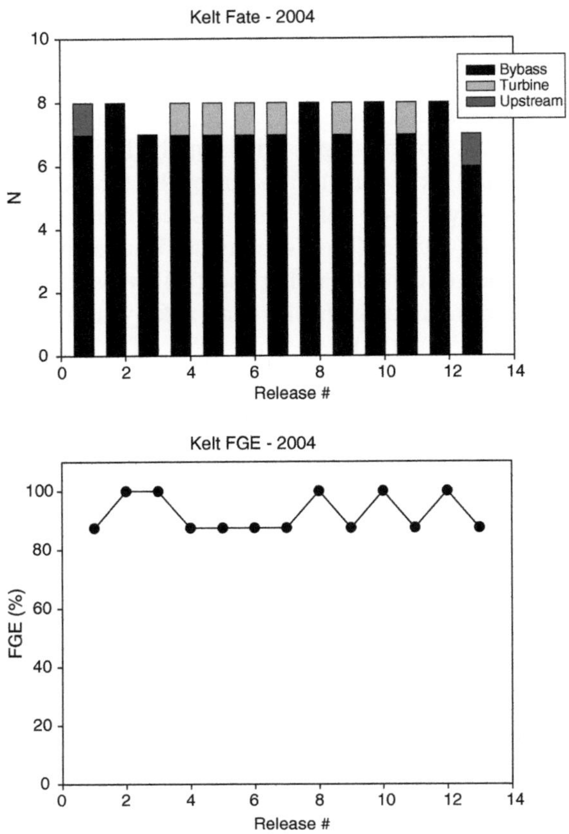

Fig. 8 The fate of radio tagged kelt in relation to passage of the Bishop's Falls hydro facility in 2004 (*upper panel*) and the fish guidance efficiency (FGE) of the Bishop's Falls bypass for kelt, by release in 2004 (*lower panel*)

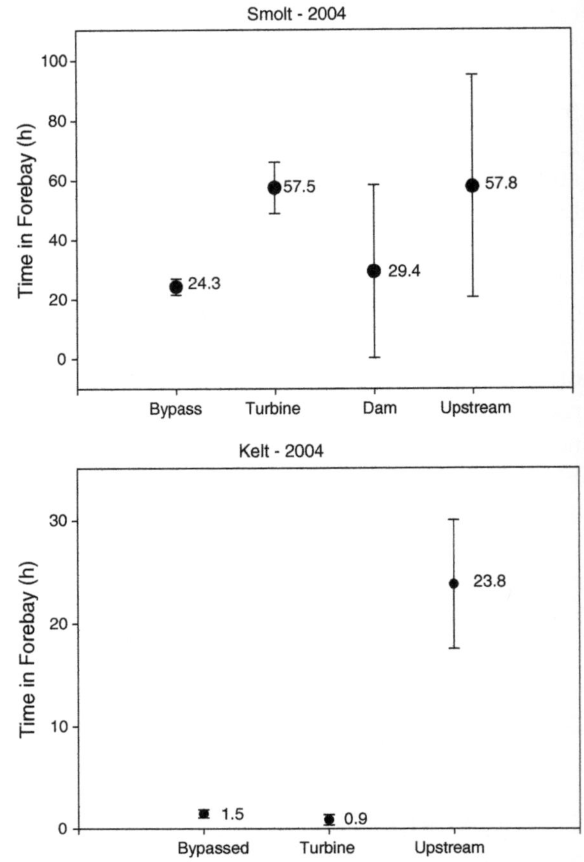

Fig. 9 The mean (±S.E.) amount of time spent in the Bishop's falls forebay by smolt (*upper panel*) and kelt (*lower panel*) for (i) bypassed fish, (ii) turbine entrained fish, (iii) fish that passed over the dam, or (iv) fish that returned upstream in 2004

57.5 h (SE ± 8.6), 29.4 h (SE ± 29.1), and 57.8 h (SE ± 37.1), respectively. The mean time spent by bypassed smolt was significant less (Mann–Whitney rank sum test, $P = 0.001$) than turbine passed fish, suggesting that the longer time smolt spent in the forebay the greater the likelihood of turbine entrainment. Similarly, the mean time spent (±SE) for bypassed kelt, turbine entrained kelt, and kelt that returned upstream was 1.52 h (SE ± 0.40), 0.90 h (SE ± 0.53), and 23.8 h (SE ± 6.23), respectively (Fig. 9 (lower)). The difference in time spent in for forebay for bypassed and turbine entrained kelt was not significantly different (Mann–Whitney rank sum test, $P = 0.83$), suggesting the propensity for turbine entrainment was unrelated to time spent in the forebay.

The distribution of time spent in the forebay by bypassed and turbine entrained smolt is provided in Fig. 10 (upper) and, as in 2003, in both groups, fish selected their passage route, actively or passively, fairly quickly, mostly within the first 10 h of release. Sixty-five percent (170 of 263) of bypassed smolt and 33% (33 of 92) turbine entrained smolt actively or passively selected their passage route in the first 10 h. For bypassed smolt, there was a secondary peak (9%, 24 fish) at 20–40 h, corresponding to evening passage in the second night, after release. For kelts, both bypassed and turbine entrained fish selected passage routes extremely quickly, within the first hour after release. Seventy five percent (71 of 94 fish) of bypassed fish and 86% (6 of 7 fish) of turbine entrained fish actively or passively selected their passage route in the first hour (Fig. 10 (lower)).

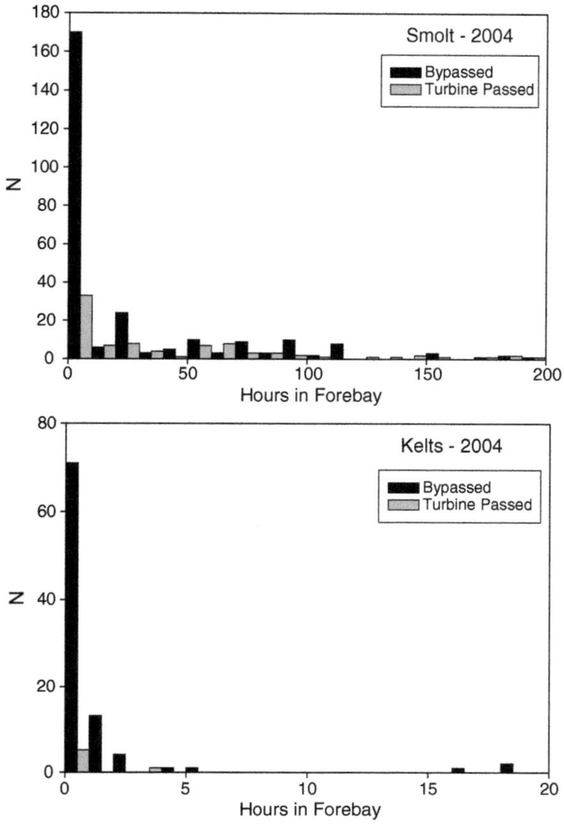

Fig. 10 The distribution of time spent in the forbay by bypassed and turbine entrained smolt (*upper panel*) and kelt (*lower panel*) in 2004

Smolts and kelt were released at about 18:00 h daily and most smolt, on a percentage basis, were either bypassed or entrained into the turbines between the hours of 19:00 and 24:00, with peak passage from 20:00 to 21:00, coincident with onset of dusk (Fig. 11). Seventy-eight percent (200 of 257) bypassed smolt and 50% (48 of 96) turbine entrained smolt left the forebay between 19:00 and 24:00, while very few smolt entered the bypass (38, 15%) or were turbine entrained (33, 34%) during daylight hours, between 05:00 h and 18:00 h (Fig. 11 (upper)). Kelts were bypassed or entrained very quickly after release and 95% (90 of 95) and 86% (6 of 7) of the bypassed and turbine entrained kelt, respectively, left the forebay during the hours of 19:00 to 24:00 (Fig. 11 (lower)). Only three kelt were bypassed during daylight hours, from dawn (05:00) to the release time (18:00) and no kelt were turbine entrained during this period.

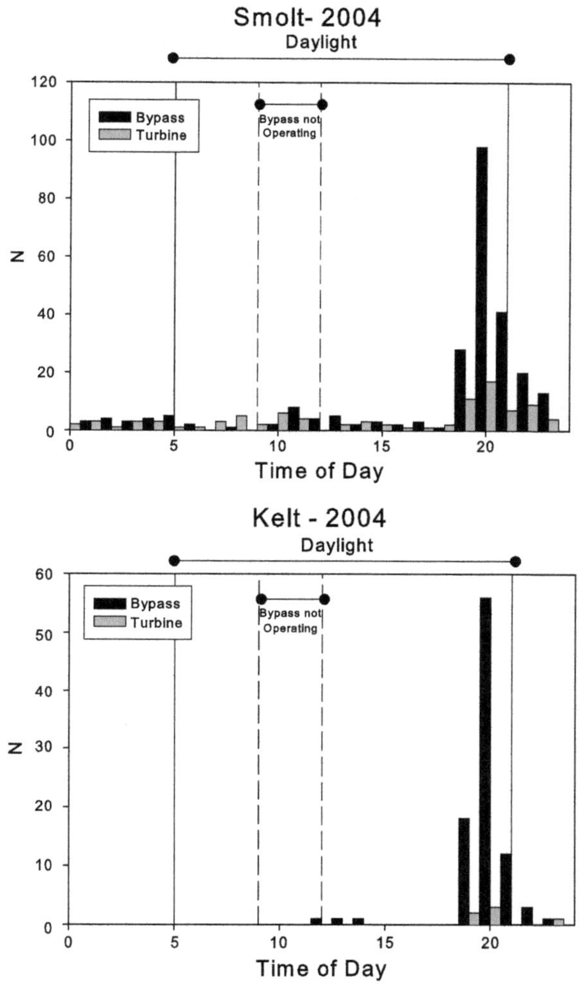

Fig. 11 Distribution of radio tagged smolt (*upper panel*) and kelt (*lower panel*) by time of day (hour) that exited via bypass and turbines. Daylight hours and the time the bypass was closed to count fish are indicated

Discussion

Bypass efficiency at the Bishop's Falls bypass for smolt in 2003 (62.3%) and 2004 (71.7%) is within the range of those published for operational hydroelectric facilities and are promising given that these tests are in the early stages of operation and testing. A release of 35 radio tagged smolt released in the Bishop's Falls headpond in 2001 resulted in an FGE of 68% (Scruton et al., 2004b) while optimized FGEs from a louver and bypass system at Grand Falls in 2000 and 2001, upstream of Bishops Falls, were 65.3% and 73.3%, respectively (Applied Biometrics Inc., 2000). FGEs at other facilities have ranged from 50% (Vernon and Bellows Falls, Connecticut River; U.S. Congress, 1995), to 80% (East River, Nova Scotia; Ducharme, 1972), to 87% (Holyoke Dam, Connecticut River; Ruggles et al., 1993). These installations had more ideal laminar flow conditions, which was not the case in the turbulent flow conditions of the Bishop's Falls forebay. There was no obvious explanation for the increase in FGE in 2004, other than the refinement of fish handling procedures and an increase in recovery time in the release cages. Release location was also varied, with cage 1 considered a possible 'worst case' scenario while cage 2 was considered more representative of naturally migrating smolt and kelt, however while FGEs were higher for cage 2, they were not significantly so.

In 2003, the time spent in the forebay by bypassed smolt was 39.2 h, significantly greater than the time spent (30.0 h) by turbine passed smolt. In 2004, bypassed fish spent significantly less time in the forebay than turbine passed fish (24.3 h versus 57.5 h for bypassed and turbine passed smolts respectively). In both years, smolt from both groups selected their passage route fairly quickly (within the first 10 h). The 2004 results suggest that the longer fish reside in the forebay the greater the likelihood they will find or select alternative passage routes such as turbine passage, returning to the headpond, or passing over the dam. In both years, bypassed smolt demonstrated a secondary peak in total time spent in the forebay, in relation to dusk in the second night after release, suggesting fish not initially bypassed may wait until dusk to attempt to locate a passage route. Forebay turbulence can determine smolt distribution, both vertically and horizontally, and smolt spend moretime higher in the water column during darkness, likely due to lower predation risk.

FGEs for radio tagged kelt were very high, averaging 92.3% and being 100% for 5 of 13 trials and there are no comparable data on kelt FGEs in the literature. Time spent in the forebay for bypassed and turbine entrained kelt was very short, averaging 1.52 and 0.90 h, respectively, with no significant difference, suggesting the propensity for turbine entrainment was unrelated to time spent in the forebay. This suggests that kelt are very efficient at locating the bypass entrance, and occupying the upper parts of the water column may facilitate this. Alternatively, kelts are stronger swimmers (Booth et al., 1997), despite their emaciated state, and may be able to navigate through the turbulent waters of the forebay better than smolt.

Conditions in the Bishop's Falls forebay are extremely turbulent and fish have been observed to be buffeted in the standing waves in the forebay. Turbulence makes fish susceptible to passive downstream displacement and, despite potential disorientation of smolt related to turbulence in the Bishop's Falls forebay, most released fish located and used the bypass system. Smolt are able to detect water velocity and acceleration gradients at dams and bypasses and sense changes in water acceleration, which may be an important cue for body orientation, change in swimming mode, and locating bypass entrances (Popper & Platt, 1993; Coutant, 1999). Fish use turbulent flow cues in a natural stream and exploit features of turbulent flow (i.e. bursts, vortices, waves) to locate regions of relatively higher velocity to speed their migration and this behavior may be exploited to facilitate fish locating fish bypasses (Coutant, 1998).

In both years, there was a distinct diel pattern in the time of day that fish used bypass and turbine passage routes. In 2004, 78% of bypassed and 50% of turbine entrained fish left the forebay between 18:00 and 24:00 h with a clear peak coincident with the onset of dusk (20:00–21:00 h). Similarly, 95% of bypassed and 86% of turbine entrained kelt exited the forebay during these hours (18:00 to 24:00). This was similar to observations of smolt in 2003, except that the peak movement period in 2003, extended to 02:00. Conversely very few radio tagged smolt (15% of bypassed and 34% of turbine entrained fish) used either of these passage routes during the day (05:00 h to 18:00 h). Results are consistent with observations on timing of smolt passage at the Grand Falls louver and bypass system, 10 km upstream from Bishop's Falls, in studies in 1998 and 1999 (Applied Biometrics Inc., 1998, 1999) and with known smolt migratory behavior and patterns, as smolt tend to move more after dark (Ruggles & Murray, 1983;

McCormick et al., 1998). Residency times in the forebay in 2003 were greater than in 2001 (average of 14.1 h; Scruton et al., 2004b). In 2004, there was a slight, but insignificant, trend for radio-tagged bypassed smolt to spend more time in the forebay as the study (smolt run) progressed. This is however an important observation and may be related to decreasing smoltification later in the migration and significant delays at this stage could lead to greater likelihood of desmoltification. Effectiveness of downstream bypass has been found to be species specific, strongly related to fish physiological condition (e.g. stage of smoltification) and very dependent on natural and regulated hydraulic conditions, and owing to these consideration, is largely considered to be site specific (Taft & Bazarian, 1983; U.S. Congress, 1995).

Atlantic salmon smolt are considered to have a limited 'window' whereby their physiological readiness, environmental conditions, and the timing of entry into salt water is critical to survival and migration delays can potentially cause desmoltification or increase mortalities (Ruggles, 1980; McCormick et al., 1998). Telemetry studies in 2003 and 2004 determined that some smolt did spend variable time in the forebay, however bypassed and turbine passed fish spent less time than fish using other exit routes. Generally, these results suggest some minor disruption in the migration process. As fish spent some time trying to locate an alternative passage routes, these delays, in a watershed the size of the Exploits, would not be considered to have any significant effect on migration delays to the extent of missing the smolt window for entry into salt water. However, there may be cumulative effects from delays at other locations in the watershed and reductions in the delay that smolt experience as they encounter obstacles in downstream migration is considered a critical element in enhancing passage and survival (Haro et al., 1998).

Conclusions

It is a continuous challenge for industry and government to minimize potential negative impact of hydroelectric power generation on downstream migrating Atlantic salmon. This is particularly complex for installations that have been constructed and operated for many years. In 2002, the refurbishing of the Bishops Falls hydro plant included the development of a surface spill type fish bypass in a forebay spill gate. Initial studies in 2003 and 2004 have documented very promising fish guidance efficiencies during initial operations and testing. Telemetry studies have documented residency time in the forebay, time of day of passage and entrainment, and post turbine passage survival of migrating fish and this information has been used to assist in improvement of the function of the bypass to optimize fish passage and survival. It is apparent that retrofit solutions can be found to improve downstream fish passage conditions at existing power plants and biological studies, involving telemetry, are important to document and improve the effectiveness of these systems.

Acknowledgements The authors would like to acknowledge the assistance of Fisheries and Oceans staff (C. Kelly, K. Smith, J. Murray, G. Clarke) and in particular the support and guidance of Mr. W. Eddy with the Abitibi Consolidated Company of Canada. Funding for this research was provided by the Exploits River Hydro Partnership, Fisheries and Oceans' Environmental Sciences Strategic Research Fund (ESSRF) and Natural Resources Canada's Panel for Energy Research and Development (PERD).

References

Adams, N. S., D. W. Rondonf, S. D. Evans, J. E. Kelly & R. W. Perry, 1998. Effects of surgically and gastrically implanted radio transmitters on swimming performance and predator avoidance of juvenile chinook salmon (*Oncorhynchus tshawytscha)*. Canadian Journal of Fisheries and Aquatic Sciences 55: 781–787.

Applied Biometrics Inc., 1998. Atlantic salmon outmigration program, 1998. Report prepared for Abitibi Consolidated Inc., Grand Falls, Newfoundland 122 pp.

Applied Biometrics Inc., 1999. Atlantic salmon outmigration program, 1999. Report prepared for Abitibi Consolidated Inc., Grand Falls, Newfoundland 73 pp.

Applied Biometrics Inc., 2000. Atlantic salmon outmigration program, 2000. Report prepared for Abitibi Consolidated Inc., Grand Falls, Newfoundland 93 pp.

Booth, R. K., E. B. Bombardier, R. S. McKinley, D. A. Scruton & R. F. Goosney, 1997. Swimming performance of post spawning adult (kelts) and juvenile (smolts) Atlantic salmon, *Salmo salar.* Canadian Manuscript Report of Fisheries and Aquatic Sciences 2406:v + 18 pp.

Bourgeois, C. E., R. F. Goosney, N. Hefford, J. Murray, C. Pennell, R. Porter & D. Scruton. 2004. Evaluation of the Efficiency of the Downstream Fish Bypass Facility, Bishops Falls Hydro Generating Station: Year 1: 2003. Prepared by Fisheries and Oceans Canada for Abitibi Consolidated Company of Canada, Grand Falls-Windsor, NL. vii + 122 pp.

Bourgeois, C. E., R. F. Goosney, N. Hefford, J. Murray, C. Pennell, R. Porter & D. Scruton, 2005. Evaluation of the Efficiency of the Downstream Fish Bypass Facility, Bishops Falls Hydro Generating Station: Year 2: 2004. Prepared by Fisheries and Oceans Canada for Abitibi Consolidated Company of Canada, Grand Falls-Windsor, NL, 108 pp.

Bourgeois, C. E., J. Murray & V. Mercer, 1994. Status of the Exploits River stock of Atlantic salmon (*Salmo salar* L.) in 1993. Canadian Department of Fisheries and Oceans Atlantic Fisheries Research Document 94/73. 13 p.

Coutant, C. C., 1998. Turbulent attraction flows for juvenile salmonid passage at dams. Oak Ridge National laboratory, Report ORNL/TM-13608. Oak Ridge, Tennessee, 29 pp.

Coutant, C. C., 1999. Think like a fish! Emphasizing the 'behaviour' in behavioural guidance systems. Hydro Review XVIII: 18–24.

Coutant, C. C. & R. R. Whitney, 2000. Fish behavior in relation to passage through hydropower turbines: a review. Transactions of the American Fisheries Society 129: 351–380.

Ducharme, L. J. A., 1972. An application of louver deflectors for guiding Atlantic salmon (*Salmo salar*) smolts from power turbines. Journal of the Fisheries Research Board of Canada 29: 1397–1404.

Haro, A., M. Odeh, J. Noreika & T. Castro-Santos, 1998. Effect of water acceleration on downstream migratory behaviour and passage of Atlantic salmon smolts and juvenile American shad at surface bypasses. Transactions of the American Fisheries Society 127: 118–127.

Hvidsten, N. A. & B. O. Johnsen, 1997. Screening of descending Atlantic salmon (*Salmo salar* L.) smolts from hydropower intakes in the River Orkla, Norway. Nordic Journal of Freshwater Research 73: 44–49.

McCormick, S. D., L. P. Hansen, T. P. Quinn & R. L. Saunders, 1998. Movement, migration, and smolting of Atlantic salmon (*Salmo salar*). Canadian Journal of Fisheries and Aquatic Sciences 55: 77–92.

Moore, A., I. C. Russell & E. C. E. Potter, 1990. The effects of intreperitoneally implanted dummy acoustic transmitters on the behaviour and physiology of juvenile Atlantic salmon, (*Salmo Salar* L). Journal of Fish Biology 37: 713–721.

Mullins, C. C., C. E. Bourgeois & T. R. Porter, 2003. Opening up new habitat: Atlantic salmon (*Salmo salar* L.) enhancement in Newfoundland, pp. 200–221. *In* Mills, D. [ed.] Salmon at the edge 307 p.

O' Connell, M. F. & C. E. Bourgeois, 1987. Atlantic salmon (*Salmo salar* L.) Enhancement of the exploits River, Newfoundland, 1957–1984. North American Journal of Fisheries Management 7: 207–214.

O'Connell, M. F., J. B. Dempson, C. C. Mullins, D. G. Reddin, C. E. Bourgeois, T. R. Porter, N. M. Cochrane & D. Caines, 2003. Status of Atlantic Salmon (*Salmo salar* L.) Stocks of Insular Newfoundland (SFAs 3–14A), 2002. Canadian Department of Fisheries and Oceans, Canadian Science Advisory Secretariat Research Document 2003/002.

Popper, A. N. & C. Platt, 1993. Inner ear and lateral line, pgs. 99–139, In Evans D. H. (eds). The physiology of fishes. CRC Press, Boca Raton, Florida, 544 pp.

Porter, T. R., 1975. Biology of Atlantic salmon in Newfoundland and Labrador. Fisheries and Marine Service. Newfoundland Region. Resource Development Branch. Information Report Series NEW/N. 75–2.

Ruggles, C. P., 1980. A review of the downstream migration of Atlantic salmon. Canadian Technical Report of Fisheries and Aquatic Sciences 952: 39 pp.

Ruggles, C. P. & D. G. Murray, 1983. A review of fish response to spillways. Canadian Technical Report of Fisheries and Aquatic 1172: 31 pp.

Ruggles, C. P., D. A. Robinson & R.J. Stira, 1993. The use of floating louvers for guiding Atlantic salmon smolts from hydroelectric turbine intakes. *In* Proceedings of the workshop on fish passage at hydroelectric developments, 26–28 March 1991. St. John's, Newfoundland. *Edited by* U. P. Williams, D. A. Scruton, R. F. Goosney, C. E. Bourgeois, D. C. Orr and C. P. Ruggles. Canadian Technical Report of Fisheries and Aquatic 1905. 87–94.

Scruton, D. A., C. J. Pennell, R. F. Goosney, C. E. Bourgeois, T. R. Porter & K. D. Clarke, 2004a. Initial assessment of a retrofitted downstream fish bypass system for wild Atlantic salmon (*Salmo salar*) smolts on the Exploits River, Newfoundland, Canada. P. 1026–1030. In Garcia de Jalon Lastra, D. & Vizcaino Martinez, P. (eds), Fifth International Conference on Ecohydraulics. Aquatic Habitats: Analysis and Restoration. Madrid, Spain, 2004. International Association of Hydraulic Research Congress Proceedings. Volume II, 1453 pp.

Scruton D. A., C. J. Pennell, M. J. Robertson, K. D. Clarke, W. Eddy & R. S. McKinley, 2004b. Telemetry studies of the passage route and entrainment of downstream migrating wild Atlantic salmon (*Salmo salar*) smolts at two hydroelectric installations on the Exploits River, Newfoundland, Canada. In Spedicato, M. T., G. Lembo, G. Marmulla (eds), *Aquatic telemetry: advances and applications*, 91–101 pp., *FAO/COISPA*, Rome, 296 pp.

Taft, E. P. & E. Bazarian, 1983. State-of-the-art in preventing turbine mortality at hydro facilities. Stone and Webster Report, TP 83–79. 16 pp.

Taylor, V. R. & B. R. Bauld, 1973. A program for increased Atlantic salmon (*Salmo salar*) production on a major Newfoundland River. International Atlantic Salmon Foundation Special Publication Series 4: 339–347.

U.S. Congress, Office of Technologies Assessment, 1995. Fish Passage Technologies: Protection at Hydropower facilities. OTA-ENV-641. Washington, DC, United States Government Printing Office, 167 pp.

Hydrobiologia (2007) 582:171–185
DOI 10.1007/s10750-006-0551-z

FISH TELEMETRY

PIT telemetry as a method to study the habitat requirements of fish populations: application to native and stocked trout movements

Amílcar Teixeira · Rui M. V. Cortes

Abstract Passive integrated transponder (PIT) technology was used to study the behaviour of fishes during the summer season in two headwater streams of northeastern Portugal. A total of 71 PIT tags (12 mm long × 2.1 mm diameter) were surgically implanted in 1^+ stocked (39) and native (32) brown trout of two size classes (<20.0 and ≥20.0 cm). Eight independent antennae, connected to a multi-point decoder (MPD reader) unit, were placed in different microhabitats, selected randomly every 3 days during the observation period (29 August–9 September in Baceiro stream and 19 September–4 October in Sabor stream). The results confirmed this method as a suitable, labour efficient tool to assess the movement and habitat use of sympatric stocked and native trout populations. About 76.9% of stocked and 59.4% of native PIT tagged trouts were detected. Multivariate techniques (CCA, DFA and classification tree) showed a separation in habitat use between the two sympatric populations. Stocked trout mainly used the microhabitats located in the middle of the channel with higher depths and without cover. Furthermore, these fishes displayed a greater mobility and a diel activity pattern different to native trout populations.

Keywords PIT tag · Brown trout · Stocking · Habitat use · Movement

Guest editors: P. R. Almeida, B. R. Quintella, M. J. Costa and A. Moore
Developments in Fish Telemetry

A. Teixeira (✉)
CIMO-Escola Superior Agrária Bragança, Campus Sta Apolónia, Apt. 1172, 5301-855 Bragança, Portugal
e-mail: amilt@ipb.pt

R. M. V. Cortes
Dep. Florestal, Universidade de Trás-os-Montes e Alto Douro, Apt. 1013, 5001-911 Vila-Real, Portugal
e-mail: rcortes@utad.pt

Introduction

Stocking of hatchery-reared salmonids into natural water systems is still a fishery management tool frequently used to mitigate loss of stocks and enhance or restore fishery activities (Cowx, 1999). Increasing concerns have been expressed about the negative ecological and genetic effects of stocking on wild populations (White et al., 1995; Pearsons & Hopley, 1999). For this reason, several studies were developed in order to understand the potential impacts resulting from the competitive interactions established (McMichael et al., 1999; Weber & Fausch, 2005) and distinct methods were applied to study the habitat requirements, movements and fish behaviour (reviewed by Weber & Fausch, 2003). But, all these methods, broadly grouped by Lucas & Baras (2000) in two main categories: (1) capture-independent (e.g. underwater observations) and (2) capture-dependent (e.g. mark-recapture and telemetry techniques), present some limitations (Heggenes et al., 1990; Bridger & Booth, 2003). Recent technology based on radiotelemetry has been widely used, providing a high-resolution, in temporal and spatial scale, of information at individual level. Applications of these studies may be the definition of the home range of a target species, specifying the diel (Young, 1999; Belanger & Rodriguez, 2001) and seasonal movements (Burrell et al., 2000), the influence of environmental factors (Ovídio et al., 1998) and the efficacy of fish pass programs (Scruton et al., 2002). Nevertheless, the high costs of individual tags and the detection equipment restrict the number of fishes utilized in the studies. Furthermore, the small-bodied fishes and fine scale movements are also other limitation that constrains the use of radiotelemetry (Prentice et al., 1990a; Lucas & Baras, 2000). Alternatively, passive integrated transponder (PIT) technology was developed (Prentice et al., 1990b; Barbin-Zydlewski et al., 2001) for tracking even small aquatic animals in shallow waters, involving low equipment costs and the possibility of addressing numerous questions in fields of animal behaviour, habitat use and population dynamics (Roussel et al., 2000; Quintella et al., 2005). Besides, other advantages could be assigned to this technique, like the theoretically indefinite life span added to high tag retention and no apparent effects on growth and survival of tagged animals (Ombredane et al., 1998; Bubb et al., 2002). In addition, PIT telemetry equipment options (e.g. stationary versus portable detectors) were developed and applied with similar objectives of radiotelemetry technology, for example, the evaluation of fish pass efficiency (Prentice et al., 1990b; Castro-Santos et al., 1996) or habitat use and behavioural studies of fish populations (Riley et al., 2003; Roussel et al., 2004).

The objective of the present study was to examine the comparative behaviour and habitat use of sympatric native and stocked brown trout in two streams of northeastern Portugal using PIT telemetry technology and multiple antennae. In detail, the detector system allowed continuous surveys on the movements, habitat use and activity pattern of both populations in two stream segments. PIT tags were selected because, besides having a comparatively low cost, their dimensions have a negligible effect on fish movements, which is a crucial aspect in fine-scale movements. Further, a comparison between PIT telemetry and a direct observation method (snorkelling) was also made.

Materials and methods

Study area

The experiments were carried out in the Baceiro and Sabor headwater streams (3rd order), tributaries of the River Douro basin, located in the Montesinho Natural Park, northeastern Portugal. These streams are subjected to a reduced human pressure, which contributes to the low impact on water composition (conductivity <70 $\mu S\ cm^{-1}$, alkalinity <25 mg $HCO_3^{-1}\ l^{-1}$, hardness <15 mg $CaCO_3\ l^{-1}, NO_3^- < 0.50$ mg $l^{-1}, PO_4^{3-} < 0.02$ mg l^{-1}). Banks are covered by dense riparian vegetation dominated by alder (*Alnus glutinosa*, Gaertn), limiting the primary production in these systems where the food webs are energetically dependent on allochthonous inputs of organic matter. The fish community is dominated by native brown trout populations, but endemic cyprinid species are also present, such as the Iberian chub *Squalius carolitertii* (Doadrio, 1987) and the Iberian nase *Chondrostoma duriensis* Coelho, 1985.

PIT telemetry equipment

The PIT technology used was based on a multi-point decoder (MPD) unit (UKID Systems Ltd, Preston, U.K.). This unit consists of DC integrated MPD/antenna multiplexer (8-channel) powered by a 24 V (18 Ah) rechargeable lead-acid battery pack, which provided more than 24 h of continuous use, and eight black circular panel antennae connected to the PIT-tag reader by cable lengths of 10 m. Each panel antenna (22 mm deep and 300 mm in diameter) operates at a frequency of 134 kHz and has a detection range of approximately 90 mm for the 12.0 mm long and 2.1 mm in diameter PIT tags (122IJ;

UKID Systems) used in this study. This system enables logging up to 1000 time-stamped events from an onboard Real Time Clock and the Battery Backed-up Memory. In order to reduce the number of repetitive events, resulting from a fish that remained over the same antenna, a data repeated filter precluded the repeat reading of the same tag code within each 25 s period. The identification data (ID) output was further downloaded from the MPD (via RS232) to a personal computer. The battery pack and the MPD were protected by a special enclosure (Peli-Plastic equipment case).

Field surveys

Two representative stream segments (30 m long by a mean width of 6 m), with riffle and pool habitats, were selected in the Baceiro and Sabor streams. Aquatic habitat was assessed, before PIT telemetry experiments, based on transects (starting point randomly chosen), made perpendicular to the stream, with 5 m intervals throughout each stream segment. Point measurements were done at 0.5 m intervals across each transect for the variables of total depth, surface velocity (measured 10 cm below the surface), bottom velocity (10 cm above the streambed) and mean water column velocity (0.6 of total depth), substrate composition and cover. Substrate composition was classified according to a modified Wentworth scale, adopting the following categories: (1) organic detritus; (2) silt and sand (<2 mm); (3) gravel (2–16 mm); (4) pebble (17–64 mm); (5) cobble (65–256 mm); (6) boulder (>256 mm) and (7) bedrock. Cover types were divided into five categories: (1) objects >15 cm (substrate emerging from the streambed); (2) overhanging vegetation; (3) roots, undercut banks and submerged woody debris; (4) surface turbulence and (5) no cover. Total depth was directly measured with a stick meter and the velocities were measured with a Valeport electronic flowmeter. For the available habitat surveyed, total depth was 56 cm (±0.08 SE; maximum depth = 110 cm) in the Baceiro and 42 cm (±0.15 SE; maximum depth = 95 cm) in the Sabor. Mean water column velocity was higher (0.071 ± 0.03 m s^{-1} SE; maximum velocity (riffle)—1.10 m s^{-1}) in the Sabor than in the Baceiro (0.041 ± 0.09 m s^{-1} SE maximum velocity (riffle)—0.90 m s^{-1}). Substrate composition was dominated by pebbles and cobblestones in the Sabor and sand, cobblestones and boulders in the Baceiro streams. Overhanging vegetation, undercut banks and boulders were the main types of available cover for fish. During field experiments the water temperature displayed a low variation in both streams (14–19°C in the Baceiro and 13–18°C in the Sabor).

A total of 39 stocked and 32 native brown trout were marked with PIT-tags in the two experiments realized in the Baceiro and Sabor streams. Before the operation, each fish was individually anesthetized in a solution of 2-phenoxy-ethanol (0.4 ml l^{-1}) and the abdominal region disinfected with an iodine solution (Betadine). The PIT-tag was surgically implanted in the peritoneal cavity with a sterilised needle linked to a special tagging gun. Stocked trout were tagged in the hatchery and held for eight days (at low density) in raceway tanks to recover from this operation. No tag loss or mortalities were registered during this period. Between 24 August and 9 September 2004, the entire study section in the Baceiro was closed with barrier nets until the end of the experiment, and the study area was sampled to depletion with several electrofishing sweeps (Hans Grassl DC, 1.5 W, 300/600 volts). Nineteen native trout (size range between 12.0 and 26.0 cm), were captured and also anaesthetized, the total length measured (±1 mm) and marked with the PIT-tags following the procedure mentioned above. After a recovery period of 2 h, they were released, together with 26 stocked fishes (size range between 16.2 and 23.0 cm), in the stream segment selected. Between 19 September and 4 October 2004 the same procedures were conducted along the segment selected in the Sabor. Here, the trout sample tagged consisted of 13 native (size range between 13.7 and 28.5 cm) and 11 stocked trout (size range between 17.0 and 26.5 cm).

The MPD unit and the eight panel antennae were installed in the streambed according to the diagram showed in Riley et al. (2003), although with a random antenna positions. Five sections (6 m long) were considered in the selected segment, since the maximum length of each cable

(10 m) connecting the antenna to MPD unit limited the sampling, at the same time, of the whole study site. The experiments began systematically from the upstream to the downstream section, and each antenna was randomly positioned and every 3 days repositioned in a new section. Immediately after each antenna installation, the following microhabitat variables were recorded: total depth, dominant and subdominant substrate, surface, water column and bottom (near the antenna) velocities, aquatic cover, overhanging vegetation (shading), distance to riffle and distance to nearest stream bank. During study experiments the weather conditions remained quite stable (no substantial variations were detected for water level regimes) and the values of microhabitat measurements assumed constant for every 3-day period. In order to minimize the visual effect of the antennae on fish, they were covered, when possible, with a thin layer of the adjacent substrate and the first recording considered only 30 min (normal activity of fish re-established) after the perturbation promoted by the operator with the antennae repositioning and habitat measurements.

Habitat use by stocked and native trout was also studied, during September 2002, by snorkelling observations. This is a direct method usually employed in the evaluation of habitat use by fish in small streams (Thurow & Schill, 1996). The stream segments (defined for telemetry experiments) were surveyed and the data observations distributed by 140 small native (<20.0 cm), 26 big native (≥20.0 cm) and 208 stocked trout to assess the microhabitat used. To the data collection a snorkeler moved systematically in upstream direction and after locating an undisturbed fish (only considered when maintaining a position for at least during 2 min), the TL-total length (estimated to the nearest cm through comparison with substrate particles) and the trout identification (stocked versus native) were registered. Several characteristics allowed the snorkeler to correctly identify a stocked fish, such as the presence of VIE *elastomer* (Northwest Marine Technology©) located in the adipose fin and post-ocular tissue, the body's color (uniform and gray) and the pectoral fins (smaller). Then, each fish position was marked with a numbered lead-weighted float (surface and focal position identified) and, after the whole stream segment was snorkeled, the measurements were made for the variables of focal velocity (velocity in the holding fish position), water column velocity (measured at 0.6 of total depth), surface velocity, focal elevation (vertical distance from bottom to fish's snout), total depth, dominant and subdominant substrate (25 × 25 cm) in a vertical below the fish and cover (objects that could provide shelter for, at least, of 50% fish's body).

Data analyses

The statistical analyses were based on the information obtained in terms of repeated (all events recorded in the overall antennae by the different individuals) and non-repeated (the continuous repeated records of each fish in the same antenna were not considered) frequency data recording by de MPD unit during the sampling period in both streams.

Multivariate analyses techniques were applied to the PIT-telemetry data. A direct ordination technique—canonical correspondence analysis (CCA), was performed for both streams through the CANOCO software package (Version 4.02, Ter Braak & Smilauer, 1998) to evaluate the links between environmental variables and microhabitat used by fish. Data were standardized for microhabitat variables and logged transformed [log $(x + 1)$] for the non-repeated frequency of the detected fish in all antennae positions. Only those variables with a variation inflation factor (VIF) of less than 20 were included to avoid multicollinearity (Ter Braak, 1986). In addition, a Monte Carlo permutation test (199 permutations) was performed to test the significance of stocked and native trout and the environmental variables.

A classification tree analysis was used to predict the membership of cases in the three classes of a categorical dependent variable, defined as: (1) stocked (age 1+ and mean size of 18.0 cm); (2) small native <20.0 cm and big native ≥20.0 cm trout, from the following predictor variables: maximum values of water column velocity, cover, distance to streambank, both maximum and minimum values of depth, dominant substrate and distance to riffle. These variables were selected after being eliminated those with low variance. Fish without registered occurrences, at least, in three distinct antennae were not included in this analysis in order to eliminate the outlier data resulting from fish detected only in the first days. The classification tree performed was based on discriminant univariate splits, considering estimated prior probabilities, equal misclassification costs and a FACT-style direct stopping with fraction of objects of 0.30. This is an exploratory technique known for its flexibility to examine the effects of the predictor variables one at a time, which display a graphic presentation of easy interpretation. However, when more stringent theoretical and distributional assumptions are met, traditional methods may be preferable (Breiman et al., 1984). In this way, another multivariate technique was carried out—Discriminant Function Analysis (DFA), with a statistical base similar to the classification tree, for the same variables and fish groups (stocked, small and big native trout) considering a forward stepwise analysis. Analogous analysis (DFA) was also applied to direct observation data, taking in account the same trout classes pre-defined in order to find the variables with the most discrimination power and to establish comparisons between both methods (PIT telemetry versus snorkelling).

Trout activity was analysed considering intervals of 3 h and the antennae non-repeated frequency records and, subsequently, polynomial regressions were fitted to the data. Differences among all trout classes for the microhabitat variables defined were assessed using Kruskall–Wallis H-tests because data did not fit to the assumptions of normality (Bartlet test). All statistical analyses were performed using STATISTICA 7.0 package (STATSOFT, 2004).

Results

A total of 19,326 fish records (tag codes) were successfully registered by the MPD unit in the experiments realized in both streams, and only 0.07% corresponded to unidentified tag codes. From 71 PIT-tagged fishes, 76.9% (30 individuals) of stocked and 59.4% (19 individuals) of native trout were detected at least once by the reader unit. Comparatively, a higher number of fish detections were obtained in the Sabor than in the Baceiro (84.6 vs. 73.1% for stocked and 76.9 vs. 47.4% for native trout). Dominant native trout showed a higher activity (based on the frequency of the recordings) during the dawn period (6.01–9.00 h) in both streams. Small native trout had similar behaviour to big native trout in the Baceiro, but avoided the same period in the Sabor, probably related to the greater mobility exhibited by bigger native and stocked trout found in the major activity in the evening. No apparent reasons justify the distinct stocked trout behaviour detected between the two streams (Figs. 1, 2). Stocked fishes displayed greater mobility than native trout, since a higher frequency was observed (66.3% for total events and 68% for non-repeated events).

Habitat use by stocked and native trout differed markedly for the majority of variables considered (Table 1). Total depth observed by small and big native trout was significantly lower than the total depth used by stocked fish in both streams ($P < 0.001$, H test). Similarly, the velocities measured at the three vertical levels were significantly higher for the big native trout than for small native and stocked trout, in both streams ($P < 0.05$, H test). Bigger native trout more often used the upstream area closest to the riffle, meanwhile stocked trout were located in the middle of the channel, since, comparatively with native trout, they occupied habitats significantly more distant from the stream bank ($P < 0.05$, H test). For hatchery-reared fish the aquatic cover provided by undercut banks and roots was, in both streams, the category type less used. We must mention that an important proportion of stocked fish remained in areas without any type of cover. Conversely, the habitat used by the dominant native trout displayed a preference for coarse particles (pebbles, cobblestones and boulders). Meanwhile for small native trout the undercut banks and the overhanging vegetation (especially in the Sabor) was also another elected shelter type. The links between the microhabitat variables and stocked and native trout for both streams are shown in the CCA ordinations (Figs. 3, 4). The first two CCA axes had eigenvalues of 0.750 and 0.314 for the Baceiro and

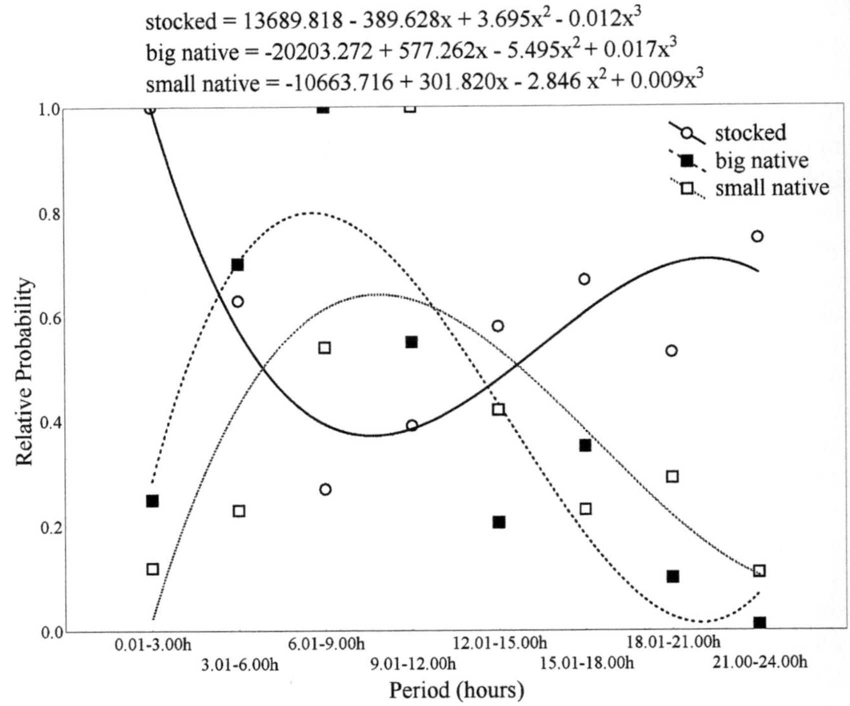

Fig. 1 Activity pattern, using polynomial regressions, performed for stocked (n = 527), small native (n = 148) and big native trout (n = 215) relative to eight diel periods and 3-h classes, in the Baceiro stream. The dependent variable represents the relative probability of use (standardized to a 0–1 scale). Stocked trout are represented by a solid line, big native trout by a dashed line and small native trout by a dotted line

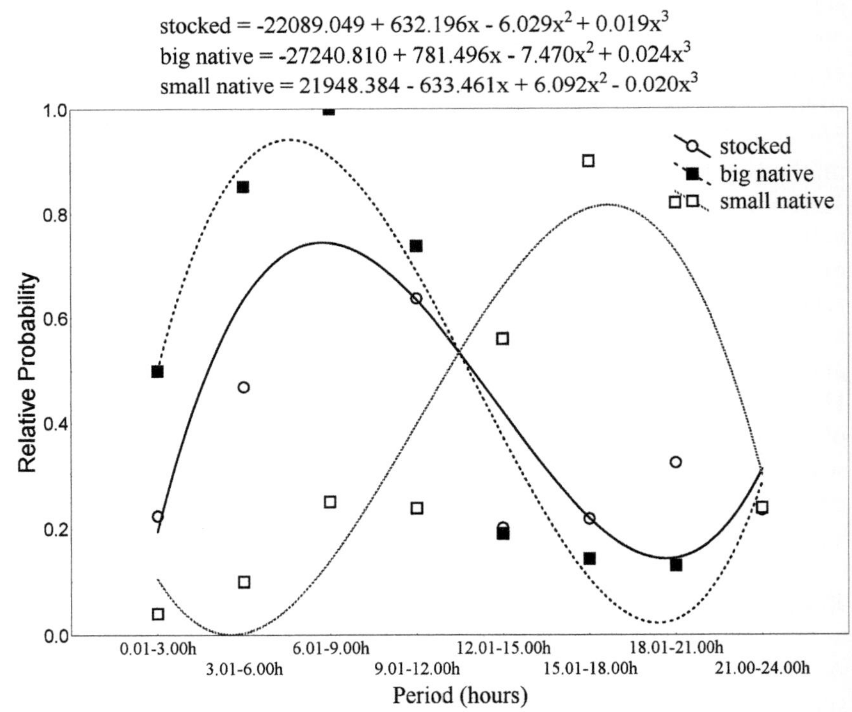

Fig. 2 Activity pattern, using polynomial regressions, performed for stocked (n = 585), small native (n = 88) and big native trout (n = 72) relative to eight diel periods and 3-h classes, in the Sabor stream. The dependent variable represents the relative probability of use (standardized to a 0–1 scale). Stocked trout are represented by a solid line, big native trout by a dashed line and small native trout by a dotted line

0.298 and 0.224 for the Sabor, explaining together, respectively, 65.2% and 53.7% of variation in the relationship between native and stocked trout and environmental variables (Table 2). Furthermore, the overall Monte Carlo randomization tests showed significant results for the sum of all eigenvalues (199 permutations, $P < 0.05$). The environmental variables most

Table 1 Mean (±1 S.E.) total depth (cm), distance to nearest stream bank (cm), distance to riffle (cm) surface velocity (m s^{-1}), water column velocity (m s^{-1}) and bottom velocity (m s^{-1}) from the 40 antennae positions of small native, big native and stocked trout in the Baceiro and Sabor streams

Microhabitat variables	Baceiro stream			Sabor stream		
	Small native (n = 178)	Big native (n = 185)	Stocked (n = 527)	Small native (n = 18)	Big native (n = 142)	Stocked (n = 585)
Total depth	56.0 ± 1.16	59.2 ± 1.75	65.9 ± 0.98	34.6 ± 2.88	34.3 ± 1.22	62.3 ± 1.19
Distance to stream bank	93.0 ± 9.51	100.7 ± 5.51	117.8 ± 4.40	58.4 ± 12.12	83.6 ± 8.53	211.7 ± 6.05
Distance to riffle	1393.6 ± 64.23	314.2 ± 10.28	445.5 ± 18.59	1477.2 ± 142.9	648.9 ± 46.98	1236.1 ± 32.94
Surface velocity	0.017 ± 0.003	0.073 ± 0.004	0.065 ± 0.003	0.055 ± 0.030	0.223 ± 0.02	0.123 ± 0.01
Water column velocity	0.014 ± 0.003	0.077 ± 0.004	0.057 ± 0.003	0.044 ± 0.024	0.165 ± 0.01	0.083 ± 0.01
Bottom velocity	0.014 ± 0.003	0.088 ± 0.006	0.056 ± 0.003	0.034 ± 0.020	0.134 ± 0.01	0.057 ± 0.01
Dominant substrate						
Organic detritus	21.5	4.9	18.0	0.0	0.0	0.0
Silt and sand	0.6	0.0	10.2	0.0	0.0	0.0
Gravel	28.1	9.2	6.3	33.3	2.8	23.6
Pebble	5.6	31.4	19.7	38.8	3.5	2.1
Cobble	21.2	43.2	27.8	5.6	41.5	13.2
Boulder	9.0	11.3	17.3	16.7	41.5	53.3
Bedrock	14.0	0.0	0.8	5.6	10.7	7.8
Aquatic cover						
Substrate particles (>15 cm)	60.7	64.3	35.1	11.1	45.8	28.4
Overhanging vegetation	7.3	3.2	5.3	44.4	12.0	13.2
Roots, undercut banks, woody debris	10.7	0.5	0.4	38.9	12.7	0.7
Surface turbulence	2.8	17.8	16.7	0.0	3.5	6.0
No cover	18.5	14.1	42.5	5.6	26.1	51.8

Cover and dominant substrate type used by fishes are shown as relative frequency (%). Numbers of observations (n) are presented in parentheses and correspond to non-repeated frequency data for each trout class

important in the stocked and native trout distribution among antennae sites were the distance to riffle and the distance to nearest stream bank for both streams and subdominant substrate (Baceiro), total depth and overhanging vegetation (Sabor) (Table 3, Figs. 3, 4).

The classification tree diagram is shown on Fig. 5. Initially the 36 trout considered are assigned to the root node (node 1, top-left corner) classified as *stocked* (top-right corner) because there is a prevalence of stocked (23 individuals) compared to native trout (small-7 and big-6 individuals). Two new nodes (nodes 2 and 3) were formed by the split of the root node. The computational method selected the maximum depth as the predictor variable that produced the smallest P-level and the cluster analysis (2-means) defined the separation value of 86.5 cm between nodes. Similarly, subsequent node splits were successively made and different predictor variables were selected (maximum cover, minimum and maximum depths). In the end, five splits and six terminal nodes allowed the separation between stocked and native trout. Just one terminal node (node 8) reached "pure" classification; nevertheless, the percentage of objects misclassified was 16.7. Additionally, the discriminant-based univariate split option allowed for the identification of the main predictor variables. The best rankings (0–100 scale) were obtained by the following variables: maximum depth (rank 100), maximum distance to riffle (rank 69) and minimum dominant substrate (rank 62).

In addition, another multivariate technique, DFA was applied to these data, considering the same nine variables and the predefined trout groups. The results, confirmed that the maximum depth was the variable that most contributed to the separation between stocked, small and big native trout (Table 4). However, all stocked trouts were correctly classified (100%), unlike the big (33.3%) and the small (14.3%) native

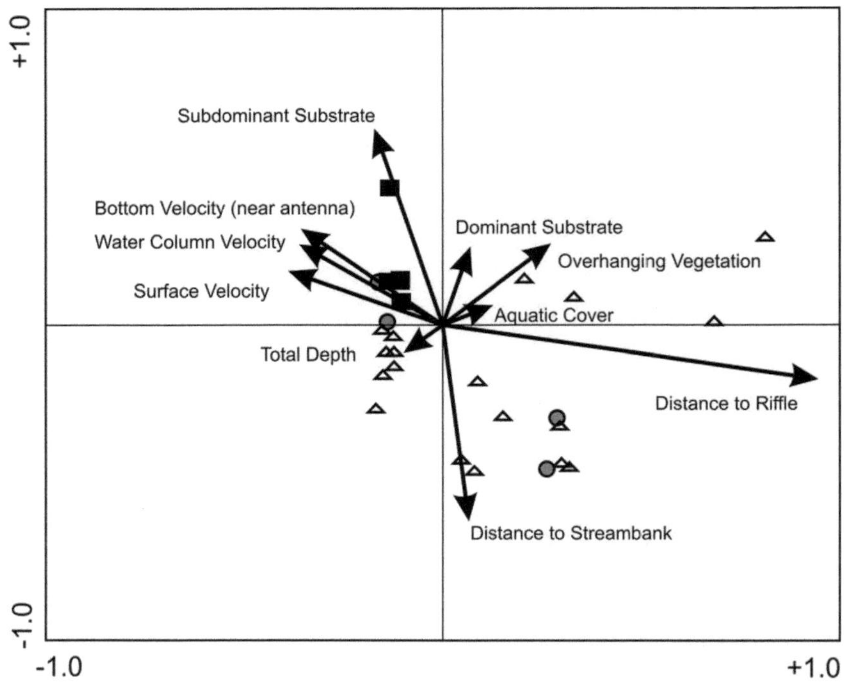

Fig. 3 CCA ordination diagram—Baceiro stream: distribution of native and stocked trout according to the selected microhabitat variables for the two first axes. The arrows represent the microhabitat variables and the symbols are the trout identification: (A) arrows—total depth; aquatic cover; overhanging vegetation; surface, water column, and bottom (near antenna) velocities; distance to riffle; distance to the nearest streambank; (B) Symbols: ● = small native; ■ = big native; Δ = stocked trout. The length of the arrow is a measure of the importance of the environmental variable and the arrowhead points at the direction of increasing influence

trout. The total percentage of corrected cases obtained by the DFA (72.2%) was lower when compared with classification tree analysis (83.3%), which is more flexible. The scatterplot of the two principal roots (Fig. 6) exhibits, in fact, a less clear separation between the groups. Table 4 presents the summary output of DFA analysis, where no significant functions (canonical roots) were found.

Finally, when this analysis (DFA) was applied to direct observational data (i.e. snorkelling), focal elevation, was identified as one of the most important variables in the discrimination detected between the predefined groups (Table 5). In fact, stocked trout occupied water column positions (focal elevations) more distant from the streambed when compared with native trout. In the DFA analysis two significant functions were calculated and an evident separation between stocked and native trout can be shown along the root 1 (Fig. 7). Besides, comparatively with DFA made with PIT telemetry data, a superior percentage (94.1) of corrected cases was detected (Table 5).

Discussion

Distinct habitat use was identified for the stocked and the two native trout classes by PIT-telemetry. These results corroborate the direct observations (snorkelling) that took place in the same streams considering the two sympatric populations mentioned. Multivariate analyses (CCA, DFA and classification tree) performed with data collected by both methods (PIT-telemetry and snorkel observations) found the total depth and the aquatic cover as being important microhabitat variables that contributed to the separation of predetermined groups. Despite this, some important differences between both methods were registered, resulting from the limitations assigned

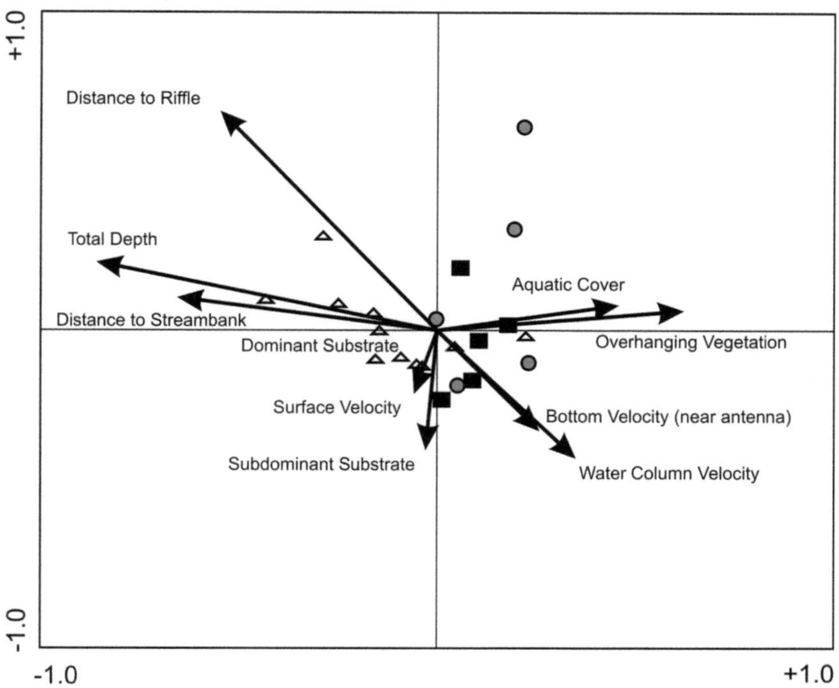

Fig. 4 CCA ordination diagram—Sabor stream: distribution of native and stocked trout according to the selected microhabitat variables for the two first axes. The arrows represent the microhabitat variables and the symbols are the trout identification: (A) arrows—total depth; aquatic cover; overhanging vegetation; surface, water column, and bottom (near antenna) velocities; distance to riffle; distance to the nearest streambank; (B) Symbols: ● = small native; ■ = big native; Δ = stocked trout. The length of the arrow is a measure of the importance of the environmental variable and the arrowhead points at the direction of increasing influence

Table 2 Summary of the canonical correspondence analysis (CCA) for Baceiro and Sabor streams

	Baceiro stream			Sabor stream		
	Axis 1	Axis 2	Total inertia	Axis 1	Axis 2	Total inertia
Eigenvalues	0.750	0.314	4.090	0.298	0.224	2.900
Species–environment correlations	0.963	0.816		0.792	0.748	
Cumulative percentage variance						
of species data	18.4	26.0		10.3	18.0	
of species–environment relations	46.0	65.2		30.6	53.7	
Sum of all canonical eigenvalues			1.633			0.973

Monte Carlo permutation tests run for the sum of all eigenvalues were significant (199 permutations, $P = 0.005$ for Baceiro stream and $P = 0.015$ for Sabor stream)

to each method. For example, the focal elevation (vertical distance of fish position to the bottom of the stream) was the most discriminant microhabitat variable in underwater observation method, but can not be considered in the PIT-telemetry method applied in this study, since the detection range of each panel antenna is approximately 90 mm for the PIT-tags used (12.0 mm long). For this reason, biased data could be related to the efficiency of detection especially when the fish holds a position, or swims, just above an antenna and over the detection range. It is important to mention that in this study a considerable area is out of the detection range of the PIT antennae and underestimate data were obtained essentially by stocked trout. Indeed, it was observed during

Table 3 Interset-correlations (Pearson correlation coefficient) of environmental variables with the CCA axes for Baceiro and Sabor streams

Environmental variables	Baceiro stream		Sabor stream	
	Axis 1	Axis 2	Axis 1	Axis 2
Total depth	–0.092	–0.063	–0.681**	0.156
Surface velocity	–0.373*	0.135	–0.049	–0.155
Water column velocity	–0.349*	0.211	0.278	–0.304*
Bottom (antenna) velocity	–0.340*	0.249	0.207	–0.240
Dominant substrate	0.066	0.201	0.004	–0.091
Subdominant substrate	–0.164	0.507**	–0.024	–0.286
Aquatic cover	0.116	0.049	0.361*	0.054
Distance to nearest streambank	0.061	–0.503**	–0.520**	0.075
Distance to riffle	0.908**	–0.137	–0.433**	0.515**
Overhanging vegetation	0.257	0.210	0.491**	0.046

*$P < 0.05$, **$P < 0.01$

the snorkelling survey that native trout normally stay close to the streambed, except in particular moments like the feeding activity, in contrast with the behaviour displayed by stocked trout, which often occupy higher focal elevations (superficial and middle water column positions).

Similarly to other studies (Heggenes 1988; Rincón & Lobón-Cerviá, 1993; Maki-Petays et al., 2002) where different sampling methods (e.g. snorkelling, electrofishing, seining) were used, PIT-telemetry proved to be an effective method to study the habitat requirements of fish populations. However, all methods present different limitations (Heggenes et al., 1990; Dolloff et al., 1996; Joyce & Hubert, 2003). Generally, the main limitations cited for PIT-telemetry in field surveys are the stream depth (this method is more adapted to shallow streams), antenna range detection, species behaviour and the alterations promoted in the substrate composition due to the installation of the antenna (Riley et al., 2003; Cucherousset et al., 2005). On the other hand,

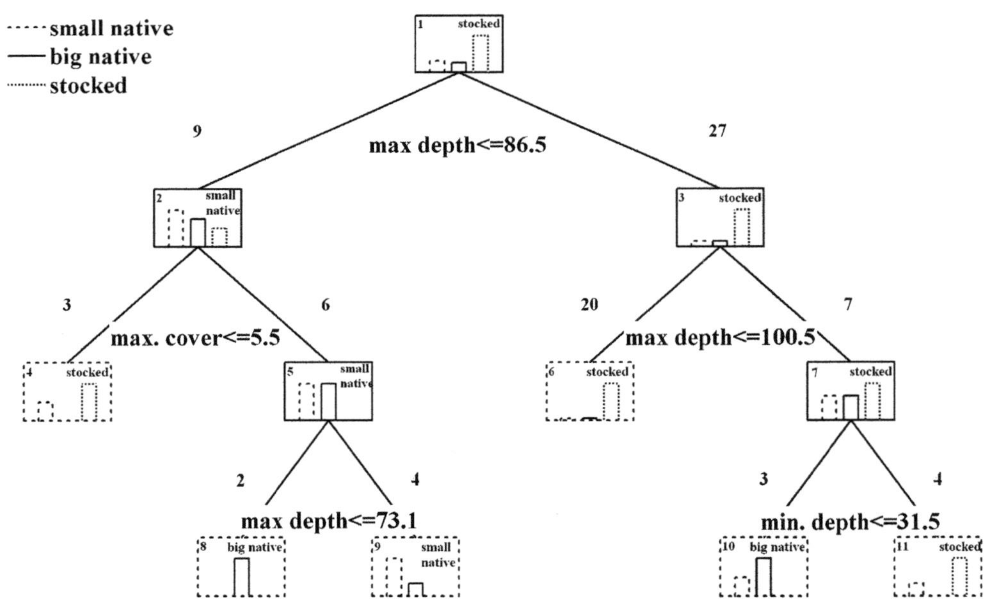

Fig. 5 Classification Tree—Tree structure graph for small native, big native and stocked trout. Five splits and six terminal nodes were found, at the end, with a global misclassification of 16.7%. Non-repeated frequency data (n = 1635) of 36 trout were used in this analysis excluding all fishes without, at least, recorded events on three different antennae. Terminal nodes are outlined with dashed lines

Table 4 Discriminant function analysis (DFA) summary applied to PIT-telemetry data-Partial Wilks' Lambda for each variable and the classification functions according to different trout classes (small native, big native and stocked trout)

Discriminant function analysis

n = 1635 Variables	Wilks' lambda	Classification functions		
		Small native P = 0.194	Big native P = 0.167	Stocked P = 0.639
Maximum depth	0.701	–0.231	–0.734	0.262
Max. water col. velocity	0.666	0.384	–0.260	–0.050
Max. dominant substrate	0.695	0.475	–0.796	0.063
Maximum cover	0.715	–0.333	1.007	–0.161
Max. distance to bank	0.661	–0.060	–0.087	0.041
Max. distance to riffle	0.665	–0.233	0.203	0.018
Minimum depth	0.701	0.809	–0.460	–0.126
Min. dominant substrate	0.704	–0.600	0.730	–0.008
Min. distance to riffle	0.663	–0.001	–0.300	0.078
Constant	–	–1.933	–2.607	–0.487
Predicted classifications (% corrected cases)		14.3	33.3	100

Variables are listed in the order in which they were included in forward stepwise analysis

several advantages are linked to this method, such as the continuous remote monitoring even of small pit-tagged fishes with high spatial and temporal resolution and a minimum disturbance of aquatic environment, when compared, for example, with direct methods. In addition, the shallow and fast flowing riffle areas, impossible to sample with underwater observation, are easily surveyed through the PIT-telemetry (Riley et al., 2003). Finally, many researchers emphasize the

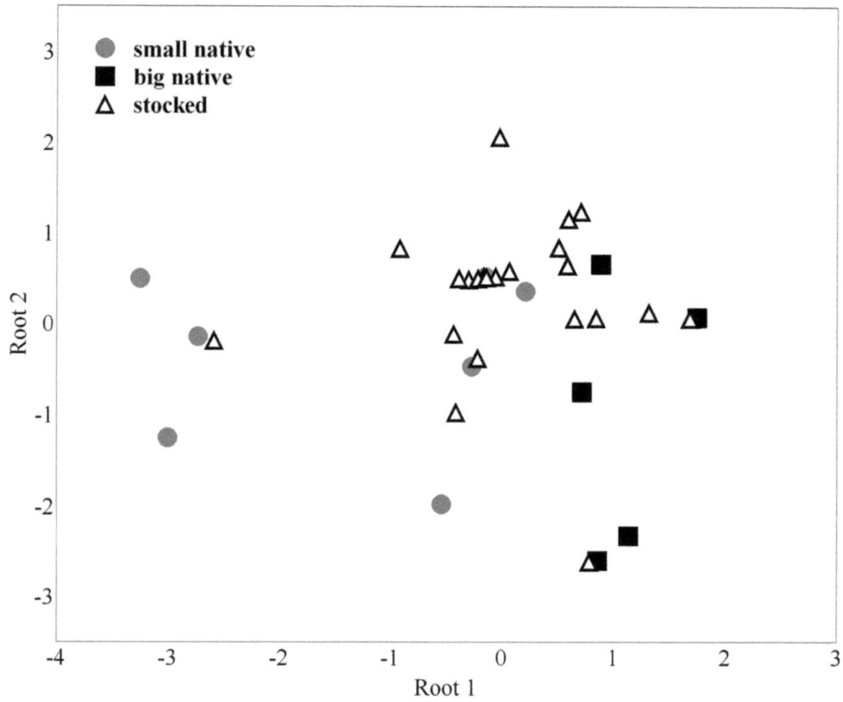

Fig. 6 Discriminant multivariate ordination (DFA) of predefined groups (classes are indicated by symbols: ● = small native; ■ = big native; Δ = stocked trout), considering the PIT-Telemetry data (n = 1635) for both stream segments—Scatterplot of the first two canonical roots

Table 5 Discriminant function analysis (DFA) summary applied to underwater observation data—Partial Wilks' lambda for each variable and the classification functions according to different trout classes (small native, big native and stocked trout)

Discriminant function analysis				
n = 374	Wilks' lambda	Classification functions		
Variables		Small native P = 0.409	Big native P = 0.035	Stocked P = 0.556
Focal elevation	0.671	–1.198	–1.241	0.959
Total depth	0.501	–0.447	0.838	0.276
Subdominant substrate	0.485	0.298	0.410	–0.245
Aquatic cover	0.481	0.066	0.897	–0.104
Dominant substrate	0.480	–0.245	0.001	0.180
Surface velocity	0.476	3.391	–0.607	–2.456
Water column velocity	0.475	–3.171	0.570	2.297
Constant	–	–1.558	–4.370	–0.993
(% corrected cases)	–	94	0	82.7

Variables are listed in the order in which they were included in forward stepwise analysis

PIT-tagging system's very efficient monitoring of the fish movements and its relative low costs (Castro Santos et al., 1996).

A greater proportion of stocked trout were detected in the different antennae of PIT-telemetry system, confirming the high mobility displayed by these fishes and suggesting the lack of a capacity to define a territory and a specific resting area, usually found in wild salmonid species (Fausch, 1984). Certainly, the erratic swimming activity and the shoaling behaviour, more visible in the first days after their release in

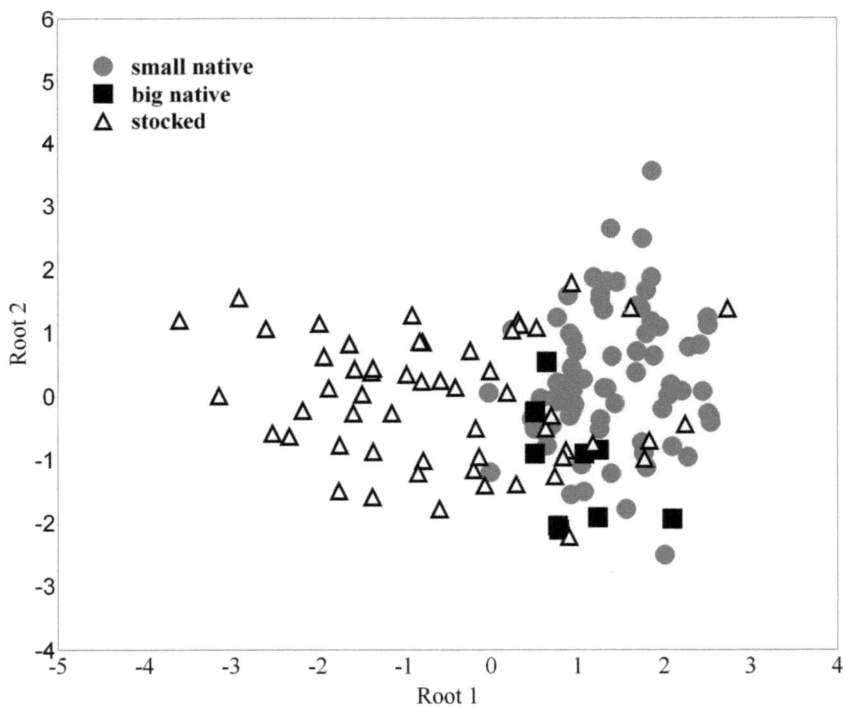

Fig. 7 Discriminant multivariate ordination (DFA) of predefined groups (classes are indicated by symbols: ● = small native; ■ = big native; Δ = stocked trout), considering the underwater observation method applied for both stream segments (n = 374)—Scatterplot of the first two canonical roots

the stream, are related to this behaviour. The high densities and the feeding regimes that promote a scramble-for-food environment are habits acquired by fish in hatcheries and contribute to suppress the establishment of social hierarchies that commonly occur in wild salmonid populations (Steward & Bjornn, 1990). For these reasons, aggressive behaviour and more intense agonistic encounters were associated by many researchers with stocking activity (Mesa, 1991; Deverill et al., 1999). An immediate consequence is the displacement from energetically profitable positions of the wild fish, leading to a negative impact on growth and survival of the natural populations (Weber & Fausch, 2005). However, our study detected a temporal (distinct maximum activity periods were observed, except for big native trout in the Sabor) and spatial segregation (significant differences were detected for the habitat used) between stocked and native trout, suggesting resource partitioning and probably, a distinct behaviour between the sympatric trout populations. Furthermore, the habitat used by dominant fishes (big native class) also differed from the small native ones. Comparatively, bigger trout were located, more often, by the PIT system in areas that provided refuge (resting activity), considering the aquatic cover and overhanging vegetation variables, near the upstream section (riffle area), normally associated with superior invertebrate drift rates. Indeed, the dominance in salmonids are linked to the profitable feeding areas and related to visible benefits in terms of survival and growth (Fausch, 1984). It is important to mention that, despite the presence of stocked trout, similar mean values for the variables of water column velocity (<0.30 m s^{-1}, Rincón & Lobón-Cerviá, 1993; Vismara et al., 2001) and depth (20–60 cm, Horton & Cochnauer, 1978; Roussel et al., 1999) were found by other studies related to habitat used by native trout, in spite of a wide range of values being referred to in the bibliography (Bunt et al., 1999; Heggenes, 1988; Vismara et al., 2001). All comparisons of habitat used by PIT-tagged fishes with literature data must be, however, carefully analysed taking into account not only the different stream characteristics, namely the habitat availability, but also the limitations of each survey method utilized.

The knowledge of the ecological processes that regulate these aquatic systems will be essential to define the best management strategies of trout populations in Portuguese salmonid streams. In conclusion, the detailed information obtained with PIT-telemetry allowed us to consider this technology as a useful alternative, in low order streams, to several methods currently available in the study of behaviour and habitat use by fishes. However, further research is needed for a better understanding of the stocked and native trout behaviour in the study area. Experimental designs with this type of equipment must be applied in shallower habitats (riffle zones) and bigger PIT tags (e.g. 23–34 mm long) should be used to improve the detection range. Moreover, an extended field survey is essential to define the adaptation period and survival of stocked fish.

Acknowledgements We would like to thank Jorge Bochechas (Direcção Geral de Recursos Florestais), who made possible this collaboration, Vitor Rego, Julieta Sampaio (Circunscrição Florestal do Norte) and Graça Barreira (Serviços Florestais-Núcleo de Bragança) who provided all the conditions in the fishfarms, and Tiago Martins and Paula Escalda for their assistance on field work. Ray Vogensen provided helpful suggestions, which led to linguistic improvements of this manuscript.

References

Barbin-Zydlewski, G., A. Haro, K. G. Whalen & S. D. McCormick, 2001. Performance of stationary and portable passive transponder detection systems for monitoring of fish movements. Journal of Fish Biology 58: 1471–1475.

Belanger, G. & M. A. Rodriguez, 2001. Homing behaviour of stream-dwelling brook charr following experimental displacement. Journal of Fish Biology 59: 987–1001.

Breiman, L., J. H. Friedman, R. A. Olshen & C. J. Stone, 1984. Classification and regression trees. Wadsworth & Brooks/Cole Advanced Books & Software. Monterey, CA.

Bridger, C. J. & R. K. Booth, 2003. The effects of biotelemetry transmitter. Presence and attachment procedures on fish physiology and behavior. Reviews in Fisheries Science 11: 13–34.

Bubb, D. H., M. C. Lucas, T. J. Thom & P. Rycroft, 2002. The potential use of PIT telemetry for identifying and tracking crayfish in their natural environment. Hydrobiologia 483: 225–230.

Bunt, C. M., S. J. Cooke, C. Katopodis & R. S. McKinley, 1999. Movement and summer habitat of brown trout (*Salmo trutta*) below a pulsed discharge hydroelectric generating station. Regulated Rivers: Research and Management 15: 395–403.

Burrell, K. H., J. J. Isely, D. B. Bunnell Jr, D. H. Van Lear & C. A. Dolloff, 2000. Seasonal movement of brown trout in a Southern Appalachian River. Transactions of the American Fisheries Society 129: 1373–1379.

Castro-Santos, T., A. Haro & S. Walk, 1996. A passive integrated transponder (PIT) tag system for monitoring fishways. Fisheries Research 28: 253–261.

Cowx, I. G., 1999. An appraisal of stocking strategies in the light of developing country constraints. Fisheries Management and Ecology 6: 21–34.

Cucherousset, J., J. M. Roussel, R. Keeler, R. A. Cunjak & R. Stump, 2005. The use of two new portable 12-mm PIT tag detectors to track small fish in shallow streams. North American Journal of Fisheries Management 25: 270–274.

Deverill, J. E., C. E. Adams & C. W. Bean, 1999. Prior residence, aggression and territory acquisition in hatchery-reared and wild brown trout. Journal of Fish Biology 55: 868–875.

Dolloff, C. A., J. Kershner & R. Thurow, 1996. Underwater observations. In Murphy, B. R. & D. W. Willis (eds), Fisheries Techniques, American Fisheries Society, Bethesda, Maryland, 533–554.

Fausch, K. D., 1984. Profitable stream positions for salmonids: relating specific growth rate to net energy gain. Canadian Journal of Zoology 62: 441–451.

Heggenes, J., 1988. Physical habitat selection by brown trout (*Salmo trutta*) in riverine systems. Nordic Journal of Freshwater Research 64: 74–90.

Heggenes, J., A. Brabrand & S. J. Saltveit, 1990. Comparison of three methods for studies of stream habitat use by young brown trout and Atlantic salmon. Transactions of the American Fisheries Society 119: 101–111.

Horton, W. D. & T. Cochnauer, 1978. Instream flow methodology evaluation, biological criteria determination, and water quality needs for selected Idaho streams. Report 14–16-0001-77090. Idaho Department of Fish and Game, Jerome, IA.

Joyce, M. P. & W. A. Hubert, 2003. Snorkeling as an alternative to depletion electrofishing for assessing Cutthroat and brown trout in stream pools. Journal of Freshwater Ecology 18: 215–222.

Lucas, M. C. & E. Baras, 2000. Methods for studying spatial behaviour of freshwater fishes in natural environment. Fish and Fisheries 1: 283–316.

Maki-Petays, A., A. Huusko, J. Erkinaro & T. Muotka, 2002. Transferability of habitat suitability criteria of juvenile Atlantic salmon (*Salmo salar*). Canadian Journal of Fisheries and Aquatic Sciences 59: 218–228.

McMichael, G. A., T. N. Pearsons & S. A. Leider, 1999. Behavioral interactions among hatchery-reared steelhead smolts and wild *Onchorhynchus mykiss* in natural streams. North American Journal of Fisheries Management 19: 948–956.

Mesa, M. G., 1991. Variation in feeding, aggression and position choice between hatchery and wild cutthroat trout in an artificial stream. Transactions of the American Fisheries Society 120: 723–727.

Ombredane, D., J. L. Baglinière & F. Marchand, 1998. The effects of Passive Integrated Transponder tags on survival and growth of juvenile brown trout (*Salmo trutta* L.) and their use for studying movement in a small river. Hydrobiologia 371: 99–106.

Ovidio, M., E. Baras, D. Goffaux, C. Birtles & J. C. Philippart, 1998. Environmental unpredictability rules the autumn migration of brown trout (*Salmo trutta* L.) in the Belgian Ardennes. Hydrobiologia 371/372: 263–274.

Pearsons, T. N. & C. W. Hopley, 1999. A practical approach for assessing ecological risks associated with fish stocking programs. Fisheries 24: 16–23.

Prentice, E. F., T. A. Flagg & C. S. McCutcheon, 1990a. Feasibility of using implantable passive integrated transponder (PIT) tags in salmonids. American Fisheries Society Symposium 7: 317–322.

Prentice, E. F., T. A. Flagg, C. S. McCutcheon & D. F. Brastow, 1990b. PIT-Tag monitoring systems for hydroelectric dams and fish hatcheries. American Fisheries Society Symposium 7: 323–334.

Quintella, B. R., N. O. Andrade, R. Espanhol & P. R. Almeida, 2005. The use of PIT telemetry to study movements of ammocoetes and metamorphosing sea lampreys in river beds. Journal of Fish Biology 66: 97–106.

Riley, W. D., M. O. Eagle, M. J. Ives, P. Rycroft & A. Wilkinson, 2003. A portable passive integrated transponder multi-point decoder system for monitoring habitat use and behaviour of freshwater fish in small streams. Fisheries Management and Ecology 10: 265–268.

Rincón, P. A. & J. Lobón-Cerviá, 1993. Microhabitat use by stream-resident brown trout: bioenergetic consequences. Transactions of the American Fisheries Society 122: 575–585.

Roussel, J. M., A. Bardonnet & A. Claude, 1999. Microhabitats of brown trout when feeding on drift and when resting in a lowland salmonid brook: effects on Weighted Usable Area. Archiv fur Hydrobiologie 146: 413–429.

Roussel, J. M., A. Haro & R. A. Cunjak, 2000. Field test of a new method for tracking small fishes in shallow rivers using passive integrated transponder PIT technology. Canadian Journal of Fisheries and Aquatic Sciences 57: 1326–1329.

Roussel, J. M., R. A. Cunjak, R. Newbury, D. Caissie & A. Haro, 2004. Movements and habitat use by PIT-tagged Atlantic salmon parr in early winter: the influence of anchor ice. Freshwater Biology 49: 1026–1035.

Scruton, D. A., R. S. McKinley, N. Kouwen, W. Eddy & R. K. Booth, 2002. Use of telemetry and hidraulic modelling to evaluate and improve fish guidance efficiency at a louve rand bypass system for downstream-migration Atlantic salmon (*Salmo salar*) smolts and kelts. Hydrobiologia 483: 83–94.

STATSOFT, Inc. 2004. STATISTICA (Data Analysis Software System). Version 7. www.statsoft.com. Tulsa, USA.

Steward, C. R. & T. C. Bjornn, 1990. Supplementation of salmon and steelhead stocks with hatchery fish: a synthesis of published literature. Technical Report n° 90-1. U.S. Fish and Wildlife Service, and Bonneville Power Administration.

Ter Braak, C. J. F., 1986. Canonical correspondence analysis: a new eigenvector technique for multivariate direct gradient analysis. Ecology 67: 1167–1179.

Ter Braak, C. J. F. & P. Smilauer, 1998. CANOCO. Reference manual and user's guide to Canoco for Windows: Software for Canonical Community Ordination (version 4). Microcomputer Power, Ithaca, NY, USA.

Thurow, R. F. & D. J. Schill, 1996. Comparison of day snorkeling, night snorkeling, and electrofishing to estimate bull trout abundance and size structure in a second-order Idaho stream. North American Journal of Fisheries Management 16: 314–323.

Vismara, R., A. Azzellino, R. Bosi, G. Crosa & G. Gentili, 2001. Habitat suitability curves for brown trout (*Salmo trutta fario* L.) in the river Arda, Northern Italy: comparing univariate and multivariate approaches. Regulated Rivers: Research & Management 17: 37–50.

Weber, E. D. & K. D. Fausch, 2003. Interactions between hatchery and wild salmonids in streams: differences in biology and evidence for competition. Canadian Journal of Fisheries and Aquatic Sciences 60: 1018–1036.

Weber, E. D. & K. D. Fausch, 2005. Competition between hatchery-reared and wild juvenile Chinook salmon in enclosures in the Sacramento River, California. Transactions of the American Fisheries Society 134: 44–58.

White, R. J., J. R. Karr & W. Nehlsen, 1995. Better roles for fish stocking in aquatic resource management. In Schramm, H. L. Jr. & R. G. Piper (eds), Uses and Effects of Cultured Fish in Aquatic Ecosystems, American Fisheries Society, Bethesda, Maryland, 527–547.

Young, M., 1999. Summer diel activity and movement of adult brown trout in high-elevation streams in Wyoming, U.S.A. Journal of Fish Biology 54: 181–189.

Hydrobiologia (2007) 582:187–197
DOI 10.1007/s10750-006-0556-7

FISH TELEMETRY

Introducing a method for extracting horizontal migration patterns from data storage tags

Bjørn Ådlandsvik · Geir Huse · Kathrine Michalsen

Abstract While data storage tags typically provide information about depth and temperature, the horizontal position of fish is difficult to obtain. The objective of this study is therefore to introduce a method for reconstructing horizontal migration patterns of fish tagged with DSTs. The method works by: establishing a database on bathymetry and environmental information about the target area, moving a large number of virtual fish between the release and recapture positions using a biased random walk procedure, and then terminate trajectories where the information at position is in disagreement with the tag information. For example a trajectory is terminated if the tag depth is greater than the bottom depth. The method is exemplified with two tag recordings from Northeast Arctic cod (*Gadus morhua* L.) in the Barents Sea. The results show that termination of impossible trajectories limits the number of potential trajectories between release and recapture positions, and in particular the usage of multiple termination criteria (depth and temperature) proved to be effective in reducing the number of possible trajectories. The method is general, simple to use, and can also be used in combination with geolocation methods.

Guest editors: P. R. Almeida, B. R. Quintella, M. J. Costa and A. Moore
Developments in Fish Telemetry

B. Ådlandsvik (✉) · G. Huse · K. Michalsen
Institute of Marine Research, P.O. Box 1870, Nordnes, 5817 Bergen, Norway
e-mail: bjorn.adlandsvik@imr.no

Keywords Geolocation · Horizontal migration · Data storage tags · Movements · Random walk

Introduction

Tagging has become established as the main technology for acquiring information about the movement of animals. A range of different tagging technologies exist, from simple tags without any stored information to the more advanced electronic data storage tags (DST), which can save information about the immediate surroundings of the animal (Arnold & Dewar, 2001). DSTs have lately been applied in several studies of fish migrations and distribution (e.g. Arnold et al., 1994; Gunn et al., 1994; Karlsson et al., 1996; Metcalfe & Arnold, 1997; Godø & Michalsen, 2000; Sibert & Nielsen, 2001; Neat et al., 2006). Typically such tags store the depth and temperature at each time step, but also tags storing salinity and light are available. While a lot of information about fish behaviour can be gathered from stored patterns of depth use, it is generally desirable to get patterns of horizontal positions as well in order to establish migration patterns. Acoustic tags can provide information about the three dimensional distribution of fish (Arnold & Greer Walker, 1992; Arnold et al., 1994; Godø, 1995). However, such tags are dependent on listening devices with limited space-time coverage and are restricted to a low number of tagged fish. The geographic position of aquatic animals that spend a considerable time at the surface, such as marine mammals and turtles, can be recorded using satellite tags. Newly developed DSTs store the position of vessels, transmitted by sonars, that come within a range of 1 nautical mile of the tagged fish (Furdal et al., 2003). While this technology can be very useful for getting the geographic position of fish, such recordings will typically occur irregularly and possibly not at all during a tagging period. Acquiring information about horizontal migration patterns of fish therefore remains non-trivial.

Alternative methods for estimating horizontal position are therefore valuable. Light measurements can be used to estimate horizontal position (e.g. Hill, 1994; Block et al., 2001), but the errors associated with such estimates can be considerable (Keating, 1995). Several attempts have been made to utilize the recorded information about depth, tidal patterns, temperature, or salinity from DSTs to make 'virtual recaptures' of individuals throughout the time they were free (e.g. Arnold & Holford, 1995; Metcalfe & Arnold, 1997; Hunter et al., 2003; Neuenfeldt et al., 2004). These approaches try to identify fish movement trajectories by geolocation and thus focus on identifying areas where the fish has been at certain times by making use of specific environmental features.

Even though the Barents Sea is a shelf sea, there is considerable variation in bottom topography with both relatively deep trenches of more than 400 m depth and banks shallower than 100 m (Fig. 1). The Barents Sea is characterised by a warm and saline Atlantic inflow to the southwest and cold Arctic water masses to the north and east (Loeng, 1991). The polar front, where these water masses meet, is characterised by distinct horizontal and vertical patterns in temperature. While the tidal difference is rather small and it can be difficult to identify unique locations in the Barents Sea using the geolocation

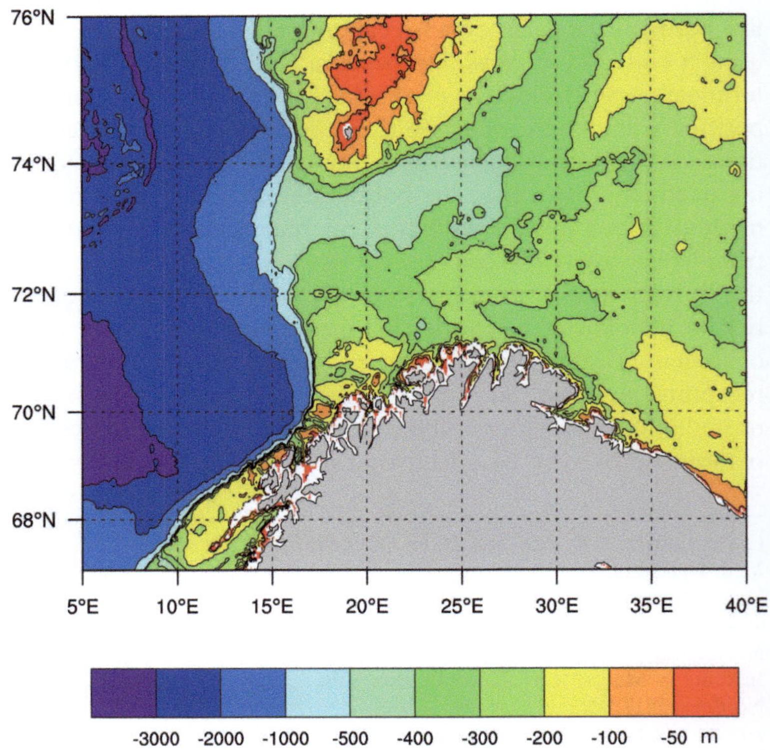

Fig. 1 Bottom topography for the southwestern Barents Sea

methods discussed above, it might be possible to combine information on depth and temperature over the year to restrict the number of possible locations where a tagged fish can have been.

Here, we present a novel method for generating possible horizontal movement trajectories of tagged fish between release and recapture positions. The approach is based on initiating a large number of virtual fish, move them about by using a biased random walk procedure and then use information from the tag confronted with an environmental database to terminate impossible trajectories. The procedure is general, and it is exemplified here by using two tag recordings from Northeast Arctic cod (*Gadus morhua* L.) in the Barents Sea.

Material and methods

Description of method

From a DST time series, environmental variables are recovered from the animal's path between the known release and the recapture positions. If these environmental variables can be mapped, areas with mismatch between tag and map information can be masked out as regions where the animal could not possibly have been at that time. As the animal has a limited propagation speed, subsequent impossible areas are not independent of each other. However, it is not trivial to combine these maps to produce a time evolving probability distribution of the animal's location.

A simple method is presented here to generate an ensemble of trajectories consistent with the information in the tag. That is, trajectories that join the tag and capture events and avoid the excluded areas where tag and environmental information disagree. The method is illustrated in Fig. 2, and a more formal description is given in Table 1. By using a biased random walk approach, the method tries to adequately sample the space of possible trajectories. The method is described as it applies to fish, but it could be useful for other kinds of animals where precise geolocation is not possible.

A large number of trajectories are started from the tag release position. Using a time step Δt the positions are updated by moving forward with a combination of a deterministic velocity pointing towards the recapture position, and an unknown velocity modelled as an isotropic random walk. The heart of the method is then to terminate impossible trajectories. This is done if the fish cannot reach the recapture position in the remaining time or if the tag information at this time step is inconsistent with environmental

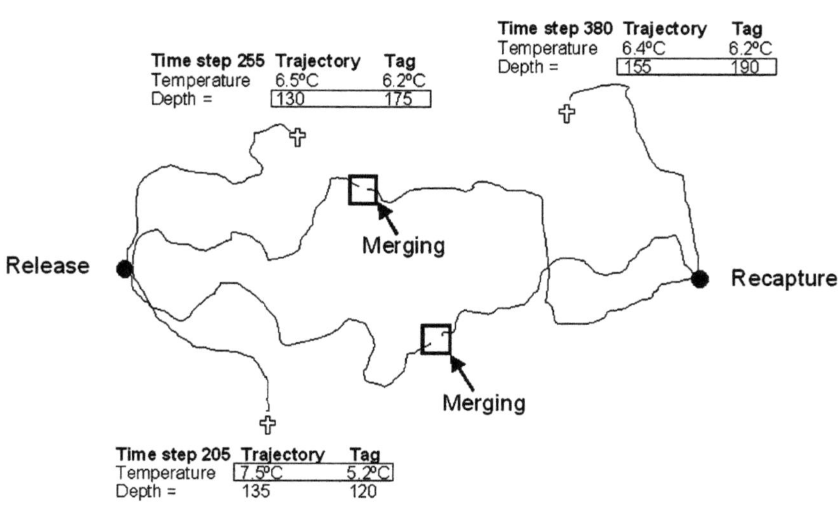

Fig. 2 Schematic outline of the principle behind the presented method for recreating horizontal movement trajectories of fish. The trajectories are arbitrary and illustrate termination due to impossible depth (time steps 255 and 380) and too large temperature discrepancy (time step 205) and merging of potential (unterminated) trajectories that come sufficiently close

Table 1 Pseudocode for the trajectory algorithm

```
for each time step
    for each trajectory
        compute the deterministic velocity u_det(X^n, t_n) = (X^final − X^n)/(t_final − t_n)
        if the speed ||u_det|| is too large, terminate the trajectory
        compute the random velocity u_rnd, with speed u_0 and random direction
        compute the new position X^(n+1) = X^n + (u_det + u_rnd)Δt
        estimate environmental data with error bars at X^(n+1) at time t_n
        if environmental data from tag are outside error bars then
            backtrack, set X^(n+1) = X^n
            estimate environmental data at old position at the new time t_(n+1)
            if environmental data from tag are outside error bars then
                terminate the trajectory
```

A trajectory is given by its position vector $\vec{X}^n$ at time t_n

(Table formulas: $\vec{u}_{\mathrm{det}}(\vec{X}^n, t_n) = \frac{\vec{X}^{final} - \vec{X}^n}{t_{final} - t_n}$; $\|\vec{u}_{\mathrm{det}}\|$; $\vec{u}_{\mathrm{rnd}}$, u_0; $\vec{X}^{n+1} = \vec{X}^n + (\vec{u}_{\mathrm{det}} + \vec{u}_{rnd})\Delta t$; $\vec{X}^{n+1} = \vec{X}^n$; t_{n+1})

information at this point in space and time. Repeating this process for the unterminated trajectories creates an ensemble of possible trajectories. At the start of the trajectory, the deterministic part of the movement is weak and randomness dominates. Towards the end of the trajectory the deterministic velocity component becomes dominating. The procedure is therefore modified to treat the release and recapture positions similarly. This is done by starting a set of backwards trajectories at the position and time of recapture, with deterministic velocity towards the release position and the same random walk implementation. The same termination criteria are applied for these trajectories. At mid-time between tagging and capture, unterminated forward and backward trajectories that are close enough (<5 km) become merged.

Termination criteria

Most DSTs used for fish, record the pressure, which gives the approximate depth. This immediately gives a minimum bottom depth at the position of the fish. As bathymetric data have a limited horizontal resolution, an uncertainty interval must be included to account for non-resolved bathymetric variability. Thus trajectories that pass positions where the bottom is clearly shallower than the concurrent tag depth, are terminated.

If the fish is known to stay *near* the bottom, the tag provides an estimate of the bottom depth leading to more efficient termination of trajectories. If the fish is known to stay *at* the bottom, the pressure data contain tidal information. In areas with well-known and strong tidal signal, this may be enough to give a precise geolocation (Hunter et al., 2003). If these conditions are not met, the method presented here can still use the tidal information to rule out trajectories in areas with clearly wrong tidal amplitude or phase.

Typically DSTs also record temperature. To use this we need an estimate of the 3D time developing temperature field in the actual ocean area. For shorter periods data from field programs can be used. For longer periods, climatologies give a coarser but still useful estimate of the temperature conditions. If available, operational ocean models with data assimilation will provide the best estimates of the temperature conditions. A trajectory can then be terminated if its temperature differs too much from the concurrent tag temperature.

Topography and temperature data

The gridded bottom topography is taken from the global $2' \times 2'$ global dataset ETOPO2, provided by US National Geophysical Data Center, http://www.ngdc.noaa.gov. The subset used in this study is shown in Fig. 1. The topography in the area is dominated by the wide Barents Sea shelf, with Bear Island on the shallow Svalbard Bank and the deeper Bear Island Trench further south. The figure also shows the narrow continental shelf off northern Norway and parts of the deep Norwegian Sea.

A temperature atlas for the Barents Sea has been developed at the Institute of Marine Research (IMR) (Ottersen and Ådlandsvik, personal communication). The atlas consists of quarterly fields of temperature, interpolated to a grid with resolution 0.5° west-east and 0.2° south-north. The vertical, resolution is 5 m. The atlas is interpolated from all CTD-stations taken by IMR in the area, from 1970 to 1996. Based on this dataset a quarterly climatology with standard deviation is calculated. Figure 3 shows the climatological temperature for the September-November quarter at 50 m depth. The map shows warmer Atlantic water masses to the south and west and colder Arctic water to the north and east. The estimation of temperature along the trajectories uses linear interpolation in space and the actual quarter in time.

Tag data

For the examples presented here, data from two tagged cod has been used. The specific tags are referenced as M106 and M235. The fish were tagged and released along the coast of Finnmark, position 25.35° E, 71.34° N, on 16 March 2004. M106 was recovered in the southeastern part of the Barents Sea, position 44.10° E, 69.37° N, on 26 November 2004, and M235 was recaptured southwest of Bear Island (19.10° E, 74.30° N) on 16 June 2004. The tag type used was a Star-Oddi, Data Storage Tag, which recorded depth and temperature every 2 h for 6 days and then every 12 h for 1 day, during the recording period. The 12 h recordings were interpolated linearly to 2 h values. For further information on the specifications of the data storage tags and the method used for tagging, see the web site of Star-Oddi http://www.star-oddi.com and Godø & Michalsen (2000), respectively.

Results

Tag M106

The cod with tag M106 moved from the coast of Northern Norway to the shallow south-east Barents Sea in 255 days. For the model experiment the time step used is 2 h and a random speed of 0.7 m/s. The parameter values in termination criteria are tuned and represent a

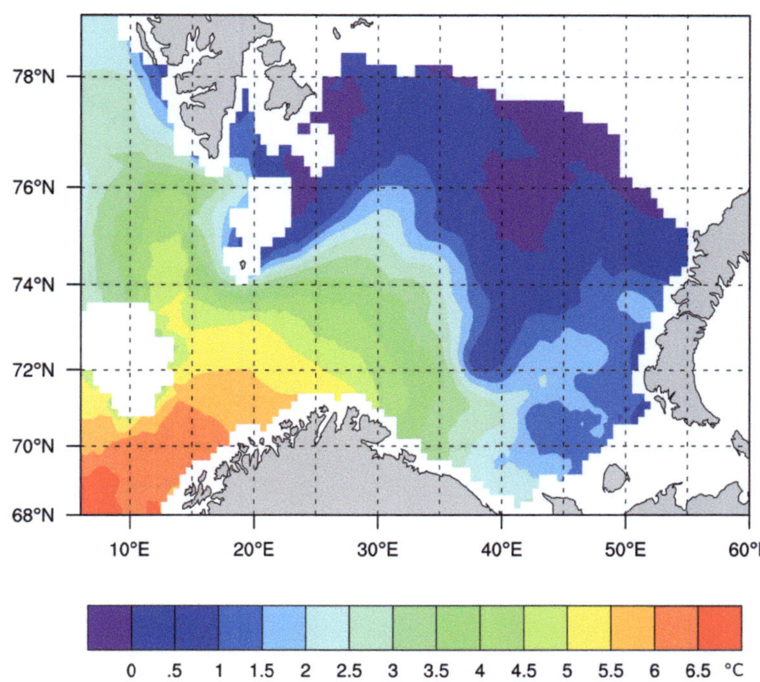

Fig. 3 Climatological temperature at 50 m depth for the quarter September–November

compromise between staying close to the gridded data sources and keeping the ensembles large enough for meaningful statistics. The following termination criteria were used:

(1) If the new position had bottom depth greater than 1000 m, the trajectory was not moved forward.
(2) If the tag depth was greater than 25 m plus the bottom depth multiplied with 1.3, the trajectory was terminated.
(3) If the tag temperature differed from the climatology with more than 3.0 standard deviations, the trajectory was terminated.

From a total of 100,000 trajectories backwards and forwards, the method gave a total of 2,854 merged trajectories without the temperature condition (3), and 110 trajectories with the temperature condition.

The time series from the tag is shown in Fig. 4 together with the horizontal standard deviation. At both ends it has the square root profile that can be predicted theoretically. The smoothest curve shows the spreading without the temperature condition. Here the standard deviation increases to 200 km and decreases slowly thereafter. In the first half of the period, the temperature condition leads to decreased variance. However, after 140 days the bimodal distribution of the trajectories increases the standard deviation beyond the case without the temperature based termination. The period with extensive temperature changes, around day 180, is in the middle of this period.

Figure 5 shows subsamples of 50 tracks from both ensembles together with the mean trajectory obtained from all the ensembles. In both cases the mean trajectory starts out towards the east before turning more towards southeast. Imposing the temperature condition leads to a splitting of the ensemble with most of the trajectories following a northern route, but with a considerable fraction following a southern route close to the coast. The trajectories avoid the Skolpen bank area (Fig. 5b) between the two routes. With such a bimodal distribution the mean path is not a likely path for the fish.

Tag M235

The cod with tag M235 moved from the coast of Norway to a location west of Bear Island where it

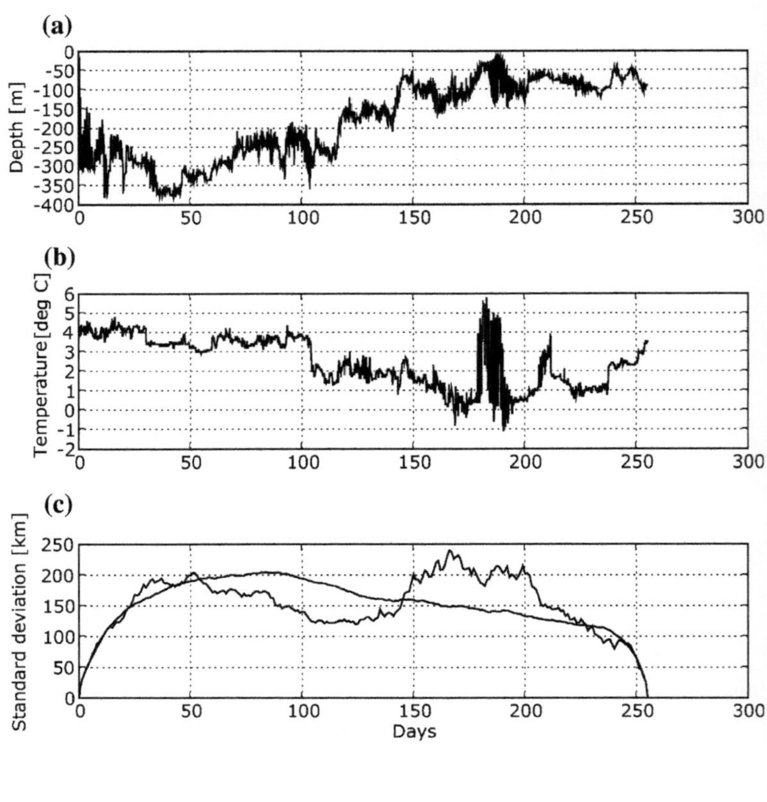

Fig. 4 Time series of depth (**a**), temperature (**b**) and horizontal standard deviation (the root mean square distance from the centre of gravity) (**c**) for tag M106

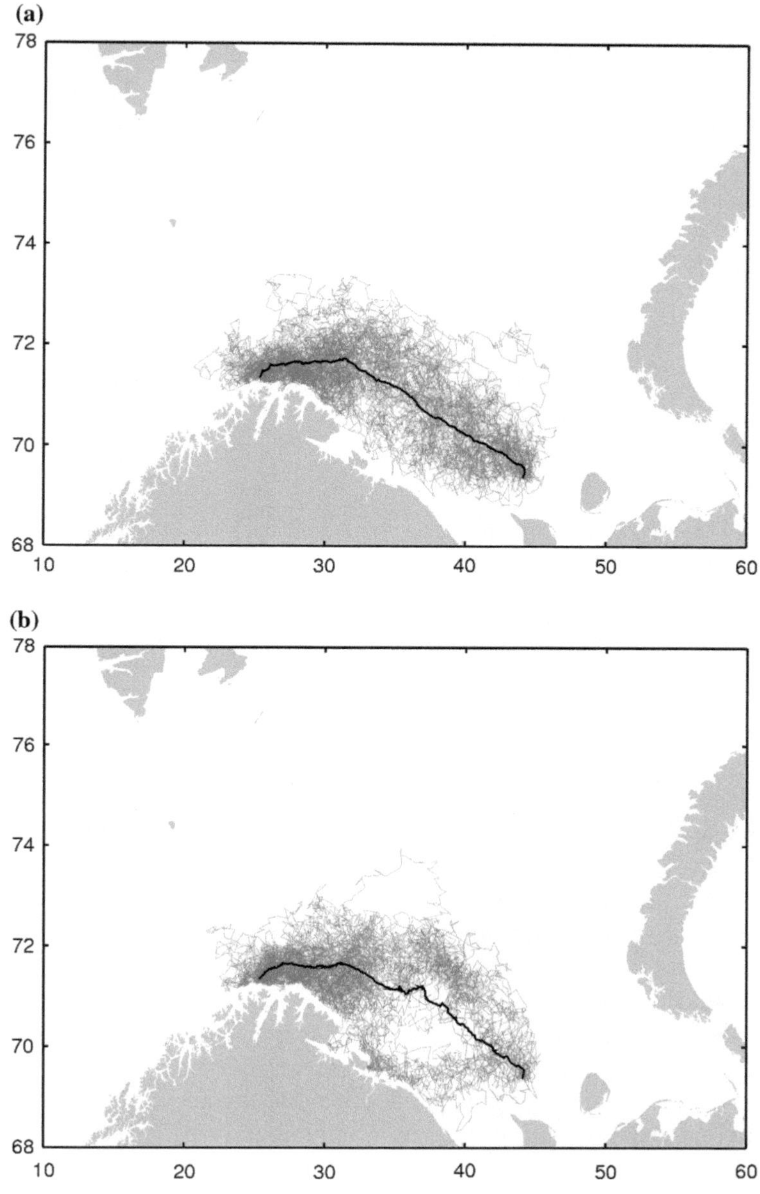

Fig. 5 Spaghetti plots with potential movement trajectories for tag M106 given termination with depth only (**a**) and with both depth and temperature (**b**). The *bold line* shows the means trajectory and the thin lines show 50 randomly selected potential trajectories

was recaptured after 91 days free. The model used the same time stepping and random speed as above.

The criteria for this tag are:

(1) If the new position had bottom depth larger than 1000 m, the trajectory was not moved forward.
(2) If the tag depth was larger than the bottom depth multiplied with 1.3, the trajectory was terminated.
(3) If the tag temperature differed from the climatology with more than one standard deviation, the trajectory was terminated.

Starting with 10,000 trajectories in each direction, 5,190 merged trajectories were obtained without the temperature condition, and 1,113 trajectories with the condition. The time series of depth, temperature and horizontal standard deviation is given in Fig. 5. The standard deviations have a maximum of about 150 km around day 30.

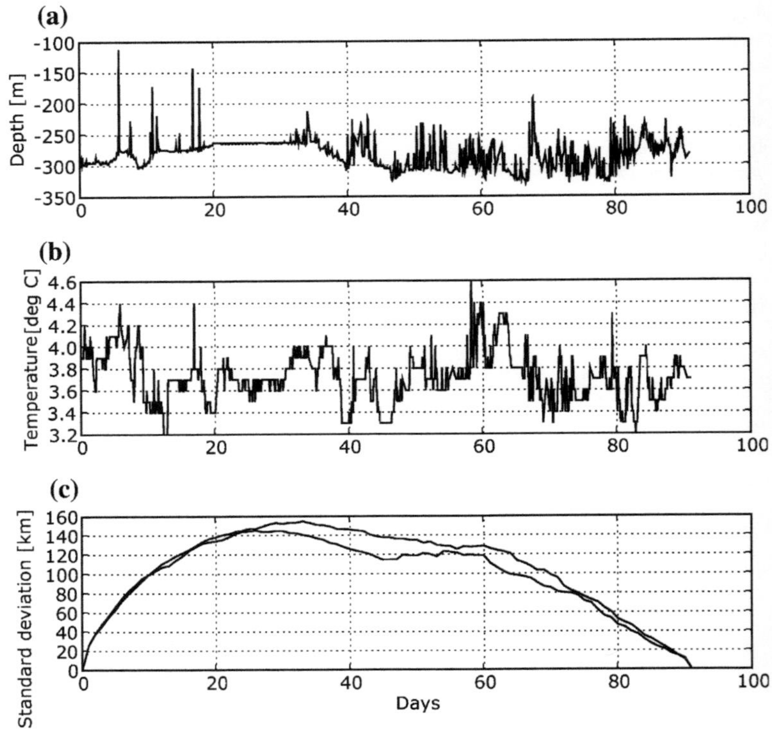

Fig. 6 Time series of depth (**a**), temperature (**b**) and horizontal standard deviation (the root mean square distance from the centre of gravity) (**c**) for tag M235

The temperature condition reduces the standard deviation only slightly (Fig. 6). Note that the decay of the standard deviations from day 60 follows a linear profile instead of the square root profile expected from diffusion theory. The backwards trajectories are limited by the requirement to stay deep close to the shallow areas around Bear Island.

Subsamples of 50 trajectories from both ensembles are shown in Fig. 7. The tag series have low variability, the mean depth is 282 m and the standard deviation is 23 m. For temperature the mean is 3.7°C with a standard deviation of 0.2°C. Both ensemble means start out towards the northwest before turning northeast towards Bear Island. The temperature conditions rule out some trajectories following a north-easterly route getting into too cold waters. Another effect is to make the initial phase of the mean movement more towards the north.

Discussion

We have presented a novel method for generation of possible horizontal movement trajectories in fish tagged by DSTs. The example applications on cod given above show that termination of impossible trajectories is an effective method for limiting the number of potential trajectories between release and recapture positions, and in particular the usage of multiple termination criteria proved effective in reducing the time-space of possible trajectories.

Understanding the extent and times of mixing between different stocks, and knowledge of seasonal distribution and migration, is crucial to effective fisheries management (Cushing, 1981). Information on the general geographical distribution of fish may be available, but tracking single fish or shoals in time or space, other than over short ranges in lakes, and within fjords, has been almost impossible. Knowledge of fish distribution gained from the fishing fleet might not be representative for the total distribution area for the different fish species, and scientific abundance surveys on the population level might not fully reflect the dynamic behaviour of individual fish. Tagged fish, and especially those tagged with DSTs, can therefore provide substantial information on migration and distribution of the stock. The use of DSTs provides excellent opportunities

Fig. 7 Spaghetti plots with potential movement trajectories for tag M235 given termination with depth only (**a**) and with both depth and temperature (**b**). The *bold line* shows the means trajectory and the thin lines show 50 randomly selected potential trajectories

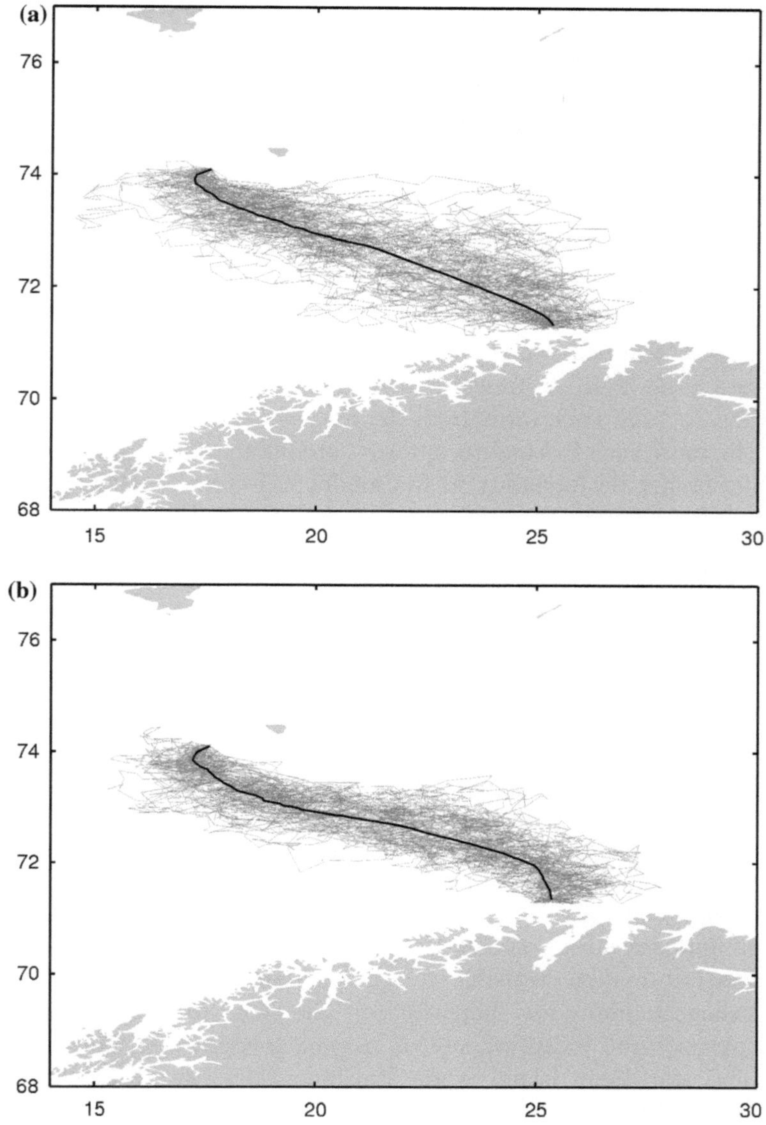

to get long term, high-resolution observations of individuals that can elucidate the factors affecting fish movements. In order to put vertical movements into an ecological context, it is also important to know the geographical location of individual fish. These factors have motivated a number of tagging studies on cod (Godø & Michalsen, 2000; Righton & Metcalfe, 2002; Palsson & Thorsteinsson, 2003; Neat et al., 2006).

Several attempts have been made to utilize the recorded information about depth, tidal patterns, temperature, salinity or light information from tags to make 'virtual recaptures' of individuals throughout the time they were at liberty (Hill, 1994; Metcalfe & Arnold, 1997; Block et al., 2001; Hunter et al., 2003; Neuenfeldt et al., 2004). Most models trying to identify fish migrations from DSTs, focus on identifying areas where the fish has been at a certain time, or recognising movement vectors (Stensholt & Stensholt, 1999). The present method focuses on excluding areas, or terminating trajectories where the fish definitely could not have been. The remaining trajectories then represent possible migration routes for the fish from the release to recapture site. Strength of the termination method is that "virtual recaptures" can be used in the same manner as real recaptures. This means that if the fish position can be estimated from a different method, for example a GPS tag or through the tidal geolocation method as used in the North Sea (Hunter et al., 2003), this position can be used as a recapture position as well as a release position. The model will then be run forwards from the real release and backwards from the "virtual recapture", as well as forwards from the "virtual release" and backwards from the real recapture. The more such "virtual recaptures" are obtained, the higher the certainty in the model trajectory will be. Similarly, bottom contact can sometimes be gleaned from the tag information. This can be applied as a termination criterion, by removing trajectories that are at locations with depths different from the bottom contact depth.

The example applications demonstrate the method's ability to extract information from the tag data. In the experiment with tag M106 we see that combining the depth and temperature data we are able to identify two possible routes in the eastern part of the area. This splitting is a consequence of the strong temperature signal around day 180. The data from tag M235 do not contain any strong signals and therefore, there is less information available for the method. The strongest limitation here is for the fish to approach the recapture site on sloping topography near Bear Island while staying at depths above 250 m.

The biased random walk is a stochastic process. Without any termination of trajectories, the growth rate of the horizontal variance is $u_0^2 \Delta t$, where u_0 is the random speed. This gives a square root profile for the standard deviation, as seen in the initial phase of both examples and also in the final phase with M106. The formula also shows how the time step and the random speed together, control the spreading of trajectories. The method can be generalized to include an element of autocorrelation between the steps (Vabø et al., 2004). This will reduce the dependency on the length of the time step.

The method uses some unknown parameters that can be tuned. This includes the random speed u_0 and the maximum distance allowed for merging forwards and backwards trajectories. The speed is most important as it determines the spreading of the trajectories. This spreading must be reasonable in view of existing knowledge of the distribution of the stock. The tolerance factors for uncertainties in the gridded bathymetry and temperature data must also be estimated. Depending on data quality a larger tolerance may be required to not falsely terminate possible trajectories. However, larger tolerance makes the termination criteria less effective. In practice, the values must represent a compromise, terminating impossible trajectories while keeping a large enough ensemble to do meaningful statistics.

The model approach presented here can be developed further. The procedure as presently described has some limitations, mostly due to geometry. For instance, the straight line between the release and capture positions may pass through land. The deterministic velocity may therefore lead to mass termination of trajectories. In such situations it may be necessary to increase the number of trajectories and/or increase the random speed to obtain merging trajectories. To keep up a reasonable number of trajectories, it is also possible to implement branching of the remaining trajectories. Another type of solution is to change the deterministic velocity to follow a predefined curve around the obstacle instead of following a straight line. The method presented here is meant to be a part of the toolbox for the experienced fish biologist in analysing results from DSTs. The method is best used interactively, where the user can tune parameters, identify bottom contact, and combine it with geolocation methods. There is need for further validation of the approach to investigate the efficiency of different model formulations and assumptions. Such a validation should ideally be performed both for known trajectories of tagged marine animals and by using artificial tags with given 3D behaviour. This will be pursued in future studies.

Acknowledgements This work was supported by the European Commission through the project "Cod spatial dynamics and vertical movements in European waters and implications for fishery management" (QLRT-2001-00813).

References

Arnold, G. P. & M. Greer Walker, 1992. Vertical movements of cod (*Gadus morhua* L.) in the open sea and the hydrostatic function of the swimbladder. ICES Journal of Marine Science 49: 357–372.

Arnold, G. P. & B. H. Holford, 1995. A computer simulation model for predicting rates and scales of movement of demersal fish on the European shelf. ICES Journal of Marine Science 52: 981–990.

Arnold, G. P. & H. Dewar, 2001. Electronic tags in marine fisheries research: a 30 year perspective. In Sibert, J. R. & J. L. Nielsen (eds), Electronic Tagging and Tracking in Marine Fisheries. Kluwer Academic Press, 7–64.

Arnold, G. P., M. Greer Walker, L. S. Emerson & B. H. Holford, 1994. Movements of cod (*Gadus morhua* L.) in relation to the diel streams in the southern North Sea. ICES Journal of Marine Science 51: 207–232.

Block, B. A., H. Dewar, S. B. Blackwell, T. D. Williams, E. D. Prince, C. J. Farwell, A. Boustany, S. L. H. Teo, A. Seitz, A. Walli & D. Fudge, 2001. Migratory movements, depths preferences, and thermal biology of Atlantic bluefin tuna. Science 293: 1310–1314.

Cushing, D. H., 1981. Fisheries Biology. The University of Wisconsin Press, London, 295 pp.

Furdal, A., S. Gudbjørnsson, F. R. Knudsen & O. R. Godø, 2003. A GPS-receiver for free-swimming fish. Fifth Conference on Fish Telemetry, Ustica, Italy.

Godø, O. R., 1995. Transplantation-tagging-experiments in preliminary studies of migration of cod off Norway. ICES Journal of Marine Science 52: 955–962.

Godø, O. R. & K. Michalsen, 2000. Migratory behaviour of north-east Arctic cod, studied by use of data storage tags. Fisheries Research 48: 127–140.

Gunn, J. S., T. Polacheck, T. L. O. Davis, M. Sherlock & A. Betlehem, 1994. The application of archival tags to study the movement, behaviour and physiology of southern bluefin tuna, with comments on the transfer of the technology to groundfish research. ICES CM 1994/Mini:21, 23 pp.

Hill, R., 1994. Theory of geo-location by light levels. In Le Boeuf B. J. & R. M. Laws (eds), Elephant Seals Population Ecology, Behaviour and Physiology. University of California Press, Los Angeles, 227–236.

Hunter, E., J. N. Aldridge, J. D. Metcalfe & G. P. Arnold, 2003. Geolocation of free-ranging fish on the European continental shelf as determined from environmental variables. I. Tidal location method. Marine Biology 142: 601–609.

Karlsson, L., E. Ikonen, H. Westerberg & J. Sturlaugsson, 1996. Use of data storage tags to study the spawning migration of Baltic salmon (*Salmo salar* L.) in the Gulf of Bothnia. ICES CM 1996/M:9, 15 pp.

Keating, K. A., 1995. Mitigating elevation-induced errors in satellite telemetry locations. Journal of Wildlife Management 59: 801–808.

Loeng, H., 1991. Features of the physical oceanographic conditions of the Barents Sea. Polar Research 10: 5–18.

Metcalfe, J. D. & G. P. Arnold, 1997. Tracking fish with electronic tags. Nature 387: 665–666.

Neat, F. C., P. J. Wright, A. F. Zuur, I. M. Gibb, F. M. Gibb, D. Tulett, D. A. Righton & R. J. Turner, 2006. Residency and depth movements of a coastal group of Atlantic cod (*Gadus morhua* L.). Marine Biology 148: 643–654.

Neuenfeldt, S., H. H. Hinrichsen & A. Nielsen, 2004. A method to geolocate eastern Baltic cod by using Data Storage Tags (DSTs). ICES C.M. 2004/L:06., pp. 1–14.

Palsson, O. K. & V. Thorsteinsson, 2003. Migration patterns, ambient temperature, and growth of Icelandic cod (*Gadus morhua*): evidence from storage tag data. Canadian Journal of Fisheries and Aquatic Sciences: 1409–1423.

Righton, D. & J. Metcalfe, 2002. Multi-torsking: simultaneous measurements of cod behaviour show differences between North Sea and Irish Sea stocks. Hydrobiologia 483: 193–200.

Sibert, J. R. & J. L. Nielsen, 2001. Electronic Tagging and Tracking in Marine Fisheries. Kluwer Academic Press, 468 pp.

Stensholt, E. & B. K. Stensholt, 1999. Fish movement vectors and the temperature gradient: a geometric analysis method for the depth-temperature time series from data storage tags. ICES Journal of Marine Science 56: 537–544.

Vabø, R., G. Huse, A. Fernö, T. Jørgensen, S. Løkkeborg & G. Skaret, 2004. Simulating search behaviour of fish towards bait. ICES Journal of Marine Science 61: 1224–1232.

Hydrobiologia (2007) 582:199–207
DOI 10.1007/s10750-006-0554-9

FISH TELEMETRY

Feeding activity and opercular pressure transients in Atlantic salmon (*Salmo salar* L.): application to feeding management in fish farming

Jo Arve Alfredsen · Bård Holand · Torfinn Solvang-Garten · Ingebrigt Uglem

Abstract Ecological and economical sustainability of marine aquaculture operations depend on proper feeding management. Feed wastage from overfeeding is a source of pollution, represents futile use of precious marine resources, and undermines the economic viability of operations. Additionally, underfeeding reduces growth and fish welfare. Finding an optimal feeding regime in terms of temporal and spatial distribution of the feed ration require intimate knowledge of the individual feeding behaviour of fish sustaining intensive culturing conditions. Fish telemetry has proved to be a valuable tool for studying spatial behaviour in sea cages, however there are currently no practical methods available with respect to detection of actual feed intake in fish on the individual level. The present study investigates pressure transients arising in the opercular cavity of farmed Atlantic salmon (*Salmo salar* L.) in connection with feeding, and whether such measurements can serve as an indication of feed ingestion. A technical solution to the sensing problem based on a differential pressure transducer is presented along with typical pressure signal traces obtained during feeding in a hard wire tank experiment. Measurements showed considerable variation of sub-ambient pressure transients (1.5 kPa ± 0.95) and their duration (519 ms ± 117), suggesting that the fish modulates its strike intensity depending on the particular feeding situation. Despite variations in scale, opercular pressure waveforms have distinct structural features that repeat between feeding instants. From a signal processing point of view waveforms provide sufficient information with respect to isolation and detection of feeding incidents, which is important with respect to a potential implementation of the sensing principle in a telemetry tag design. Issues regarding development and application of a telemetry system based on this sensing principle are discussed.

Keywords Feeding behaviour · Feeding management · Opercular pressure · Telemetry · Atlantic salmon

Introduction

Fish farming, using primarily sea cages, usually involves massive aggregations of fish in relatively

Guest editors: P. R. Almeida, B. R. Quintella, M. J. Costa and A. Moore
Developments in Fish Telemetry

J. A. Alfredsen (✉) · B. Holand · T. Solvang-Garten
Department of Engineering Cybernetics, The Norwegian University of Science and Technology, NO-7491 Trondheim, Norway
e-mail: jo.arve.alfredsen@itk.ntnu.no

I. Uglem
Norwegian Institute for Nature Research, Tungasletta 2, NO-7485 Trondheim, Norway

small volumes, where living conditions and fish behaviour deviate significantly from the conditions experienced and the behaviour exhibited in the natural environment. For example, outgrowth of Atlantic salmon (*Salmo salar* L.) is typically conducted in sea cages. Investigation of farm fish behaviour is important because it reflects critical information about fish welfare and represents the premises for efficient farm management in terms of optimum feeding regimes and sustainable production. The behaviour of intensively cultured fish can not be extrapolated from knowledge about free-ranging fish, which suggests that effort should be invested in adaptation and development of suitable methods for monitoring fish behaviour under such conditions. As an effective tool for studying fish behaviour on the individual level, fish telemetry has already found several useful applications within aquaculture research in general (Juell & Westerberg, 1993; Baras & Lagardère, 1995; Begout Anras, Kadri, Juell & Hansen, 2000; McFarlane et al., 2002; Cooke, Thorstad & Hinch, 2004).

Feeding management in fish farms encompasses the control of ration size and type as well as the temporal and spatial delivery of feed. Improper feeding management has serious impact on both productivity and environment, and the industry is reported to have considerable potential for improvement (Houlihan, Boujard & Jobling, 2001). Overfeeding weakens the feed conversion ratio (FCR = kg. feed used/kg. biomass gained), pollutes the environment through microbial spoilage of waste feed and spread of feed particles to populations of wild fish, and causes waste of precious marine protein and lipid resources. On the other hand, underfeeding results in reduced growth, elevated competition and stress among the fish, and a weakened FCR (Ang & Petrell, 1997; Talbot, Corneillie, & Korsøen, 1999). Optimum feeding management is a complicated decision process where the amount of information available for making the decisions may be limited. The ability of the farmer to assess the size and amount of fish in the cages in order to determine an adequate pellet size and ration (enough feed to support growth potential) is obviously of great importance. However, the decision regarding the temporal and spatial delivery of feed and the corresponding feeding behaviour of fish during a particular meal, depending on the hunger level of the fish and environmental conditions such as light level and water temperature, is what ultimately determine feeding efficiency (Alver, Alfredsen, & Sigholt, 2004).

The problem of finding an optimum delivery regime represents an active field of research and technological development. Most investigations are based on the population level perspective of fish feeding, such as through controlled large scale feeding experiments (Jobling et al., 2001). However, observations made at the population level will primarily provide aggregated data on how the fish respond to feed, i.e. the relationship between gross feed consumption and the average response to a particular feature of the feeding strategy or environmental condition. These are average responses where signs of individual variation and interactions within the population are essentially hidden. In order to fully understand the underlying mechanisms that govern feeding efficiency, observations should also be obtained at the individual level (Jobling et al., 2001). Observations of individuals or specific groups of fish can provide valuable information on important matters such as intrinsic feeding rates, feeding histories, and consequences of social hierarchy. This suggests that the possibilities of fish telemetry should be assessed with respect to observing individual feeding behaviour and feed intake in fish cages.

Key to the development of a "feeding sensitive" telemetry system is the identification and detection of a distinct signal that somehow reflects the act of feed ingestion (feeding signal). Once the detection principle is established its further implementation in a conventional acoustic transmitter based system should be attainable. Ideally, the system should be able to reliably detect every (and only those) feed pellets passing into the tagged fish' stomach. However, detection of alternative signals giving indirect indications of feed ingestion should also be considered, particularly if such feeding signals can lead to a robust solution with respect to the sensing principle at an acceptable signal to noise ratio. In any case, the balance between measurement accuracy, complexity, and invasiveness of the required technology need to be assessed.

Fish use three different modes of prey capture: ram feeding, suction feeding, and manipulation (Norton & Brainerd, 1992). Suction feeding, in which the predator causes prey to be dragged into the mouth through rapid buccal expansion, is the principal strike mode among many predatory fish (Alexander, 1967; Nemeth, 1997). In experiments conducted with rainbow trout, *Oncorhynchus mykiss* (Van Leeuwen, 1984), it was demonstrated that suction is an important component of the prey capture process. However, most fish exercise a combination of suction and ram feeding (Norton & Brainerd, 1992), as is probably the case for salmon when presented with formulated feed pellets. As regards the question of identifying possible feeding signals, the production of pressure transients in the fish's mouth during feeding may turn out as a possible indicator. The buccal and opercular cavities are hydrodynamically connected through the gill arches (Van Leeuwen, 1984) and the opercular cavity is readily accessible for pressure measurements by means of cannulation through the anterior wall of the cleithrum (Shelton, 1970; Norton & Brainerd, 1992). It is probable that the invasiveness and complexity of this kind of instrumentation is feasible with respect to implementation in a future telemetry system. However, the information present in the pressure signals must be analysed and reliable methods of signal interpretation must be developed. The present study investigates the characteristics of the pressure signal in the opercular cavity of Atlantic salmon during ingestion of typical feed pellets. The feasibility of this kind of instrumentation in a fish telemetry system and some technical considerations with respect to its implementation are also discussed.

Materials and methods

Animals

Atlantic salmon are the most important species employed in Norwegian aquaculture. Two specimens (A: standard length 52 cm, weight 1540 g; B: 50 cm, 1388 g) were selected and prepared for the investigations of this study by installing a pressure sensing device in their opercular cavity. In addition, a number of similar fish were employed for the purpose of experimental control. The fish were acquired from a tank reared population of the Salmobreed-strain at Akvaforsk, Sunndalsøra, and transferred to Brattøra Research Centre, Trondheim, where they were kept in rectangular 1,000 l flow-through tanks. In order to facilitate acclimation the fish were fed formulated feed pellets of the same size and quality (Skretting Nutra, 4 mm) as they were fed prior to transport. Feed delivery was administered by use of automatic feed dispensers three times per day (2.5% of body weight). Consistent feed intake at the level prior to transport was considered as evidence of acclimation, which occurred after four weeks. During experiments instrumented fish were accompanied by unaltered fish as control and to facilitate feeding behaviour. Temperatures during acclimation and the experiments were in the range 8.0–8.2°C. Animals were housed and maintained according to national regulations and protocols applying to animal use and care.

Surgery

Prior to surgery fish were anaesthetized by immersion in a solution of 2-phenoxyethanol (2-PE, 0.65 ml/l). Following anaesthesia the fish were transferred to specialized cradle where surgery could be performed with the fish' gills partially submerged in pure sea water. Water was poured over the fish repeatedly to keep their skin moist during surgery, and the surgical procedure was timed to last about six minutes. After surgery the fish were revived in their respective holding tanks. The fish returned to a stable position and normal respiration rate approximately three hours after revival, and feeding resumed after 24 h. The experiment employed a "hard wired" experimental setup where the fish were physically connected to the data recording equipment through a signal cable (diameter 2.8 mm, 3 wire + shield). Low friction four terminal slip rings (Poly-Scientific, AC6373) were introduced between the tagged fish and the recording equipment in order to avoid excessive twisting of the cable as the fish moved around. After an initial period of habituation the cable did not seem to affect the behaviour of the fish noticeably.

Opercular pressure was measured by inserting a medical grade plastic cannula (Venflon, BD Medical Systems, 1.1 mm outer diameter, 0.8 mm inner diameter) into the opercular cavity with the distal tip of the cannula protruding approximately 5 mm past the anterior wall of the cleithrum. This technique is similar to the technique described e.g. by Shelton (1970). The cannula was positioned by inserting a hypodermic needle (2.4 mm outer diameter, 1.7 mm inner diameter) through the tissue beneath the cleithrum. The needle carried the plastic cannula into place and was subsequently removed. The external portion of the cannula was fixed to the skin by four sutures (2/0 silk) 15 mm dorsally of the pectoral fin.

The end of the cannula was flanged to 11 cm of flexible Tygon R3603 tubing (2.4 mm outer diameter, 0.8 mm inner diameter). The remote end of the tubing was attached to the positive pressure port of a miniature differential pressure sensor (Honeywell 26PC01 SMT, pressure range ± 6.9 kPa, frequency response 1 kHz). The negative port of the sensor was left open making ambient water pressure reference for the pressure measurements. Interior parts of cannula, tubing and pressure port were filled completely by degassed fluid to assure adequate incompressible coupling to the pressure sensitive membrane of the sensor. Prior to surgery, the pressure sensor had been connected to an IC instrumentation amplifier (Texas Instruments, INA321) for amplification of the pressure signal, and both components were mounted on a purpose built printed circuit board and subsequently moulded into a solid epoxy cylinder of length 32 mm and diameter 13 mm (denoted as the tag). The tag was then attached to the fish by passing two fine surgical steel wires guided by two hypodermic needles through the epaxial musculature just below the dorsal fin. The steel wires were tightly wrapped around the tag prior to surgery causing firm attachment of the tag to the fish body. An outline of the arrangement can be seen in Fig. 1.

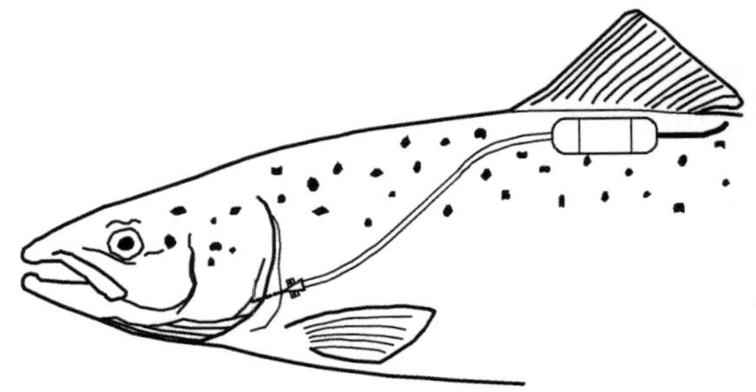

Fig. 1 Outline of the instrumentation of the fish used in the experiment

Observations

After the surgical procedure the pressure signal of the opercular cavity was readily accessible through the flexible cable extending from the tag. The cable was connected to dedicated computer hardware for AD conversion of the pressure signal (National Instruments, DAQCard 6062E, sampling rate 100 Hz, resolution 12 bit). Purpose built computer software using the LabVIEW programming system (National Instruments) was developed permitting real time visualization of pressure waveforms and data recording. Pressure data time series were recorded continuously during feeding episodes. Measurements were adjusted for the difference in depth between the tip of the cannula and the reference pressure port in the tag on the fish' back.

Feeding behaviour was observed by employing multiple underwater cameras (Type WATEC 107LH) in the fish tank. The camera permitted a close view of the feeding fish without causing any disturbance. Video sequences of the feeding fish were recorded to DVDs for subsequent analysis and correlation with pressure recordings. Synchronization of video recordings and time series of opercular pressure was secured by adding accurate time information to both videos and data files. Analysis was performed by identifying positive occurrences of feed ingestion on the video recordings and inspecting the associated pressure signals looking for features of the waveforms that distinguish feeding events.

Results

The results presented here are based on measurements and observations of fish A. Fish B was treated in a similar manner, but served mainly as a backup in case of technical failure and as an indicator with respect to detection of potential sources of variability arising from the tag attachment procedure. However, variability between two individuals appeared to be substantially lower than the intra-individual variability experienced.

Feeding related opercular pressure data on salmonids are sparse. In order to get an impression of the validity of the pressure measurements, it was decided to observe the fish's respiratory waveform first. Respiration in salmonids, and fish in general, is relatively well documented by pressure waveforms of the buccal and opercular cavities, EMG recordings from the muscles involved, and visual observations of moving skull structures (Alexander, 1967; Shelton, 1970; Oswald, 1978). Figure 2 shows three representative periods of the respiratory cycle as measured by the tag. The most conspicuous property of the waveform are the regular subambient drops in pressure which are known to correspond to the phase of the respiratory cycle where the opercular suction pump is active (sucking water over the gills). The waveform also contains a relatively flat region where the buccal pressure pump is active (pressing water over the gills). The flat region is preceded by a "knee" and followed by a minor peak which indicates the transitional phase between the two distinct pumping actions. Some noise is present in the signal, but its magnitude is too small to obscure the main features of the waveform. The noise is apparently quantization effects due to the small signal amplitude of the respiration signal. Respiration is actually below the pressure range for which the tag electronics is designed to work optimally.

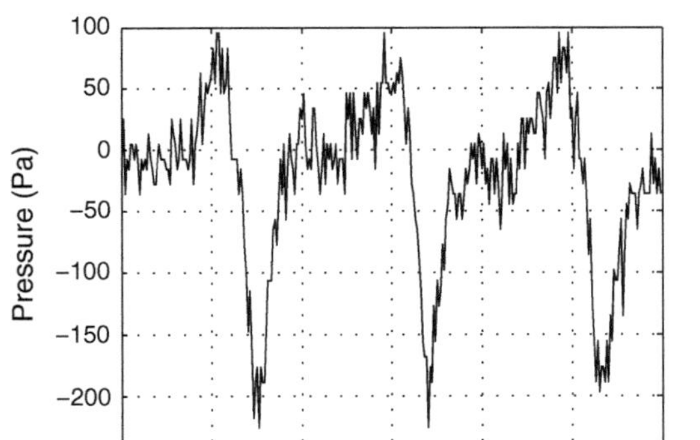

Fig. 2 Pressure measurements in the opercular cavity during three cycles of respiration

The fish commenced feeding within 24 h of surgery. 4 mm feed pellets were delivered from an automatic feed dispenser and the fish's response was recorded on video while the corresponding pressure signal was recorded by the computer software. Positive feeding events were noted for subsequent analyses. Generally, the opercular pressure during feed capture and ingestion showed high variability with respect to the magnitude of the subambient pressure transient (P = [0.8, 4.6] kPa, $\bar{P}$ = 1.5 kPa, s.d. = ±0.95 kPa, N = 16). The transients were, however, of considerably greater magnitude than corresponding subambient transients of the respiration signal. Also, the shape of the waveform that followed after the initial transient endured substantial fluctuations. Such fluctuations are probably caused by irregular and rapidly changing water flows in the fish mouth as a result of manipulation and ingestion of the captured feed pellet. The duration of the subambient transient was also investigated and it was found that this parameter endured less variation than what was observed with the magnitude of the pressure transient (t = [360, 829] ms, $\bar{t}$ = 519 ms, s.d. = ±117 ms, N = 16). Compared to other species that have been investigated (non-salmonids) the subambient opercular pressure of salmon seems to be of longer duration.

Figure 3 shows a selection of four pressure signatures of varying magnitude recorded during the experiment. The waveforms show common features with respect to their shape when differences in scale are disregarded. The transient phase seems to be preceded by a short period of elevated pressure, presumably caused by a preparatory phase of buccal compression leading the strike. Similar observations have been done with other species, e.g. *Micropterus salmoides* and *Cichla ocellaris* of the Centrarchidae family

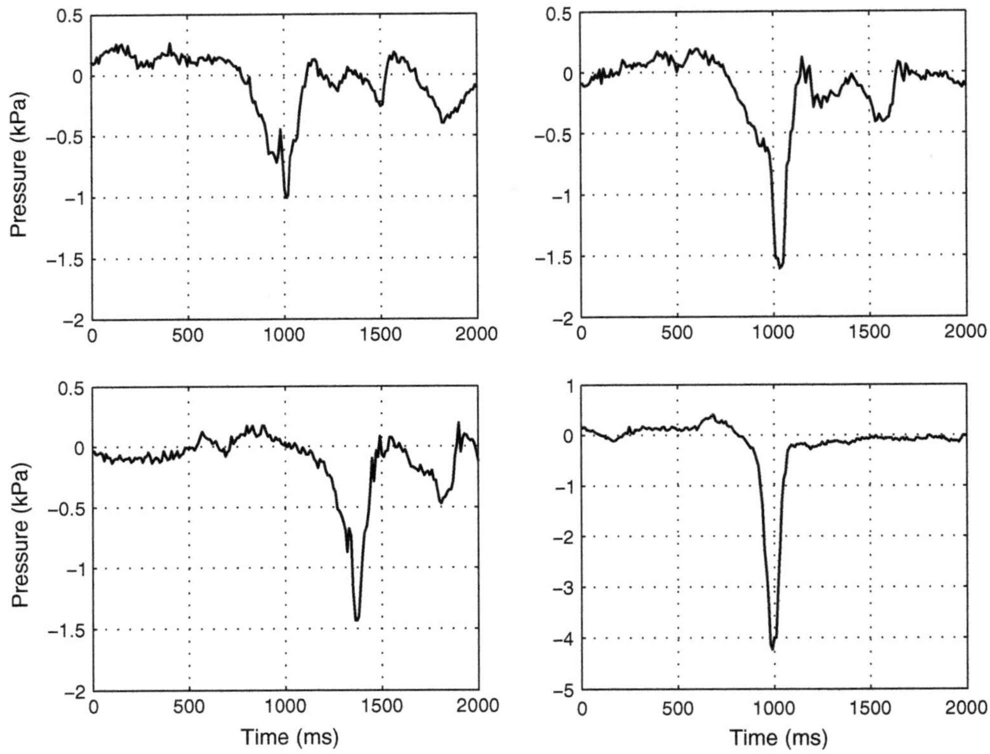

Fig. 3 Pressure waveforms measured in the opercular cavity during four consecutive feeding incidents

(Norton & Brainerd, 1992). The preparatory phase culminates with the development of an almost monotonically increasing negative pressure gradient during the abduction phase causing build up of suction. When the point of minimum pressure is achieved an even more characteristic positive gradient ensues which corresponds to the adduction phase of the strike where pressure is brought back to the ambient level. The two gradient phases constitute the most apparent features of the opercular pressure signal.

The features of the opercular pressure signal that relate to waveform are clearly recognizable, despite large variations in scale, and this fact is important with respect to the applicability of the signal as the source of information in a potential "feeding sensitive" telemetry system. In this context it is a matter of developing a robust signal processing algorithm that is able to equalize the effects of scale variation at the same time as it recognizes features of the pressure waveform that reflect feed ingestion. On the basis of the pressure traces in Fig. 3, it seems logical that the algorithm should employ some form of band pass filtering strategy where the pass band is selected to amplify signals in the frequency range containing the relatively powerful transient that represents the subambient pressure drops of feed ingestion. Other ranges of the signal spectrum containing sources of slow pressure variations of relatively low power (such as respiration and fluctuations caused by swimming) and high frequency noise would be suppressed with this type of filter. Spectral analysis of the signals recorded during the study showed that the most of the power associated with the subambient pressure drops appeared in the frequency range 5–20 Hz. Based on this information a 4th order band pass filter of the Butterworth type was designed and tested on the signals. As can be seen from Fig. 4, the output of the filter indicates feeding episodes by corresponding ripples in the signal while other features of the signal are effectively suppressed.

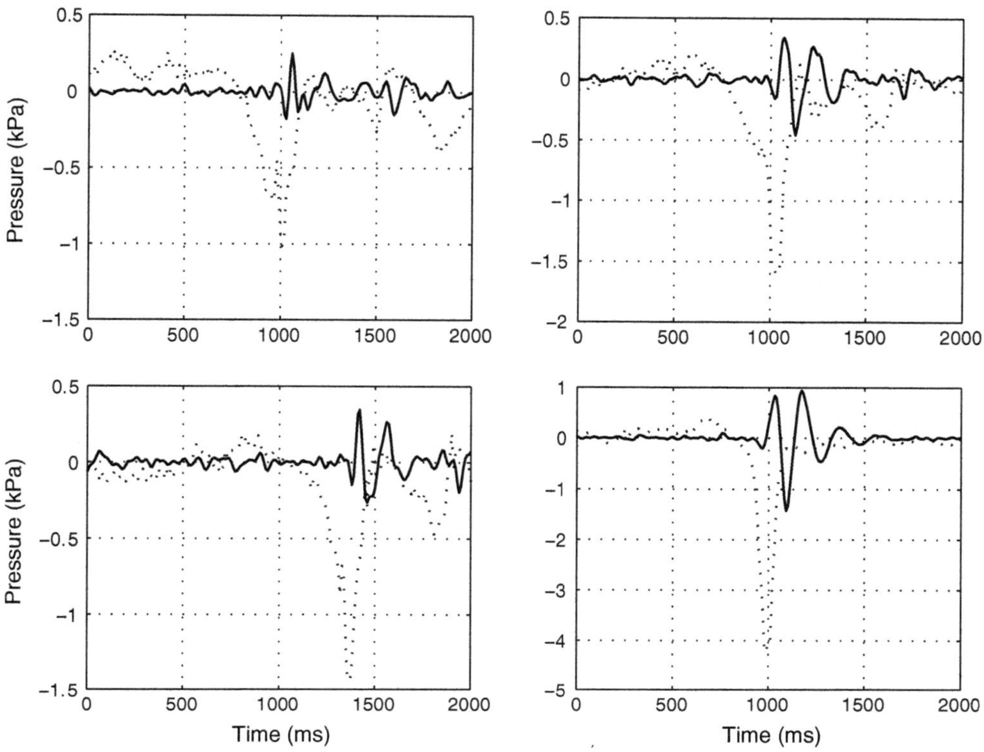

Fig. 4 Original pressure signal (dotted) and corresponding band pass filtered signal (solid). Pass band frequencies 5 and 20 Hz. Substantial amounts of noise are removed while feeding incidents are indicated as ripples in the filter output

Discussion

Comparison of the time course of respiration pressure measured by the tag with data acquired from trout (*Salmo trutta* L.) (Shelton, 1970: 296) shows that the waveforms are similar in structure. This was taken as an indication that the method of instrumentation is adequate and that measurements will give qualitatively representative data on the pressure conditions in the opercular cavity during feeding. Quantitatively, differences are seen in pressure amplitudes and respiration rate. The trout employed in the study was 70 g with peak-to-peak opercular pressure amplitude of 200 Pa compared to 320 Pa on the 1.5 kg salmon employed in the current experiment. Differences are most likely consequences of animal size discrepancies, deviating conditions under which the recordings were taken, and differences in the instrumentation. It is reasonable to expect that a larger fish of similar morphological design exhibits its higher respiratory pressure, perhaps even higher than the difference measured here. However, attenuation of the pressure signal within reasonable limits will not affect results qualitatively.

Multiple pressure signatures were recorded as the fish was feeding on pellets, and it was observed that the magnitude of the pressure transients showed substantial variation between feeding incidences and the duration of the transients were typically longer than similar measurements on non-salmonid species (Muller, 1989, Norton & Brainerd, 1992). In order to understand this variation, it should be observed that rather than being limited to some constant strike level, fish are known to operate in a continuum of different strike intensities. Strike intensity has for example been shown to depend on prey type, and non-elusive prey normally evokes less powerful strikes in terms of suction magnitude and duration than what is the case for elusive prey

(Nemeth, 1997). Passively drifting feed pellets must be regarded as notoriously non-elusive prey by a salmon that has fed on this kind of food its entire lifetime. Besides, the pellet size employed, which was a compromise for the sake of acclimation time, may be regarded as undersized considering the size of the fish. These factors may suggest that the magnitude of pressure transients in the present study could be expected to lie in the lower range of possible strike intensities. However, the fact that the instrumented fish was accompanied by control fish that were feeding on the same pellets suggest that episodes of competition could easily increase strike intensity. Visual observations indicate indeed that competition occurred occasionally.

With respect to the development of a detection system for feed ingestion, it is likely that information related to the spectral properties of the signal will be essential. Band pass filtering of the feeding signal in the 5–20 Hz range as shown in Fig. 4 provides a new signal where the feeding-specific parts of the signal are boosted. Subsequent equalization and logical manipulation on the filtered signal waveform will extract the required information, e.g. by exploiting the fact that feeding signals always consist of an ordered set of two consecutive gradients of opposite direction within a limited time period. Moreover, multiple feeding incidents will at least be separated by some minimum time period which can be used to eliminate signal ripples following the main transients. Implementation of a signal processing algorithm of this complexity in small acoustic telemetry tags is feasible with the availability of modern mixed signal microprocessors. The differential pressure sensor used in the experiment also permits easy integration into a fish tag due to its "tag friendly" specifications (low power, miniature design, low cost). Measuring the pressure differentially is important in order to cancel the effect of varying swimming depth and to maintain sensitivity at all depth ranges. Knowledge of the exact pressure is not essential as far as the detection algorithm operates on relative changes in signal amplitude.

The motivation of this study was based on the urgency for better researching tools and solutions to important issues regarding feeding management of aquaculture facilities. However, the results presented here were obtained in a controlled tank environment with a limited number of fish and should merely be regarded as an initial contribution to a particular part of a more extensive problem. Opercular pressure transients represent just one candidate feeding signal whose properties should be investigated further under realistic cage conditions. Several questions remain to be answered concerning the effect of the cage environment on fish behaviour and the feasibility of the measurement principle under such conditions. Alternative feeding signals exist and must be explored carefully before decisions can be made with respect to up-scaling to a fully functioning telemetry-based monitoring system for individual feeding behaviour in sea cages.

Conclusion

The objective of the present study was to investigate the properties of the pressure signal arising in the opercular cavity of farmed Atlantic salmon during feeding, and to discuss the measurement principle with respect to its potential for application in a telemetry system for remote monitoring of feed intake and feeding history of fish kept in sea cages. Acquisition of differential pressure data from a fish fed with 4 mm feed pellets in a tank environment showed that feed ingestion indeed influenced pressure conditions in the opercular cavity. Pressure transients showed considerable variation in terms of signal magnitude and duration between feeding events. However, the signal waveform contains distinct features that can be related to feed ingestion, and it is regarded as feasible from a signal processing point of view to develop an algorithm that will be able to recognize these features while suppressing the influence from other sources of pressure variation. Nonetheless, the approach must be put to test under realistic conditions in sea cages in order to be conclusive about its ability to mediate individual feeding behaviour in farmed fish.

Acknowledgements This work is partially financed by research project 165197/S40, "FeedTag", granted by The Research Council of Norway.

References

Alexander, R. McN., 1967. Functional Design in Fishes. Hutchinson & Co, London.

Alver, M. O., J. A. Alfredsen & T. Sigholt, 2004. Dynamic modelling of pellet distribution in Atlantic salmon (*Salmo salar* L.) cages. Aquacultural Engineering 31: 51–72.

Ang, K. P. & R. J. Petrell, 1997. Control of feed dispensation in seacages using underwater video monitoring: effects on growth and food conversion. Aquacultural Engineering 16: 45–62.

Baras, E. & J. P. Lagardère, 1995. Fish telemetry in aquaculture: review and perspectives. Aquaculture International 3: 77–102.

Begout Anras, M. L., S. Kadri, J. E. Juell & T. Hansen, 2000. Measuring individual and group swimming behaviour under production densities: test of a 3D multiple fish acoustic positioning system in a sea cage. In Moore, A & I. A. Russel (eds), Advances in fish telemetr. Centre for Environment, Fisheries and Aquaculture Sci., Lowestoft Laboratory, UK, 75–78.

Cooke, S. J., E. B. Thorstad & S. G. Hinch, 2004. Activity and energetics of free-swimming fish: insights from electromyogram telemetry. Fish and Fisheries 5: 21–52.

Houlihan, D., T. Boujard & M. Jobling, 2001. Food Intake in Fish. Blackwell Science Ltd., London.

Jobling, M., D. Covès, B. Damsgård, H. R. Kristiansen, J. Koskela, T. E. Petursdottir, S. Kadri & O. Gudmundsson, 2001. Techniques for measuring feed intake. In Houlihan, D., T. Boujard & M. Jobling (eds), Food Intake in Fish. Blackwell Science Ltd, London, 49–87.

Juell, J. E. & H. Westerberg, 1993. An ultrasonic telemetric system for automatic positioning of individual fish used to track Atlantic Salmon (*Salmo salar* L.) in a Sea Cage. Aquacultural Engineering 12: 1–18.

McFarlane, W. J., H. Williams, D. Roswell, C. White, R. Gosine, B. Finstad & R. S. McKinley, 2002. Intelligent monitoring systems for aquaculture: enhancing the efficiency of feeding systems. Aquaforum Conference Proceedings January 24–25, Science Council of British Columbia, Vancouver, BC.

Muller, M., 1989. A quantitative theory of expected volume changes of the mouth during feeding in teleost fishes. Journal of the Zoological Society of London 217: 639–662.

Nemeth, D. H., 1997. Modulation of buccal pressure during prey capture in Hexagrammos decagrammus (Teleostei: Hexagrammidae). Journal of Experimental Biology 200: 2145–2154.

Norton, S. F. & E. L. Brainerd, 1992. Convergence in the feeding mechanics of ecomorphologically similar species in the Centrarchidae and Cichlidae. Journal of Experimental Biology 176: 11–29.

Oswald, R. L., 1978. The use of telemetry to study light synchronization with feeding and gill ventilation rates in Salmo trutta. Journal of Fish Biology 13: 729–739.

Shelton, G., 1970. The regulation of breathing. In Hoar, W. S. & D. J. Randall (eds), Fish Physiology IV, Academic Press Inc., New York, 293–258.

Talbot, C., S. Corneillie & Ø. Korsøen, 1999. Pattern of feed intake in four species of fish under commercial farming conditions: implications for feeding management. Aquaculture Research 30:509–518.

Van Leeuwen, J. L., 1984. A quantitative study of flow in prey capture by rainbow trout, Salmo gairdneri with general consideration of the actinopterygian feeding mechanism. Transactions of the Zoological Society of London 37: 171–227.

Hydrobiologia (2007) 582:209–220
DOI 10.1007/s10750-006-0539-8

FISH TELEMETRY

Laboratory protocol to calibrate sea lamprey (*Petromyzon marinus* L.) EMG signal output with swimming

P. R. Almeida · I. Póvoa · B. R. Quintella

Abstract A correct application of electromyogram (EMG) telemetry in the field can be a powerful tool to evaluate activity patterns and swimming strategies of fishes. We evaluated the swim performance of seven untagged sea lampreys (*Petromyzon marinus* L.) with critical swim speed (U_{crit}) tests. The average U_{crit} observed was *c.* 1.03 ms^{-1} (i.e., 1.14 BL s^{-1}). The strongest reotaxic response was observed during tests using water velocities between 0.4 ms^{-1} and 0.8 ms^{-1}. During two consecutive years (i.e., 2004 and 2005), in order to model upstream migration of sea lampreys with CEMG transmitters (Lotek Wireless), we calibrated EMG signal with swim speed. A high correlation between EMG records and swim speed was observed in both years (r^2 = 0.74–0.93). However, in spite of methodology improvements and standardization in the second year of study, differences in intercepts and slopes were observed between individuals, making the determination of a unique calibration equation for all tagged animals unfeasible. Therefore, it appears to be necessary to obtain the relationship between EMG signals and swimming speed for each lamprey using laboratory procedures, prior to release in the wild. It is unknown whether this variability results from individual locomotor behaviour, physiological state and/or variation in placement and functioning of the EMG transmitters. The results of five laboratory calibrated lampreys, released in the River Mondego, revealed considerable differences between swim speeds calculated with EMG signal (calibration equation) and ground speed therefore it was not possible to successfully calibrate the EMG signal output with swimming speed. In order to accomplish this, longer continuous swimming tests in laboratory are necessary. Nevertheless, the calibrated swimming effort gives reliable information about the swimming behaviour and permits comparison of the results between animals.

Keywords Electromyogram · CEMG transmitters · Biotelemetry · Calibration · Critical swimming speed

Guest editors: P. R. Almeida, B. R. Quintella, M. J. Costa and A. Moore
Developments in Fish Telemetry

P. R. Almeida (✉)
Department of Biology, University of Évora, 7002-554 Evora, Portugal
e-mail: pmraposo@mail.telepac.pt

P. R. Almeida · I. Póvoa · B. R. Quintella
Institute of Oceanography, Faculty of Sciences of the University of Lisbon, Campo Grande, 1749-016 Lisboa, Portugal

Introduction

Sea lampreys (*Petromyzon marinus* L.) are considered a gastronomical delicacy in Portugal, with ecological, economical and cultural significance. During their anadromous migration, sea lampreys do not feed and therefore have a finite amount of energy reserves not only to complete migration, but also for sexual maturation and spawning (Beamish, 1980). The physiological and behavioural processes involved in spawning success can be compromised by human and/or natural obstacles responsible for the waste of excessive amounts of energy (Mesa et al., 2003; Quintella et al., 2004).

The swimming performance is considered a golden rule that determines the survival in many aquatic species (Plaut, 2001). The anguilliform type of locomotion used by lampreys is recognized to be less efficient than the sub-carangiform motion used by salmonids (Beamish, 1974, 1978; Mesa et al., 2003; Quintella et al., 2004). The poor swimming capacity of sea lampreys already observed in the laboratory (Beamish, 1974; Hanson, 1980; Bergstedt et al., 1981) is again explored in this approach by the application of the standard measurement critical swimming speed, similar to the one applied by Mesa et al. (2003) to the Pacific lamprey *Lampetra tridentata* (Richardson, 1836).

Electromyogram (EMG) telemetry has been used for several years in a variety of free-swimming fishes, providing information on movement and activity at a resolution that is not possible using more conventional techniques (Cooke et al., 2004). The EMG signals are used as a quantitative measure of activity, representing bioelectrical voltage changes strongly related with strength and duration of muscle contraction (Kaseloo et al., 1992). From several studies, this methodology has proven to be useful to recognize behaviour and activity patterns of fish in response to their environment and anthropogenic perturbations (Booth et al., 1997; Geist et al., 2002; Murchie & Smokorowski, 2004). In sea lampreys in particular, EMG telemetry provided important information on migratory behaviour either in distinct river stretches as in difficult passage areas (Quintella et al., 2004).

The EMG signals by themselves only provide a relative index of activity unless they are calibrated to swimming speed (Geist et al., 2002). The laboratory calibration of EMG signal with the swimming speed, can promote a more reliable interpretation of the migratory behaviour of fish in the field, once it provides an accurate measure of the real swimming speed, instead of the commonly used ground speed which does not take into account the influence of different current velocities nor the distinct routes followed by the animals during the migratory movement. The calibration curve resulting from this procedure can be useful to compare swimming behaviours between species with practical application on ecology, ethology, fisheries and aquaculture studies. There already exist a considerable number of calibration studies applied to various species (Demers et al., 1996; Booth et al., 1997; Økland et al., 1997; Thorstad et al., 2000; Brown et al., 2002; Geist et al., 2002; Standen et al., 2002) most of them with salmonids (see Cooke et al., 2004). However, as far as is known by the authors, there are no studies in which calibrated EMG transmitters are used to describe the lamprey anguilliform swimming motion.

Currently, one of the biggest debates among EMG telemetry users is the need for individual calibrations for all transmitter-implanted fish (Demers et al., 1996; Økland et al., 1997; Thorstad et al., 2000; Geist et al., 2002; Cooke et al., 2004). The cause for this variability is not well known however there are some hypotheses based on differences in behaviour, electrodes position, physiological state, total length or even manufactory differences of the transmitters (Thorstad et al., 2000). In some studies, it was possible to recognize some factors, like sex and dimensions of the individuals, as responsible for these differences (Hinch & Rand, 1998; Standen et al., 2002).

The main objective of this study was to standardize the application of electromyogram (EMG) telemetry for the monitoring of sea lamprey anadromous migration. Two experiments were outlined; the first performed in the spawning season of 2004 was directed to test the influence of parameters like sex, river basin of capture and differences between transmitters' assemblage in EMG signal output and possible constraints to the definition of a unique calibration equation. In the second experiment, an effort was made to standardize the sample, in order to give particular emphasis to the relation between EMG signal output and swimming speed. As a complementary objective, the swimming capacity of the sea lamprey was assessed with a critical swimming speed test. Although some laboratory procedures already exist which refer to the swimming performance of sea lampreys (Beamish, 1974; Hanson, 1980; Bergstedt et al., 1981), the increasing velocity procedure described by Brett (1967) was apparently never applied. Being a preliminary test, two prescribed increments of water velocity and swimming intervals were applied, in order to determine a laboratory procedure more appropriate for this species.

Materials and methods

Telemetry equipment and tagging

To record the EMG signals (i.e., changes in bioelectrical voltage related to the muscle contraction) we used coded electromyogram radio-transmitters (CEMG-R11-25) manufactured by Lotek Wireless, Newmarket, Ontario. These cylindrical transmitters are covered by an epoxy cap and are 61 mm long by 11 mm in diameter and weigh 12 g in the air (<1% of sea lamprey body mass). The transmitters measured the same signal (muscle electrical impulse) as the old EMG transmitters first described by Kaseloo et al. (1992) however it was processed in a different manner. The voltage corresponding to the muscle activity was rectified, sampled at a very high rate (1000s of samples per second), summed and averaged for the duration of the burst interval. After the amplification of the signals, a mean value of a sample interval of 2 s (i.e., burst rate interval) was obtained. This result was then transmitted to a radio receiver SRX_400W 32C (Lotek Wireless) through a coaxial antenna. The EMG signals detected were displayed in an increasing activity index, ranging from 0 to 50. Data were downloaded to a computer via a RS-232 serial communication port, using the software WINHOST.

The surgical procedure is described with specific details in Quintella et al. (2004). EMG transmitters were surgically implanted into the peritoneal cavity, and a pair of flexible teflon-coated stainless steel electrodes with 19 carat gold muscle-anchoring tips (dimensions 7×1 mm), were fixed in the axial red musculature. During the first experiment (i.e., 2004 study), tagged animals were revived in a 2 m^3 holding tank, and left to recover under a photoperiod of 12L: 12D for one week. This procedure was changed for the second experiment (i.e., 2005 study); the recovery period was reduced to 2 days, and to avoid unnecessary manipulation the lampreys were kept in the swim tunnel used for the calibration procedures, instead of the holding tank used in the first year.

Swimming performance

During the first year of experiment, in order to assess the swimming performance of adult sea lampreys, the critical swimming speed (U_{crit}) tests were conducted on seven untagged sea lampreys, captured in the Tagus river basin during the spawning migration of 2004. The procedure was applied according to Brett (1964, 1967), (see more details in the calibration procedure) differing only in the swim apparatus. The animals presented a mean total length (L_t) of 90 cm and displayed an average total weight (W_t) of 1.4 kg. Prior to the tests the lampreys underwent a 30 min conditioning period at a low water velocity of 0.2 ms^{-1} with water temperature around 15°C. After the conditioning period, the animals were subjected to prescribed increments of water velocity (U_{ii}) for a determined duration of the swimming interval (T_{ii}), according to the methodology applied: Method A—U_{ii} = 0.1 ms^1, T_{ii} = 15 min; or Method B—U_{ii} = 0.2 ms^{-1}, T_{ii} = 10 min (Table 1). This procedure continued until the animals presented a fatigue state characterized by the "U" position. Fatigue was determined when lampreys failed to leave the downstream screen despite some attempts to induce their movement by means of mechanical stimuli. Critical swimming speed (U_{crit}) was calculated using Brett's (1964, 1967) equation:

$$U_{crit} = U_i + [U_{ii}(T_i/T_{ii})]$$

where, U_i is the highest water velocity maintained for the prescribed period (ms^{-1}), U_{ii} is the water

Table 1 Data on sea lampreys used in the critical swimming speed ($U_{crit.}$) determination

M	ID	L_t (mm)	W_t (g)	sex	U_i (ms^{-1})	T_i (min)	$U_{crit.}$ (ms^{-1})	$U_{crit.}$ (BL s^{-1})
A	LpI	940	1651	M	1.0	9.0	1.18	1.01
	LpII	927	1606	F	1.0	5.0	1.10	1.07
	LpIII	885	1390	M	1.0	9.5	1.19	0.93
	LpIV	810	1147	F	0.7	9.0	0.82	1.20
B	LpV	957	1495	M	1.0	2.0	1.03	1.26
	LpVI	900	1301	F	0.8	3.0	0.84	1.18
	LpVII	885	1453	M	0.9	13.0	1.07	1.34

M: methodology, ID: animal identification, L_t: total length, W_t: total weight, U_i: maximum water velocity, T_i: time elapsed at maximum water velocity, U_{crit}: critical swimming speed

velocity increment (ms^{-1}), T_i is the time elapsed at fatigue velocity (min), and T_{ii} is the prescribed interval of swimming (min).

Calibration of EMG signal with swimming speed

This work extended over two consecutive spawning seasons (i.e., 2004 and 2005) from February to May. During the first year of study, we used eight adult sea lampreys, five females and three males, captured in Mondego, Vouga and Minho river basins, and due to costs constrains only three CEMG transmitters were used. In 2005, to standardize the procedure the tests were performed with five adult lampreys belonging to the same gender (i.e., females), captured in the same watershed (i.e., near the mouth of River Mondego), and each animal was tagged with a different transmitter (Table 2). In the first year of experiment, the calibration procedure was performed during five consecutive days after a recovery period of one week. In the second year, the protocol was changed to reduce the experimental time and the calibration was performed during 3 days after 2 days of recovery period. The animals presented a mean total length (L_t) of 93 cm and displayed a mean weight of 1.5 kg. The relation between muscle activity and swimming speed was obtained by the correlation between the EMG signal output and the water velocity (ms^{-1}) measured in the test section. The procedure followed was according to what described in Booth et al. (1997), Thorstad et al. (2000), Brown et al. (2002), and others. However, for this study a swim tunnel was preferred to adjust the dimensions of the test section to the width amplitude of the sea lamprey anguilliform movement, and to

Table 2 Data on sea lamprey used in the EMG calibration procedure

ID	L_t (mm)	W_t (g)	SEX	BC	TF	C	S	BC	EC
Lp2	910	1547	F	Mondego	142.020*	11-Mar-04	15-Mar-04	22-Mar-04	26-Feb-04
Lp3	900	1534	F	Minho	142.040	18-Mar-04	22-Mar-04	29-Mar-04	2-Apr-04
Lp4	925	1443	M	Minho	142.000	18-Mar-04	29-Mar-04	5-Apr-04	9-Apr-04
Lp5	995	1846	F	Vouga	142.040	2-Apr-04	5-Apr-04	12-Apr-04	16-Apr-04
Lp6	100	1658	F	Vouga	142.000	8-Apr-04	12-Apr-04	19-Apr-04	23-Apr-04
Lp7	860	1157	M	Vouga	142.040	16-Apr-04	19-Apr-04	26-Apr-04	30-Apr-04
Lp8	965	1696	M	Vouga	142.000	23-Apr-04	26-Apr-04	3-Mai-04	7-Mai-04
Lp9	880	1465	F	Vouga	142.040	30-Apr-04	3-Mai-04	10-Mai-04	14-Mai-05
Lp10	920	1432	F	Mondego	142.100	4-Feb-05	7-Feb-05	9-Feb-05	11-Feb-05
Lp11	980	1853	F	Mondego	142.150	18-Feb-05	21-Feb-05	23-Feb-05	25-Feb-05
Lp12	885	1375	F	Mondego	142.200	2-Mar-05	7-Mar-05	9-Mar-05	11-Mar-05
Lp13	895	1545	F	Mondego	142.250	23-Mar-05	24-Mar-05	28-Mar-05	31-Mar-05
Lp14	945	1512	F	Mondego	142.300	5-Apr-05	11-Apr-05	13-Apr-05	15-Apr-05

ID: animal identification; L_t: total length; W_t: total weight; BC: basin of capture; TF: transmitter frequency; C: capture date; S: surgery date; BC: beginning of calibration; EC: end of calibration

* This transmitter was only used once due to a malfunction caused by a broken electrode

facilitate manipulation due to the extended period of swimming trials (i.e., 3–5 days).

This swim tunnel had a maximum capacity of 4.5 m^3, was 3 m long by 1.5 m wide (Fig. 1). The test section was located in one of the two divisions of the swim tunnel, were there was a stainless steel cage (mesh 2 × 2 cm) 2 m long by 0.7 m wide and 0.9 m high. Inside this, an acrylic structure was set to guarantee a uniform laminar flow. The test section was restricted to an area with 150 cm long by 40 cm wide, interior lined with plastic mesh (2 × 2 cm) to prevent the lampreys from holding with the sucker mouth. The water height in the test section was 20 cm making a total water volume available for the lamprey to swim of *c.* 0.12 m^3 (Fig. 1). It was not necessary to correct the blocking effect (see Beamish, 1978), because the maximum cross-sectional area of the lampreys, assumed as a circle, was about 5% of the cross-section of the experimental area. The anterior part of the test section was covered by a black acrylic plate to minimize the negative fototaxic response (see Almeida et al., 2000, 2002) and reduce stress. The water velocities (ms^{-1}) were induced by a system composed of three submersible motors, two 1 hp (Model Brio 1 from AcQua & Co) and one 2.3 hp (Model SR4620.410 HG from Flyght) propeller motors.

Adjustable frequency drives (model *Yaskawa VS mini J7*) were used to control the frequency of rotation of the motor propellers. The maximum water velocity generated by this system was *c.*1.2 ms^{-1}.

After the surgical procedure and subsequent recovery period, the animals were introduced in the test section for an acclimation period of 30 min at a low water velocity *c.* 0.2 ms^{-1}. In order to guarantee a swimming behaviour of the sea lampreys similar to that observed in their natural environment, a flat stone was added to the test section as a resting support. In this way, lampreys could alternate between voluntary swimming trials with periods of rest while fixed by the oral sucker. Resting EMG values were recorded while the animal was attached to the stone subjected to null water velocity. In the first year of study, the EMGs were also recorded while the animals were fixed to the resting support in two different positions: lateral and horizontal, subjected to distinct water velocities (i.e., 0.4, 0.7 and 1.0 ms^{-1}). This record was important to test, if monitoring the activity patterns in the natural environment, because the resting periods can be confused with activity periods due to the body oscillation when animals are fixed resting and subjected to different water velocities. Movement EMG values were recorded at three water velocities: 1.0, 0.7 and 0.4 ms^{-1}. The swimming tests were performed alternating the water velocity in a decreasing way from 1.0 ms^{-1} to 0.4 ms^{-1}, until the animal showed signs of fatigue. Between swimming trials the animals were left to rest 45–90 min at low water velocity (0.2 ms^{-1}). Only the swimming trials characterized by a unidirectional movement at a constant speed with the minimum duration of 10 s were considered valid.

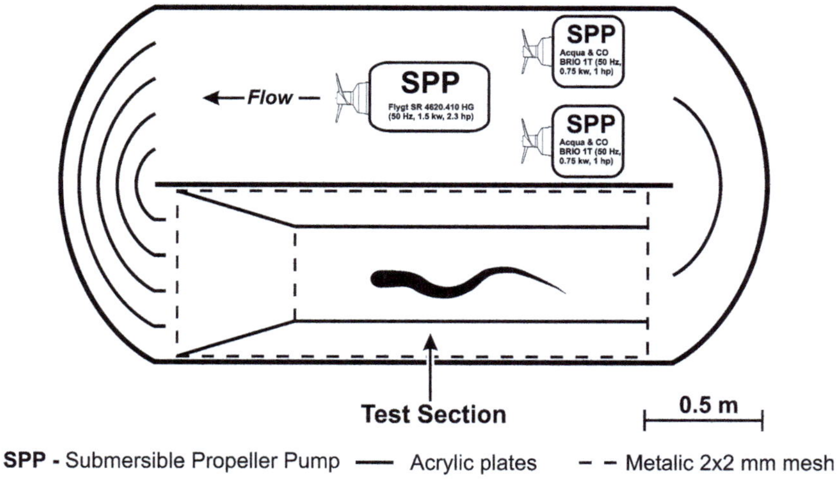

Fig. 1 Schematic diagram (top view) of the swim tunnel used for the calibration swimming trials

At the end of the calibration procedures of the eight lampreys tagged in 2004, a lethal dose of anesthesia (2-phenoxyethanol) was administrated and the final total length (L_t), final total weigh (W_t) and distance between electrodes registered. The five lampreys tagged in 2005, were released in the field (River Mondego) to test the reliability of this methodology in situ but the results are only briefly discussed in this study.

Statistical analysis

Prior to the statistical analysis, the outliers and extreme EMG values obtained from the swimming trials were removed, with the application of the boxplot procedure performed with the statistical program SPSS for Windows (version 11.0). Regression analysis was used to model the relationship between EMG signals (variable) and swimming speed (covariate). In order to eliminate the influence of length variation among animals, the swimming speed (SW) was converted to body lengths per second (BL s^{-1}) according to the equation:

$$SW_{BL/s} = \frac{SW_{m/s}}{L_t}$$

An analysis of covariance (ANCOVA) was used to compare the linear slopes and intercepts between individuals and in particular to test the influence of sex, basin of capture and transmitters in calibration equation. This statistical procedure was performed with the program BIOMstat for Windows (version 3.0) (Sokal & Rohlf, 1981).

To test the influence of the resting position (i.e., horizontal and lateral) and water velocity (i.e., 0.4 ms^{-1} and 1.0 ms^{-1}) in the EMG records, while lampreys were fixed to the resting support during recovering periods, a paired-samples *t*-test was applied. This analysis was also performed with the statistical program SPSS for Windows (version 11.0). Inconsistent EMG records obtained during the experimental trials of the lamprey Lp8, within the same resting position and subjected to the same water velocity, prevented the data collected with this animal from being included in the statistical analysis.

Results

All animals successfully completed the critical swimming speed (U_{crit}) tests. Sea lampreys, tested at water velocities ranging from 0.2 ms^{-1} (acclimatization velocity) to 1.0 ms^{-1}, showed a more satisfactory reotaxic response for velocities between 0.4 ms^{-1} and 0.8 ms^{-1}. Despite the reduced number of animals used, the U_{crit} estimated independently of the methodology applied (method A or B, see Material and methods section) was similar, with a mean value of 1.03 ms^{-1} (1.14 BL s^{-1}) (Table 1).

No mortality occurred as a result of the surgical procedure used to implant the transmitters. After a postmortem observation of the first tagged lampreys (i.e., Lp2-Lp9), it was possible to confirm the absence of infections around the antenna and incision wound, and that during the two weeks of experiment (first week of transmitter implantation recovery plus the second week of calibration trials) they lost on average 2.4% of the total length and 4.4% of the total weight. The gold-tipped electrodes were found between 1–10 mm (minimum distance) apart in the dissected animals.

In the 2004 experiment, all eight lampreys exhibited a strong relationship between EMG signal and swimming speed ($r^2 = 0.74$–0.90, $P < 0.001$) (see Fig. 2). The ANCOVA analysis revealed differences in intercepts and slopes of the linear regressions between individuals (intercepts $F = 106.74$, $P < 0.001$ and slopes $F = 40.914$, $P < 0.001$). Furthermore, the ANCOVA analysis conducted according to sex revealed the same differences between females and males (females: intercepts $F = 186.09$, $P < 0.001$ and slopes $F = 37.78$, $P < 0.001$; males: intercepts $F = 29.47$, $P < 0.001$ and slopes $F = 50350$, $P < 0.001$). Individuals from Vouga river basin, also revealed the same differences (intercepts $F = 194.72$, $P < 0.001$ and slopes $F = 19.68$, $P < 0.001$). There were no results from Rivers Mondego and Minho because there were not enough individuals for the ANCOVA analysis to be conducted. The individuals tagged with the same transmitter, also revealed differences for both intercepts and slopes (transmitter 142.040: intercepts $F = 239.23$, $P < 0.001$ and slopes $F = 56.82$, $P < 0.001$, transmitter 142.000: inter-

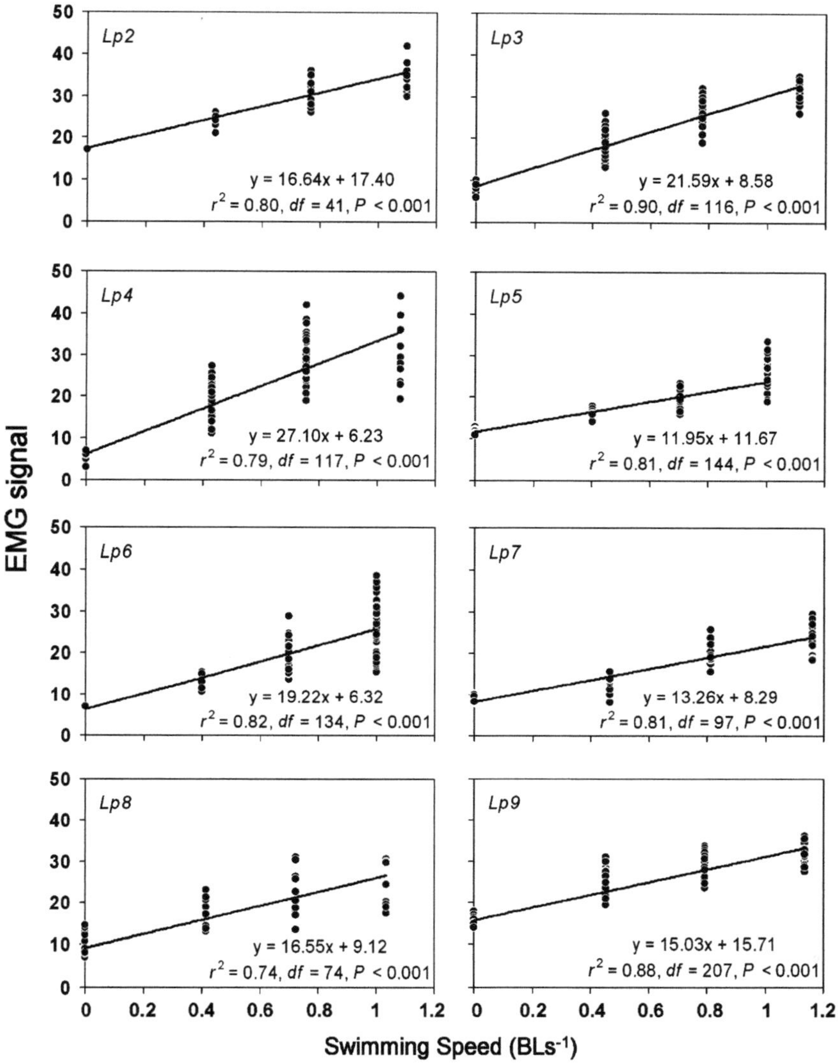

Fig. 2 Relationship between EMG signal and swimming speed expressed in body lengths per second (BL s^{-1}) during the calibration trials in the 2004 experiment. EMG data is represented by an average value of each swimming trial performed at each of the predefined water velocities (i.e., 0 ms^{-1}—resting attached to the stone, 0.4, 0.7 and 1.0 ms^{-1}). Regression parameters are also presented

cepts $F = 20.08$, $P < 0.001$ and slopes $F = 24.79$, $P < 0.001$). We did not perform an ANCOVA analysis for transmitter 142.020 because this transmitter was only used to tag one individual (see Table 2).

In the 2005 experiment, the strong correlation between EMG signal and swimming speed was again confirmed ($r^2 = 0.85$–0.93, $P < 0.001$) (see Fig. 3). Although eliminating the influence of sex, basin of capture and reducing the calibration swimming trials period, the ANCOVA analysis revealed once more statistical differences between intercepts and slopes (intercepts $F = 277.371$, $P < 0.001$ and slopes $F = 6.627$, $P < 0.001$) of the five tested lampreys.

The paired-samples t-test showed that the resting position, assumed by the lampreys (i.e., lateral and horizontal) when they are attached by means of the oral sucker to the support stone and subjected to three distinct water velocities (i.e., 0.4, 0.7 and 1.0 ms^{-1}), did not represent a significant influence in the variability of the registered

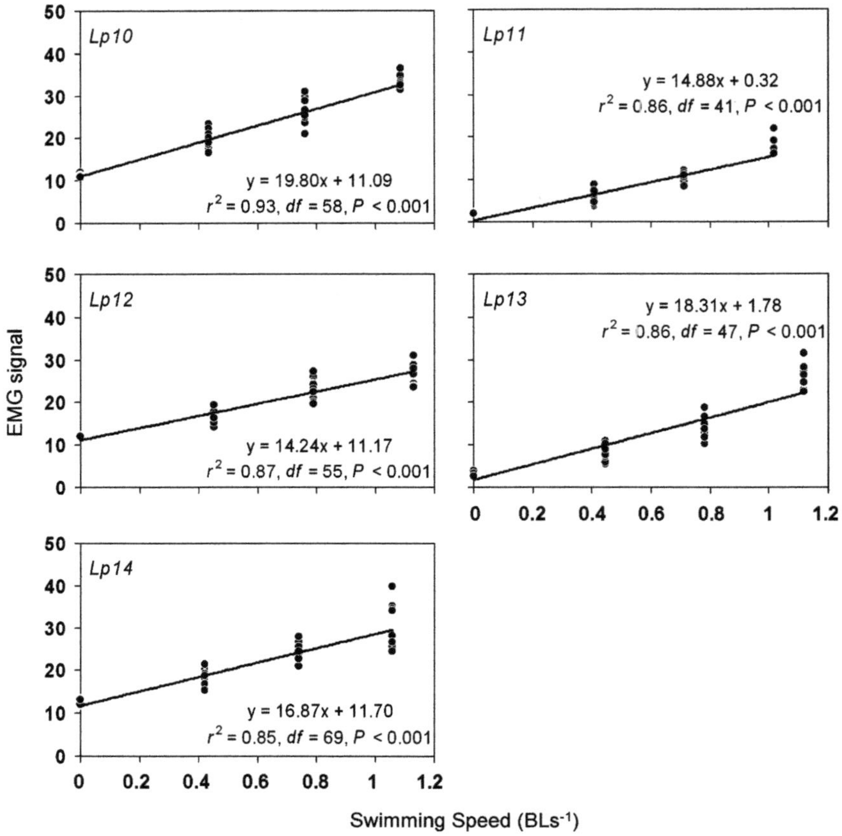

Fig. 3 Relationship between EMG signal and swimming speed expressed in body lengths per second (BL s^{-1}) during the calibration trials in the 2005 experiment. EMG data is represented by an average value of each swimming trial performed at each of the predefined water velocities (i.e., 0 ms^{-1}—resting attached to the stone, 0.4, 0.7 and 1.0 ms^{-1}). Regression parameters are also presented

EMG values (Table 3). Nevertheless, when animals were fixed at the same position and subjected to different water velocities (i.e., comparing EMG values between lower and higher velocities of 0.4 ms^{-1} and 1.0 ms^{-1}, respectively), a significant statistical variability in EMG data was detected (see Fig. 4 and Table 3).

The five lampreys tagged in 2005 were also released in the field (River Mondego) after the laboratorial calibration procedure. Data from this particular experiment are only briefly discussed here in order to verify and validate the swimming speed obtained with the EMG signal records with the measured ground speed in the field. When

Table 3 Significant levels of the paired-samples *t*-test used to determine the influence of the resting position and water velocity on the variability of EMG signal output

SV	G	Lp2	Lp3	Lp4	Lp5	Lp6	Lp7	Lp9
Position (H/L)	0.4 ms^{-1}	n.s.	n.s.	***	***	n.s.	*	**
	0.7 ms^{-1}	n.s.	n.s.	*	n.s.	n.s.	n.s.	n.s.
	1.0 ms^{-1}	*	n.s.	n.s.	n.s.	n.s.	n.s.	n.s.
Velocity (Min/Max)	H	***	***	**	**	***	***	***
	L	***	***	***	***	***	***	***

SV: source of variation, G: group of sample, Position: position of the stone support during rest (H: horizontal and L: lateral), Velocity: water velocity (Min: Minimum = 0.4 ms^{-1}, Max: Maximum = 1.0 ms^{-1}), n.s.: not significant

* Significant at $P \le 0.05$; ** Significant at $P \le 0.01$; *** Significant at $P \le 0.001$

comparing the ground speed with swimming speed it was noticed that they were considerably different. During stretches free from obstructions (i.e., 1 km long straight channel with no islands, no narrow areas, no shallow areas, no dense vegetation), the swimming speed of sea lampreys were considerably inferior when compared to the ground speed. Conversely, when lampreys were transposing obstacles (i.e., 100 m long submerged blockstone weirs with a 4.5 m drop and a 20° slope in the downstream direction) the ground speed was lower than the swimming speed (Fig. 5—for simplicity this figure shows only the average values for the five released lampreys).

Discussion

The U_{crit} estimated in this study (i.e., 1.14 BL s^{-1}), was similar to the one observed for the Pacific lamprey (i.e., 1.36 BL s^{-1}) by Mesa et al. (2003), being the first quantitative measure with this standard method for the sea lamprey. Critical swimming speed is a useful operational term for comparing swimming capacity of different species, or of individual fish of the same species but distinct swimming abilities caused by differing physiological states (Hoar & Rand, 1978). The estimated swimming capacity for sea lampreys is considerably low when compared with the U_{crit} of

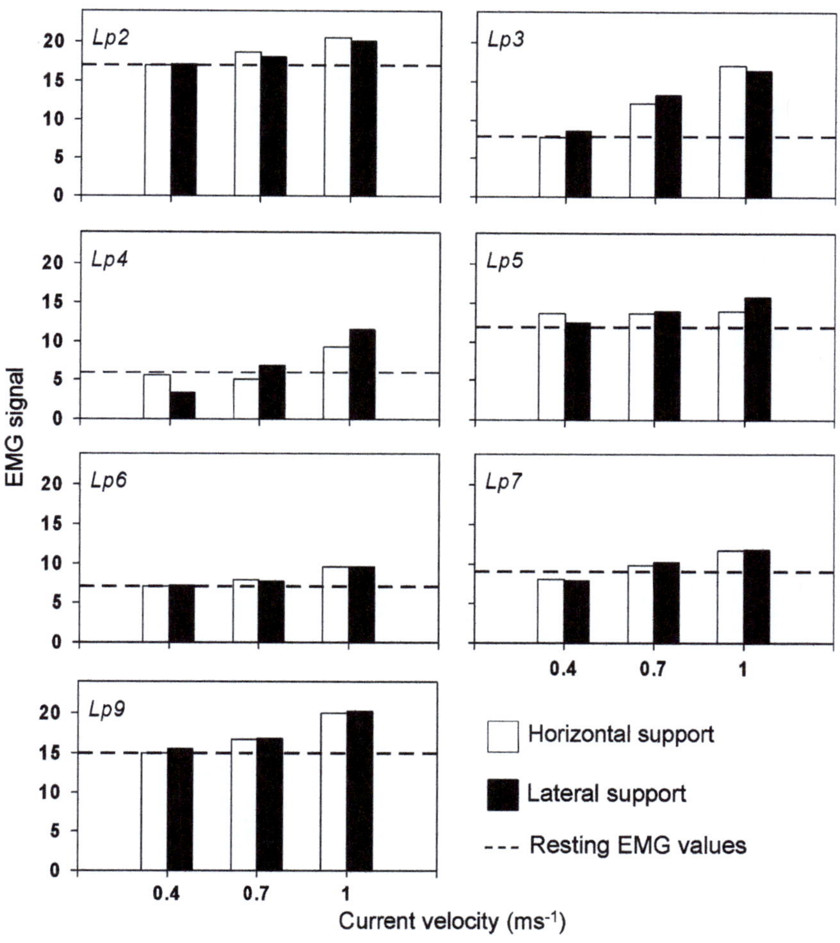

Fig. 4 Average values of EMG data when lampreys are fixed with the oral sucker to two different resting positions, horizontal (□) and lateral (■), and subject to three distinct water velocities (i.e., 0.4, 0.7 and 1.0 ms^{-1}). Also represented is the EMG resting value (---) when the lamprey is attached to the horizontal support (i.e., stone) and subjected to null water velocity. Problems during Lp8 trials did not allow for test to be carried out on the influence of the resting position and water velocity in the EMG records

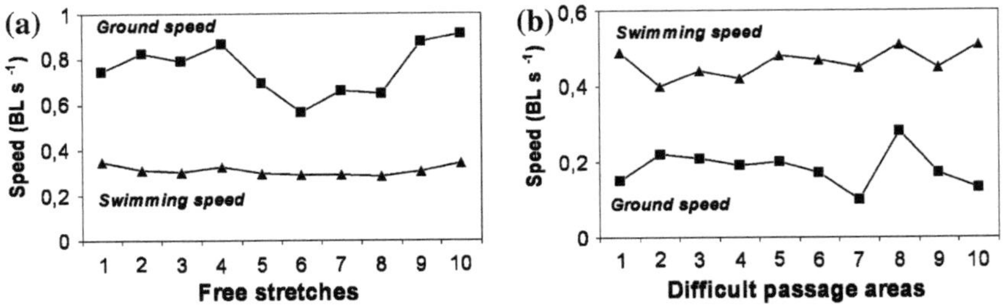

Fig. 5 Average swimming speed (-▲-) and average ground speed (-■-) expressed in body lengths per second (BL s^{-1}), determined during the migration of five calibrated sea lampreys released in River Mondego (2005), while migrating trough: (**a**) stretches free from obstructions and (**b**) difficult passage areas (i.e., submerged blockstone weirs)

c. 2.0 BL s^{-1} of most salmonids (Brown et al., 2002; Farrell et al., 2003).

A strong relationship between EMG signal and swimming speed was achieved, as observed in other studies using similar methodologies but with different species (Booth et al., 1997; Økland et al., 1997; Geist et al., 2000; Thorstad et al., 2000; Brown et al., 2002; Standen et al., 2002). This high correlation showed that EMG signal can be a good indicator of the swimming activity of sea lampreys in their natural environment. EMG records were less homogeneous at the higher water velocity (i.e., 1.0 ms^{-1}). Similar results have been shown by McFarlane et al. (2004) with rainbow trout (*Oncorhynchus mykiss* Walbaum, 1792). This observation is probably related to a less uniform swimming behaviour near the U_{crit} (i.e., burst swimming). Variability in EMG data caused by differences in swimming behaviour was also referred in anterior studies (Kaseloo et al., 1992; Økland et al., 1997; Thorstad et al., 2000; Hinch et al., 2002).

During the two years of experiment, a variability between individuals based on differences in intercepts and slopes from linear regressions of the calibration procedures was observed. In the first year, some factors like basin of capture, sex or transmitter were tested to see if they were responsible for this variability, which apparently were not, even though we only used a reduced number of individuals in this analysis. In the second year, we standardize the samples by using only females captured in the same watershed (i.e., River Mondego), and we also reduce the time for the calibration procedures, but the individual variability remained. According to Plaut (2001) the variability inter-individual is a repeatable phenomenon that reflects a real inter-individual variability in the swimming capacity of fishes. The different distances between electrodes, observed for all individuals in the post-mortem observations, revealed the need to standardize the implantation of the electrodes in the musculature, in order to maintain an identical distance in all of the tagged animals. Electrode position must guarantee that EMG signals are collected from the red muscle and the distance between them must be similar for all individuals, preferably around 10 mm apart (Cooke et al., 2004). Although according to Kaseloo et al. (1992), modifications in electrodes position are not responsible for the discrepancy between EMG resting signals of individuals, other authors are still not convinced and consider the importance of more thoroughly exploring this matter (Demers et al., 1996; Thorstad et al., 2000). Inter-variability can be related with heterogeneity of physiological responses between individuals, recently recognized as one important issue to consider (Kieffer, 2000). In fact, EMG transmitters provided different results when the same transmitter was applied to different individuals and between individuals presenting different transmitters. More studies should be addressed to analyze the ecomorphological and physiological factors underlying the muscle activity during swimming (Cooke et al., 2004). Due to this variability, the best approach will be to calibrate individually all the lampreys before releasing them in the wild. With this calibration procedure we can achieve a better interpretation of the EMG data and at the same time we are able to compare the results obtained by different tagged animals.

The calibration procedure is essential to describe the behaviour performed by the animals in the wild. One important issue to be addressed, when studying sea lampreys migratory behaviour, is their unique movement strategy. Sea lampreys alternate between swimming activity and resting, fixed by means of the oral sucker to stones or other convenient support structures. According to the test conducted in the first year of study, we found significant differences when we compared the EMG signal obtained when lampreys were fixed in the same position, but subjected to different water velocities. Body oscillation when lampreys are fixed to a support, and submitted to relatively high water velocities, can mimic the EMG activity signals. When applying the EMG telemetry in the field, a careful interpretation of the EMG records must be made, mainly in difficult passage areas, to ensure the correct distinction between activity and resting periods.

The five lampreys released in the second year in River Mondego, after one week of laboratory calibration, proved that it is possible to calibrate each animal that, as most migratory fishes, are only a few weeks from the terminus of their lives. However, when comparing the ground speed of the lampreys, with the swimming speed calculated with the EMG signal, we notice that they were considerably different. If the lamprey is migrating in a straight route (i.e., no swimming path curvatures) the swimming speed should be equal to the ground speed plus the current velocity. During the difficult passage areas, the ground speed was lower than the estimated swimming speed, probably due to the high current velocities in those areas. But when comparing the ground speed of the lampreys with the swimming speed, while migrating in the stretches free from obstacles, this was considerably higher. For some reason the lampreys presented a more efficient swimming in the wild when compared with the laboratory trials. Prolonged swimming trials were difficult to obtain in a restricted experimental area. In the laboratory we only registered small periods of continuous swimming (a mean value of 16 s for each trial), which probably results in registering the beginning of the movements on which the lampreys detached the oral sucker and have to overcome the inertia. For that reason, in the beginning of the movement, lampreys are experiencing a muscular effort higher than it would be expected for the observed swimming speed, when compared to the muscular effort needed to maintain a certain speed in a longer continuous movement. For that reason, we consider that the EMG signal measured in the laboratory trials corresponds to an effort index, and therefore it cannot be properly related to a real swimming speed.

The spawning migration of sea lampreys is one of the most critical periods in the life cycle of this species. This study proved that calibrated EMG telemetry may be a valuable tool used to recognize activity patterns and swimming strategies of adult sea lampreys in their natural environment. It also opened the possibility of future studies involving the determination of the efficiency of fishways and swimming strategies during their negotiation, location of difficult passage areas during the migratory route, quantification of energy expenditure or even definition of spawning activity. This study was important to the progress of the calibration procedure of EMG with swimming activity in sea lampreys, in order to assure a correct and useful application of this technology in future studies.

Acknowledgements This research was financially supported by two Portuguese institutions, the Water Institute (INAG) of the Ministry of Environment and by the Foundation for Science and Technology (FCT) with Project funding (reference PNAT/1999/BIA/15019/P) and a Ph.D. grant to B.R. Quintella (reference SFRH/BD/9125/2002).

References

Almeida, P. R., H. T. Silva & B. Quintella, 2000. The migratory behaviour of the sea lamprey *Petromyzon marinus* L., observed by acoustic telemetry in the River Mondego (Portugal). In Moore, A. & I. Russel (eds), Advances in Fish Telemetry. CEFAS, Suffolk, 99–108.

Almeida, P. R., B. R. Quintella & N. M. Dias, 2002. Movement of radio-tagged anadromous sea lamprey during the spawning migration in the River Mondego (Portugal). Hydrobiologia 483: 1–8.

Beamish, F. W. H., 1974. Swimming performance of adult sea lamprey, *Petromyzon marinus*, in relation to weight and temperature. Transactions of the American Fisheries Society 2: 355–358.

Beamish, F. W. H., 1978. Swimming capacity. In Hoar, W. S. & D. J. Randall (eds), Fish Physiology, Vol. 7. Academic Press, New York, 101–187.

Beamish, F. W. H., 1980. Biology of the North American anadromous sea lamprey, *Petromyzon marinus*. Canadian Journal of Fisheries and Aquatic Sciences 37: 1924–1943.

Bergstedt, R. A., D. V. Rottiers & N. R. Foster, 1981. Laboratory determination of maximum swimming speed of migrating sea lampreys: a feasibility study. US Fish and Wildlife Service Ann Arbor, Great Lakes Fishery Laboratory, administrative report no 81-3, Michigan.

Booth, R. K., R. S. McKinley, F. Økland & M. Sisak, 1997. *In situ* measurement of swimming performance of wild Atlantic salmon (*Salmo salar*) using radio transmitter electromyogram signals. Aquatic Living Resources 10: 213–219.

Brett, J. R., 1964. The respiratory metabolism and swimming performance of young sockeye salmon. Journal of the Fisheries Research Board of Canada 21: 1183–1226.

Brett, J. R., 1967. Swimming performance of sockeye salmon (*Oncorhynchus nerka*) in relation to fatigue time and temperature. Journal of the Fisheries Research Board of Canada 24: 1731–1741.

Brown, R. S., D. R. Geist & M. G. Mesa, 2002. The use of electromyogram (EMG) telemetry to assess swimming activity and energy use of adult spring chinook salmon migrating thought the tailraces, fishways, and forebays of Bonneville Dam, 2000 and 2001. US Army Corps of Engineers, Portland, PNNL-14080, Oregon.

Cooke, S. J., E. B. Thorstad & S. G. Hinch, 2004. Activity and energetics of free swimming fish: insights from electromyogram telemetry. Fish and Fisheries 5: 1–21.

Demers, E., R. S. McKinley, A. H. Weatherley & D. J. McQuenn, 1996. Activity patterns of largemouth and smallmouth bass determined with electromyogram biotelemetry. Transactions of the American Fisheries Society 125: 434–439.

Farrell, A. P., C. G. Lee, K. Tierney, A. Hodaly, S. Clutterham, M. Healey, S. Hinch & A. Lotto, 2003. Field-based measurements of oxygen uptake and swimming performance with adult Pacific salmon using a mobile respirometer swim tunnel. Journal of Fish Biology 62: 64–84.

Geist, D. R., C. S. Abernethy, S. L. Blanton & V. I. Cullinan, 2000. The use of electromyogram telemetry to estimate energy expenditure of adult fall chinook salmon. Transactions of the American Fisheries Society 129: 126–135.

Geist, D. R., R. Brown, K. Lepla & J. Chandler, 2002. Practical application of electromyogram radiotelemetry: the suitability of applying laboratory-acquired calibration data to field data. North American Journal of Fisheries Management 22: 474–479.

Hanson, L. H., 1980. Study to determine the burst swimming speed of spawning-run sea lamprey (*Petromyzon marinus*). Research completion report. Michigan, US Fish and wildlife service Hammond bay biological station, Millersburg.

Hinch, S. G. & P. S. Rand, 1998. Swim speeds and energy use of upriver-migrating sockeye salmon (*Oncorhynchus nerka*): role of local environment and fish characteristics. Canadian Journal of Fisheries and Aquatic Sciences 55: 1821–1831.

Hinch, S. G., E. M. Standen, M. C. Healey & A. P. Farell, 2002. Swimming patterns and behaviour of upriver-migrating adult pink (*Oncorhynchus gorbuscha*) and sockeye (*O. nerka*) salmon as assessed by EMG telemetry in the Fraser River, British Columbia, Canada. Hydrobiologia 483: 147–160.

Hoar, P. A. & D. J. Randall, 1978. Locomotion, Fish physiology Vol. 7. Academic Press Inc., New York.

Kaseloo, P. A., A. H. Weatherley, J. Lotimer & M. D. Farina, 1992. A biotelemetry system recording fish activity. Journal of Fish Biology 40: 165–179.

Kieffer, J. D., 2000. Limits to exhaustive exercise in fish. Comparative Biochemistry and Physiology 126: 161–179.

Mcfarlane, W. J., K. F. Cubitt, H. Williams, D. Rowsell, R. Moccia, R. Gosine & R. S. McKinley, 2004. Can feeding status and stress level be assessed by analyzing patterns of muscle activity in free swimming rainbow trout (*Oncorhynchus mykiss* Walbaum)? Aquaculture 239: 467–484.

Mesa, M. G., J. M. Bayer & J. G. Seelye, 2003. Swimming performance and physiological responses to exhaustive exercise in radio-tagged and untagged Pacific lampreys. Transactions of the American Fisheries Society 132: 483–492.

Murchie, K. & K. E. Smokorowski, 2004. Relative activity of brook trout and walleye in response to flow in a regulated river. North American Journal of Fisheries Management 24: 1050–1057.

Økland, F., B. Finstand, R. S. McKinley, E. B. Thorstad & R. K. Booth, 1997. Radio transmitted electromyogram signals as indicators of physical activity in Atlantic salmon. Journal of Fish Biology 51: 476–488.

Plaut, I., 2001. Critical swimming speed: Its ecological relevance. Comparative Biochemistry and Physiology (Part A) 131: 41–50.

Quintella, B. R., N. O. Andrade, A. Koed & P. R. Almeida, 2004. Behavioural patterns of sea lampreys spawning migration during difficult passage areas studied by electromyogram telemetry. Journal of Fish Biology 65: 1–12.

Standen, E. M., S. G. Hinch, M. C. Healey & A. P. Farrell, 2002. Energetic costs of migration trough the Fraser River Canyon, British Columbia, in adult pink (*Oncorhynchus gorbuscha*) and sockeye (*Oncorhynchus nerka*) salmon as assessed by EMG telemetry. Canadian Journal of Fisheries and Aquatic Sciences 59: 1809–1818.

Sokal, R. E. & F. Rohlf, 1981. Biometry: The Principles and Practice of Statistics in Biological Research, 2nd edn. W.H. Freeman, New York.

Thorstad, E. B., F. Økland, A. Koed & R. S. McKinley, 2000. Radio transmitted electromyogram signals as indicators of swimming speed in lake trout and brown trout. Journal of Fish Biology 57: 547–561.

Hydrobiologia (2007) 582:221–229
DOI 10.1007/s10750-006-0543-z

FISH TELEMETRY

Effects of antenna length and material on output power and detection of miniature radio transmitters

John W. Beeman · Neil Bower · Steve Juhnke · Laura Dingmon · Mike van den Tillaart · Tom Thomas

Abstract The optimal antenna of transmitters used in small aquatic animals is often a compromise between efficient radio wave propagation and effects on animal behavior. Radio transmission efficiency generally increases with diameter and length of the conductor, but increased antenna length or weight can adversely affect animal behavior. We evaluated the effects of changing antenna length and material on the subsequent tag output power, reception, and detection of tagged fish. In a laboratory, we compared the relative signal strengths in water of 150 MHz transmitters over a range of antenna lengths (from 6 to 30 cm) and materials (one weighing about half of the other). The peak relative signal strengths were at 20 and 22 cm, which are approximately one wavelength underwater at the test frequency. The peak relative signal strengths at these lengths were approximately 50% greater than those of 30 cm antennas, a length commonly used in fisheries research. Few significant differences were present in distances for the operator to hear or the telemetry receiver to decode transmitters from a boat-mounted receiving system based on antenna length, but the percent of tagged fish detected passing a hydroelectric dam fitted with an array of receiving systems was significantly greater at the antenna length with peak output power in laboratory tests. This study indicates careful choice of antenna length and material of small transmitters can be used to reduce weight and possible antenna effects on animal behavior, to maximize tag output power and detection, or to balance these factors based on the needs of the application.

Keywords Radiotelemetry · Antenna · Detection · Tag · Power · Fish

Guest editors: P. R. Almeida, B. R. Quintella, M. J. Costa and A. Moore
Developments in Fish Telemetry

J. W. Beeman (✉) · S. Juhnke · L. Dingmon
United States Geological Survey, Western Fisheries Research Center, Columbia River Research Laboratory, 5501A Cook-Underwood Road, Cook, Washington 98605, USA
e-mail: john_beeman@usgs.gov

N. Bower · M. van den Tillaart · T. Thomas
Lotek Wireless, Inc., 115 Pony Drive, Newmarket, Ontario, Canada L3Y 7B5

Introduction

The use of radio telemetry in studies of fish behavior has become commonplace and its application is growing through advances in technology and miniaturization. The use of smaller internal components has decreased the size and weight such that tags weighing less than 1.0 g are readily available, enabling studies of smaller fishes than ever before. Radio telemetry has proven useful in many studies, including evaluations of dam passage behavior, habitat use, and responses to environmental conditions (Cheek et al., 1985; Martinelli & Shively, 1997; Johnson et al., 2000; Moser et al., 2002). Evaluations of the effects of radio transmitter (tag) presence on fish performance and behavior have been conducted on a variety of species, providing recommended tag-weight to body-weight ratios, attachment procedures, and detection methods (Winter, 1996; Martinelli & Shively, 1997; Beeman et al., 2004). However, little information has been published to guide researchers in the choice of antenna length or material.

Perhaps the most critical assumption in studies of tagged animals is that the application or presence of a tag does not affect the animal's behavior (Guy et al., 1996). Despite availability of tags weighing less then 1.0 g, this assumption may be easily violated in studies of some fishes due to their small size relative to the tag. This is a common problem and will probably always be present as tags become miniaturized further and their use is extended to smaller fishes. However, while tags have been continually reduced in size, antenna configuration has changed little. Thus, as tags are miniaturized further, a greater proportion of the tag effect on fish behavior will be attributable to the antenna, making antenna choice important.

The length and construction material of antennas can be modified to reduce their contribution to tag weight, but the effects on tag power must be evaluated to ensure the resulting product is still useful for its intended purpose. The lengths of efficient radio antennas are inversely proportional to the frequency of operation, typically in some fraction of the wavelength. For example, the elements of aerial and underwater dipole antennas are commonly 0.5 wavelengths (DeMaw, 1987; Beeman et al., 2004). Lengths of efficient antennas are also a function of circuit design, the properties of the antenna material, and the medium in which it will be used. However, little is known about the effects of modifying the length of tag antennas on the output power or detection probability. Thus, the lengths of tag antennas in studies of aquatic animals have been variable, ranging from about 10 cm to 44 cm, and are rarely mentioned in the literature.

The objectives of this study were to (1) explore the relation between antenna material, length, and tag output power in a laboratory setting and (2) test the effects of two antenna materials and lengths on tag reception in a field setting. The goal of the study was to lower overall tag weight without compromising detection.

Materials and methods

All tests were performed by altering antennas on Lotek Wireless model NTC-3-1 pulse-coded tags with an operating frequency near 150 MHz. The tags were 6.3 mm in width, 4.5 mm in height, 14.5 mm in length, and weighed 0.84 g in air with a 30 cm Sava (Type 1) antenna and 0.72 g with a 30-cm Minfish (Type 2) antenna. The Type 1 antenna was composed of a 49-strand stainless steel braided wire with an outer Teflon sheath, had a diameter of 0.58 mm, and weighed 0.254 g per 30 cm length. The Type 2 material was composed of a 63-strand stainless steel braided wire with an outer Teflon sheath, had a diameter of 0.41 mm, and weighed 0.135 g per 30 cm.

Laboratory tests of tag output power

Tests of tag output power were conducted indoors at Lotek Wireless in Newmarket, Ontario, Canada in July 2004. The output powers of five 150.460 MHz tags of each antenna type were measured with antenna lengths ranging from 30 cm to 10 cm in 2 cm decrements. Tests were begun with 30 cm antennas and were repeated each time the antennas were cut 2 cm shorter. Each tag, held in a plastic jig to maintain the antenna in a straight and horizontal position, was tested individually in a plastic cylindrical tank (75 cm diameter, 77 cm water depth) in tap water with a temperature of 24°C and a conductivity of 780 μS/cm. The water was allowed to stand for at least 48 h prior to the experiments to permit chlorine to dissipate. The tags were placed at approximately 30 cm of water depth. The tag output power was measured using a Hewlett Packard 8560E spectrum analyzer with a 30 kHz bandwidth resolution connected to a Cushcraft 4 element Yagi antenna placed 2.73 m above and pointed toward the water surface. The Yagi had a gain of 6.5 decibels relative to an isotropic radiator (6.5 dBi); the elements of the receive antenna were held parallel to the transmit antenna. A 32 dB preamplifier was used to increase the received signal to improve the measurement accuracy across the dynamic range of the spectrum analyzer by shifting the level of the readings sufficiently above its noise floor.

Detection distance of tags in controlled field conditions

The effects of antenna length and material on detection distances from a boat mounted receiving system were evaluated in the Columbia River near Bingen, Washington, USA (river kilometer 274) from 18 October through 28 October, 2004. The tests were conducted by placing five tags at a time in a polyvinyl chloride (PVC) rack to hold the antennas in a consistent orientation and to expedite the tests. The tags were held 23 cm apart from each other (about 1 wavelength underwater) to reduce the impact of the presence of the other tags on transmission properties of individual tags. The lack of an effect of this procedure was confirmed prior to testing by comparing detection distances of individual tags in the rack with the same tags in the rack together (S. Juhnke, unpublished data). The rack was affixed to a line anchored at a depth of 12.5 m, enabling adjustment to the test depths of 2, 6 and 10 m. The rack was oriented in such a way that antennas were parallel to the thalweg of the river. Daily water temperatures at the study site ranged from 14.4 to 16.8°C during the trials and conductivity was 146 μS/cm on at a nearby monitoring station on 20 October 2004. (Warrendale, Oregon, river kilometer 227; J. Morace, U.S. Geological Survey, personal communication.)

Three lengths of each antenna material were tested based on results of the laboratory trials. These were the length with the greatest output power (medium), the greatest length tested (30 cm; control), and the shortest length with an output power similar to the 30 cm antenna (short). Five tags of each material and antenna length were tested at each of 2, 6 and 10 m depths. To mimic conditions when using a gastric implantation method, which is commonly used in studies of juvenile salmonid survival in the Columbia River, we bent the tag antennas in a 180 degree arc 5 cm from the end of the tag body, which allows the antenna to protrude from the mouth and trail back along the body of the fish. The effect of bending the antenna wire back on itself in this manner was assessed in the laboratory using the materials and methods described earlier. The outcome was a loss of signal strength roughly equivalent to a 2 cm reduction in straight antenna length, so we added 2 cm to the medium and short antennas to offset this loss. The control length was unchanged although it too was bent, so the control antenna represented a 28 cm straight antenna. Antenna lengths were measured from the union of the antenna and the tag body to the terminus of the antenna.

To determine the detection distances of tags at several water depths, we followed the methods of Beeman et al. (1998). The receiving system consisted of a six-element Yagi antenna (Cushcraft Model PLC1506) about 3 m above water surface connected to a Lotek SRX-400 receiver with approximately 6 m of Belden 9311 coaxial cable. The receiver gain was set at 50 (on a range of 0–99). The location of the tag rack was recorded using a Trimble model TSC1 global positioning system (GPS) for reference to subsequent positions. The boat was moved beyond the range of audible detection and advanced slowly toward the tag rack with the receiver antenna perpendicular to the tag antennas and the location was recorded with the GPS when the operator could hear the tag, or when the receiver could decode the tag identity, depending on the objective. After the location of the first audible detection was obtained, the boat retreated out of audible range and the process was repeated until the audible detection distance of each tag was recorded three times, after which the distance required for the telemetry receiver to decode the tag was determined using the same process. This procedure was repeated with the tag rack at each of the three depths.

Analyses of the detection distances were conducted using a repeated measures analysis of variance (ANOVA). The analyses were based on data from the three repeated measurements from each of the five tags at each antenna length and type at each of the three depths. The analyses were conducted using the SAS Mixed procedure specifying the Kenward-Roger method of calculating the denominator degrees of freedom for *t*-tests, the first-order autoregressive covariance matrix to model within-subjects variation with the REPEATED statement, and the Bonferroni adjustment to control Type-I error rates in comparisons of least squares means (Littell et al., 1996). Two-way ANOVAs were specified with the distance to either hear a tag or decode the signal as the dependent variables and depth and antenna length along with their interaction term as independent variables.

Percent detection of tagged fish at a hydroelectric dam

The effects of antenna length and material on detection of tagged fish at a receiving system mounted to a hydroelectric dam were evaluated at The Dalles Dam on the Columbia River (river kilometer 274). Subyearling Chinook salmon (*Oncorhynchus tshawytscha* Walbaum) implanted with tags with short Type 2 antennas and those near the medium length of Type 1 antennas (23 cm rather than the medium length of 24 cm) from this study were released at John Day Dam along with those implanted with control Type 1 antennas as part of a concurrent study of passage and survival at The Dalles Dam 39 km downstream. Tags were gastrically implanted using methods of Martinelli et al. (1998) and were held in tanks at John Day Dam for approximately 24 h to allow recovery from the procedure prior to release. Approximately equal numbers of tagged fish were released mid-river about 3.5 km downstream from John Day Dam at 1800 h on 27 July 2004 and at 0600 h and 1800 h on 28 July, 2004. The proportion of each group detected by an array of aerial and underwater receiving systems, which enabled detection of fish within about 100 m of the forebay and tailrace sides of The Dalles Dam, were compared. The percent detections of the tags with the medium Type 1 or short Type 2 antennas were compared to those of the control Type 1 antennas using a logistic regression procedure with Chi-square tests of odds ratios (Allison, 1999; Stokes et al., 2000). Statistical significance throughout the study was assumed when $P \le 0.05$.

Results

Laboratory tests of tag output power

The relationship between tag output power and antenna length was curvilinear and slightly different between antenna types (Fig. 1). The maximum output powers of Type 1 and Type 2 antennas were at 20 and 22 cm, respectively, which is similar to the 21.8 cm wavelength of a 150 MHz radio wave underwater. The powers of the 8 cm Type 1 antenna and the 14 cm Type 2 antenna were similar to their respective 30 cm antennas, so these lengths, plus 2 cm to account for the bend for gastric implantation, were used as "short" lengths in subsequent tests.

Detection distance of tags in controlled field conditions

Significant differences were present in distances to hear Type 1 antennas at the 2 m depth, but none were present at 6 m or 10 m (Fig. 2). The effects of antenna length were assessed at each depth separately due to a significant length*depth interaction term in the 2-way ANOVA model ($F = 19.00$, df = 4, 114, $P < 0.0001$). A significant difference in the distance to hear tags was present among antenna lengths at 2 m ($F = 4.90$, df = 2, 12, $P = 0.0279$), but none were present at 6 m ($F = 2.32$, df = 2, 12, $P = 0.1403$), or 10 m ($F = 3.27$, df = 2, 12, $P = 0.0737$). The distance tags at 2 m were heard was significantly greater from the medium than the short antenna, and the control length was similar to them both (Table 1).

Few significant differences were present in distances to decode tags with Type 1 antennas (Fig. 2). There was a significant difference among lengths at 10 m ($F = 5.49$, df = 1, 8, $P = 0.0473$), but not at 2 m ($F = 2.00$, df = 2, 12, $P = 0.1779$) or 6 m ($F = 1.65$, df = 2, 12, $P = 0.2333$). At 10 m the medium length was decoded at a greater distance than the control length, and the short length could not be decoded (Table 1).

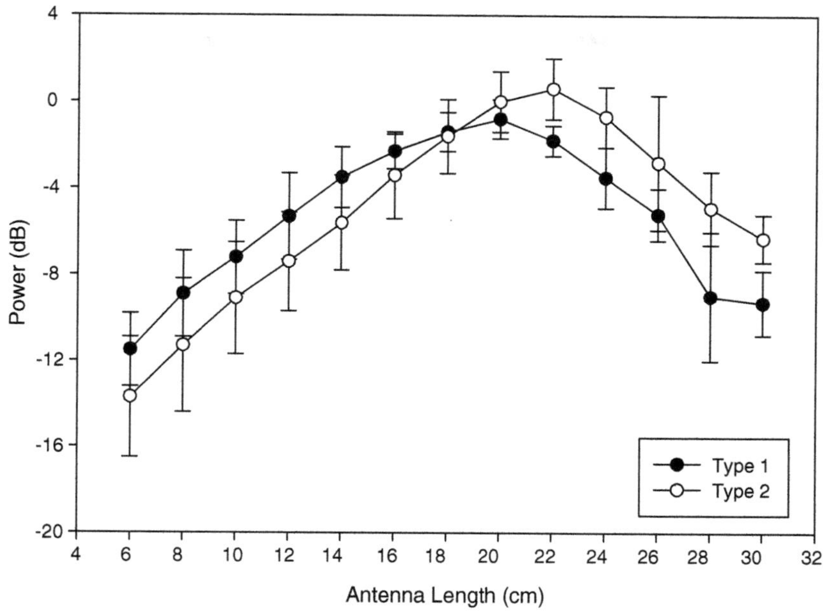

Fig. 1 Mean and 95% confidence intervals of relative output power (decibels, dB) of tags with two antenna materials measured in a laboratory

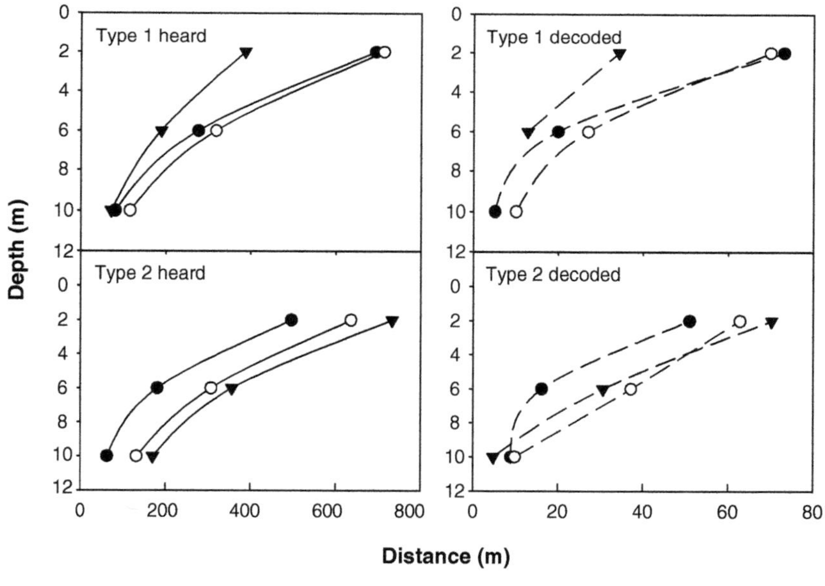

Fig. 2 Distances to hear and decode tags with Type 1 and Type 2 antennas of short (closed triangle), medium (open circle), and control (closed circle) lengths from a boat-mounted receiving system in the Columbia River, USA. Smoothed lines drawn between points are for clarity and may not represent the relation between tag depth and reception distance

Modified Type 2 antennas were generally heard from greater distances than the control (Fig. 2). A significant interaction between antenna length and depth was present in the 2-way ANOVA, so the effect of antenna length was tested at each depth separately ($F = 5.02$, df = 4,

Table 1 Distances (m) tags were heard or decoded using a boat-mounted receiving system. Values represent means of three measurements from each of five tags of each antenna material and length. Within an antenna material and distance category (Hear or Decode), distances within each depth with letters in common were not significantly different from one another ($P > 0.05$). The short Type 1 antenna could not be decoded at 10 m

Antenna material	Antenna length	Depth (m) 2	6	10
Distance to hear				
Type 1	Short	386.6 z	188.5 z	80.7 z
	Medium	712.8 y	318.0 z	115.3 z
	Control	694.2 zy	275.6 z	71.1 z
Type 2	Short	733.4 z	356.9 z	170.1 z
	Medium	636.4 z	307.0 z	132.1zy
	Control	496.4 y	180.5 y	63.1 y
Distance to decode				
Type 1	Short	34.2 z	19.9 z	Na
	Medium	69.9 z	27.0 z	10.1 z
	Control	73.1 z	12.8 z	5.1 y
Type 2	Short	70.3 z	30.6 zy	4.8 z
	Medium	62.8 z	37.2 z	9.9 z
	Control	51.0 z	16.2 y	9.0 z

114, $P = 0.0009$). Significant differences among lengths were present at 2 m ($F = 12.70$, df = 2, 12, $P = 0.0011$), 6 m ($F = 10.74$, df = 2, 12, $P = 0.0021$), and 10 m ($F = 9.44$, df = 2, 12, $P = 0.0034$). At 2 m and 6 m, the distances from the control antennas were significantly shorter than the medium and short antennas, which were not significantly different from one another (Table 1). At 10 m, the distance of the control was significantly less than the short antenna and the medium was not significantly different from the others.

Few significant differences among lengths were present in distances to decode tags with Type 2 antennas (Fig. 2). A significant interaction between length and depth was present in the 2-way ANOVA ($F = 6.61$, df = 4, 106, $P < 0.0001$), so the effect of antenna length was tested for each depth separately. No difference among lengths was present at 2 m ($F = 3.65$, df = 2, 12, $P = 0.0577$) or 10 m ($F = 1.30$, df = 2, 9, $P = 0.3184$), but there were differences at the 6 m depth ($F = 4.47$, df = 2, 12, $P = 0.0299$). At this depth, the distance of the medium length was significantly greater than the control length, and the short length was similar to them both (Table 1).

Percent detection of tagged fish at a hydroelectric dam

Detection rates of tags with modified antennas were similar or greater than tags with control antennas. There was no significant difference between detection rates of tags with short Type 2 antennas and the control ($\chi^2 = 0.69$, df = 1, $P = 0.4061$), but the tags with 23 cm Type 1 antennas were detected at a significantly greater rate than the control ($\chi^2 = 4.20$, df = 1, $P = 0.0405$; Table 2). There were no significant differences among the replicates, or in the replicate*antenna type interaction terms of either test.

Table 2 Summary of radio-tagged subyearling Chinook salmon tagged and released to test effects of antenna length and material on detection percentage as fish passed a hydroelectric dam. Overall detection percentages with letters in common were not significantly different from one another ($P > 0.05$)

Antenna type and length	Replicate	N	Fork length (mm) Mean	Range	Fish weight (g) Mean	Range	Detection (%)
Type 1 Medium	1	23	109.3	103–129	14.7	13.0–23.0	82.6
	2	24	114.9	104–130	17.4	13.0–22.8	95.8
	3	25	111.9	103–133	17.2	13.8–27.5	80.0
	Overall	72	112.1	103–133	16.4	13.0–27.5	86.1 z
Type 2 Short	1	24	111.9	103–136	16.2	13.3–16.2	75.0
	2	24	114.0	104–130	17.1	13.0–25.8	79.2
	3	22	109.2	103–126	16.6	13.7–22.5	81.8
	Overall	70	111.8	103–136	16.7	13.0–26.0	78.6 zy
Type 1 Control	1	35	110.4	103–130	15.5	13.2–25.7	65.7
	2	32	112.4	102–136	16.4	13.0–27.8	78.1
	3	46	109.5	101–130	16.5	13.0–26.8	76.1
	Overall	113	110.6	101–136	16.2	13.0–27.8	73.4 y

Discussion

Our results indicate that the output power and subsequent reception of radio tags was a function of the antenna material and length. The greatest reception ranges and detection rates were generally at the antenna lengths predicted from laboratory measurements of tag output power. These optimal antenna lengths were considerably shorter than the 30 cm many researchers have used in the past, so by simply reducing the length of the tag antenna to the proper length one can increase output power and reception and decrease tag weight and potential effects of the antenna on fish behavior.

The primary goal of this study was to determine if tag weight could be reduced by changes in antenna material or weight without compromising detectability. The result that best met this need during our tests was the short (16 cm) Type 2 antenna, though results from the medium (24 cm) Type 2 antennas were similar. This was 2 cm greater than the short length indicated in the laboratory test due to the bend in the antenna for use with a gastric implant method; use with a straight antenna, such as in a surgical or external attachment, would be optimal at a length of 14 cm. This configuration resulted in a reduction in tag weight in air from 0.84 g for a tag with the 30 cm Type 1 antenna, which is commonly used in studies of juvenile salmon in the Columbia River, to 0.65 g with the 16-cm Type 2 antenna. This is a substantial reduction in overall weight relative to the tag used, was achieved at little cost compared to redesign of internal components to achieve the same result, and resulted in no significant difference in detection (78.6%) than the tag with the 30 cm Type 1 antenna (73.4%). Use of the medium Type 1 antenna resulted in significantly greater detection (86.1%) than the 30 cm Type 1 antenna, indicating it would be a logical choice if the goal was to maximize detection of tags with this antenna material. Our study was based on external antennas, but it is conceivable that the output power and weight of tags with internal antennas could also be improved by altering antenna length and material.

Some differences between results of field and lab tests were evident. Based on lab tests, we expected the field performance of tags with medium antennas to be greater than the control and short antennas, which were expected to be similar to one another. In field trials of Type 1 antennas the control medium lengths were similar and the short length had the shortest range. In field trials with Type 2 antennas the short and medium distance were similar to one another and the control length had the shortest range. One difference between lab and field studies was control of the relative orientation of transmit and receive antennas. These were tightly controlled in lab tests, but variation in river current, wind and other environmental factors surely affected the consistency of the antenna orientations during field tests, introducing an additional source of error. Received signals are of greatest powers when transmit and receive antennas are parallel (Beeman et al., 2004). Differences between lab and field results may also have been due to the conductivity and temperature of the water in the two settings. The conductivity and temperature of the water used in lab tests in Ontario, Canada was 780 μS/cm and 24°C and those during field tests in Washington, USA were 146 μS/cm and 14.4–16.8°C. The propagation of radio waves through water is affected by both conductivity and temperature, and the differences in our study may have affected the results by altering the velocity factor, and therefore, wavelength of the transmitted signal.

We suggest that the benefits of understanding the relation between antenna material, length, and output power are threefold. First, weight of small tags can be simply reduced. This could allow smaller fish to be tagged, enabling telemetry to be applied to smaller species, or a greater proportion of the population of species already studied, such as subyearling Chinook salmon, which range in weight from about 5 to 23 g in the Columbia River (S. Juhnke, unpublished data). Decreases in tag weight through changes in antennas will become more important as tags are further reduced in size, because antenna weight will represent a larger portion of the overall tag weight. Second, the increase in tag output power and the resulting increase in detection could result in increased precision of survival or passage estimates, a means to decrease sample sizes while maintaining existing precision in ongoing studies, or to simply gather more data from tagged fish. These considerations are important in recent studies of juvenile salmon passage and survival at dams throughout the Columbia River basin, which have recently required about 40,000 tagged fish annually. Lastly, fish behavior may be less affected by lighter tags with shorter antennas. Little direct study of the effects of antenna length on fish behavior or performance has been published, but Anglea et al. (2004) described fewer effects of acoustic tag (i.e., no antenna) implantation on swimming performance and susceptibility to predation of juvenile Chinook salmon than typically seen in studies of tags with antennas (Adams et al., 1998; Martinelli et al., 1998), suggesting external antennas may affect fish performance or behavior. If so, a shorter or lighter antenna may impart less of an effect on the fish. In some studies, using radio tags with antennas coiled within the embedding material could obviate the need for "shorter" trailing antennas. Coiled antennas generally result in lower output power than trailing antennas (Cooke & Bunt, 2001) and may be inappropriate in cases where output power is important, such as for determining accurate pulse intervals or where reception distances need to be maximized due to large study areas or small detection fields (Beeman et al., 1998; Johnson et al., 2000).

There can be no universally-applied "optimum" antenna length for radio tags, as many biological and technical factors must be considered. We only tested antenna lengths ranging from 6 to 30 cm, because our goal was to use the tags in small fish. The antenna lengths resulting in the greatest output power were near 1 wavelength of the tag frequency underwater, but this may not be true with all tags or antenna materials. The wavelength of a 150 MHz radio wave underwater is approximately 22 cm (at temp = 10 °C and conductivity = 150 μS/cm), which is near the peak output power of the two antenna types we tested (20 and 22 cm). However, we have since tested other antenna materials in the laboratory and found one with peak output power at 16 cm. The effect of the antenna on tag output is controlled by many factors, including the size and shape of the tag, and the properties of the antenna wire. Tag size and shape are important, because the body of the transmitter (including its battery) is a significant part of the total radiating aperture. The conductivity of the antenna wire depends on the material, number and thickness of strands, as well as wire diameter. These factors determine the distributed capacitance between the antenna and the surrounding medium. The thickness and dielectric properties of the insulating material also contribute to the antenna properties. In addition to preventing leakage of conducting charges from the antenna, the insulation material reduces the sensitivity of the distribution of current in the antenna to the electrical properties of the ambient medium. Hence, the optimal antenna length for any given frequency may vary for different antenna materials. Finally, it must be noted that the propagation properties of a wire antenna will vary considerably with the conductivity of water. We suggest that tests of the effects of antenna length and material on tag output power should be conducted for specific tag/antenna combinations, or that further studies be conducted to determine the relations among these factors.

Future studies using radio telemetry may be improved by careful selection of antenna material and length. Increases in output power and detection can be realized through the use of studies of the effects of tag material and length on output power. We found the output power of tags with 30 cm antennas and those with antennas as short as 8 cm were similar, depending on the antenna material, and that intermediate lengths provided the greatest detection range and percentage. Information about the effect of antennas on tag output power enables researchers to balance needs for tags with low weight, short antennas, or optimal detection depending on study requirements. Decisions about antenna lengths and materials will be of increasing importance as tags are miniaturized and used in smaller fish. Results from laboratory studies were generally indicative of performance in field studies. However, we found the relation between antenna length and output power was a function of the antenna material and is likely affected by the tag itself, making tests with other tag types or models warranted.

Acknowledgements We thank our colleagues at the US Geological Survey and at Lotek Wireless for their assistance in this research, and the staff of the Pacific States Marine Fisheries Commission at John Day Dam for providing test fish. The manuscript was improved by insightful comments of two anonymous reviewers. This work was funded in part by the Portland District of the US Army Corps of Engineers, contract W66QKZ51376463.

References

Adams, N. A., D. W. Rondorf, S. D. Evans, J. E. Kelly & R. W. Perry, 1998. Effects of surgically and gastrically implanted radio tags on swimming performance and predator avoidance of juvenile Chinook salmon (*Oncorhynchus tshawytscha*). Canadian Journal of Fisheries and Aquatic Science 55: 781–787.

Allison, P. D., 1999. Logistic Regression using the SAS System: Theory and Application. SAS Institute Inc, Cary, North Carolina, USA.

Anglea, S. M., D. R. Geist, R. S. Brown & K. A. Deters, 2004. Effects of acoustic tags on swimming performance and predator avoidance of juvenile Chinook salmon. North American Journal of Fisheries Management 24: 162–170.

Beeman, J. W., P. V. Haner & A. G. Maule, 1998. Evaluation of a new miniature pressure-sensitive radio transmitter. North American Journal of Fisheries Management 8: 458–464.

Beeman, J. W., C. Grant & P. V. Haner, 2004. Comparison of three underwater antennas for use in radio telemetry. North American Journal of Fisheries Management 24, 275–281.

Cheek, T. E., M. J. Van den Avyle & C. C. Coutant, 1985. Influences of water quality in distribution of striped bass in a Tennessee River impoundment. Transactions of the American Fisheries Society 114: 67–76.

Cooke, S. J. & C. M. Bunt, 2001. Assessment of internal and external antenna configurations of radio tags implanted in smallmouth bass. North American Journal of Fisheries Management 21: 236–241.

DeMaw, D., 1987. W1FB's Antenna Notebook. The American Radio Relay League, Newington, Connecticut, USA.

Guy, C., H. Blankenship & L. Nielsen, 1996. Tagging and marking. In Murphy, B. W. & D. W. Willis (eds), Fisheries Techniques, 2nd edn. American Fisheries Society, Bethesda, Maryland, USA, 353–384.

Johnson, G. E., N. A. Adams, R. L. Johnson, D. W. Rondorf, D. D. Dauble & T. L. Barila, 2000. Evaluation of the prototype surface bypass for salmonid smolts in spring 1996 and 1997 at Lower Granite Dam on the Snake River, Washington. Transactions of the American Fisheries Society 129: 381–397.

Littell, R. C., G. A. Milliken, W. W. Stroup & R. D. Wolfinger, 1996. SAS System for Mixed Models. SAS Institute Inc., Cary, North Carolina, USA.

Martinelli, T. L. & R. S. Shively, 1997. Seasonal distribution, movements and habitat associations of northern squawfish in two lower Columbia River reservoirs. Regulated Rivers: Research and Management 13: 543–556.

Martinelli, T. L., H. C. Hansel & R. S. Shively, 1998. Growth and physiological responses to surgical and gastric radio tag implantation techniques in subyearling Chinook salmon (*Oncorhynchus tshawytscha*). Hydrobiologia 371/372: 79–87.

Moser, M. L., A. L. Matter, L. C. Stuehrenberg & T. C. Bjorn, 2002. Use of an extensive radio receiver network to document Pacific lamprey (*Lampetra tridentata*) entrance efficiency at fishways in the Lower Columbia River, USA. Hydrobiologia 483: 45–53.

Stokes, M. E., C. S. Davis & G. G. Koch, 2000. Categorical Data Analysis Using the SAS System, 2nd edn. SAS Institute Inc, Cary, North Carolina, USA.

Winter, J., 1996. Advances in underwater biotelemetry. In Murphy, B. W. & D. W. Willis, (eds), Fisheries Techniques, 2nd edn. American Fisheries Society, Bethesda, Maryland, USA, 550–590.

Hydrobiologia (2007) 582:231–242
DOI 10.1007/s10750-006-0562-9

Comparison between PIT and radio telemetry to evaluate winter habitat use and activity patterns of juvenile Atlantic salmon and brown trout

Eva C. Enders · Keith D. Clarke · Curtis J. Pennell · L. M. Neil Ollerhead · David A. Scruton

Abstract Winter habitat use and activity patterns of juvenile Atlantic salmon and brown trout were analysed in a comparative study between Passive Integrated Transponder (PIT) technology, radio telemetry and underwater observation by snorkelling. Two study periods were conducted in Stoney River, Newfoundland, Canada. During Study period I, 49 juvenile Atlantic salmon (fork length: 11.0–18.0 cm) and 7 brown trout (11.0–17.3 cm) were tagged with PIT tags and/or radio transmitters in late winter of 2004. During Study period II, 18 juvenile Atlantic salmon (fork length: 12.0–18.4 cm) and 23 brown trout (10.9–20.8 cm) were tagged and tracked twice a day at 10:00 h and 22:00 h on five consecutive days in late winter of 2005. From the 56 fish released during Study period I, on average 19.6 ± 6.0% of the PIT tagged fish and 99.3 ± 2.2% of the radio tagged fish were relocated during any given survey. Over the Study period II, 39% of fish emigrated from the study site. PIT technology had an efficiency of 39.2 ± 14.1% to detect the remaining fish. In contrast, radio telemetry relocated on average 96.9 ± 6.5% of the tagged fish whereas by snorkelling on average only 4.1 ± 5.6% of the tagged fish were observed. PIT telemetry may however be more efficient in smaller, less heterogeneous streams. The advantage of PIT technology over radio telemetry is clearly that it is relatively less costly permitting higher numbers of individuals to be tagged and there is no limit in the operational life of the transponder. In winter, juvenile salmonids preferred low flow velocity and no preferences were observed for any specific water depth over the range of available water depths. Fish selected preferentially boulder habitat over other substrates in the environment. Habitat utilisation did not differ between day and night. The use of winter preference indices may be important for future habitat modelling.

Keywords PIT technology · Radio telemetry · Winter habitat · Activity · *Salmo salar* · *Salmo trutta*

Guest editors: P. R. Almeida, B. R. Quintella, M. J. Costa and A. Moore
Developments in Fish Telemetry

E. C. Enders (✉) · K. D. Clarke · C. J. Pennell · L. M. N. Ollerhead · D. A. Scruton
Ecological Sciences Section, Fisheries and Oceans Canada, 80 East White Hills Road, P.O. Box 5667, St. John's, NL, Canada A1C 5X1
e-mail: enderse@dfo-mpo.gc.ca

Introduction

Protection and management of suitable habitat for migration, rearing and survival of juvenile fish is thought to be one of the key factors for the conservation of salmonid populations (Armstrong et al., 1998). Availability of suitable overwintering habitat for juvenile fish may play a central role in regulating salmonid populations in northern streams and rivers (Cunjak, 1996; Quinn & Peterson, 1996). Fish habitat models are often used to describe and predict the effect of natural and anthropogenic perturbations on quality, quantity and distribution of fish habitat (Bovee, 1982). Biological models implemented in fish habitat models assess habitat quality using preference indices that indicate fish preference for a specific range of flow velocities, water depths and substrate sizes over the complete range available (deGraaf & Bain, 1986; Morantz et al., 1987; Guay et al., 2000). Preference indices are often estimated by visual surveys of fish distribution (Scruton et al., 1998; Guay et al., 2000). Hence, variables that evoke behavioural changes of fish from actively foraging and holding position to cryptically hiding in the interstices of the substrate may change the fish visibility to observers. Changes in the visibility to observers during surveys may consequently have an affect on the resulting preference indices and thus significantly affect the predictions of the biological models.

Juvenile salmonids are known to exhibit complex diel activity patterns. During summer, juvenile salmonids are commonly holding position on the substrate and are actively foraging during daytime (Valdimarsson et al., 1997; Amundsen et al., 1999). In winter, juvenile salmonids display a cryptic behaviour concealed in the interstices of the substrate during daytime but emerging from interstices at night (Rimmer et al., 1983; Cunjak, 1988; Bremset, 2000). However, nocturnal behaviour may also be observed during summer (Gries et al., 1997; Johnston et al., 2004) and diurnal behaviour during winter (Hiscock et al., 2002). Information on habitat quality obtained from visual observation may therefore provide only a partial perspective of the temporal variability in habitat use and activity patterns of juvenile salmonids.

In the past decades, new technologies have been developed allowing the remote localisation of fish exhibiting cryptic behaviour in natural flow environments. Firstly, Passive Integrated Transponder (PIT) technology has been developed as a valuable method to track individually tagged juvenile salmonids (Armstrong et al., 1996; Roussel et al., 2000, 2004). Secondly, the development of small radio transmitters allowed the application of radio telemetry to increasingly smaller (i.e. juvenile) fish. Recent radio telemetry studies have provided new insights into fish behaviour including that juvenile salmonids may not always utilise restricted home ranges, as previously suggested, but may display a high level of mobility (Hiscock et al., 2002; Berland et al., 2004; Økland et al., 2004).

The objectives of this study were to (1) compare the feasibility and efficiency of PIT and radio telemetry and underwater observation by snorkelling to detect and locate juvenile salmonids, (2) develop preference indices for winter habitat use and (3) study activity patterns of juvenile Atlantic salmon (*Salmo salar* L.) and brown trout (*Salmo trutta* L.) in late winter.

Materials and methods

Study site

Observations were carried out in Stoney River, Newfoundland, Canada (46°43′ N; 53°20′ W; Fig. 1). Observations were made during two periods (4–9 March 2004 and 10–15 March 2005) further defined as Study periods I and II, respectively. In Study period I, the studied river reach consisted of a 220 m section including a riffle-pool sequence and was located 1.2 km upstream from the river mouth (Fig. 1). River width was on average 9 m. Average water depth was 32.0 cm and maximum depth was 84.0 cm. The substratum of the river bed ranged from gravel to boulder. Water temperature in the riffle was recorded every hour with a temperature sensor (model Minilog TR; Vemco, Shad Bay, Nova Scotia, Canada). Water temperature ranged from 0.0 to 3.5°C and water discharge was 1.2 m s^{-1} during the observations.

In Study period II, the study site consisted of a 200 m section that was located 250 downstream from study site I (Fig. 1). The river width was on average 10 m. Average water depth was 31.7 cm (±13.3 cm SD, maximum depth 72.0 cm). River

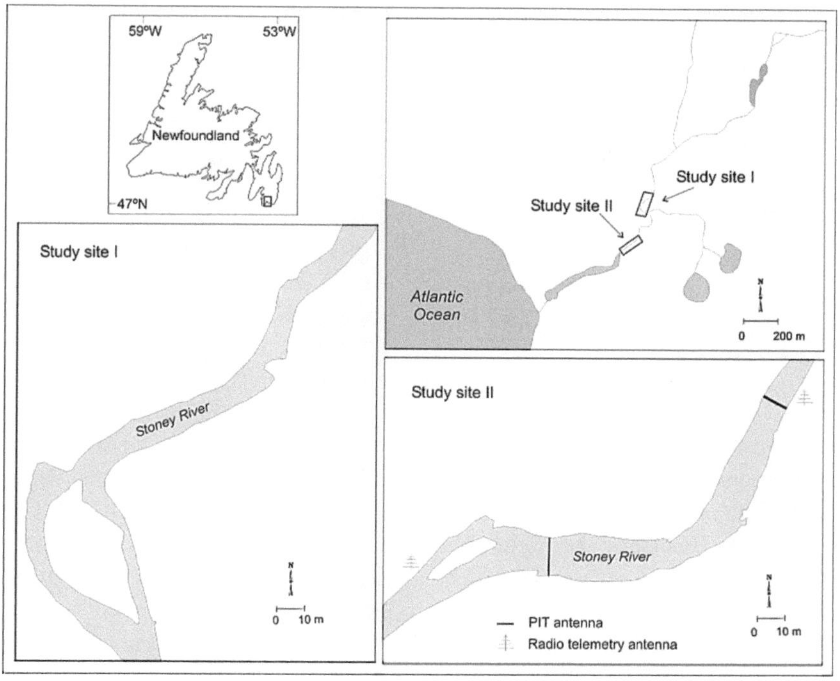

Fig. 1 Map of study sites located in Stoney River, in insular Newfoundland Canada

slope in the study site was 1%. Mean flow velocity was 50.9 cm s^{-1} (SD: 29.4 cm s^{-1}, maximum flow velocity: 185.0 cm s^{-1}) and water discharge was 1.2 m s^{-1}. The substratum of the river bed ranged from gravel to boulder. Water temperature in riffle and pool section was recorded every hour with two temperature sensors and ranged from –0.1 to 4.3°C. Atlantic salmon (*Salmo salar* L.) and brown trout (*Salmo trutta* L.) were the dominate species in the lower sections of Stoney River, while brook charr (*Salvelinus fontinalis* Mitchell) and American eel (*Anguilla rostrata* Lesueur) were also commonly found.

Fish surgery

During Study periods I and II, 56 and 41 juvenile salmonids respectively were caught with a Smith-Root Inc.® backpack electrofisher (model 12-B; Vancouver, Washington, USA). After capture, fish were individually anaesthetised (clove oil; 0.6 mg l^{-1}), weighed (g wet) and measured (fork length L_F; cm; Table 1).

Captured fish were tagged with PIT tags (model RI-TRP-WRHP; Texas Instruments Inc., Dallas, Texas, USA; 23.1 mm in length and 3.9 mm in diameter, mass in air of 0.6 g) and/or coded radio transmitters (model NTC-4-2; Lotek Wireless Inc., Newmarket, Ontario, Canada; 17.0 mm in length and 7.5 mm in diameter, mass in air of 1.65 g, guaranteed battery life time of 9 d). Fish were either tagged with a pit tag, a radio transmitter or double-tagged with a PIT and a radio transmitter (Table 2). PIT tags and radio

Table 1 Number *n*, mean and range of body mass (*M*; g), and mean and range of fork length (L_F; cm) of tagged Atlantic salmon (AS) and brown trout (BT) during Study period I and II

Species		Study period	
		I	II
AS	*n*	49	18
	Mean *M*	25	29
	Range *M*	13–51	17–63
	Mean L_F	13.2	13.9
	Range L_F	11.0–18.0	12.–18.4
BT	*n*	7	23
	Mean *M*	30	30
	Range *M*	14–54	12–90
	Mean L_F	13.9	14.0
	Range L_F	11.0–17.3	10.9–20.8

Table 2 Number *n* of Atlantic salmon (AS) and brown trout (BT) tagged with only PIT tag, only radio transmitter and double tagged with both PIT tag and radio transmitter during Study period I and II

n of tagged fish	Study period			
	I		II	
PIT tagged	36 AS	4 BT	12 AS	15 BT
Radio tagged	4 AS	2 BT	4 AS	4 BT
Double tagged	9 AS	1 BT	2 AS	4 BT
Total	49 AS	7 BT	18 AS	23 BT

transmitters were surgically implanted. For the PIT tags, a 0.6 cm incision was made on the ventral surface, posterior to the pelvic girdle using a scalpel. The tag was inserted through the incision and smoothly pushed forward into the body cavity. The incision was closed using one silk suture (4–0 SofSilk^{TM}). For the radio transmitters, a 1.0 cm incision was made. The Teflon-coated antenna was threaded through the body wall using a hollow needle and the incision was closed with two independent silk sutures. In double tagged fish, the radio transmitter was first implanted during the surgery and gently pushed forward in the body cavity whereas the PIT tag was implanted secondly and pushed backward in the body cavity. In this position, detection and reading distance of the PIT tag was not affected by the radio transmitter. Implanted transmitters represented on average 3.5% of fish body mass in air. Additionally, in Study period II, fish were marked with a Polyethylene streamer tag (Hallprint Pty Ltd., Adelaide, Australia), passed through the dorsal surface, to facilitate visual observations. After surgery, fish were placed overnight in flow-through cages within the river for recovery before they were released in the middle of the river section.

Fish tracking

Fish were not located for the first day after surgery to exclude potential tagging effects. Tagged fish were manually tracked during Study period I nine times from 5 to 9 March 2004. Five day (8:00 h) and four night (19:00 h) surveys were carried out using both PIT and radio telemetry. During Study period II, a loop antenna with a PIT tag reader (model Series 2000 RI-CTL-MB2A; Texas Instruments Inc.) and a radio telemetry receiver (model SRX_400; Lotek Wireless Inc.) were each installed at the up- and downstream end of the study site to act as gates to monitor fish emigrating from the study site. Fish were tracked ten times from 11 to 15 March 2005. Five day (10:00 h) and five night (22:00 h) surveys using PIT and radio telemetry were carried out to assess diurnal changes in habitat use and activity patterns of juvenile salmonids. During Study period II, to obtain additional information on fish behaviour, direct visual observations were made by snorkelling the entire study site during one day and five night time surveys. The diver swam upstream to minimise fish disturbance (Cunjak et al., 1988).

A portable PIT tag detector was constructed after the description of Roussel et al. (2000). The radio frequency identification system (model Series 2000 RI-CTL-MB2A; Texas Instruments Inc.) powered by a 12 V battery was carried in a backpack and connected to an antenna (60 cm diameter of coil) and to a handheld PDA computer using custom software. The antenna was mounted on a 2 m pole. The reading distance of system was up to 75 cm when the PIT tag was horizontally orientated. When tracking, the operator wadded the site in an upstream direction sweeping the antenna just above the water surface from river bank to river bank. In the deeper river sections, the antenna was moved under the water surface to better detect the fish close to the river bed. Radio tagged fish were tracked using an H-antenna (AF Antronics Inc., Urbana, Illinois, USA) and a handheld radio telemetry receiver (model SRX_400; Lotek Wireless Inc.). No fish mortality was recorded. Three PIT tagged fish from Study period I were located and tracked a second time in the Study period II.

Habitat survey

Once a fish was detected by manual tracking or snorkelling, fish location was marked with a numbered fluorescent marker nut. Fish locations were then geo-referenced and microhabitat variables at the fish location measured including mean velocity (0.6 of water depth) and nose velocity (model 2000 portable flowmeter, Marsh-McBirney), water depth and substrate [gravel (0.004–3.2 cm), cobble (3.2–25 cm), boulder (>25 cm) modified after Scruton et al., 1998]. Additionally, during Study period II, the fluvial habitat available to fish was assessed based on 350 reference points along a stratified random sampling grid, which were later used to develop habitat preference curves.

Data analysis

For data from Study period II, preference curves were estimated for three abiotic habitat variables: mean flow velocity, water depth and substrate. Preference curves illustrate the tendency of fish to choose specific ranges of available habitat characteristics that represent fish habitat quality (Bovee, 1982; Morantz et al., 1987). Data on the three habitat variables, collected at the fish location, were classified in categories and the percentage of utilisation of each category was determined. The same approach was applied to habitat variables randomly sampled in the river reach to quantify habitat availability. For the three habitat variables, Jacobs' preference indices D were computed to reveal potential habitat preference (Jacobs, 1974)

$$D = (r - p) / (r + p - 2rp) \quad (1)$$

where D is the habitat preference index, r is the proportion of fish located in a specific habitat category and p the proportion of the specific habitat category available in the study site. Positive values for D indicate selection for a particular habitat category, $D = 0$ indicates no preference and negative values suggest habitat avoidance.

Fish positions were mapped in ArcView GIS 3.2 (Environmental Systems Research Institute Inc., Redlands, California, USA). The mean distances travelled by fish of a given species were compared between the two telemetry techniques using Mann–Whitney tests. The significance level for all statistical tests were $P \leq 0{\cdot}05$. All statistical analyses were performed using SPSS 11.0.

Results

Fish tracking

From the 56 fish released during Study period I, on average 19.6 ± 6.0% (range: 11.8–27.5%) of the PIT tagged fish were found during any given survey. Conversely, on average 99.3 ± 2.2% (range: 93.3–100.0%) of the radio tagged fish were relocated during any given survey. The tracking efficiency was significantly different between the methods (Mann–Whitney test, $n = 18$, $P < 0.001$; Fig. 2a).

During Study period II, loop antennas monitored fish migrating from the study site. From the

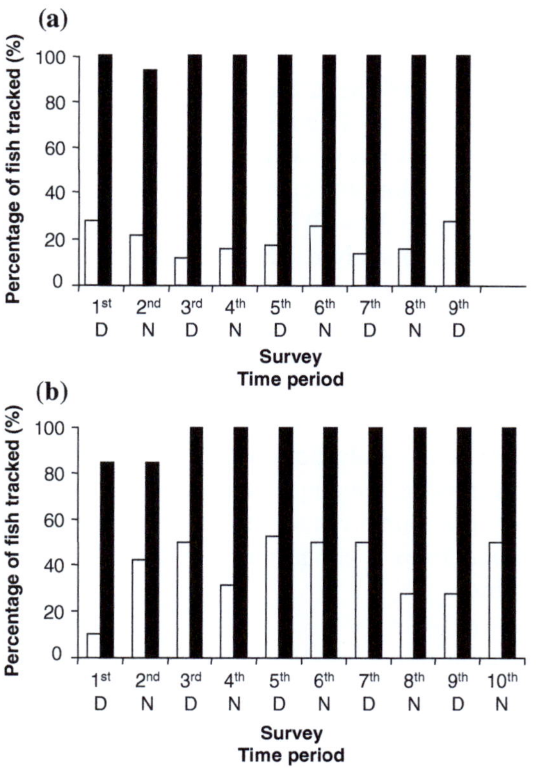

Fig. 2 Percentage of juvenile salmonids tracked during a given survey by PIT (*open bars*) and by radio-telemetry (*solid bars*) during (**a**) Study period I and (**b**) Study period II

41 fish released during Study period II, 16 fish left the study site during dusk (34.7% of the emigrating fish), around midnight (50%) and at dawn (14.3%). Interestingly, a high post-release emigration rate of 15 fish was observed within the first 24 h after release. From these fish, 12 (44.4% of all PIT tagged fish) were PIT, one (12.5% of all radio tagged fish) radio and two (33.3% of all double tagged fish) double tagged fish. Interestingly, of the 16 fish emigrating from the study site, 12 fish migrated upstream and 4 fish moved downstream. The efficiency of tracking the remaining fish in the study site differed significantly between PIT and radio tagged fish (Mann–Whitney test, $n = 20$, $P < 0.001$; Fig. 2b). From the PIT tagged fish only 39.2 ± 14.1% (range: 10.5–52.6%) were found during any given survey. In contrast, on average 96.9 ± 6.5% (range: 84.6–100.0%) of the radio tagged fish were located during any given survey. Temporal effort for both methods was similar ranging from 0:55 h to 1:30 h (t-test, $n = 20$, $P = 0.10$). In comparison to both telemetry techniques, snorkelling displayed much lower detection rates. On average, only 4.1 ± 5.6% of the tagged fish (range: 0–5) were observed during visual survey in Study period II. Snorkelling surveys lasted on average 0:57 h.

The distribution of water depths in fish locations obtained by PIT and radio telemetry were compared to analyse if PIT telemetry successfully detected fish in deeper waters (Fig. 3). No difference in the distributions of water depths was found between the fish locations detected by PIT and radio telemetry (Kolmogorov–Smirnov test, $n = 108$, $P = 0.32$ in Study period I and $n = 184$, $P = 0.34$ in Study period II). This suggests that both tracking techniques allowed detection of fish over the same range of water depths, in the studied river sections and the associated water depths.

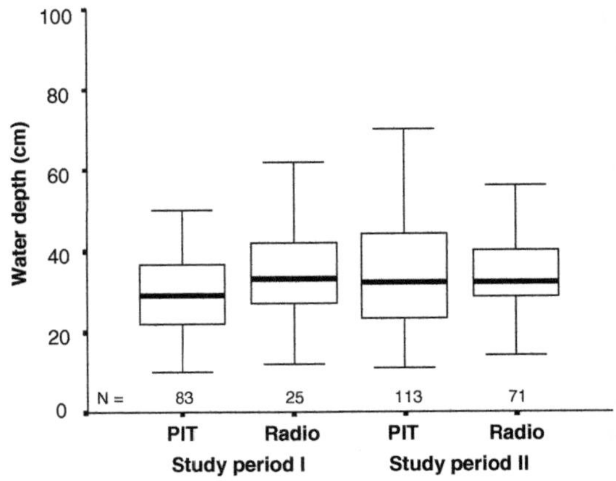

Fig. 3 Water depth of fish positions detected by PIT technology and radio-telemetry during Study period I and Study period II. Box-Whisker plot displays the median water depth, lower and upper quartiles, and minimum and maximum water depth for given study period and tracking technology

Preference indices for winter habitat use

In comparison to the availability in the environment, juvenile salmonids expressed distinct preferences for low velocity (Kruskal–Wallis test, $n = 265$, $P < 0.001$; Fig. 4a, 5a); where brown trout occupied significant slower flowing microhabitats (mean and nose flow velocity 29.7 cm s^{-1} and 9.0 cm s^{-1}) than Atlantic salmon (mean and nose flow velocity of 48.7 cm s^{-1} and 12.6 cm s^{-1}; Tamhane Statistic, $n = 125$, $P < 0.001$; Table 3). No differences were observed between the water depth available and used by fish (Kruskal–Wallis test, $n = 265$, $P > 0.05$, Fig. 4b, 5b). Mean water depth in fish locations were similar for Atlantic salmon (34 cm) and brown trout (33 cm). Both species selected boulder habitat over the availability of substrates in the environment (Kruskal–Wallis test, $n = 265$, $P < 0.001$). Furthermore, brown trout selected gravel habitat over its availability in the study site (Fig. 4c, 5c). For both species, no differences in habitat utilisation were observed between day and night (Mann–Whitney tests, always $P > 0.05$; Fig. 4, 5).

Activity pattern

Mean distance travelled between two consecutive surveys was higher for radio/double tagged (mean: 10.2 m, range: 0.0–64.2 m) than for PIT tagged (mean: 3.4 m, range: 0.0–31.8 m) Atlantic salmon (Mann–Whitney test, $n = 41$, $P = 0.003$). No differences were observed in the mean distance travelled between radio/double tagged (mean: 3.2 m, range: 0.0–40.4 m) and PIT tagged (mean: 2.8 m, range: 0.0–2.9 m) brown trout (Mann–Whitney test, $n = 74$, $P = 0.10$). Over

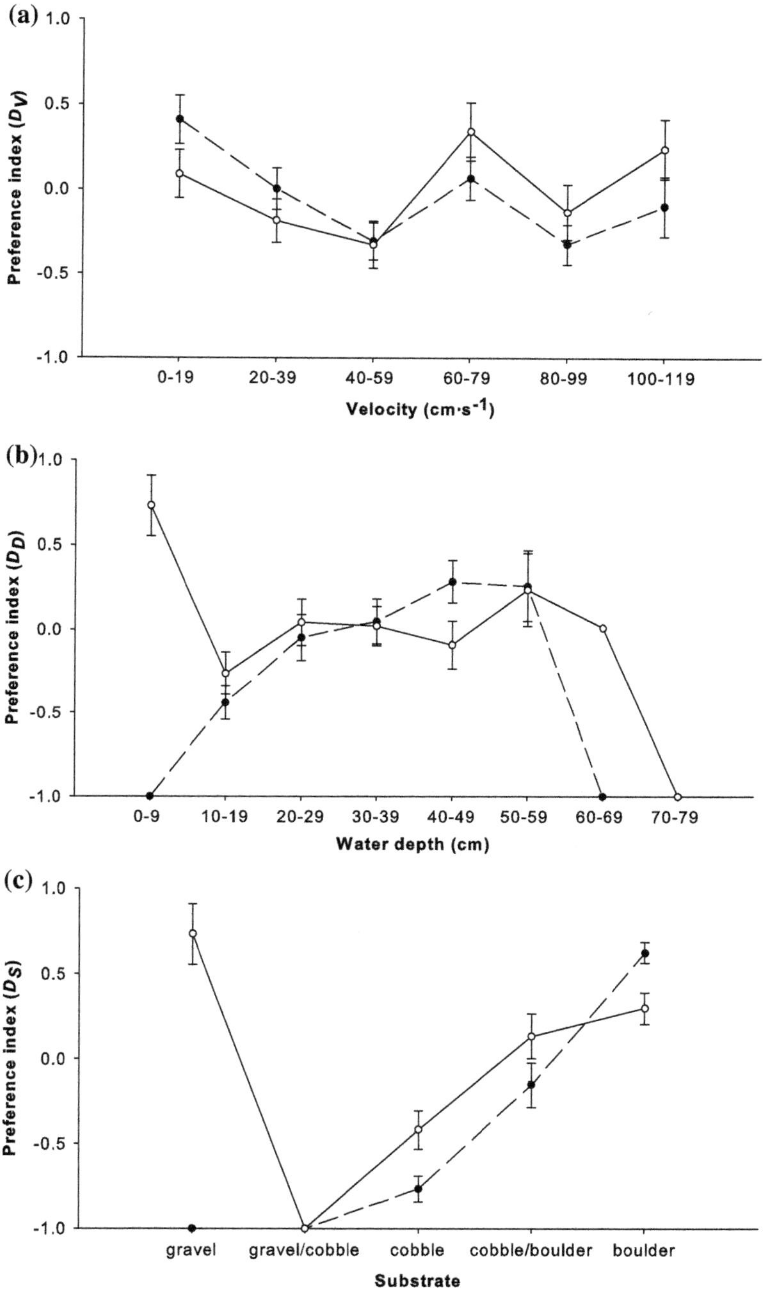

Fig. 4 Day (in *open triangle*) and night (in *closed triangle*) preference indices of (**a**) mean flow velocity D_V, (**b**) water depth D_D and (**c**) substrate D_S for Atlantic salmon from Study period II

the study period, individual Atlantic salmon moved on average a cumulative distance of 48.1 m (range: 3.1–138.1 m) in comparison to individual brown trout that moved only 21.6 m (range: 0.6–104.1 m).

Discussion

Efficiency of PIT and radio telemetry to locate individual juvenile salmonids under winter conditions was compared. The results showed that

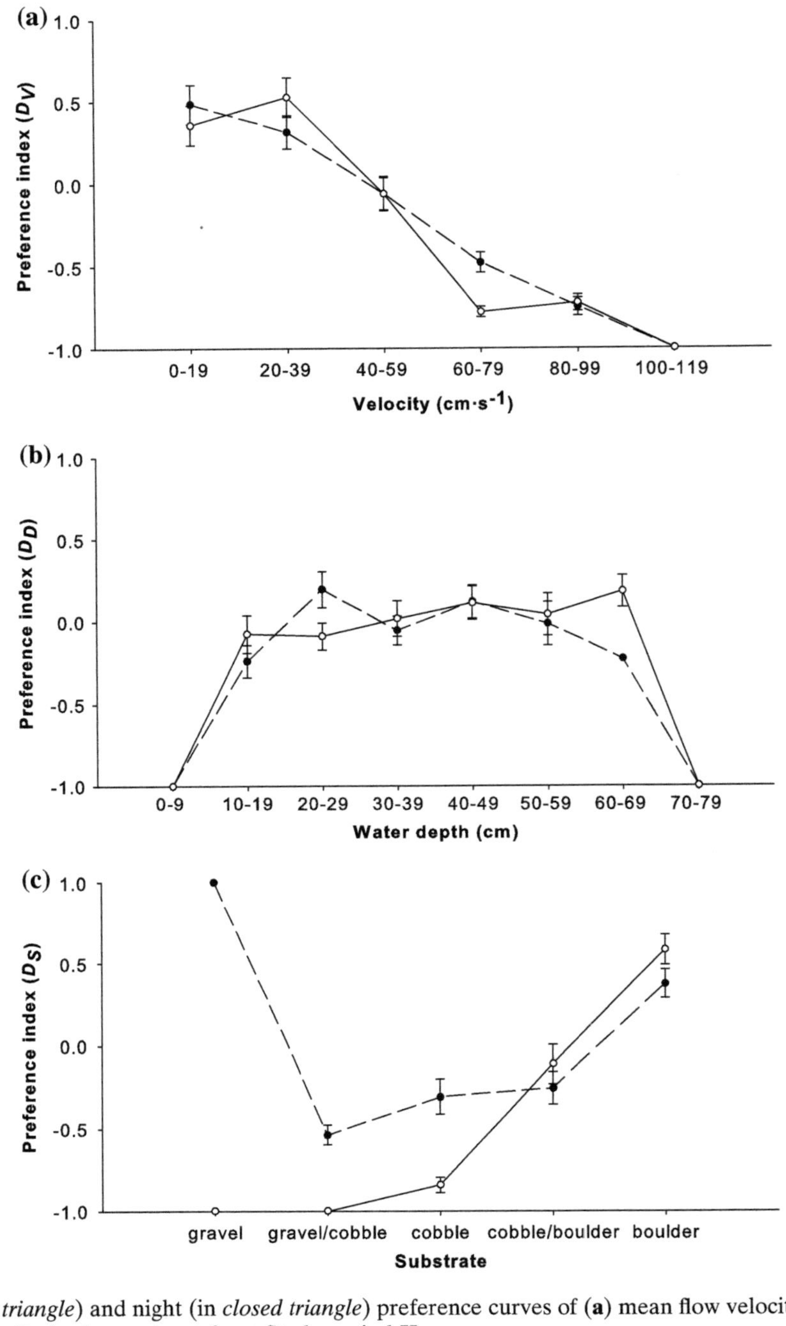

Fig. 5 Day (in *open triangle*) and night (in *closed triangle*) preference curves of (**a**) mean flow velocity D_V, (**b**) water depth D_D and (**c**) substrate D_S for brown trout from Study period II

both techniques provide a similar accuracy in the prediction of the fish position within 1.5 m^2. The comparison of frequency distributions of water depths obtain by PIT and radio telemetry revealed no differences. This suggests that in the study sites (water depth < 1 m) PIT telemetry was not limited within respect to water depth. While using a comparable temporal effort, radio telemetry provided an essentially higher detection rate than PIT telemetry (Fig. 2). The potential reasons of the low detection efficiency of PIT telemetry in our study are twofold. First, habitat complexity in our study site was relatively high. However, PIT telemetry may be more efficient

Table 3 Mean, standard deviation and ranges of three abiotic habitat variables used by Atlantic salmon (AS) and brown trout (BT) and available in study site II

Abiotic variable	Habitat utilisation of AS				Habitat utilisation of BT				Habitat availability			
	Mean	SD	Min	Max	Mean	SD	Min	Max	Mean	SD	Min	Max
Mean flow velocity (cm s^{-1})	48.7	31.9	0.0	105.0	29.7	22.2	0.0	92.0	52.2	29.0	0.0	105.0
Bottom flow velocity (cm s^{-1})	12.6	18.3	0.0	60.0	9.0	15.5	0.0	66.0				
Water depth (cm)	34	14	0	62	33	12	11	60	33	13	6	72
Substrate			Cobble	Boulder			Gravel	Boulder			Gravel	Boulder

than evaluated in this study when employed in smaller streams with less heterogeneous habitat. For example, in a small and shallow stream (7.5 m in width, 21 cm in mean depth), Roussel et al. (2000) found on average 81.9% (range: 73.9–95.6%) of the PIT tagged fish. Second, in order to compare the techniques, we kept the sampling effort similar. Increased PIT tracking effort would potentially lead to higher detection rates of PIT tagged fish. The clear advantage of PIT tags in comparison to radio transmitters is that they are less costly allowing tagging of high numbers of individuals. Recent developments in the portable PIT readers detect even smaller PIT tags (12 mm in length, 2.1 mm in diameter) allowing to monitor the movements of 0+ fish (Cucherousset et al., 2005; Quintella et al., 2005). Furthermore, unlike radio transmitters, PIT tags do not necessitate batteries to operate, allowing an operational life of several years. Thus, PIT tags provide a valuable tool in long-term studies on life-history of individuals or population dynamics. In comparison to snorkelling, both telemetry techniques displayed much higher detection rates whereas fish were seldom visible to the observer during visual surveys. On average, only 4.1% fish were observed during a visual survey. The limitation of underwater observation is clearly due to the cryptic behaviour of juvenile salmonids during winter. Even when the observer knew where a fish was located due to PIT and radio telemetry information, fish were in most cases hidden in the interstitial space of the substrate and invisible for the observer. This leads to the suggestion that during winter, when juvenile salmonids exhibit predominantly cryptic behaviour, snorkelling may not be an appropriate method to obtain fish abundances or habitat preferences. However, visual observation may produce additional information on fish behaviour or may be useful in validating telemetry derived fish positions.

Numerous studies have analysed the habitat preferences of juvenile salmonids during summer (deGraaf & Bain, 1986; Morantz et al., 1987; Guay et al., 2000; Heggenes & Dokk, 2001). Although, studies were conducted in different regions using different salmonid populations, there is good agreement in the observed ranges of flow velocity (20–60 cm s^{-1}), water depth (15–60 cm) and substrate (gravel-cobble) used by juvenile salmonids. However, for the survival of juvenile salmonids the availability of suitable overwintering habitat is critical (Cunjak, 1996; Quinn & Peterson, 1996). The comparison of ranges of winter habitat preferences obtained in the present study to summer habitat preferences highlight essential differences in winter and summer habitat use. The results revealed that fish preferred low flow velocities and boulder substrate. Furthermore, snorkelling surveys revealed that fish seek shelter between the interstices in the boulder. The use of winter habitat preferences indices may therefore be important for future habitat modelling that requires the description of all critical life stages. The proximity of suitable winter habitat and rearing habitat is therefore essential for the fitness and survival of juvenile salmonids and consequentially to their populations.

Despite early evidence that individual juvenile Atlantic salmon undertake important up- and downstream migration over several hundred meters (Saunders & Gee, 1964), juvenile salmonids have generally been described as station-holding territorial fish with restricted home ranges (Kalleberg, 1958; Keenleyside & Yamamoto, 1961). Recently, movements of juvenile salmonids and understanding why fish move have received much attention (Van Winkle et al., 1998; Railsback & Rose, 1999; Gowan & Fausch, 2002). Radio telemetry studies have demonstrated that juvenile salmonids may be much more active, displaying large home ranges than previously thought (Armstrong et al., 1999; Berland et al., 2004; Økland et al., 2004; Robertson et al., 2004). Gowan & Fausch (2002) suggested that summer movements of salmonids are related to maximisation of energy intake by the fish at a reach scale (100 s of m). The present study underlines the fact that even during winter, juvenile salmonids are more mobile then previously thought. It may be hypothesised that although food intake rates decrease at low water temperature due to slower metabolic and digestion rates (Brett & Groves, 1979), juvenile salmonids may still need to feed during winter months, especially during late winter when energy reserves are exhausted. They may only emerge from their shelter in the interstices at night time in order to minimise predation risk from diurnal homeothermal predators (Fraser et al., 1993; Valdimarsson & Metcalfe, 1998). An alternative hypothesis for the increased mobility of juvenile salmonids during winter is that they leave shelters at night to escape anchor and frazil ice formation that often forms during decreasing water temperatures at night (Heggenes et al., 1993).

The present study demonstrated that juvenile salmonids displayed considerable mobility. However, some of the observed mobility might be artificially introduced due to the displacement. Fish were captured over 200 m river section and released in the middle of the river section. Therefore, fish could have experienced a displacement of up to 100 m. Consequently, dislocated fish may initially attempt to return to their original home range and the observed mobility may represent attempted homing. Armstrong et al. (1997) displaced Atlantic salmon parr and examined exploratory and settlement behaviour. Most of the displaced fish showed only for a short period, of less than 2.5 days, mobile exploratory behaviour before resettling to restricted home ranges. Once establishment of new home ranges occurred, Atlantic salmon parr rarely displayed exploratory behaviour outside the home range (Armstrong et al., 1997). Similarly, in the present study, highest rates of up- and downstream movements occurred during the first 24 h after release. Levels of mobility then decreased over time as fish settled into distinct home ranges. The high post-release emigration is most likely explainable by the displacement of the fish within the river reach rather than by the tagging or type of tag. High mobility and emigration rates within the first hours after fish survey have been frequently observed (Armstrong et al., 1999; Peterson et al., 2005).

This study is one of the first attempts to compare the efficiency of PIT and radio telemetry. The results show that the relocation rate of radio telemetry is significantly higher than PIT tracking and snorkeling may considerably underestimate fish abundance. Furthermore, winter habitat preferences of juvenile salmonids may differ from summer habitat preferences as fish displayed a strong preference to boulder substrate in order to seek shelter in the interstice. Therefore, successful habitat conservation and management requires information on temporally variable habitat preferences at key life stages, which include the winter period.

Acknowledgements We thank C. Kelly for assistance in the field. This research was funded by Natural Resources Canada's (NRCan) Panel for Energy Research and Development (PERD). E.C. Enders was supported by a Visiting Fellowship in Canadian Government Laboratories from the Natural Sciences and Engineering Research Council of Canada (NSERC).

References

Amundsen, P. A., R. Bergersen, H. Huru & T. G. Heggberget, 1999. Diel feeding rhythms and daily food consumption of juvenile Atlantic salmon in the River Alta, northern Norway. Journal of Fish Biology 54: 58–71.

Armstrong, J. D., V. A. Braithwaite & P. Rycroft, 1996. A flat-bed passive integrated transponder antenna array for monitoring behaviour of Atlantic salmon parr and other fish. Journal of Fish Biology 48: 539–541.

Armstrong, J. D., V. A. Braithwaite & F. A. Huntingford, 1997. Spatial strategies of wild Atlantic salmon parr: exploration and settlement in unfamiliar areas. Journal of Animal Ecology 66: 203–211.

Armstrong, J. D., F. A. Huntingford & N. A. Herbert, 1999. Individual space use strategies of wild juvenile Atlantic salmon. Journal of Fish Biology 55: 1201–1212.

Armstrong, J. D., J. W. A. Grant, H. L. Forsgren, K. D. Fausch, R. M. DeGraaf, I. A. Fleming, T. D. Prowse & I. J. Schlosser, 1998. The application of science to the management of Atlantic salmon (*Salmo salar*): integration across scales. Canadian Journal of Fisheries and Aquatic Sciences 55(Suppl. 1): 303–311.

Berland, G., T. Nickelsen, J. Heggenes, F. Økland, E. B. Thorstad & J. H. Halleraker, 2004. Movements of wild atlantic salmon parr in relation to peaking flows below a hydropower station. River Research and Applications 20: 957–966.

Bovee, K. D., 1982. A guide to stream habitat analysis using the instream flow incremental methodology. FWS/OBS 82/26, U.S. Fish and Wildlife Service.

Bremset, G., 2000. Seasonal and diel changes in behaviour, microhabitat use and preferences by young pool-dwelling Atlantic salmon, *Salmo salar*, and brown trout, *Salmo trutta*. Environmental Biology of Fishes 59: 163–179.

Brett, J. R. & D. D. Groves, 1979. Physiological energetics. In Hoar, W. S., J. R. Randall & J. R. Brett (eds), Fish Physiology – Bioenergetics and Growth. Academic Press, New York, 279–352.

Cucherousset, J., J. M. Roussel, R. Keeler, R. A. Cunjak & R. Stump, 2005. The use of two new portable 12-mm PIT Tag detectors to track small fish in shallow streams. North American Journal of Fisheries Management 25: 270–274.

Cunjak, R. A., 1996. Winter habitat of selected stream fishes and potential impacts from land-use activity. Canadian Journal of Fisheries and Aquatic Sciences 53(Suppl. 1): 267–282.

Cunjak, R. A., 1988. Behaviour and microhabitat of young Atlantic salmon (*Salmo salar*) during winter. Canadian Journal of Fisheries and Aquatic Sciences 45: 2156–2160.

deGraaf, D. A. & L. H. Bain, 1986. Habitat use by and preferences of juvenile Atlantic salmon in two Newfoundland rivers. Transactions of the American Fisheries Society 115: 671–681.

Fraser, N. H. C., N. B. Metcalfe & J. E. Thorpe, 1993. Temperature-dependent switch between diurnal and nocturnal foraging in salmon. Proceedings of the Royal Society of London – Series B: Biological Sciences 242: 135–139.

Gowan, C. & K. D. Fausch, 2002. Why do foraging stream salmonids move during summer? Environmental Biology of Fishes 64: 139–153.

Gries, G., K. G. Whalen, F. Juanes & D. L. Parrish, 1997. Nocturnal activity of juvenile Atlantic salmon (*Salmo salar*) in late summer: evidence of diel activity partitioning. Canadian Journal of Fisheries and Aquatic Sciences 54: 1408–1413.

Guay, J. C., D. Boisclair, D. Rioux, M. Leclerc, M. Lapointe & P. Legendre, 2000. Development and validation of numerical habitat models for juveniles of Atlantic salmon (*Salmo salar*). Canadian Journal of Fisheries and Aquatic Sciences 57: 2065–2075.

Heggenes, J. & J. G. Dokk, 2001. Contrasting temperatures, waterflows, and light: seasonal habitat selection by young Atlantic salmon and brown trout in a boreonemoral river. Regulated Rivers: Research and Management 76: 623–635.

Heggenes, J., O. M. W. Krog, O. R. Lindas, J. G. Dokk & T. Bremnes, 1993. Homeostatic behavioural responses in a changing environment: brown trout (*Salmo trutta*) become nocturnal during winter. Journal of Animal Ecology 62: 295–308.

Hiscock, M. J., D. A. Scruton, J. A. Brown & C. J. Pennell, 2002. Diel activity pattern of juvenile Atlantic salmon (*Salmo salar*) in early and late winter. Hydrobiologia 483: 161–165.

Jacobs, J., 1974. Quantitative measurement of food selection: a modification of the forage ratio and Ivlev's electivity index. Oecologia 14: 413–417.

Johnston, P., N. E. Bergeron & J. J. Dodson, 2004. Diel activity patterns of juvenile Atlantic salmon in rivers with summer water temperature near the temperature dependent suppression of diurnal activity. Journal of Fish Biology 65: 1305–1318.

Kalleberg, H., 1958. Observations in a stream tank of territoriality and competition in juvenile Atlantic salmon and trout (*Salmo salar* L. and *S. trutta* L.). Fisheries Board of the Swedish Institute of Freshwater Research Drottingholm Report 39: 55–98.

Keenleyside, M. H. A. & F. T. Yamamoto, 1961. Territorial behaviour of juvenile Atlantic salmon (*Salmo salar* L.). Behaviour 19: 139–169.

Morantz, D. L., R. K. Sweeney, C. S. Shirvell & D. A. Longard, 1987. Selection of microhabitat in summer by juvenile Atlantic salmon (*Salmo salar*). Canadian Journal of Fisheries and Aquatic Sciences 44: 120–129.

Økland, F., E. B. Thorstad & T. F. Næsje, 2004. Is Atlantic salmon production limited by number of territories? Journal of Fish Biology 65: 1047–1055.

Peterson, J. T., N. P. Banish & R. F. Thurow, 2005. Are block nets necessary? Movement of stream-dwelling salmonids in response to three common survey methods. North American Journal of Fisheries Management 25: 732–743.

Quinn, T. P. & N. P. Peterson, 1996. The influence of habitat complexity and fish size on over-winter survival and growth of individually marked juvenile coho salmon (*Oncorhynchus kisutch*) in Big Beef Creek, Washington. Canadian Journal of Fisheries and Aquatic Sciences 53: 1555–1564.

Quintella, B. R., N. O. Andrade, R. Espanhol & P. R. Almeida, 2005. The use of PIT telemetry to study movements of ammocoetes and metamorphosing sea lampreys in river beds. Journal of Fish Biology 66: 97–106.

Railsback, S. F. & K. A. Rose, 1999. Bioenergetics modeling of stream trout growth: temperature and food consumption effects. Transactions of the American Fisheries Society 128: 241–256.

Rimmer, D. M., U. Paim & R. L. Saunders, 1983. Autumnal habitat shift of juvenile Atlantic salmon (*Salmo salar*) in a small river. Canadian Journal of Fisheries and Aquatic Sciences 40: 671–680.

Robertson, M. J., C. J. Pennell, D. A. Scruton, G. J. Robertson & J. A. Brown, 2004. Effect of increased flow on the behaviour of Atlantic salmon parr in winter. Journal of Fish Biology 65: 1070–1079.

Roussel, J. M., A. Haro & R. A. Cunjak, 2000. Field test of a new method for tracking small fishes in shallow rivers using passive integrated transponder (PIT) technology. Canadian Journal of Fisheries and Aquatic Sciences 57: 1326–1329.

Roussel, J. M., R. A. Cunjak, R. Newbury, D. Caissie & A. Haro, 2004. Movements and habitat use by PIT-tagged Atlantic salmon parr in early winter: the influence of anchor ice. Freshwater Biology 49: 1026–1035.

Saunders, R. L. & J. H. Gee, 1964. Movements of young Atlantic salmon in a small stream. Journal of the Fisheries Research Board Canada 21: 27–36.

Scruton, D. A., J. Heggenes, S. Valentin, A. Harby & T. H. Bakken, 1998. Field sampling design and spatial scale in habitat-hydraulic modelling: comparison of three models. Fisheries Management and Ecology 5: 225–240.

Valdimarsson, S. K. & N. B. Metcalfe, 1998. Shelter selection in juvenile Atlantic salmon or why do salmon seek shelter in winter. Journal of Fish Biology 52: 42–49.

Valdimarsson, S. K., N. B. Metcalfe, J. E. Thorpe & F. A. Huntingford, 1997. Seasonal changes in sheltering: effect of light and temperature on diel activity in juvenile salmon. Animal Behaviour 54: 1405–1412.

Van Winkle, W., H. I. Jager, S. F. Railsback, B. D. Holcomb, T. K. Studley & J. E. Baldrige, 1998. Individual-based model of sympatric populations of brown and rainbow trout for instream flow assessment – Model description and calibration. Ecological Modelling 110: 175–207.

Hydrobiologia (2007) 582:243–256
DOI 10.1007/s10750-006-0549-6

FISH TELEMETRY

Assessment of largemouth bass (*Micropterus salmoides*) behaviour and activity at multiple spatial and temporal scales utilizing a whole-lake telemetry array

K. C. Hanson · S. J. Cooke · C. D. Suski · G. Niezgoda · F. J. S. Phelan · R. Tinline · D. P. Philipp

Abstract A whole-lake acoustic telemetry array was utilized to monitor the three-dimensional position of 20 largemouth bass (*Micropterus salmoides*). Code division multiple access (CDMA) technology enabled the simultaneous monitoring of the 20 transmitters (equipped with pressure and temperature sensors) at 15 s intervals with sub-meter accuracy. Fish were monitored between November 2003 and April 2004 to evaluate the behaviour of fish across different temporal and spatial scales. The distance moved by largemouth bass, assessed both on a daily and hourly basis, varied by season and was positively correlated with water temperature. For example, daily movement rates were 2.69 ± 1.45 km/day in mid November (average daily water temperature 5.9°C), 2.24 ± 0.73 km/day in early January (5.1°C), and 7.28 ± 2.62 km/day in mid April (7.7°C). Interestingly, daily movement rates varied by as much as 25 fold among individual fish. Visualization of fish swimming paths revealed that whereas some fish occupied discrete areas and made only localized movements, other individuals made lengthier journeys covering much of

Guest editors: P. R. Almeida, B. R. Quintella, M. J. Costa and A. Moore
Developments in Fish Telemetry

K. C. Hanson · D. P. Philipp
Program in Ecology and Evolutionary Biology, University of Illinois, and Center for Aquatic Ecology and Conservation, Illinois Natural History Survey, 607 E. Peabody Drive, Champaign, IL 61820, USA

S. J. Cooke
Department of Biology and Institute of Environmental Science, Carleton University, 1125 Colonel By Drive, Ottawa, ON, Canada K1S 5B6

C. D. Suski
Harkness Laboratory of Fisheries Research, Ontario Ministry of Natural Resources, 300 Water St., Peterborough, ON, Canada K9J 8M5

G. Niezgoda
Lotek Wireless Inc, 115 Pony Dr., Newmarket, ON, Canada L3Y 7B5

F. J. S. Phelan
Queen's University Biological Station, Department of Biology, Queen's University, Kingston, ON, Canada K7L 3N6

R. Tinline
Geographic Information Systems Laboratory, Department of Geography, Queen's University, Kingston, ON, Canada K7L 3N6

K. C. Hanson (✉)
Ottawa-Carleton Institute of Biology, Carleton University, 1125 Colonel By Drive, Ottawa, ON, Canada K1S 5B6
e-mail: kchanson@uiuc.edu

the lake in periods of as little as one day. Analysis of fish behaviour at a finer temporal scale revealed that during the winter, fish spend more than 95% of their time swimming at speeds less than 0.1 m/s (0.07 ± 0.24 m/s). During late fall, and especially in spring, swimming speeds were higher with mean swimming speeds of 0.11 ± 0.27 m/s and 0.19 ± 0.29 m/s, respectively. When the telemetry dataset was queried to simulate 24 h manual tracking intervals, it was clear that manual tracking data would not have been representative of actual daily movement rates, underestimating daily movement and swimming speeds by at least 75 fold. This study identifies the importance of evaluating fish activity at multiple spatial (whole lake to sub-meter position) and temporal (seasonal to seconds) scales and illustrates the potential of CDMA telemetry to yield such data.

Keywords Biotelemetry · Behaviour · Scale · Winter biology · Activity

Introduction

Studies of fish movement and activity have increased in both number and importance in recent years (Lucas & Baras, 2000), reflecting the need to understand the spatial ecology of fish when developing effective management and conservation strategies. Underlying the explosion of studies evaluating movement and activity have been major advances in research techniques for monitoring the behaviour of fish in the wild. The advent of telemetry techniques has enabled researchers to obtain detailed activity and movement data remotely across different temporal and spatial scales (Lucas & Baras, 2000; Cooke et al., 2004a, b). Rarely, however, have multiple scales been considered within a single study. Although mobile tracking of telemetered individuals has become a standard technique in fisheries science, it still fails to provide researchers with fine scale activity data, and if fish are to be tracked continuously, it requires large investments in labour (Cooke et al., 2004a, b). To overcome those deficiencies, researchers have turned to either fixed telemetry stations to continuously monitor movements (e.g., Cooke et al., 2001a; Cote et al., 2002) or physiological telemetry devices to assess fine scale patterns of swimming activity (e.g., Cooke et al., 2004a, b). Fixed monitoring stations can provide information on the two- and three-dimensional position and movement of fish continuously and in a manner that is independent of the presence of a researcher. With sub-meter accuracy, an array of fixed stations also has the potential to provide detailed information on fish swimming speeds and energetics, data previously limited to physiological telemetry systems. Fixed stations have become even more robust when coupled with code division multiple access (CDMA) technology, enabling numerous animals to be tracked simultaneously in a confined area without the signal collision problems evident in previous deployments of fixed telemetry stations (see Niezgoda et al., 2002; Cooke et al., 2005). Niezgoda et al. (2002) utilized such an acoustic CDMA system to continuously monitor the movements of blue crab (*Callinectes sapidus*) in a salt marsh pond for several weeks. For the first time, we use a whole-lake three-dimensional CDMA acoustic telemetry observatory to assess fish behaviour and activity at multiple spatial and temporal scales.

The use of a remote, three dimensional positioning system allows us to generate a continuous and detailed dataset of fish behaviour for the entire life of each transmitter, a period that can span multiple seasons. Such a complete dataset is one of the first of its kind. In their review of biotelemetry, Cooke et al. (2004a) noted that desirable characteristics of remotely monitored telemetry studies included the partnering of physiology with field data, individual based focus, continuous monitoring over multiple scales, and ability to work with unrestrained animals. This study incorporated all of those characteristics. This initial exploratory analysis is intended to provide an assessment of system capabilities and to identify future detailed research questions that are possible to address using this dataset.

Using largemouth bass (*Micropterus salmoides*) as the test organism, we investigated fish behaviour and activity at multiple spatial and temporal scales using a three dimensional CDMA acoustic whole-lake telemetry array in an eastern

Ontario lake. Experiments focused on largemouth bass as a model species for several reasons. First, a long-term population ecology study of largemouth bass in the study lake (i.e., Warner Lake) provided baseline information (see Suski, 2000). In addition, because largemouth bass exhibit different seasonal movement patterns, with some individuals occupying discrete home ranges, while others being more transient, it provides the opportunity to assess the level of individual variation (Demers et al., 1996). Largemouth bass are also thought to vary their activity among seasons, being active in the warmer seasons and quiescent in the winter (Demers et al., 1996; Cooke et al., 2003b). Finally, largemouth bass are a heavily managed species owing in part to their popularity as a recreational sport fish (Pullis & Laughland, 1999), thereby providing much opportunity for applied research.

Methods

Study site

In 2003 a freshwater lake (Warner Lake, Queen's University Biological Station [QUBS]) was instrumented with a CDMA acoustic telemetry system as the backbone of an aquatic "ecological observatory". The observatory extends traditional telemetry studies involving manual triangulation of a few representative animals to a fully automated all-season community and species-interactive multi-dimensional view. Warner Lake is a private research reserve located in eastern Ontario (44°31′ N, 76°22′ W) that is wholly enclosed on QUBS property, enabling the deployment and field testing of equipment in an undisturbed setting. The small lake (18.2 hectare surface area) is composed of two basins; a smaller shallow basin (max depth = 2 m) and a slightly larger deeper basin (max depth = 7 m). The entire shallow basin and the near shore regions of the deeper basin have extensive littoral zone characterized by emergent and submergent vegetation. The deeper areas have dense weed beds with the majority of the lake bottom covered by *Chara* spp. The entire shoreline has substantial amounts of fallen timber. Warner Lake is a closed system for fish and has been the focus of a long-term ecological study on the reproductive dynamics of largemouth bass (see Suski, 2000). Beginning in 1996, researchers have conducted underwater nesting surveys of largemouth bass and have implanted nesting males with passive integrated transponders (PITs) to monitor the growth, survival, and reproductive activity of individual fish over time, providing ample background data for this current research activity. Other documented species include white sucker (*Catostomus commersonii*), pumpkinseed (*Lepomis gibbosus*), yellow perch (*Perca flavescens*), brown bullhead (*Ameiurus nebulosus*) and golden shiner (*Notemigonus crysoleucas*).

Telemetry array

The underwater acoustic telemetry array was installed in Warner Lake during November 2003 (Detailed description of the equipment, configuration, and system performance is provided in Cooke et al., 2005). To meet the acoustic telemetry requirements of the observatory, a CDMA-based telemetry system (MAP_600, Lotek Wireless, Newmarket, ON) was installed at the observatory. The telemetry equipment consists of two multi-port MAP_600 receivers monitoring a total of 13 hydrophones distributed in such geometry as to provide coverage of the entire lake, including the littoral zones. Equipment was configured to monitor eight hydrophones (large basin) on one receiver, with the remaining five hydrophones (small basin) on the other receiver. A CDMA temperature-pressure sensing tag (burst rate 15 s) was also placed at each moored hydrophone location. Cabling (i.e., connecting hydrophones to shore-based receiving equipment) was routed along the bottom of the lake to one point on the shoreline and brought out of the water through conduit. Hydrophones were moored from fixed posts (steel piping driven into the lake bottom) at an approximate depth of 2 m from the water surface to ensure that lake ice conditions would not damage or move the hydrophones. To facilitate sub-meter positioning of instrumented fish, all hydrophones were surveyed using differential GPS (±0.2 m). Data were transferred from each receiver to a personal computer for further processing by use of flash storage cards.

Study animals

Largemouth bass were collected by angling between October 14 and October 16, 2003. We attempted to target individuals that had previously been implanted with PITs. These PIT tagged fish were part of a long-term study of the reproductive ecology of largemouth bass in the lake and provided information on reproductive history, sex, age, and capture history. In total, we implanted 14 males and 8 females with CDMA temperature-pressure sensing acoustic transmitters (Lotek CTP-M11-55, 11 mm × 55 mm, burst rate 15 s, life expectancy of one year). After surgery, two telemetered individuals either died or suffered from transmitter failure. These individuals were removed from all analyses in this study. The burst rate of the tags was selected to ensure transmitter longevity through the fall-winter and winter-spring transitions. Depth and temperature resolution for the fish tags was ±0.7 m and ±0.5°C, respectively.

To implant fish, they were first anesthetized in a 60 ppm induction bath of clove oil solution (emulsified in ETOH, clove oil:ETOH, 1:9; Anderson et al., 1997). When fish lost equilibrium, they were measured (total length in mm) and weighed (mass in g). Fish were then placed on a foam surgery table with a recirculating supply of water containing a maintenance dose of anesthetic (20 ppm clove oil solution) irrigating the gills. An initial laparotomy incision of 12 mm was made slightly off center from the ventral midline behind the pelvic girdle of each fish (Cooke et al., 2003a). Cleaned transmitters (ETOH and betadine) were inserted into the body cavity, and the incision was closed with two simple interrupted sutures (3/0 PDS II, absorbable monofilament sutures; Ethicon Inc.). All surgeries were conducted by the same experienced individual to eliminate variance associated with multiple surgeons (Cooke et al., 2003a). Fish were then placed in coolers containing lake water and allowed to recover from anesthesia. Equilibrium was usually regained within 5 min, and the fish were then released in the lake at one central location.

Data processing and analysis

Raw data were loaded into the BioMAP program (v. 2.1.12.1, Lotek Wireless, Newmarket, ON), where temperature information was retrieved to compute sound speed on a day-to-day basis to enable hyperbolic position determination. Raw position solutions were generated after running flash card files through the two-dimensional positioning engine within the BioMAP program. Depth information (i.e., the 3rd dimension) was determined from the pressure information relayed by each tag. To remove outliers, a series of filtering processes within BioMAP were applied to the data. At the time a position solution is recorded, the array assigns a reliability number (RN) and condition number (CN) to assess the numerical stability of the position (Niezgoda et al., 2002). If these numbers showed that a point may have been an outlier, it was filtered out of the data set (RN > 0.75 , CN < 10). Remaining position solutions that did not fall within sub-meter degree of precision (GDOP) were filtered out of the dataset as well. Lastly, impossible position solutions, defined as areas the fish could not possibly access (i.e., land) as well as movements that were not physically possible (large distance movements in impossibly short periods of time; i.e., swim speeds that exceeded 6 body/lengths per second) were also filtered out. Overall, more than 75% of solutions were deemed valid locations.

This paper sought to assess the behaviour of largemouth bass on varying scales across seasons. As such, we first plotted the seasonal trends of the data (on a weekly basis) and then selected one representative day within each season to analyze movements in detail (fall, November 11, 2003; winter, January 1, 2004; spring, April 13, 2004). Fully filtered data were then queried to generate tables of information for each fish on each particular day. To determine daily distance traveled for the seasonal trend line, position estimates were averaged hourly, and then the difference was found between successive X, Y positions assuming that the fish maintained the same elevation (Z) between positions. For the representative days in each season, distance traveled was determined as the distance between each X,

Y position recorded. The Pythagorean Theorem was then utilized to determine linear distance between the points. Daily distances were measured as the sum of all distances moved across a day for each individual. Hourly distances were measured as the sum of all distances moved across a single hour of a day by each fish. Swim speeds were determined by dividing each instantaneous distance moved by the time required to move that distance (represented by the time between signal transmissions). To simulate manual tracking, a daily distance was determined by the above procedure, but only the first and last points of the data set were utilized (i.e., the closest point to 00:00 h and almost 24 h later (i.e., closest signal to 00:00 h the following day). Statistical analysis was performed in JMP IN v. 4.0. All values presented represent means ± SD unless otherwise noted.

Results

Seasonal trends

At the broadest scale, we observed trends across the fall and winter seasons. Activity, measured as daily distance swam, decreased from November 2003 to early December 2003, corresponding to a drop in water temperature (Fig. 1) and a period when fish were observed in deep water. As temperature steadied from December 2003 to early March 2004, activity level and average depth utilization remained relatively constant (Fig. 1). In spring, however, fish were found shallower in the water column, and activity estimates declined dramatically (Fig. 1). Both depth and activity seemed to be correlated with water temperature, though this was not quantified. We used these general trends to identify representative days for further in depth analysis.

Daily movements

Focusing our efforts on three representative 24 h periods, we observed that daily swimming activity varied among the three seasonal periods examined. Overall, mean total daily swimming activity across all individuals was lowest in winter, but was 1.2 fold higher in fall and 3.25 fold higher in spring (Table 1, Fig. 2). This range of daily swimming activity seemed to be positively correlated with water temperature (fall: November 11, 2003 = 5.9 ± 0.32°C, winter: January 1, 2004 = 5.1 ± 0.62°C, spring: April 13, 2004 = 7.7 ± 0.82°C), but this was not quantified statistically. Although individual activity levels were generally consistent with this pattern, there were instances in which certain individuals swam more during periods of cooler temperatures, revealing individual variation in behaviour.

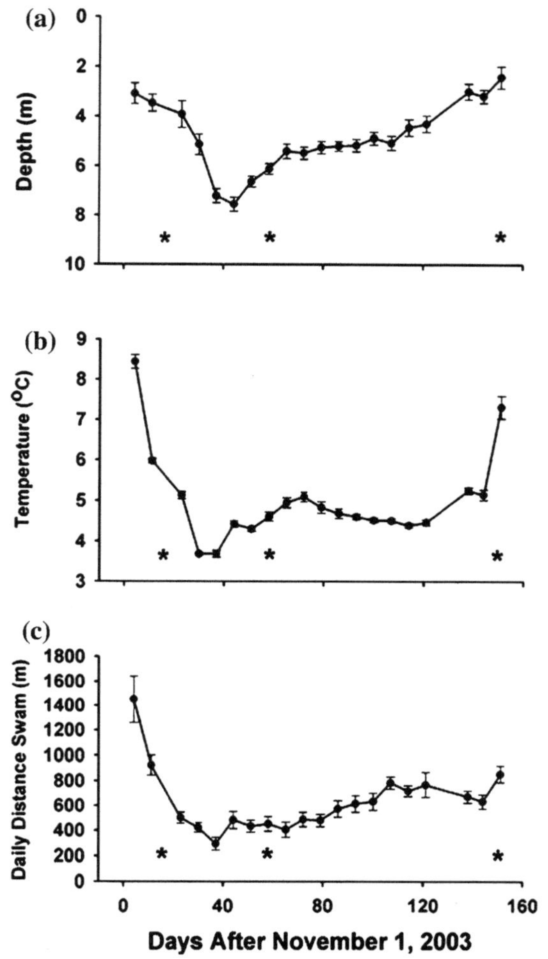

Fig. 1 Seasonal patterns of activity, depth, and water temperature of largemouth bass. (**a**) details the mean depth utilized by all individuals weekly across the seasons. (**b**) details the mean temperature utilized by all individuals weekly across the seasons. (**c**) details the mean daily distance swam by all individuals weekly across the seasons. Error bars represent ±1 SD. Asterisks indicate representative 24 h periods analyzed

Table 1 Mean total daily swimming distances for all 20 telemetered largemouth bass in Warner Lake over the three representative dates. All numbers are shown mean ± SD, with minimum and maximum values in parentheses below

Date	Daily water temperature (°C)	Total daily swimming distance, array tracking (m)	Total daily swimming distance, manual tracking (m)
11/11/3	5.9 ± 0.3	2,700 ± 1,500 (920/6,800)	76 ± 59 (1.5/200)
1/1/4	5.1 ± 0.6	2,200 ± 730 (1,040/3,400)	68 ± 71 (2.4/202)
4/13/4	7.7 ± 0.8	7,300 ± 2,600 (450/11,000)	61 ± 93 (0.7/320)

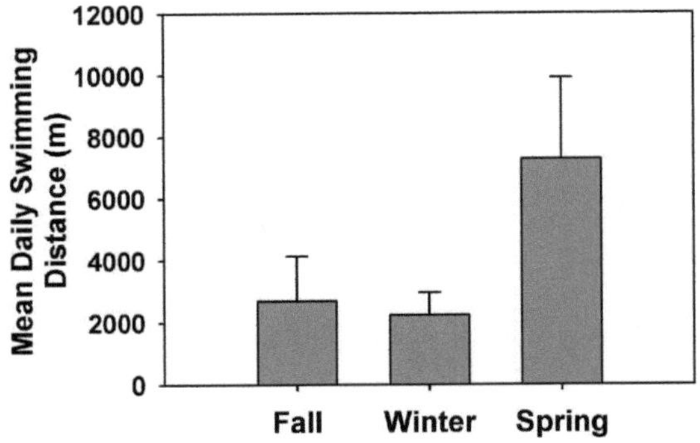

Fig. 2 Mean daily distance traveled (mean ± SD) as recorded by the hydrophone array for the three representative dates. Statistical quantification of between-date differences were not performed, as discussed in the text

Individual fish also varied extensively in daily swimming activity within a seasonal period, in some cases with differences exceeding 25 fold. For example, on April 13, LMB #*2* moved less than 1 km in 24 h whereas, LMB #*7* moved more than 11 km (Figs. 2, 3). Indeed, when daily movement traces were plotted for these two individuals, distinct differences in behaviour were evident. LMB #*2* occupied a very specific site and conducted only localized movement compared to LMB #*7* that made extensive movements within the lake, transiting numerous habitat types.

Hourly movements

For an intermediate time scale, mean hourly movement of all individuals was examined across all three seasonal periods. Hourly movement was also positively correlated with water temperature. Fish moved the longest distances during any hour in spring (150 ± 47 m/h), followed by winter (82 ± 10 m/h), and finally by fall (75 ± 9 m/h) (Fig. 4). The larger standard deviation around the

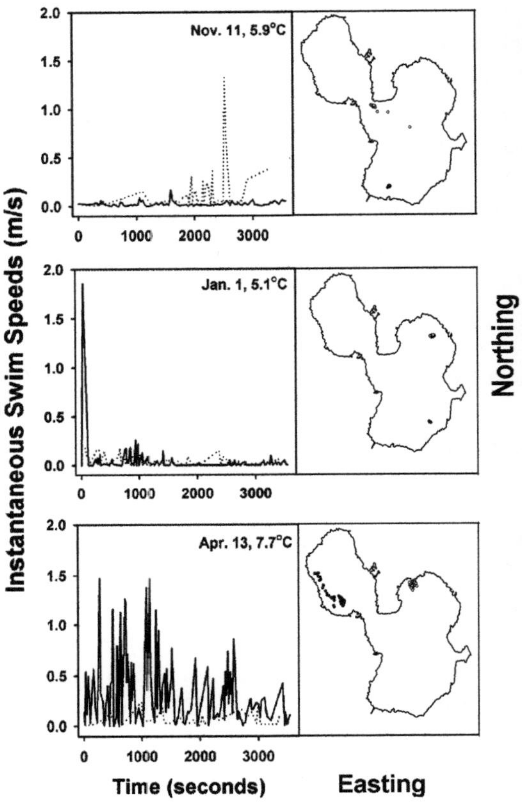

Fig. 3 Individual variation in behaviour and activity as determined by movement pattern and swim speed utilization over the three representative dates. The left panels show the instantaneous swim speed (m/s) utilized over time (s) across a one hour period of each date. The solid line represents individual LMB #*7*, and the dashed line represents individual LMB #*2*. The right panels show the two-dimensional position solutions within the lake that were used to determine the above mentioned instantaneous swim speeds. Individual LMB #*7* is represented by dark colored dots, while individual LMB #*2* is represented by light colored dots

mean hourly movement value for spring reflects the fact that fish were more active during daylight hours than at night (Fig. 4). In contrast, activity

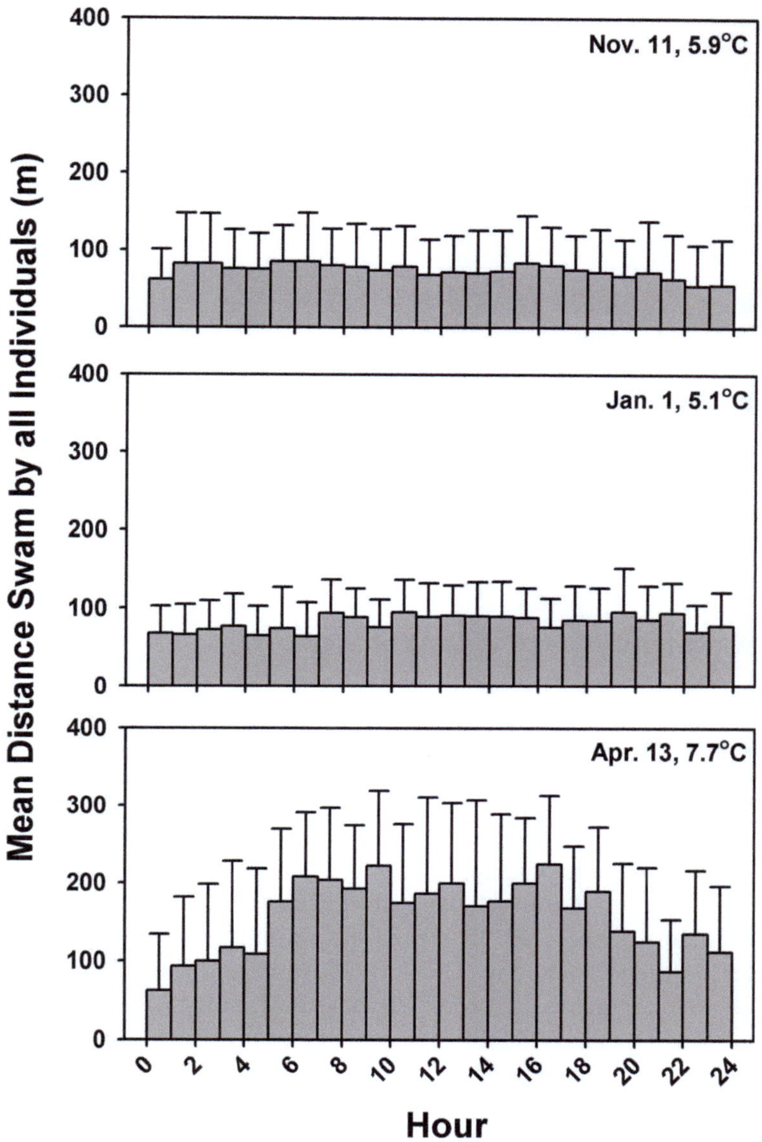

Fig. 4 Mean hourly daily distances swum, as recorded by the hydrophone array, for the three representative dates by each fish. Error bars represent ±1 SD

levels remained at a fairly consistent level during the day in both fall and winter (Fig. 4).

Instantaneous swim speeds

The final temporal scale investigated was that of instantaneous swim speeds. Again, a positive correlation was found between temperature and the frequency of faster swim speeds in both body lengths per second and meters per second. In fall, fish spent over 80% of the time swimming at a rate of 0.30 body lengths per second or less (between 0 and 0.12 m/s). The winter mean swimming speed was the lowest, with the fall and spring mean swimming speeds being 1.78 and 3.18 fold higher, respectively (Table 2, Fig. 5). In winter, fish exhibited swim speeds of 0.30 body lengths per second or less (between 0 and 0.12 m/s) over 90% of the time indicating that fish were moving much slower most of the time. During spring fish were much more active and moved faster. Approximately 55% of the observations of swimming speeds were in the 0–0.30 body lengths per second class (between 0 and 0.12 m/s) and speeds between 0.30 and 1.2 (between 0.12 and 0.60 m/s) body lengths per second accounted for

Table 2 Array tracking and manual tracking simulation instantaneous swim speeds for all 20 telemetered largemouth bass over three representative dates. All numbers are shown mean ± SD, with minimum and maximum values in parentheses below

Date	Daily water temperature (°C)	Array swim speed (m/s)	Manual mean swim speed (m/s)	Array mean swim speed (body length/s)	Manual mean swim speed (body length/s)
11/11/3	5.9 ± 0.32	0.107 ± 0.267 (0/2.67)	0.0009 ± 0.0007 (0.000017/0.0022)	0.262 ± 0.645 (0/5.99)	0.0023 ± 0.0017 (0.000042/0.0056)
1/1/4	5.1 ± 0.62	0.060 ± 0.237 (0/2.60)	0.0008 ± 0.0008 (0.000028/0.0027)	0.121 ± 0.402 (0/5.99)	0.0020 ± 0.0021 (0.000063/0.0064)
4/13/4	7.7 ± 0.82	0.191 ± 0.291 (0/2.50)	0.0008 ± 0.0011 (0.0000083/0.0041)	0.472 ± 0.715 (0/5.99)	0.0019 ± 0.0029 (0.000024/0.010)

40% of the other swim speeds. This result indicates that fish were more active during spring, a period that also had the highest water temperatures. During spring fish swam at higher speeds at least 20% more than they did during the fall and winter time periods.

Array versus manual tracking simulations

Daily swimming activity estimates were influenced dramatically by tracking interval. As described above, mean distances traveled by all individuals when tracked by the array were on the order of several km/day. When manual tracking simulations were conducted from the same empirical data set, however, daily swimming activity estimates plummeted for all seasons, in all cases to less than 0.5 km/day. These values represent a 3.5 fold decrease in estimated daily swimming activity for fall, a 3.3 fold decrease for winter, and a 12 fold decrease for spring. In addition, the positive correlation between water temperature and daily swimming activity that was present in the array tracking data was lost in the manual tracking simulation. This is not surprising considering that there was no correlation between daily swimming distance estimates generated from both techniques (Fig. 6). Mean instantaneous swimming speeds generated from manual tracking simulations were significantly lower than those empirical values from the array data in all seasons (fall, 122 fold lower; winter, 75 fold lower; and spring, 238 fold lower; Table 2, Fig. 6). Again, the positive correlation between water temperature and mean instantaneous swimming speed noted for the array tracking data was lost in the simulated manual tracking data.

Discussion

For the first time, we provide preliminary information with sub-meter accuracy on the simultaneous movement and activity of teleost fish within an entire lake. Our analyses focused on evaluating fish activity and movement at different spatial and temporal scales to yield new insights into largemouth bass behaviour as well as to assess the capabilities of the telemetry array. Hydrophone arrays that enable the real-time 3-dimensional positioning of numerous fish for extended durations, while remotely recording data on a fine scale, have enormous potential for answering biological questions (Cooke et al., 2005). Data derived from this research were incorporated into databases that can be used to form individual-based studies that continuously monitor unrestrained animals over multiple scales (Cooke et al., 2004a; Cooke et al., 2005).

Our analyses of fish movement began at a broad temporal scale, covering several seasons and eventually focused on instantaneous swimming speeds using intervals on the order of seconds. A consistent pattern was observed; i.e., there was a positive correlation between activity level of largemouth bass and water temperature at various temporal scales all derived from the same CDMA acoustic telemetry dataset. The fact that largemouth bass behaviour is influenced by temperature is not surprising and is supported by a large body of literature on largemouth bass thermal ecology (e.g., Beamish, 1970; Kolok, 1992), as well as general relationships between metabolic rate and temperature (Fry, 1971; Brett & Groves, 1979). That said, some of our findings are inconsistent with the notion that during winter, most

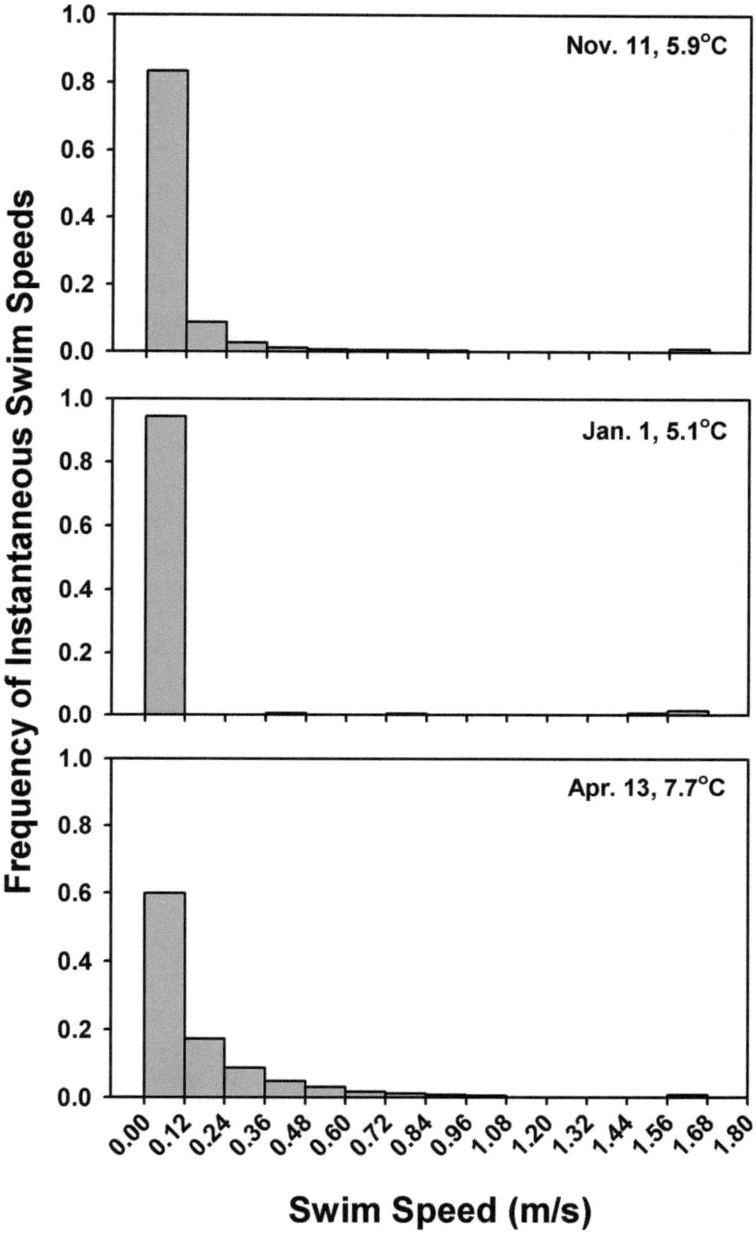

Fig. 5 Frequency histogram of instantaneous swim speeds (m/s) of 20 telemetered largemouth bass as recorded by a hydrophone array across three seasons

largemouth bass are sedentary, as has been a long standing paradigm (e.g., Heidinger, 1975). Our under-ice telemetry observations during the winter reveal that although the spatial distribution of largemouth bass is rather confined relative to the entire lake, fish still undertake significant localized movements, at times swimming at speeds up to 1.6 m/s. The majority (over 95%) of time, however, is spent swimming at speeds between zero (i.e., holding position) and 0.1 m/s. Due to the possibility of a winter energy deficit (Sullivan, 1985), coupled with thermal constraints on metabolism (Cooke et al., 2003b), when water temperatures decrease, bass may minimize energy usage by localizing movements. Restricted winter movements may also reflect environmental conditions

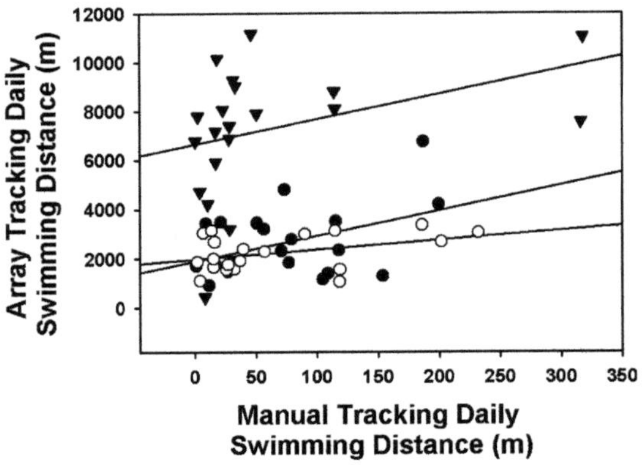

Fig. 6 Regression of array tracking vs. manual tracking daily distances swum over the three representative dates. Filled circles represent data from November 11, 2003, open circles represent data from January 1, 2004, and filled triangles represent data from April 13, 2003

that may limit distribution in a waterbody (e.g., hypoxia or anoxia). Localized movements may reflect subtle attempts to behaviourally thermoregulate (Crawshaw, 1984) or to seek or avoid other environmental conditions (Cunjak, 1996; Raibley et al., 1997). Behavioural paradigms of both warmwater fish in general and largemouth bass in particular propose that fish become quiescent in winter (Heidinger, 1975; Crawshaw, 1984). This study shows that although largemouth bass may significantly decrease activity levels during winter months, they certainly do not become completely dormant or sedentary.

In the fall, and especially in the spring, fish spent more time swimming at speeds that were between 0.2 and 1.0 m/s. Higher swim speeds during these times were positively correlated with higher water temperatures on these days. Also, fish moved up to 10.7 times farther on a daily basis in spring than winter, and up to 6.5 times farther in fall than winter. These movement patterns were also positively correlated with water temperature. Fish may show increased activity levels during fall in an attempt to obtain additional food and increase overwinter energy reserves (Miranda & Hubbard, 1994; Mackereth et al., 1999) and/or to locate overwintering regions with suitable water quality conditions (e.g., Raibley et al., 1997). Increased activity in spring may be in preparation for reproductive activities. Fish seek out food and warmer water temperatures to acquire food and enable gonad development needed for spawning. When water temperatures approach 15°C, male largemouth bass construct nests, court females, spawn, and then assume active parental care of up to one month depending on fry development (Kramer & Smith, 1962). During this season spawning and active parental care occurred during the second week of May and lasted until early June (Hanson, unpublished data). The spring date (April 13) that was analyzed in this study may have detected fish movements in preparation for spawning.

Across the three periods examined, largemouth bass activity (daily distance swam) varied considerably, and this variation may be partly due to competition and prey density. During warmer months, patches of vegetation attract both prey species seeking to avoid being consumed (Crowder & Cooper, 1979), as well as largemouth bass involved in ambush predation (Savino & Stein, 1982; Valley & Bremigan, 2002). Largemouth bass may therefore seek out areas of intermediate structural complexity that increase prey density as well as improve the probability of encountering a prey item (Crowder & Cooper, 1979). Due to the fact that vegetation growth in lakes is partly influenced by light penetration into water, many vegetated areas occur in the shallow reaches of the water column (Barko & Smart, 1981; Sheldon & Boyle, 1977) and are limited during the winter when the lake is covered by ice. In the spring, when vegetation has just begun to emerge, largemouth bass may be quite active as they search for ideal areas in which to forage. When vegetated areas are fully grown in, largemouth bass activity may decline as fish remain in localized areas while foraging. Therefore, the seasonal variation in swimming activity observed in largemouth bass may be partly driven by habitat complexity, competition, and prey behavior. Future studies to document vegetation dynamics will be needed to elucidate its potential role in mediating activity associated with competition and foraging.

Individual variation in daily distances traveled was greatest during spring, correlating with the highest water temperatures. Differences of up to 25 fold were seen between individuals on the April sampling day. In November, differences as great as 7.4 fold were seen. The lowest amount of variation occurred in January (up to 3.2 fold). This same variation occurred with hourly movements as well, indicating that individual fish behave much more similarly during winter than during either fall or spring. This finding is consistent with other work (Carlson, 1992; Karchesky & Bennett, 2004) that points to the fact that largemouth bass enter large aggregations and make localized movements during winter. In addition to variation among individuals and across seasons, diel differences were also noted. Interestingly, these differences were only apparent in the spring and were characterized as elevated activity during the day, with slightly higher peaks at dawn and dusk. Previous laboratory research has revealed that largemouth bass exhibit crepuscular activity peaks at warm water temperatures (Reynolds & Casterlin, 1978).

Data analysis on multiple scales can be used to develop energetic models of free-swimming fish, but of particular relevance are the data generated from the instantaneous swimming speeds (Lucas et al., 1993; Cooke et al., 2004b). Profiles of fish swimming speeds are typically generated by manually tracking telemetered fish to determine rates of movement (Lucas & Baras, 2000) or using physiological telemetry to assess muscle activity or tail beats (Demers et al., 1996; Cooke et al., 2004b). Our data on swim speeds represent one of the first long-term tracks of fish swimming activity at this whole-lake scale for multiple individuals. As discussed above, we noted that fish spent little time swimming faster than 1 m/s, even in the spring. This is not surprising considering both the swimming capability of bass as well as the sampling limitations associated with the transmitters that we used. Based upon laboratory swim tunnel work, field acclimated juvenile largemouth bass are believed to have maximum critical swimming speeds (i.e., the speed at which fish switch from aerobic to anaerobic activity) of 2.22 body lengths per second at 5°C (Kolok, 1992). Although we monitored fish in three seasons, the water temperature was always reasonably low. In fact, much of the current dogma on largemouth bass is that they are quiescent below about 8°C (our highest temperature analyzed here was 7.7°C), reflecting the fact that under laboratory conditions bass often fail to swim at low water temperatures (Cooke et al., 2001b). It is somewhat remarkable, therefore, that we observed the level of activity that we did. A second reason why we detected so little high speed swimming may be because the "instantaneous" speeds we generated are actually average speeds over a 15 s or greater period. Using transmitters with more frequent burst rates would greatly improve our ability to generate detailed information on swimming dynamics and energetics. Furthermore, because fish rarely swim in a straight line, more frequent pulses would also reduce underestimates of activity costs. That said, the dataset that we present is truly unique and unprecedented in its level of detail.

When the same dataset was subjected to simulated manual tracking, fish activity estimates were greatly underestimated. Multiple studies have indicated that telemetry investigations that utilize large time intervals between position solutions may in fact miss the vast majority of daily movements by fish. For example, experiments described in Demers et al. (1996) using EMG telemetry revealed that the majority of largemouth bass activity on daily basis may occur at spatial and temporal scales that are too small to be detected by current telemetry techniques that rely upon generating position solutions using manual tracking. Løkkeberg et al. (2002) showed that when artificially increasing the interval between signal transmissions, estimates of swimming speeds of cod (*Gadus morhua*) were decreased by 30–70%. Ovidio et al. (2000) also found a drastic reduction in accuracy of home range estimates of brown trout (*Salmo trutta*) when utilizing subsamples of a dataset with longer intervals between radio tracking position fixes. Our findings also indicate that manual tracking with larger intervals between position solutions may be particularly prone to underestimating movement and activity levels for fish that are highly mobile, but on a localized basis (or, in this case, during seasons when fish show increases in activity and mobility). Underestimation of fish movements seemed to be at the lowest during periods when the fish were most sedentary. During these time periods, although fish are least active and occupy the most discrete areas, longer intervals between position solutions are less likely to miss gross movements. In addition, we observed dramatic differences in daily activity estimates depending upon how these activity estimates were generated. For example, to evaluate seasonal trends at weekly intervals, we calculated distances using mean hourly position (Fig. 1). Conversely, to evaluate more fine scale activity, we summed the distances between every valid position solution (Figs. 2–5). Not surprisingly, using average hourly position solutions generated lower activity estimates than when all data points were essentially connected and summed. These findings have profound ramifications for the field of energetics modeling based upon field data, as discussed by Cooke et al. (2001a). Due to the differences between activity estimates based upon time interval between position solutions and on how data are analyzed, much of the current energetics modeling efforts likely incorporates activity estimates that are underestimations of actual fish activity. Not only is this relevant to understanding variation among studies, it is also important when considering variation and in selecting data analysis techniques within a single study or dataset.

This paper, based upon work collected during a trial period for the whole lake array, is intended to help focus future research efforts. Indeed, to refine our energetic estimates and develop seasonal sex-specific energetics models, transmitters with more frequent burst rates should be utilized (recognizing the tradeoffs between burst rate, battery life, and transmitter size). A study by Rand & Hinch (1998) determined that averaging over periods even as short as 5 s can underestimate the true costs of activity for migratory salmonids. Furthermore, this technology shows much promise for being able to detect the frequency with which fish burst at speeds that require anaerobic metabolism. Integrating this CDMA technology with other sensors such as EMG could provide information on fish energetics during periods when fish are stationary in terms of an X–Y coordinate but are turning continuously, such as those activities that occur when male largemouth bass are engaged in parental care. Future research efforts will focus on more detailed analyses of distribution, movements, depth use, habitat relationships and the factors associated with individual variation. Novel analytical software and filtering techniques currently under development will enable us to process and analyze data for the entire study period, providing power for statistical tests and enabling hypothesis driven research.

Acknowledgements The authors thank Raleigh Robertson, and the staff at QUBS for supporting the project. Caleb Hasler and Dave Ball of the Queen's GIS Laboratory assisted with data processing. Brandon Barthel and Kevin Esseltine assisted with fish capture. Glenn Crossin kindly commented on an earlier version of this MS. The array infrastructure was provided by Lotek Wireless, Inc. and maintained using funds provided by the Queen's University Biological Station and Queen's University. Funding for transmitters was partially provided by the University of Illinois Research Board. KCH and other aspects of this research project were partially supported by the Ron Ward Memorial Scholarship from the Champaign-Urbana Bass Club. SJC was supported by an NSERC and Killam Fellowship and the Queen's University Biological Station's Visiting Scientist programme. Additional support was provided by Lotek Wireless, Inc., University of British Columbia, Illinois Natural History Survey, and Queen's University.

References

Anderson, W. G., R. S. McKinley & M. Colavecchia, 1997. The use of clove oil as an anesthetic for rainbow trout and its effects on swimming performance. North American Journal of Fisheries Management 17: 301–307.

Barko, J. W. & M. Smart, 1981. Comparative Influences of light and temperature on the growth and metabolism of selected submersed freshwater macrophytes. Ecological Monographs 51: 219–236.

Beamish, F. W. H, 1970. Oxygen consumption of largemouth bass, *Micropterus salmoides*, in relation to swimming speed and temperature. Canadian Journal of Zoology 48: 1221–1228.

Brett, J. R. & D. D. Groves, 1979. Physiological energetics. In Hoar, W. S., Randall D. J. & J. R. Brett (eds), Fish Physiology, vol VIII. Academic Press, New York, 279–352.

Carlson, D. M., 1992. Importance of wintering refugia to the largemouth bass fishery in the Hudson River estuary. Journal of Freshwater Ecology 7: 173–180.

Cooke, S. J., C. M. Bunt, J. F. Schreer & D. H. Wahl, 2001a. Comparison of several techniques for mobility and activity estimates of smallmouth bass in lentic environments. Journal of Fish Biology 58: 573–587.

Cooke, S. J., T. A. Kassler & D. P. Philipp, 2001b. Physiological performance of largemouth bass related to local adaptation and interstock hybridization: implications for conservation and management. Journal of Fish Biology 58A: 248–268.

Cooke, S. J., B. D. S. Graeb, C. D. Suski & K. G. Ostrand, 2003a. Effects of suture material on incision healing, growth and survival of juvenile largemouth bass implanted with miniature radio transmitters: case study of a novice and experienced fish surgeon. Journal of Fish Biology 62: 1366–1380.

Cooke, S. J., E. C. Grant, J. F. Schreer, D. P. Philipp & A. L. Devries, 2003b. Low temperature cardiac response to exhaustive exercise in fish with different levels of winter quiescence. Comparative Biochemistry Physiology 134A: 157–165.

Cooke, S. J., S. G. Hinch, M. Wikelski, R. D. Andrews, L. J. Kuchel, T. G. Wolcott & P. J. Butler, 2004a. Biotelemetry: a mechanistic approach to ecology. Trends in Ecology and Evolution 19: 334–343.

Cooke, S. J., E. B. Thorstad & S. G. Hinch, 2004b. Activity and energetics in free-swimming fish: insights from electromyogram telemetry. Fish and Fisheries 5: 21–52.

Cooke, S. J., G. H. Niezgoda, K. C. Hanson, C. D. Suski, F. J. S. Phelan, R. Tinline & D. P. Philipp, 2005. Use of CDMA acoustic telemetry to document 3-D positions of fish: relevance to the design and monitoring of aquatic protected areas. Marine Technology Society Journal 39: 17–27.

Crowder, L. B. & W. E. Cooper, 1979. Structural complexity and fish-prey interactions in ponds: a point of view. In Johnson, D. L. & R. A. Stein (eds), Response of Fish to Habitat Structure in Standing Water. American Fisheries Society, North Central Division, Special Publication 6, Bethesda, Maryland, 2–10.

Cote, D., L. M. N. Ollerhead, R. S. Gregory, D. A. Scruton & R. S. McKinley, 2002. Activity patterns of juvenile Atlantic cod (*Gadus morhua*) in Buckley Cove, Newfoundland. Hydrobiologia 483: 121–127.

Crawshaw, L. I., 1984. Low-temperature dormancy in fish. American Journal of Physiology 246R: 479–486.

Cunjak, R. A., 1996. Winter habitat of selected stream fishes and potential impact from land-use activities. Canadian Journal of Fisheries and Aquatic Sciences. 53(Supplement 1): 267–282.

Demers, E., R. S. McKinley, A. H. Weatherley & D. J. McQueen, 1996. Activity patterns of largemouth and smallmouth bass determined with electromyogram biotelemetry. Transactions of the American Fisheries Society 125: 434–439.

Fry, F. E. J., 1971. The effect of environmental factors on the physiology of fish. In Hoar, W. S. & D. J. Randall (eds), Fish physiology, vol. VI. Academic Press, New York, 1–98.

Heidinger, R. C., 1975. Life history and biology of the basses. In Clepper, H. (ed.), Black Bass Biology and Management. Sport Fishing Institute, Washington, DC, 125–139.

Karchesky, C. M. & D. H. Bennett, 2004. Winter habitat use by adult largemouth bass in the Pend Oreille River, Idaho. North American Journal of Fisheries Management 24: 577–585.

Kolok, A., 1992. Morphological and physiological correlates with swimming performance in juvenile largemouth bass. American Journal of Physiology 263: 1042–1048.

Kramer, R. H. & L. L. Smith, 1962. Formation of year classes in largemouth bass. Transactions of the American Fisheries Society 91: 29–41.

Løkkeberg, S., A. Fernö & T. Jørgensen, 2002. Effect of position-fixing interval on estimated swimming speed and movement pattern of fish tracked with a stationary positioning system. Hydrobiologia 483: 259–264.

Lucas, M. C., A. D. F. Johnstone & I. G. Priede, 1993. Use of Physiological Telemetry as a method of estimating metabolism of fish in the natural environment. Transactions of the American Fisheries Society 122: 822–833.

Lucas, M. C. & E. Baras, 2000. Methods for studying the spatial behaviour of freshwater fishes in the natural environment. Fish and Fisheries 1: 283–316.

Mackereth, R. W., D. L. G. Noakes & M. S. Ridgway, 1999. Size-based variation in somatic energy reserves and parental expenditure by male smallmouth bass, *Micropterus dolomieu*. Environmental Biology of Fishes 56: 263–275.

Miranda, L. E. & W. D. Hubbard, 1994. Length-dependent winter survival and lipid composition of age-0 largemouth bass in Bay Springs Reservoir, Mississippi. Transactions of the American Fisheries Society 123: 80–87.

Niezgoda, G., M. C. Benfield, M. Sisak & P. Anson, 2002. Tracking acoustic transmitters by code division multiple access (CDMA) based telemetry. Hydrobiologia 483: 275–286.

Ovidio, M., J. C. Philippart & E. Baras, 2000. Methodological bias in home range and mobility estimates when locating radio-tagged trout, *Salmo trutta*, at different time intervals. Aquatic Living Resources 13: 449–454.

Pullis G. & A. Laughland, 1999. Black bass fishing in the U.S., U. S. Fish and Wildlife Service, Report 96–3. U. S. Fish and Wildlife Service, Washington, DC.

Raibley, P. T., K. S. Irons, T. M. O'Hara, K. D. Blodgett & R. E. Sparks, 1997. Winter habitats used by largemouth bass in the Illinois River, a large river-floodplain ecosystem. North American Journal of Fisheries Management 17: 401–412.

Rand, P. S. & S. G. Hinch, 1998. Swim speeds and energy use of river migrating adult sockeye salmon: simulating metabolic power and assessing risk of energy depletion. Canadian Journal of Fisheries and Aquatic Sciences 55: 1832–1841.

Reynolds, W. W. & M. E. Casterlin, 1978. Complementarity of thermoregulatory rhythms in *Micropterus salmoides* and *M. dolomieui*. Hydrobiologia 60: 89–91.

Savino, J. F. & R. A. Stein, 1982. Predator-prey interaction between largemouth bass and bluegills as influenced by simulated, submersed vegetation. Transactions of the American Fisheries Society 111: 255–266.

Sheldon, R. B & C. W. Boylen, 1977. Maximum depth inhabited by aquatic vascular plants. American Midland Naturalist 97: 248–254.

Sullivan, K. M., 1985. Physiology of feeding and starvation tolerance in overwintering freshwater fishes. Environmental Biology of Fishes 7: 259–268.

Suski, C. D., 2000. Linking reproduction with conservation for the largemouth bass (*Micropterus salmoides*) and smallmouth bass (*M. dolomieui*). M.S. Thesis, University of Illinois at Urbana Champaign, Urbana, IL.

Valley, R. D. & M. T. Bremigan, 2002. Effects of macrophyte bed architecture on largemouth bass foraging: implications of exotic macrophyte invasions. Transactions of the American Fisheries Society 131: 234–244.

Hydrobiologia (2007) 582:257–269
DOI 10.1007/s10750-006-0560-y

FISH TELEMETRY

Techniques for evaluating the spatial behaviour of river fish

Kathy H. Hodder · Jerome E. G. Masters · William R. C. Beaumont · Rodolphe E. Gozlan · Adrian C. Pinder · Carolyn M. Knight · Robert E. Kenward

Abstract Radio-tagging is widely used for studies of movements, resource use and demography of land vertebrates, with potential to combine such data for predictive modelling of populations from individuals. Such modelling requires standard measures of individual space use, for combination with data on resources, survival, dispersal and breeding. This paper describes how protocols for efficient collection of space-use data can be developed during a pilot study, and reviews the ways in which such data can be used for space-use indices that help answer biological questions, with examples from a study of riverine pike (*Esox lucius*). Analyses of diurnal activity and spatio-temporal correlation were used to assess when to record locations, and analyses of home range increments were used to define the number of location records necessary to assess seasonal ranges. We stress the importance of developing protocols that use minimal numbers of locations from each individual, so that analyses can be based on samples of many individuals. The efficacy of link-distance (e.g. cluster analysis) and location density (e.g. contouring) techniques for spatial analysis for river fish were compared, and the utility of clipping off areas to river banks was assessed. In addition, a new automated analysis was used to estimate distances along river mid-lines. These techniques made it possible to quantify interactions between individuals and their habitat: including a significant increase in core range size during floods, significant preference for deep pools, and a lack of exclusive territories.

Keywords Home range · Core range · Radio-tracking · Cluster polygon · Pike

Guest editors: P. R. Almeida, B. R. Quintella, M. J. Costa and A. Moore
Developments in Fish Telemetry

K. H. Hodder (✉) · J. E. G. Masters · W. R. C. Beaumont · R. E. Gozlan · A. C. Pinder · C. M. Knight · R. E. Kenward
Centre for Ecology and Hydrology, Winfrith Technology Centre, Winfrith Newburgh, Dorchester, Dorset DT2 8ZD, UK
e-mail: khh@ceh.ac.uk

C. M. Knight
School of Biological and Biomedical Sciences, Science Laboratories, University of Durham, South Road, Durham DH1 3LE, UK

Introduction

Radio tagging has been used in the study of home ranges of river fish for several decades, but there is little guidance available for efficient data collection and analysis in the lotic environment. At the outset of a project to study spatial behaviour, decisions need to be made about the sample size of fish, the number of locations to record for each individual, the timing of recording and the time interval between location records. In this paper, we show how initial tracking sessions can be designed as a pilot study in order to determine a tracking protocol appropriate for providing answers to the biological questions of interest. The pilot study can determine the number of location records required to describe a stable home range, an issue that has rarely been addressed in studies of river fish (exceptions are Natsumeda, 1998; Snedden et al., 1999). It can also be used to establish the optimal sampling interval (Lucas & Batley, 1996; Baras, 1998; Ovidio et al., 2000), in part by determining the degree of autocorrelation between location records (Chapman & Mackay, 1984). Here we approach these issues with a number of methods including application of the test of Time To Independence (TTI) of locations (Swihart & Slade, 1985) to river fish for the first time.

In spatial analyses of river fish behaviour, the home range has often been defined using the range span, measured between maximal upstream and downstream locations, and expressing range area as the longitudinal displacement multiplied by mean stream width (Minns, 1996; Huber & Kirchhofer, 1998; Vokoun, 2003). This approach gives a good indication of the overall area available to the fish, but oversimplifies understanding of space use. The total area used may be disproportionately influenced by the inclusion of just one excursive location and greatly overestimate the smaller areas that are favoured by fish, as by other animals, for most of their activities. This overestimation of space use has the potential to give spurious or misleading results in assessments of habitat association or of predatory and social interaction between individuals. More detailed techniques for analysis of range structure have been developed in studies of terrestrial animals (White & Garrot, 1990; Kenward, 2001), such as contouring methods (Dixon & Chapman, 1980; Worton, 1987, 1989) and cluster polygons (Kenward, 1987). These methods describe internal range structure by excluding outlying locations to give a core range, which may have one or more nuclei (activity centres). This paper explores the utility of these methods for analysing the ranges of river fish using location data for northern pike *Esox lucius* (L.). Can these methods, originally designed for studying animals in non-linear environments, adequately represent the size, shape and structure of home ranges in rivers?

Materials and methods

Study site

Pike were tracked in the River Frome, a chalk stream in Dorset, UK. The study area (NGR 367863, 382870) included a 2 km stretch of the main channel with a mean (±se) stream width of 14.2 (±0.6) m, a millstream (1.2 km), and artificial ditches (Fig. 1). It was selected on the basis of a pilot study (see below) to give the optimum trade-off between practical constraints (terrain, available trackers, distance between tagged fish) and aims (i) to maximise the sample size of individual fish while (ii) also representing different channel characteristics (straight and meandering). The latter aim provided the potential to test how home range estimators functioned in different linear environments. A weir formed a potential barrier (depending on water levels) at the upstream limit, whereas there were no barriers to fish movement at the downstream limit. The ditches were seasonally inundated, and adjacent water meadows were flooded during one of two winter seasons. Streambed topography was mapped along 1108 m of the main channel by measuring depths on transects at 5 m intervals along the channel, and at 2 m intervals across it.

Radio tagging and tracking

Twenty-seven pike with a mean fork length of 69 cm (range 52–95 cm) were implanted with Biotrack TW-5 radio tags (length 8 cm, diameter 1.6 cm, weight 22 g in air, 7 g in water), released into the river and allowed 10 days to recover prior to the commencement of data collection (Beaumont et al., 2002; Jepsen et al., 2000). Location records for the pike were determined from the riverbank, using Yaesu FT290 receivers and 3 element Yagi antennae, by triangulation and detuning the radio receiver. Blind experiments with tags in known positions indicated that a location resolution of 1–2 m^2 could be achieved with this method.

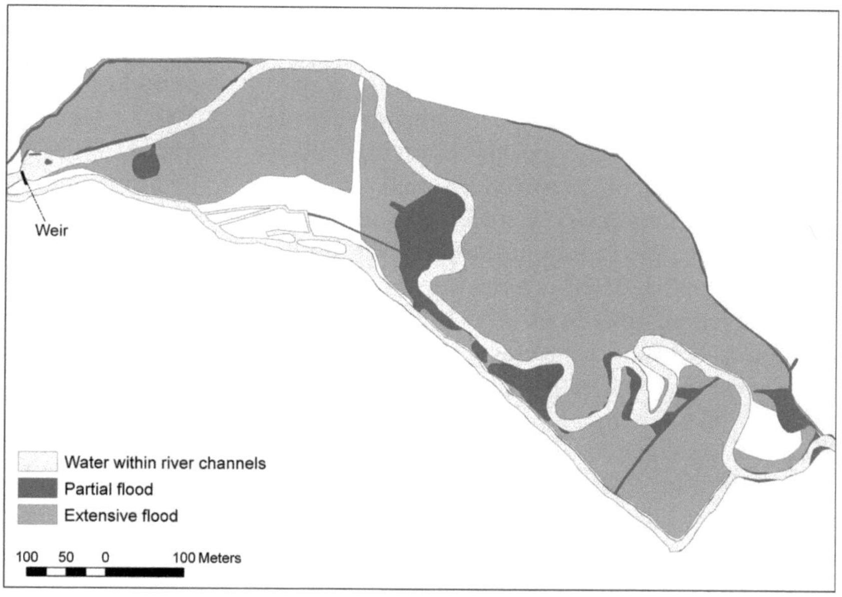

Fig. 1 The extent of area that was available to pike in the study area on the River Frome, at three different water levels

Development of tracking protocol

The protocol for estimation of pike home ranges was developed through a pilot tracking study at the very start of the project. Whilst in theory the greater the number of fish tracked in a pilot study, the greater the confidence in the results, the practicalities of capturing the fish and tracking them at short time intervals resulted in pilot data being available for only the first two pike tagged (body lengths 86 cm (A) and 71 cm (B)). Location data for these fish were recorded in a 300 m stretch of river at hourly intervals over five consecutive 24-h periods in June 2000. The aim was to determine the minimum number of location records required to estimate a home range, the optimum sampling interval, and the most appropriate time(s) of day for sampling. A minimum number of locations was required to allow simultaneous tracking of as many individuals as possible, and also estimation of ranges within relatively short time periods, avoiding range shifts following events such as floods (Masters et al., 2002), or seasonal migrations such as spawning.

Sampling interval for location records

Locations that are recorded every minute, for a fish that moves at 10 m a minute, would not be expected to be more than 10 m apart after the first minute, 20 m after the second minute etc. In other words they are spatio-temporally correlated. However, when animals move back and forth through their ranges, eventually a time interval will be reached when the first location cannot predict where the second will be; this is the TTI. A method for determining TTI between consecutive locations was developed by Swihart & Slade (1985). The procedure examines the way that the distance between location records changes with sampling intervals using Schoener's index, $V = t^2/r^2$, where t^2 is the mean squared distance between consecutive location records and r^2 is the mean squared distance from each location to the range centre (the arithmetic mean of all coordinates). TTI between locations is indicated when the first of three successive time intervals exceed $V = 2$. This is roughly equivalent to the time required for an animal to traverse its whole range (Swihart & Slade, 1985). In practice the time required to traverse half the range width ($V = 1$) may be more useful (Kenward, 2001) so the outcomes of using both $V = 1$ and $V = 2$ were examined for the pilot tracking data.

Diel activity patterns

Many animals timetable their activities; for example, they may roost in one place and move to forage in another during each day (e.g. Clough & Ladle, 1997). Such habitual behaviour may be indicated in autocorrelation analysis (see above). If an animal tended to be in the same place at the same time each day, serial correlation of its locations would be expected to peak at about 24 h. In order to design a tracking protocol that would adequately represent the locations used by the fish, more detailed information was required about levels of activity at different times of day. Distances between consecutive location records were used to investigate whether the mobility of the two pike varied on a diel basis and provide guidance on the most appropriate times of day to track the pike. The diel period was split into dawn, day, dusk and night for the analysis; crepuscular periods were defined as dawn, 1 h before and 2 h after sunrise, and dusk, 2 h before and 1 h after sunset. The hours before sunrise and after sunset were included so that first and last light would fall within the crepuscular period.

The number of location records required

It is always necessary to balance the need for sufficient locations for each individual with the need to collect data on an adequate sample of animals. So to give time for data collection on enough individuals, it is useful to estimate the minimum number of location records that are required to calculate stable home ranges for individuals in the study population. In other words, within a predefined time period (e.g. season) when does the range area cease to increase as further location records are added? The change in range area with addition of successive fixes was determined for the two original pike plus one other pike that was tagged part way through the pilot study. The channel used by the three pike was relatively straight, so it was possible to use the outer minimum convex polygon (Mohr, 1947) to delineate the ranges in this analysis without excessive expansion of the range outside of the channel (see discussion on 'clipping' range areas below).

Estimation of pike home ranges

Definition of a home range has progressed from an area traversed during the 'normal' activities of an animal (Burt, 1943) to a concept that refers to an area repeatedly traversed within a specific period of time (Kenward, 2001; Kernohan et al., 2001). This was described by Doncaster & Macdonald (1991) as the 'prevailing range', which may shift with season, life history, environmental or demographic changes. Delineation of this short-term range enables quantitative comparisons between animal categories, and investigation of habitat utilisation and interactions, both intra- and interspecific (depending on the animals tagged). There are numerous statistical techniques available for home range analysis, which can provide estimates not only of the total area used but also the structure, in terms of core and excursive areas, and the range shape. Realistic representation of the home range shape is important if spatial coincidence with habitat features or other individuals is of interest. Detailed discussion of home range estimators and their suitability for different types of data or study question are reviewed by Kenward (1992, 2001) and White & Garrot (1990). In this study, a sample of the most promising methods were investigated to determine their efficacy for analysis of animal locations in a linear environment, and also for comparison between seasons for a river that habitually floods—leading to a less restricted available area for the fish. Home ranges estimated by the distance (or area) between furthest upstream and downstream locations were compared with ranges estimated with cluster polygons and by a kernel contouring method.

A home range tracking protocol was finalised using the data collected during the pilot track (see results) and eight tracking sessions were conducted between July 2000 and March 2002 (Table 1). Over this period, sufficient data were collected on 20 pike to investigate the extent and structure of 68 seasonal ranges.

Outer ranges

The total, or outer, home ranges were assessed using two linear techniques and one contouring method:

(i) a linear range span—the distance along the midline of the channel between the maximal

Table 1 Dates and flow conditions of the radio-tracking sessions used to establish pike home ranges

Date	N pike	Flow conditions
July 2000	3	Within the channel
September 2000	3	Within the channel
December 2000	5	Extensive flood
March 2001	10	Partial flood
July 2001	9	Within the channel
September 2001	12	Within the channel
December 2001	12	Within the channel
March 2002	14	Within the channel

upstream and downstream location records was estimated using a modified version of RANGES V (Kenward & Hodder, 1996). The area of the channel between these two locations (range span area, see Fig. 2), was also calculated from maps. There were slight variations in channel width in the study area; therefore this method was more accurate than multiplying the range span by mean channel width. The range span measure can provide results that are comparable with linear home ranges from other studies e.g. (Huber & Kirchhofer, 1998; Ovidio et al., 2002; Khan et al., 2004), and the range span area gives a measure more comparable with the home range areas derived from outlines based on contours or polygons.

(ii) Kernel contours (Worton, 1989) were selected for estimating outer ranges from the density distribution of locations; this approach has previously been applied to linear ranges by (Blundell et al., 2001). The K_{100} contour was used; this just includes all the locations, differing from the minimum convex polygon (Mohr, 1947) in that it can identify separate activity nuclei.

Kernel contours (K_{100}) that used the default reference smoothing parameter tended to expand beyond the river channel (Fig. 2); therefore, the areas were corrected by clipping off regions outside of the river (Mesing & Wicker, 1986; Allouche et al., 1999). Three maps were required for this purpose, delineating different water levels during the study: within-channel, partial flood and extensive flood (Fig. 1). These three boundary maps were used to clip off parts of the range area of the K_{100} estimator, which made it possible to compare ranges at the different water levels, whereas the range span could only be calculated when the river was within its banks.

The expansion of the kernel ranges beyond the channel made it inevitable that area estimates during floods would be larger than those recorded in lower water conditions, even if fish locations

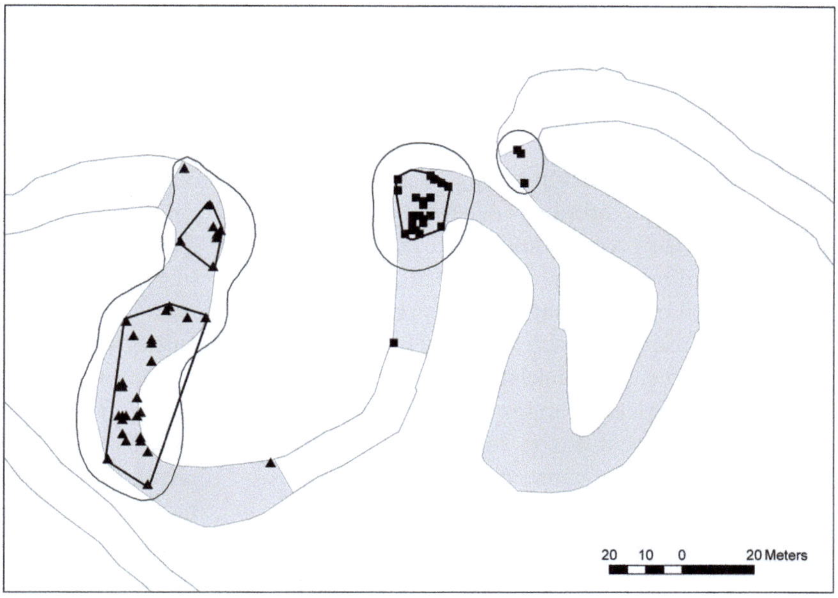

Fig. 2 Home range estimates for two pike (*E. lucius*) J (■) and P (▲) (fork lengths 93 cm and 58 cm respectively). Shaded areas represent the range span between the upstream and downstream limits; the outer contours show the K_{100} range edges; and the bold lines the C_{tx} range edges

were identical at the two water levels. In order to check for Type I statistical errors associated with this problem, range areas were estimated for 'in channel' conditions using both within-channel and extensive flood maps (Fig. 1).

Core ranges

Convex cluster polygons (Kenward, 1987) were used to investigate core areas of intensive use. The polygons are estimated round clusters of locations that represent an ascending sum of nearest-neighbour link distances. Core areas (C_{tx}) were estimated separately for each range, by excluding outlying location records through truncation of the upper 5% of the nearest neighbour distance distribution (Hodder et al., 1998; Kenward et al., 2001). Generally, only small areas of the cluster polygons expanded beyond the river channel (Fig. 2), these areas were removed by clipping, as for the K_{100} ranges. The core range estimates were also tested for methodological bias relating to the effects of using the three different channel boundaries when clipping off parts that expanded beyond the river.

In addition to estimation of core areas, the C_{tx} ranges indicated the degree of home range fragmentation by the number of separate nuclei in each core range, and also by a partial area index. This index divides the areas in all the cluster polygons by the area in a single polygon round all the clusters, and therefore tends to zero, if the nuclei are far apart, and to one if all the locations are in one nucleus (Kenward et al., 2001). As the calculation of C_{tx} was unique for each range, the number of location records included in the core could vary. Many radio-tracking studies employ a standard core for all ranges, which ensures that the same percentage of locations is used for each range (Hodder et al., 1998). However, the standard core will overestimate core areas for ranges with unusually low percentages of locations in the C_{tx} range. This is particularly likely to cause problems when statistical comparisons with small sample sizes are required; therefore, the unique C_{tx} cores, and not standard cores are presented here.

Range overlap

The degree of spatial interaction between fish during each tracking session was addressed by measuring the percentage area of each pike's range (C_{tx}, K_{100} and Range Span) overlapped by other pike, and by counting the number of pike whose ranges overlapped each individual. Overlays of ranges were performed on range outlines that had been clipped.

Spatial analyses

Estimation of spatio-temporal correlation, home range outlines and their incremental increase with addition of locations used RANGES V (Kenward & Hodder, 1996). Estimation of range span, the clipping of range outlines and their overlap initially used ArcInfo 8.1 (ESRI, Redlands, USA) but is now implemented in RANGES 6 (Anatrack, Wareham BH20 5AX, UK). All statistical tests were performed with Minitab 13, and all *t*-tests were conducted with degrees of freedom adjusted for differing sample variances. In statistical comparisons (e.g. of range area in different flow conditions), the pike, and not the range, was use as the sample unit. Where necessary average values were found from multiple ranges for the same individual.

Results

Development of the tracking protocol

Sampling interval for location records

The autocorrelation analysis provided information about TTI between consecutive locations, and also timetabling of daily activities. When values of Schoener's Index (V) were required to exceed two (equivalent to traversing the range span), TTI between consecutive locations was 11 h for pike A and 6 h for pike B. When V was required to exceed one (equivalent to crossing half the range span), TTI was 4 h for pike A and 3 h for pike B (Fig. 3). The autocorrelation plots (Fig. 3) of the two fish also indicated that spatial

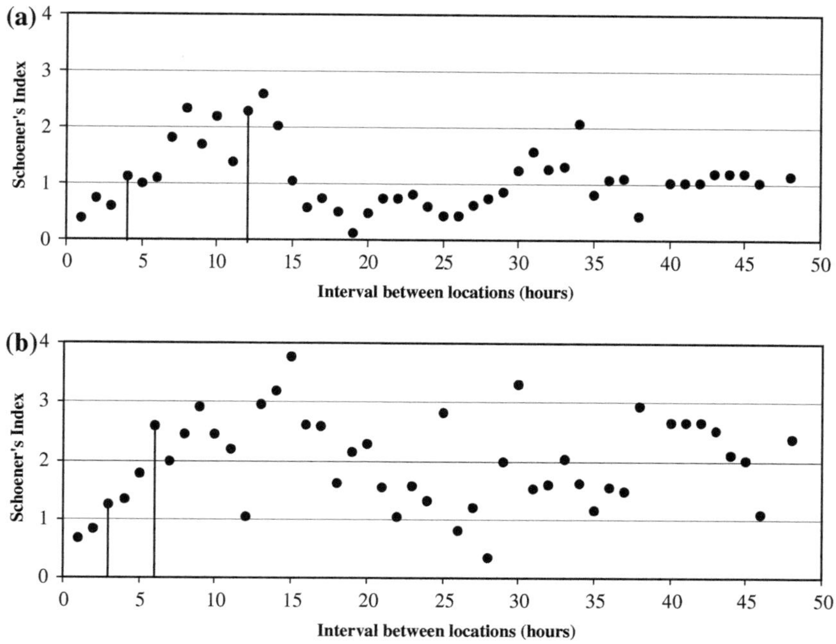

Fig. 3 Autocorrelation analysis of location data for two pike (**a, b**). Vertical lines indicate the Time To Independence (TTI) as the first of three consecutive points that exceed the critical value of Schoener's Index (V), set as $V = 1$ or $V = 2$

dependence of the location records may be related to diel behaviour patterns: V was greatest at an interval of about 15 h between fixes. However, at intervals of 20–25 h the value of V became very small again, indicating that fixes at this time interval were spatially close. Therefore, one record a day, taken at the same time, would be unlikely to describe the whole range of the fish.

Diel activity patterns

A Kruskal–Wallis test using distances between consecutive locations showed that there were significant differences between the levels of mobility recorded in the four stages of the diurnal cycle ($H = 20.6$, d.f. = 3, $P < 0.001$). The pike moved about more during crepuscular periods than during the day and night. Median distances moved were 13 m (Q1-3 = 1–37) at dawn and 13 m (Q1-3 = 0–12) at dusk, compared with 4 m (Q1–3 = 3–33) during the day, and 1 m (Q1–3 = 0–9) at night. It was therefore important to schedule tracking so that a good proportion of locations would be recorded during the active periods. During the summer months in temperate latitudes, tracking during normal office hours would underestimate the mobility of this species, and also the area used.

The number of location records required

After the 5-day pilot track, although 120 fixes had been collected for each fish, incremental analysis showed that their range areas were still increasing, suggesting that data sampling should be extended over a longer period. Additional locations were recorded three times a day: at dawn, midday and dusk. This provided records at the periods of greatest mobility, and also gave the maximum sampling interval possible during daylight hours (approximately 6 h) to minimise the spatial dependence between records. A second incremental analysis for three pike (A and B plus C), with locations from the initial hourly tracking sub-sampled to the same three times a day, showed that the range size stabilised after around 40 location records collected over a period of 13 days (Fig. 4).

Finalising the tracking protocol

For this study, three location records a day kept spatial autocorrelation to a minimum, allowed records to be made during the most active times of day and ensured that adequate data for a stable home range should be attainable after 13 days tracking (39 locations for each fish). This

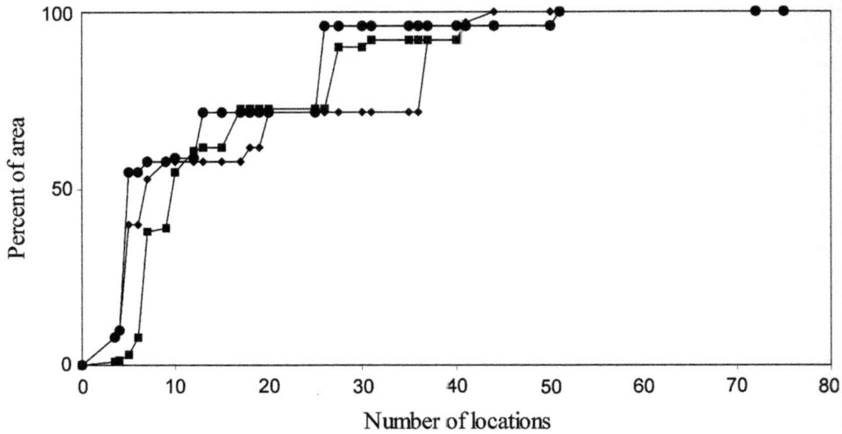

Fig. 4 Incremental analysis of the percentage of the maximum range area (outer minimum convex polygon) against consecutive locations for three pike (*E. lucius*)

maximised the opportunity to complete tracking sessions during short-term events such as floods.

Structure of pike home ranges

Data collected during the eight tracking sessions, using the protocol described above, showed that there were considerable differences between the core ($C_{tx,}$) and outer range (Range Span, K_{100}) areas (Table 2, Fig. 2). This indicated that the pike concentrated their activity in range cores, with occasional forays to locations that could be a great distance from the core.

This range structure was found in 65 of the 68 ranges, but three ranges, for two pike, had a smaller proportion of locations in the core, indicating greater mobility (Fig. 5). Excluding these three mobile ranges, a mean (±se) of 90% (±0.5) of location records were included in the core ranges.

The core areas tended to be located around pools in the channel. Streambed topography data was available for 16 ranges recorded for nine pike, and areas defined by the C_{tx} ranges for this sample were significantly deeper (mean ± se 16 ± 5 cm difference) than the area in the range span (paired $t = 3.35$, $n = 9$, $P = 0.01$).

Range fragmentation

The internal structure of the pike home ranges was investigated with the C_{tx} core ranges. These are represented by multiple nuclei where there is

Table 2 Range areas (m^2) and span distance (m) for pike in the River Frome during different water levels

Water level	Range estimator	N pike	N ranges	Median	Min	Max	Number of nuclei			Partial area index		
							Median	Min	Max	Median	Min	Max
Within-Channel	C_{tx}	19	53	515	51	3,000	2	1	6	0.40	0.05	1.00
	K_{100}	19	53	2769	127	14,602						
	Range span area	19	53	2580	171	11,231						
	Range span	19	53	205	15	899						
Partial Flood	C_{tx}	10	10	636	89	1,440	3	1	5	0.13	0.01	1.00
	K_{100}	10	10	6499	631	31,523						
Extensive Flood	C_{tx}	5	5	6174	346	8,318	3	2	4	0.26	0.04	0.56
	K_{100}	5	5	43012	14,452	63,646						

Range estimators included convex cluster polygons with outlier exclusion by truncation (C_{tx}), kernel contours around all of the locations (K_{100}), and the area of channel between the maximum upstream and downstream locations (Range Span). Range fragmentation statistics are given for core ranges estimated by convex cluster polygons (C_{tx}). A partial area index value of one indicates a single nucleus, with the value becoming smaller as multiple nuclei are more widely separated

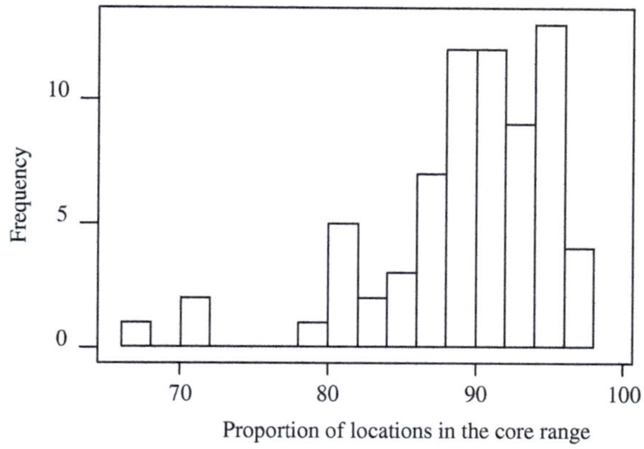

Fig. 5 The proportion of location records remaining in the C_{tx} cores of 68 pike ranges after removal of outlying locations

more than one area of intensive use in the range (Fig. 2). Whilst C_{tx} ranges had a median of two activity nuclei, there was great variation in the results (Table 2). Of the 53 ranges recorded when the river was not flooded, 11 were mononuclear, but all others had several nuclei. A wide range of partial area indices, including small values that indicated wide dispersion of nuclei, was recorded at all water levels (Table 2). When the water was within the channel, the activity centres for a single range were separated by up to 520 m. The maximum number of C_{tx} nuclei (6) was recorded when the water remained within the channel, however the sample size of ranges was far greater than at other times, and *t*-tests indicated that there were no significant differences in fragmentation between ranges recorded at different water levels ($P > 0.05$).

Area of home ranges

The area of the outer home ranges was best estimated for ranges within the channel by the range span. Although the K_{100} contours conformed well to some ranges (for instance Pike P, Fig. 2), in other cases, particularly in meandering stretches of river, large parts of the overall range were excluded (for instance Pike J, Fig. 2). The extension of K_{100} range edges beyond the channel was also problematic. The K_{100} area estimates for within-channel conditions were significantly larger when the map for the extensive flood was used to clip the range boundaries ($t = 4.5$, $n = 59$, $P < 0.001$). To avoid this methodological bias, the comparison of range areas at different water levels was repeated using the river outlines during the extensive flood for all ranges (K_{100a} Table 3). This adjustment overestimated the range areas for within-channel conditions, but ensured that any significant increase in range area during the floods would not be due to the effects of clipping. There was generally only minimal extension of C_{tx} ranges beyond the river channel, and *t*-tests indicated there was no significant increase in range area due to methodological bias.

Environmental influences on range area

Both core (C_{tx}) and outer (K_{100}) range areas were greatest during extensive flooding: the range areas of 12 pike in December, in within-channel conditions, were significantly smaller than the ranges of five pike recorded during extensive floods the previous December (Table 3). Although the median ranges during partial flooding were larger than the within-channel ranges (Table 2), there were no significant differences (Table 3). During the six tracks carried out when the water remained within the channel, discharge (recorded at the weir, Fig. 1) varied between 2 and 5.3 $m^3 s^{-1}$, with an associated 20 cm difference in river height, but there were no correlations between discharge and range areas ($r < 0.75$, $P > 0.08$).

Table 3 Comparison of range areas estimated at different water levels

	Range estimator	d.f.	*t*	*P*
In channel vs. extensive flood	C_{tx}	4	2.9	0.04
	K_{100}	4	4.4	0.01
	K_{100a}	8	2.2	0.05
In channel vs. partial flood	C_{tx}	21	–0.5	NS
	K_{100}	11	1.5	NS

Areas were estimated with convex cluster polygons (C_{tx}), kernel contours (K_{100}), and kernel contours allowing for bias (K_{100a} where the extensive flood boundary was used for estimates at the two water levels) The degrees of freedom in *t*-tests were adjusted to differing variances between samples

Individual variation

During the 20-month study period there was considerable variation between the range areas recorded for individual pike. This variation was found even when the data considered were limited to similar (i.e. within-channel) conditions. Data from nine pike, which all had at least four separate range estimates, showed that the variation between ranges collected for the same individual was very high in the outer ranges (mean CV (±se) for nine pike: K_{100} = 71% (±8), Range Span = 61% (±7)). The mean differences between largest and smallest outer range areas for a single pike was 5,193 m^2 (Range Span) and 5,998 m^2 (K_{100}), and the maximum difference was 9,429 m^2 (Range Span) 10,039 m^2 (K_{100}). The C_{tx} core range areas were also variable; (mean CV (±se) for nine pike 55% (±5)) with a maximum difference between largest and smallest areas for a single pike of 1,301 m^2.

Range overlap

There was a high degree of overlap, often with the entire outer range overlapped by other pike (Table 4). The median percentage overlap was smaller for the core ranges, but even these were sometimes completely overlapped. Medians of two or three pike overlapped their outer ranges, and there was a median of one overlap in the cores, but up to five other individuals could be found within an individual's C_{tx} core range (Table 4). The results were not suitable for analysis of seasonal differences in overlap, or variations related to river discharge, because different numbers of pike (3–14), at different distances apart, were tracked during the eight sessions. The results are also likely to underestimate the degree of overlap between the fish because other non-tagged pike may have been present. However, it was possible to demonstrate that areas where more C_{tx} ranges overlapped tended to be deeper (Pearson's correlation, $r = 0.99$, d.f. = 3, $P < 0.01$, Fig. 6). It was relatively unusual for the core ranges of more than two fish to overlap in the same position (Fig. 6), but the relationship was still significant when the highest overlap value (4 core ranges overlapping), which had a sample size of two, was excluded ($r = 0.99$, d.f. = 2, $P < 0.05$).

Table 4 Percentage overlaps and counts of overlaps between pike ranges that were estimated by three methods: convex cluster polygons with outlier exclusion by truncation (C_{tx}), kernel contours around all of the locations (K_{100}), and the area of channel between the maximum upstream and downstream locations (range span area)

	Range estimator	Median	Min	Max
Mean % overlap	K_{100}	83	0	100
	C_{tx}	31	0	100
	Range span area	100	0	100
Number of overlaps	K_{100}	3	0	9
	C_{tx}	1	0	5
	Range span area	2	0	7

Discussion

Manually tracking fish in the lotic environment has the potential to provide novel insights into fish behaviour and ecology, including habitat preferences, demography and territoriality. However, the collection of sufficient data to estimate home ranges can be time consuming, and hence has high associated staff costs. An intensive pilot study at the outset of a tracking project is therefore imperative to maximise the output from the effort applied, although logistical issues, such as the population distribution of the study species, and the nature of the terrain to be traversed by

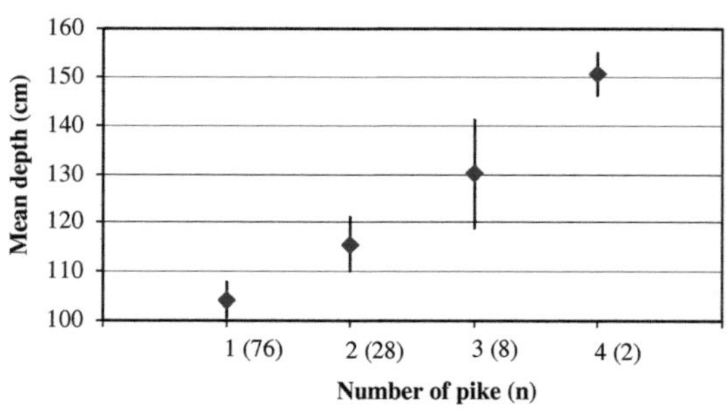

Fig. 6 The relationship between the degree of overlap between pike C_{tx} core ranges, and mean (±standard error) depth. On the *x*-axis, one pike indicates zero overlaps, two pike one overlap etc., and the number of records for each value is shown in brackets

the tracker, will inevitably impose constraints on the application of tracking protocol that is developed from the pilot work.

Both the tracking protocol and the data analyses will depend on the biological questions of interest, and various logistical or statistical constraints. In this study, the kernel contours were used for comparative purposes because the pike moved out into areas beyond the river channel during floods. Despite attempts to improve area estimation by kernels through clipping off areas not within the river, there were difficulties with this technique; namely (i) expanding *along* the channel beyond the area utilised by the fish on straight stretches, and (ii) omitting parts of ranges in meandering stretches (Fig. 2). Although the general rule would dictate that comparisons of range areas should only be made with results obtained from the same analysis technique, in this case it may be more meaningful to compare kernel contours in floods with range span areas at other times. The cluster technique for estimating core ranges; however, was much more promising as a tool for analyses that can include linear and open systems, hence lotic and lacustrine or marine environments. It does not expand along the channel, and showed a lower tendency to expand beyond the river channel, such that there was relatively little need to correct this latter inaccuracy by clipping.

The home range analysis showed that pike in the River Frome occupied stable home ranges that could be determined over a period of two weeks with the majority of individuals spending about 90% of their time in a core range with occasional excursions. These core ranges were generally located in deeper water than the outer ranges, where the pike would presumably experience slower flows, enabling them to reduce energy requirements. This behaviour is similar to that described for lake dwelling pike, which made many localised movements and fewer long distance movements (Diana, 1980; Lucas et al., 1991).

Both outer and core ranges of individual pike in the River Frome overlapped considerably, and the degree of overlap was necessarily an underestimate, given that it was not possible to ensure that all pike in the study area were tagged. The strong positive correlation between the number of pike found in an overlap area, and the mean depth of that area, showed that pike tended to congregate in the deeper pools, although it was not possible to show whether they occupied those areas at the same time. Comparative data on range overlap were not available from other studies of pike, but Larsen (1966) recorded up to four pike of similar size within 30–40 m stretches of a Danish trout stream, and overlap of home range 'activity centres' has been reported for lake dwelling muskellunge (Miller & Menzel, 1986). In this study, pike were also found very close together on several occasions, outside of spawning activity, even when they would have been clearly visible to one another. Ideally, analysis of dynamic interaction (Macdonald & Amlaner, 1980; Kenward et al., 1993) could have produced a 'social cohesion index' based on proximity of individuals. However, the data were not suitable for dynamic interaction analysis because the time taken to move between individuals during each tracking round meant that the requirement for virtually simultaneous location records could not be met. Despite this, the degree of overlap between pike in the River Frome clearly showed that they *did not hold exclusive territories* during this study.

The great variation in range size and structure suggests that these fish exhibit the high degree of behavioural versatility that might be expected of an opportunist predator (Bry, 1996), and has also been recorded for lake-dwelling pike (Jepsen et al., 2001). Large differences also occur between studies (reviewed in Lucas & Baras (2001)), some populations of pike in lakes occupying home ranges and others apparently ranging widely with no 'clear route'; while populations in brackish waters may display anadromous migration patterns (Muller, 1986).

Where an animal remains in a given area for sufficient time, perhaps a season or a life stage, delineation of its stable short-term home range is a useful analysis tool, which has been applied extensively in terrestrial studies over many decades (e.g. Burt, 1943). For the pike in this study, home range analysis showed that the activity of most individuals was concentrated in multinuclear core areas, situated in deeper areas of the channel, but that some individuals may be more mobile. When the floodplains became available to the pike, knowledge of home range structure made it possible to demonstrate that the pike expanded their *core ranges*, as opposed to making occasional forays into their extended habitat. The definition of range structure also provided a means to investigate and quantify the proximity of neighbouring pike and to demonstrate a lack of exclusive territories. So, despite the need for clipping of ranges, which was imposed by the linear habitat, the home range analysis techniques were able to provide quantitative insights into the behaviour and ecology of a river-fish.

Acknowledgements This work could not have been completed without the dedication of the field trackers: Chris Flower, Bethan Lewis, Neasa McDonnell, Luke Scott and Sarah Venis. John Cooper processed the depth data. We are also grateful to Mike Ladle and Stewart Welton for advice and practical help, to local landowners for allowing access to the river, and to the Environment Agency for providing river discharge data. Jerome Masters was in receipt of a Freshwater Biological Association Frost Scholarship. The manuscript was improved by incorporation of constructive comments from Eva B. Thorstad and an anonymous referee.

References

Allouche, S., A. Thevenet & P. Gaudin, 1999. Habitat use by chub *Leuciscus cephalus* L. 1766 in a large river, the French Upper Rhone, as determined by radiotelemetry. Archiv fur Hydrobiologie 145: 219–236.

Baras, E., 1998. Selection of optimal positioning intervals in fish tracking: an experimental study on *Barbus barbus*. Hydrobiologia 372: 19–28.

Beaumont, W. R. C., B. Cresswell, K. H. Hodder, J. E. G. Masters & J. S. Welton, 2002. A simple activity monitoring radio tag for fish. Hydrobiologia 483: 219–224.

Blundell, G. M., J. A. K. Maier & E. M. Debevec, 2001. Linear home ranges: effects of smoothing, sample size, and autocorrelation on kernel estimates. Ecological Monographs 71: 469–489.

Bry, C., 1996. Role of vegetation in the life cycle of pike. In Craig, J. F. (ed.), Pike: Biology and Exploitation. Chapman and Hall, London, 45–67.

Burt, W. H., 1943. Territoriality and home range concepts as applies to mammals. Journal of Mammalogy 24: 346–352.

Chapman, C. A. & W. C. Mackay, 1984. Versatility in habitat use by a top aquatic predator, *Esox lucius* L. Journal of Fish Biology 25: 109–115.

Clough, S. & M. Ladle, 1997. Diel migration and site fidelity in a stream-dwelling cyprinid, *Leuciscus leuciscus*. Journal of Fish Biology 50: 1117–1119.

Diana, J. S., 1980. Diel activity pattern and swimming speeds of Northern Pike *Esox lucius* in Lac Ste. Anne, Alberta. Canadian Journal of Fisheries and Aquatic Sciences 37: 1454–1459.

Dixon, K. R. & J. A. Chapman, 1980. Harmonic mean measure of animal activity measures. Ecology 61: 1040–1044.

Doncaster, C. P. & D. W. Macdonald, 1991. Drifting territorialty in the red fox *Vulpes vulpes*. Journal of Animal Ecology 60: 423–439.

Hodder, K. H., R. E. Kenward, S. S. Walls & R. T. Clarke, 1998. Estimating core ranges: a comparison of techniques using the common buzzard *Buteo buteo*. Journal of Raptor Research 32: 82–89.

Huber, M. & A. Kirchhofer, 1998. Radio telemetry as a tool to study nase *Chondrostomata nasus* L. in medium sized rivers. Hydrobiologia 372: 309–319.

Jepsen, N., S. Beck, C. Skov & A. Koed, 2001. Behavior of pike (*Esox lucius* L.) >50 cm in a turbid reservoir and in a clearwater lake. Ecology of Freshwater Fish 10: 26–34.

Jepsen, N., S. Pedersen & E. Thorstad, 2000. Behavioural interactions between prey (trout smolts) and predators (pike and pikeperch) in an impounded river. Regulated Rivers Research & Management 16: 189–198.

Kenward, R. E., 1987. Wildlife radio tagging-equipment, field techniques and data analysis. Academic Press, London.

Kenward, R. E., 1992. Quantity versus quality: programmed collection and analysis of radio-tracking data. In Priede, I. G. & S. M. Swift (eds), Wildlife Telemetry: Remote Monitoring and Tracking of Animals. Ellis Horwood, London, 231–244.

Kenward, R. E., 2001. A Manual for Wildlife Radio Tagging. Academic Press, London.

Kenward, R. E., R. T. Clarke, K. H. Hodder & S. S. Walls, 2001. Density and linkage estimators of home range: nearest-neighbor clustering defines multinuclear cores. Ecology 82: 1905–1920.

Kenward, R. E. & K. H. Hodder, 1996. RANGES V: An Analysis System for Biological Location Data. Institute of Terrestrial Ecology, Dorset, UK.

Kenward, R. E., V. Marcstrom & M. Karlbom, 1993. Post-nestling behaviour in goshawks, *Accipiter gentilis*: sex differences in sociality and nest switching. Animal Behaviour 46: 371–378.

Kernohan, B. J., R. A. Gitzen & J. J. Millspaugh, 2001. Analysis of animal space use and movements. In Millspaugh, J. J. & J. M. Marzluff (eds), Radio Tracking and Animal Populations. Academic Press, San Diego, California.

Khan, M. T., T. A. Khan & M. E. Wilson, 2004. Habitat use and movement of river blackfish (*Gadopsis marmoratus* R.) in a highly modified Victorian stream, Australia. Ecology of Freshwater Fish 13: 285–293.

Larsen, K., 1966. Studies on the biology of Danish stream fishes. II. The food of Pike (*Esox lucius* L.) in trout

streams. Meddelelser fra Danmarks Fiskeri- og Havundersogelser 4: 271–326.
Lucas, M. C. & E. Baras, 2001. Migration of Freshwater Fishes. Blackwell Science Ltd, Oxford.
Lucas, M. C. & E. Batley, 1996. Seasonal movements and behaviour of adult barbel *Barbus barbus*, a riverine cyprinid fish: Implications for river management. Journal of Applied Ecology 33: 1345–1358.
Lucas, M. C., I. G. Priede, J. D. Armstrong, A. N. Z. Gindy & L. DeVera, 1991. Direct measurements of metabolism, activity and feeding behaviour of pike, *Esox lucius* L. in the wild, by the use of heart rate telemetry. Journal of Fish Biology 39: 325–345.
Macdonald, D. W. & C. J. Amlaner, 1980. A practical guide to radio-tracking. In Amlaner, C. J. & D. W. Macdonald (eds), A Handbook on Biotelemetry and Radio-Tracking. Pergamon Press, Oxford, 143–159.
Masters, J. E. G., J. S. Welton, W. R. C. Beaumont, K. H. Hodder, A. C. Pinder, R. E. Gozlan & M. Ladle, 2002. Habitat utilisation by pike *Esox lucius* L. during winter floods in a southern English chalk river. Hydrobiologia 483: 185–191.
Mesing, C. & A. Wicker, 1986. Home range, spawning migrations, and homing of radio-tagged Florida largemouth bass in two central Florida lakes. Transactions of the American Fisheries Society 115: 286–295.
Miller, M. & B. Menzel, 1986. Movements, homing and home range of muskellunge, *Esox masquinongy*, in West Okoboji Lake, Iowa. Environmental Biology of Fishes 16: 243–255.
Minns, C. K., 1996. Allometry of home range size in lake and river fish. Canadian Journal of Fisheries and Aquatic Sciences 52: 1499–1508.
Mohr, C. O., 1947. Table of equivalent populations of North American small mammals. American Midland Naturalist 37: 223–249.
Muller, K., 1986. Seasonal anadromous migration of the pike (*Esox lucius* L.) in coastal areas of the northern Bothnian sea. Archiv fur Hydrobiologie 107: 315–330.
Natsumeda, T., 1998. Home range of the Japanese fluvial sculpin *Cottus pollux* in relation to nocturnal activity patterns. Environmental Biology of Fishes 53: 295–301.
Ovidio, M., E. Baras, D. Goffaux, F. Giroux & J. C. Philippart, 2002. Seasonal variations of activity pattern of brown trout (*Salmo trutta*) in a small stream, as determined by radio-telemetry. Hydrobiologia 470: 195–202.
Ovidio, M., J. C. Philippart & E. Baras, 2000. Methodological bias in home range and mobility estimates when locating radio-tagged trout, *Salmo trutta*, at different time intervals. Aquatic Living Resources 13: 449–454.
Snedden, G. A., W. E. Kelso & D. A. Rutherford, 1999. Diel and seasonal patterns of spotted gar movement and habitat use in the lower Atchafalaya River Basin, Louisiana. Transactions of the American Fisheries Society 128: 144–154.
Swihart, R. & N. Slade, 1985. Testing for independence of observations in animal movements. Ecology 66: 1176–1184.
Vokoun, J. C., 2003. Kernel density estimates of linear home ranges for stream fishes: advantages and data requirements. North American Journal of Fisheries Management 23: 1020–1029.
White, G. C. & R. A. Garrot, 1990. Analysis of Wildlife Radio Tracking Data. Academic Press, San Diego.
Worton, B. J., 1987. A review of some models of home range for animal movement. Ecological Modelling 38: 277–298.
Worton, B. J., 1989. Kernel methods for estimating the utilisation distribution in home range studies. Ecology 70: 164–168.

Hydrobiologia (2007) 582:271–280
DOI 10.1007/s10750-006-0538-9

FISH TELEMETRY

Use of muscle activity indices as a relative measure of well-being in cultured sea bass *Dicentrarchus labrax* (Linnaeus, 1758)

G. Lembo · P. Carbonara · M. Scolamacchia · M. T. Spedicato · R. S. McKinley

Abstract Aquaculture of sea bass is widely spread in the Mediterranean and employs a variety of husbandry protocols that need to be evaluated in terms of fish well-being. Behavioural tests can be used as operational indicators of short-term stress, because changes in swimming performance and/or muscle activity (e.g. electromyograms) can be interpreted as response associated with a wide variety of stressors. Diagnostic procedures, based on physiological telemetry, will thus enable appropriate mitigative strategies to be implemented to ensure the well-being of cultured fish. The objective of this study was to examine the contribution of two muscle types to the swimming activity of sea bass (*Dicentrarchus labrax*, L. 1758). Hard-wire technology was used to 'fine-tune' measures obtained using a physiological transmitter. Fine-tuning showed that the aerobic muscle displayed an increase in recruitment of muscle fibres relative to increasing swimming speed up to 0.6–0.7 ms^{-1} of the U_{crit}, where the anaerobic (white muscle) activity started to exponentially augment with swimming speed, reaching up to ~7 times its initial value. Intensity of electromyogram signals were described by logarithmic (red muscle): $y = 0.5922Ln(x) + 1.2251$ ($R^2 = 0.9906$) and exponential (white muscle): $y = 0.0977e^{2.4723x}$ ($R^2 = 0.9845$) relationships. Fine-tuning indicated that the two muscle types in the sea bass are involved in fuelling swimming activity below the U_{crit}. Thus, scope for activity is not supported solely using aerobic metabolism, though the red muscle powers the majority of the swimming ability. Measurement of U_{crit} displayed an average value of 3.43 BLsec^{-1} (SE = 0.12). Associated EMG values measured during the forced swim trials using an implanted bio-sensitive radio transmitter showed that EMG's intensity increased, on average, 3.2 times between 0.2 msec^{-1} and the U_{crit} velocity (~1 msec^{-1}). Above EMG values were fine-tuned using estimates obtained from direct monitoring of the red and white musculature. Overall, the results demonstrated that the scope for activity, previously thought to represent only aerobic metabolism, is composed of both aerobic and anaerobic pathways. Fine-tuning of physiological transmitters to measure activity of free ranging fish can therefore

Guest editors: P. R. Almeida, B. R. Quintella, M. J. Costa and A. Moore
Developments in Fish Telemetry

G. Lembo (✉) · P. Carbonara · M. Scolamacchia · M. T. Spedicato
COISPA Technology & Research, Bari-Torre a Mare, Italy
e-mail: lembo@coispa.it

R. S. McKinley
Centre for Aquaculture and Environmental Research, The University of British Columbia, Vancouver, Canada BC V7V IN6

be utilised to monitor the proportion of the scope of activity utilised in response to external stressors. This proportion and the level of compensatory ability remaining could represent a measure of well-being in cultured fish.

Keywords Sea bass · Activity levels · Electromyograms · Fish well-being · Physiological telemetry

Introduction

In the Mediterranean, culture of sea bass (*Dicentrarchus Labrax* L.) and sea bream (*Sparus aurata* L.) represents greater than 95% of all marine fish species farmed. Each of these species is subjected to a variety of operational protocols and handling procedures. Assessing the level of stress and subsequent energetic expenditures associated with culture protocols, would permit the development of appropriate mitigative strategies for fish, likely improving product quality and yields at farm sites.

Fish welfare requires that the animal be in good health with its biological systems functioning appropriately and with an ability to compensate stress (FSBI, 2002).

Historically, behavioural tests have been used as indicators of short-term stress, but examples are sparse. Measurement of swimming ability, such as critical swimming speed (U_{crit}) has been used as an indicator of the effects of environmental challenges on the general health of fish (Randall et al., 1987; Brauner et al., 1994), and requires the use of a calibrated swim chamber. Chatelier et al. (2005) have, for example, recently demonstrated, using salinity and temperature challenges, that sea bass have an exceptional ability to regulate plasma osmotic homeostasis and tissue water balance. Physiological telemetry is one approach where the response to culture conditions and procedures can be assessed from the perspective of free swimming fish.

Recent advances in physiological telemetry have permitted several investigators to monitor the relative activity of free swimming fish (Weatherley et al., 1996; Økland et al., 1997, 2002), fish welfare (Beddow and McKinley, 1998; McFarlane et al., 2004; Chandroo et al., 2005), and seasonal movement (Demers et al., 1996; Quintella et al., 2004). In addition, several investigators have used laboratory derived calibrations of swimming speed to EMG level and ultimately to oxygen consumption to estimate energy expenditure to activity events (Cooke et al., 2000; Geist et al., 2000). All of these studies relied on the calibration of a measure of relative activity, typically transmitter signal level to swimming speed, from the recordings of electrical activity of the aerobic muscle. Indeed, calibration of the transmitter signals level, and swim speed was believed to be derived primarily from the aerobic muscle.

Once established, investigators can construct an estimate of scope for activity (McKinley and Power, 1992) and use the overall measurement of scope as a correlate of well-being. On the other side, at the present time the technology does not include the white muscle (anaerobic muscle) activity, which is better linked with the fish's well-being state (Jain and Farrell, 2003; Chandroo et al., 2005).

The objective of this study was to examine the contribution of two muscle types (red and white) to swimming activity of sea bass. Results were used to 'fine-tune' measures obtained using a physiological transmitter. Implications for measures of 'scope for activity' and the use of this measure as an indicator of fish well-being, defined as the level of compensatory ability available to an individual fish, are also discussed.

Materials and methods

Experimental animals

Adult sea bass were reared in a 4.5 m^3 circular tank (2.1 m wide, 1.3 m deep), at a density of 10 kg $(m^3)^{-1}$ under natural light conditions. Individuals were fed once per day to satiation. Tank water was exchanged five times per day and water temperature was maintained at 18°C. Eight fish (average TL: 263.2 ± 11.8 mm; average TW: 178.4 ± 23.2 g) and twelve fish (average TL: 255 ± 12.7 mm; average TW: 163.8 ± 29.1 g) were used for the EMG and hard-wired monitoring of the electrical activity tests, respectively.

Implantation of EMG transmitter

The EMG transmitters (CEMG-R11-25, Lotek Wireless) measured 11 mm in length and 61 mm diameter. In these transmitters, the voltage corresponding with muscle activity is rectified, summed and stored over a 5-s period. Before the end of the 5-s period, average value is determined and assigned to an activity level ranging from 0 to 50. Use of radio tags in sea water has some limitations due to the strong attenuation of the radio waves. However, in our experiment, collection of signals from these transmitters was possible because of the close proximity of the receiving antenna to the transmitting antenna of the tag.

Transmitter weight (12 g) in proportion to fish body mass was approximately 6–7%.

All fish submitted to the surgical procedure were fasted for a minimum of 24 h. Implantation procedures followed those used in McFarlane et al. (2004). Fish were initially anaesthetized using 30 mgl^{-1} clove oil (Massee et al., 1995) and gills were continuously irrigated using a maintenance level of anesthetic. The EMG transmitter was inserted through 3-cm incision (4-5 cm posterior of the pelvic girdle). The gold-tipped electrodes were inserted into the red muscle band via a hollow needle (diameter 2 mm, length 50 mm). The incision was closed using four independent surgical sutures (3–0). Total surgical time, including anaesthetization, was generally 15–18 min. EMG signals from the transmitters were recorded using a receiver (SRX_400, Lotek) and then downloaded to a computer.

Electrical activity from muscle types by hard-wires

Two pairs of stainless steel (304) wire electrodes (1.3 m in length, 0.2 mm in diameter) were surgically implanted (Fig. 1) on anaesthetized fish using clove oil at the same concentration as in the EMG transmitter implantation. One pair of electrodes were inserted subcutaneously using needles (diameter 0.6 mm, length 30 mm) into the lateral red musculature at a location approximately two thirds posterior to the mouth of the fish and above the lateral line. A second pair of electrodes was implanted into the white musculature at a similar location but at a depth of 1 cm. Electrodes were sutured to the side of individuals to minimize entanglement of the electrode wires. The duration of surgical procedure including anaesthetization, was generally 25–30 min. Individuals were permitted a 1 h recovery period. The bio-sensitive electrodes monitoring electrical activity in the two muscle types were passed through a controlled opening in the swimming chamber to the pre-amplifiers. As the system measured differential signals, a reference electrode was placed in the water of the chamber.

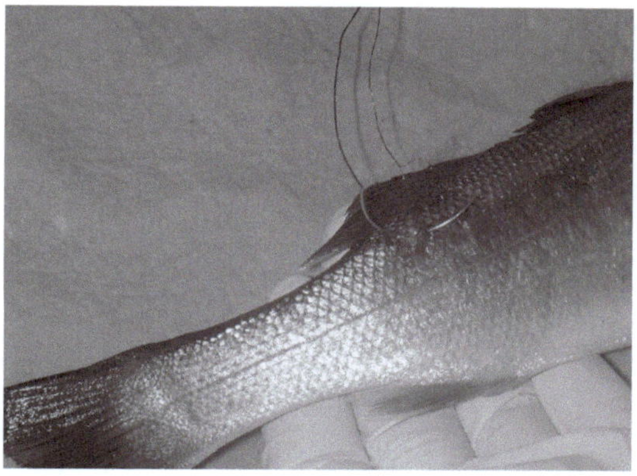

Fig. 1 Pair of electrodes inserted subcutaneously into the lateral red musculature and sutured to minimize entanglement

The wires were connected to pre-amplifiers (Grass, Model P511), where signals were filtered, amplified and transformed into non-differential signals. Data were subsequently recorded to a computer by Labview7 software, with a sample size of 5000 data per second via an analog-to-digital board (National Instrument DAQCard-AI-16E-4).

Swimming performance protocols

Swimming performance trials were conducted using a Blažka-style swimming chamber (Thorstad et al., 1997). It measured 123 cm in length with a diameter of 24 cm. The diameter of the external Plexiglas tube was 35 cm (Fig. 2). Water velocity was calculated using a Doppler flow meter (Sigma).

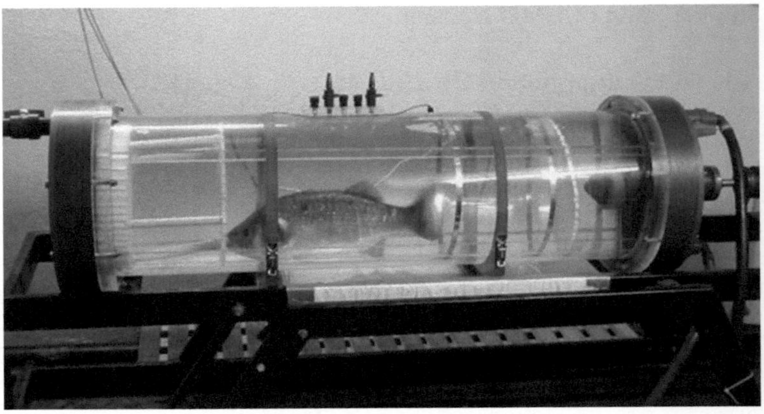

Fig. 2 Blažka-style swimming chamber measuring 123 cm in length and with an internal diameter of 24 cm

All the fish tested, were fasted for 24 h prior to swim trials to ensure a post-absorptive state (McFarlane et al., 2004). Possible stressful events, have been minimized through anesthesia, gentle manipulation, recovery period and acclimation time. All fish were transferred immediately following the surgery and allowed to recover for a minimum period of 30–40 min without flow, following an acclimation phase at a water velocity of 0.1 ms^{-1} for a minimum of 20 min.

Swimming trials followed procedures outlined in Brett (1964). Tests were conducted using increasing water flow velocity 0.1 ms^{-1} in the swimming chamber every ten minutes until individuals fatigued. Fatigue was determined if the tail of the fish maintained contact with the rear screen for five seconds. Swimming behaviour, including starting/stopping swimming style was noted.

EMG activity was recorded for each incremental increase in water velocity.

Analyses

Absolute U_{crit} (ms^{-1}) was calculated as in Brett (1964) and a correction was made for the specimens with a maximum width greater than 10% of the inner tube diameter (Smit et al., 1971).

Considering the intrinsic variability in each biological response to certain stimuli and the aim of obtaining 'average' information on the scope for activity of the fish, absolute EMG signals (mV) for each muscle type at each velocity from the 12 test fish were pooled, then averages ± SE were calculated for each speed step.

A similar conceptual approach was followed for the relative activity levels recorded from the physiological transmitter in each 10-min step. In this case, the frequency distributions of EMG signals for the eight fish were constructed and smoothed. Analysis was performed decomposing the distributions into their gaussian components using the Bhatthacharya (1967) method as implemented in Gayanilo et al. (1996). Since the slope of a gaussian curve is positive to the left of the mode, zero at the mode and negative to the right of the mode, the logarithmic derivative decreases linearly. Therefore, each gaussian component can be determined by identifying negative sloped intervals in the graph of logarithmic frequency differences against EMG midpoints. The separation of the frequency distributions into their components is an iterative process that subtracts each identified component using the Gaussian function. The separation index, i.e. the ratio between the difference of two successive means and their estimated standard deviations, allows the results of the separation process to be accepted or refused. Values of the separation index can be calculated for modes following the first one, and the implemented method (Gayanilo et al., 1996) rejects figures lower than 2.

Estimated component means (corresponding to the dominant modes), representing the EMG main signals, were plotted against water velocity steps to identify the pattern of the red muscle activity.

Results

Critical swimming speed of sea bass (Fig. 3) ranged between 0.63 and 1.00 $msec^{-1}$ (average: 0.87; SE: 0.031) equivalent to 2.58–3.98 $BLsec^{-1}$ (average: 3.43; SE: 0.12). Few fish were able to complete 1 $msec^{-1}$ velocity step, but the swimming performances of the tested individuals were only slightly lower if compared to that previously observed in fish of similar size, not surgically treated (range: 3.6–4.1 BLs¯1 Carbonara et al., 2006).

Results from the Bhattacharya's (1967) analysis of the frequency distributions of the EMG signals from eight fish samples at each water velocity stage are reported in Table 1. These results show the occurrence of 1–3 normal components (Table 1), with the first one representing between 91.7 and 99.8% of the whole distribution. Dominant modes (Fig. 4) indicate groups of EMG signals with similar values, thus we assumed the estimated average of the gaussian component as representative of the mean EMG value at each velocity step. Separation index was always higher than 2, while the coefficient of variation of the first mode was between 16.9 and 36.5%, highlighting that the EMG signals from transmitters were rather variable, probably for inherent causes (physiological differences), or technological reasons (characteristics of the transmitter) or as a consequence of the electrical activity interaction between the two muscle types. This interference is probable because of the small width of the red muscle band. As water speed increased, and the fish progressively approached the critical swimming speed, the EMG from the transmitter showed an increasing pattern similar to that observed for the red muscle monitored by hard wires (Fig. 5). The correlation ($r = 0.97$) between transmitter and red muscle hard wiring EMGs was significant ($P < 0.05$).

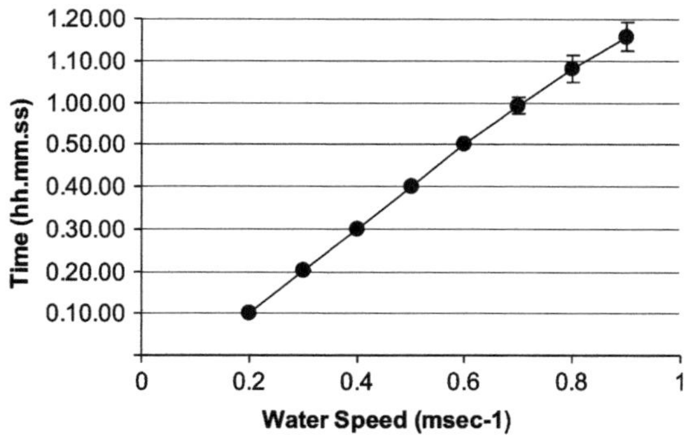

Fig. 3 Critical swimming speed (Ucrit) estimated from 12 fish tested

EMG's averages were ranging between ~5 at 0.2 $msec^{-1}$ to ~16 at 1 $msec^{-1}$ (Table 1), thus the EMG intensity increased on average 3.2 times between 0.2 $msec^{-1}$ and the U_{crit} velocity (~1 $msec^{-1}$). The calibration between forced swimming speed and average EMG was described by the following linear relationship: $y = 14.5x + 2.0778$ ($R^2 = 0.989$) where y is the EMG and x the velocity.

The averages of hard wired-red muscle electromyograms were ranging between 0.28 and 1.19 mV (Fig. 5), thus the intensity increased on average ~4 times between 0.2 $msec^{-1}$ and the U_{crit} velocity (~0.9 $msec^{-1}$). The pattern was initially rather sharp and then reached a *plateau* at higher velocity steps. It was described by the following logarithmic relationship: $y = 0.5922Ln(x) + 1.2251$ ($R^2 = 0.9906$), where y is the electrical voltage and x the velocity.

The white muscle activity showed a different pattern if compared to both transmitter and red muscle-hard wired EMGs. The averages of white muscle electromyograms varied between 0.15 and 1.00 mV (Fig. 5), thus the intensity increased exponentially and on average ~6.7 times between 0.2 $msec^{-1}$ and the U_{crit} velocity (~0.9 $msec^{-1}$). The relationship between the electrical voltage (y) and the velocity (x) was the following: $y = 0.0977e^{2.4723x}$ ($R^2 = 0.9845$).

If the swimming performance of sea bass is split in three main phases of activity, defined as (1) resting (0.1–0.3 $msec^{-1}$ speed), (2) moderate (0.4–0.6 $msec^{-1}$) and (3) intense (>0.7 $msec^{-1}$), it was observed that the EMG increased ~2 times between phase 1 and 2, and ~3 times between the phase 1 and 3. The corresponding increments of hard wired-red muscle electromyograms were ~3 and ~4, whereas electromyogram's intensity of the white muscle grew of ~2 and ~4 times, respectively.

Fine-tuning of the measures obtained using physiological transmitters by the hard-wired tests

Table 1 Results from the Bhattacharya (1967) analysis of the frequency distributions of the EMG signals from 8 fish samples at each water velocity step

Velocity	Mode	Number of EMG signals	Mean	Standard deviation	C.V. (%)	Separation index	% of 1st mode
0.2	1	531	5.2	1.9	36.5	–	96.2
	2	19	11.9	1.5	12.6	4.0	
	3	2	16.6	0.9	5.4	3.8	
0.3	1	580	6.3	2.1	33.3	–	91.8
	2	50	12.8	2.2	17.2	3.1	
	3	2	18.1	1.3	7.2	3.0	
0.4	1	573	7.5	2.4	32.0	–	91.7
	2	52	14.4	2.1	14.6	3.1	
0.5	1	591	9.0	2.7	30.0	–	92.9
	2	45	16.4	2.1	12.8	3.1	
0.6	1	501	10.6	2.9	27.4	–	99.8
	2	1	24.4	3.7	15.2	4.2	
0.7	1	511	13.1	3.0	22.9	–	99.0
	2	5	22.7	1.6	7.0	4.1	
0.8	1	568	13.9	3.5	25.2	–	97.8
	2	13	22.6	1.3	5.8	3.6	
0.9	1	332	14.8	2.5	16.9	–	89.7
	2	35	20.1	1.7	8.5	2.5	
	3	3	24.3	0.7	2.9	3.5	
1.0	1	85	16.4	3.3	20.1	–	100.0

Number of modes separated for each distribution, number of EMG signal estimated for each mode, mean of each separated mode, standard deviation, coefficient of variation and separation index are indicated. The percentage of the distribution represented by the first estimated gaussian component is also reported

(Fig. 5) indicates that the two muscle types in the sea bass are both involved in fuelling swimming activity below the U_{crit}. Thus, scope for activity in the sea-bass is not supported solely using aerobic metabolism, though the red muscle powers the majority of the swimming ability.

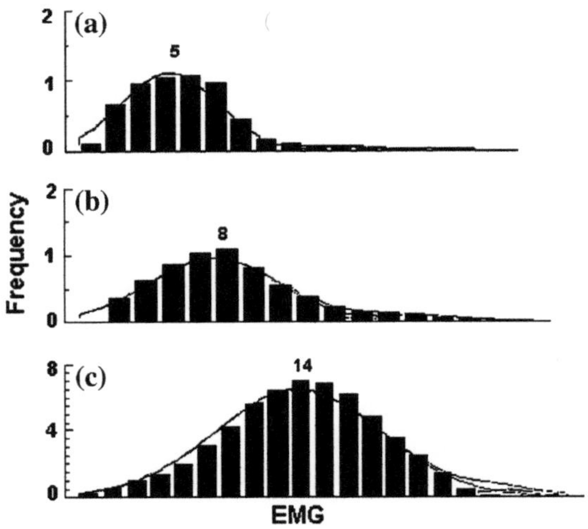

Fig. 4 Examples of the separation of the normal components of the EMG frequency distributions obtained on 8 fish samples at water speed of (**a**) 0.2, (**b**) 0.4, (**c**) 0.8 msec^{-1}. The numbers indicate the estimated average for the main component identified at each water speed step. The curves indicate the estimated distributions after the normal components separation

Discussion

The critical swimming speed is considered to be a measure of aerobically sustainable swimming capacity and has been used to define the aerobic scope of many fish species. The swimming performances observed in this study were only slightly lower if compared to that previously monitored in fish of similar size but not surgically

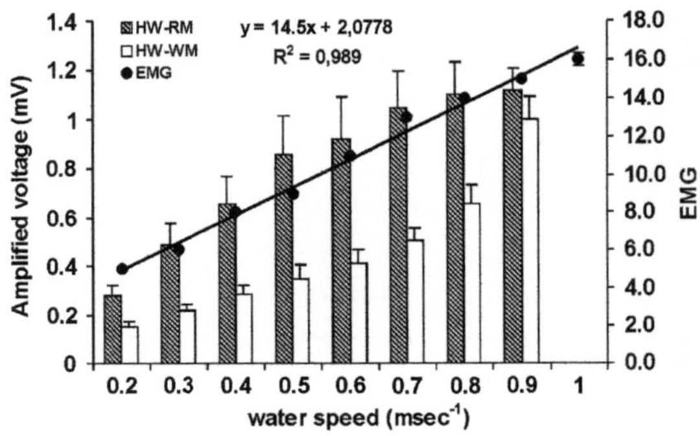

Fig. 5 Swimming activity of sea bass (average ± standard error) estimated from the contribution of two muscle types by hard-wired tests (HW-RM and HW-WM represent the electromyograms of red and white muscle, respectively), "fine-tuned" by the measures obtained using a physiological transmitter. Linear relationship describing the calibration between forced swimming speed and average EMG pulse intervals is showed

treated (Carbonara et al., 2006). Conversely, swimming ability of tested fish (average: 3.43 BLsec^{-1}; SE: 0.12) was on average, higher than reported by Chatelier et al., (2005) (average: 2.25 BLsec^{-1}; SE: 0.08), but results are not strictly comparable, as consequence of differences in protocol exercises.

Large variations in swimming ability exist among teleosteans: the maximum sustainable speed in tunas is believed to be fuelled solely by aerobic metabolism (Deware and Graham, 1994, Ellerby et al., 2000), while for other species the recruitment of both aerobic and anaerobic muscle fibres is required, as for example in Atlantic cod (Reidy et al., 2000). The results of our study show that in the sea bass, despite the low mass of red, compared to white muscle, the former fuelled the majority of the swimming capability below the U_{crit}. The pattern of the recruitment of red muscle activity increased with speed until a maximal level was attained. The plateau in the mass recruitment was maintained despite further increases in swimming speed, a result similar to that reported by Rome et al. (1992), whereas white mass recruitment continued to exponentially increase up to the U_{crit} velocity.

The beginning of the recruitment of white muscle appears to be somewhat species-specific (Burgetz et al., 1998). It has been suggested, for example, that anaerobic metabolism must be taken into account also at speeds above 70% U_{crit} for *O. mykiss* (generalist swimmer). This fact is supported by the different placement and amount of slow-twitch aerobic muscle (for sustained swimming) and fast-twitch fibres (recruited during fast starts) in different groups (Ellerby et al., 2000). Scombrids, for example, are specialized for sustained swimming by the means of elongated myotomes and the presence of connective tendons which lead to an efficient mean of power transmission caudally along the length of the fish, though loosing manoeuvrability. Salmonoids generate the majority of power by posterior musculature, while eels (high body curvature) use primarily their anterior myotome to power steady swimming.

The Scope For Activity (SFA) or energetic cost of activity can be defined as the increase in oxygen uptake in a swimming fish, while SMR (Standard Metabolic Rate) as the rate of oxygen uptake measured from a notionally stationary fish (Fry, 1971). Scope for activity is variable over time and among fish (Briggs and Post, 1996).

Results of the hard wiring exercise clearly showed that SFA in the sea-bass is not supported solely using aerobic metabolism and that the two muscle types are both involved in fuelling swimming activity below the U_{crit}. Indeed, 'fine-tuning' indicates a very clear pattern, with increasing aerobic and anaerobic activity according to the swimming speed. Results also highlight that the estimate of the overall scope of activity should take into account the energy of the white muscle, and thus the budget available is lower than expected on the basis of the red muscle metabolic pathway. In fact if sub-maximal swimming results in the recruitment of anaerobic fibres, the energy budget will be underestimated (Burgetz et al., 1998).

To evaluate the energetic expenditure of a swimming fish (Cooke et al., 2000), the available physiological transmitter technology allows only the monitoring of the red muscle and thus of the aerobic activity. This gives a rough estimate of the SFA and consequently of the fish physiological state. Correction of the SFA for the anaerobic component would provide an accurate accounting of a fish ability to compensate to stressors, which act as consumers of the fish reserves affecting well-being.

Monitoring of free swimming sea bass is particularly important, not only for introducing mitigative strategies in aquaculture practices, but also for giving insights into the migratory behaviour of this species, entering coastal lagoons and spending part of the inter-genetic phase of the life cycle there. Calibration of swimming speed through swimming chambers to EMG signals is considered by many authors as a baseline for elucidating swimming behaviour of free swimming fish (e.g. Cooke et al., 2000; Thorstad et al., 2000; Geist et al., 2002; McFarlane et al., 2004). Some authors suggest that the relationship between EMG signal and swimming speed would be constructed for each fish before it is released, but usually calibration curves are made on a separate group of fish and then applied to the ones in the field (Brown and Geist, 2002).

However a high variability of EMG among fish, stocks, or in the same individual at a different time has also been stressed (e.g. Thorstad et al., 2000). Similarly, in our study EMG signals from transmitters were rather variable, probably for inherent causes (physiological differences), or technological reasons (characteristics of the transmitter) or as consequence of electrical activity interaction between the two muscle types. Indeed, the large gold-tipped electrodes may record a large proportion of the myotome, and sometimes multiple myomeres. Variation in depth from the skin and distance from the lateral line, where the gold rods are positioned, may affect to what extent they record EMGs from red or white muscle fibres, or both (Beddow and McKinley, 1999; Thorstad et al., 2000). Also in our study, variability of EMG at each velocity stage can be interpreted considering the possible cross-interference between the signals emitted by the two muscle types.

Considering the intrinsic variability in the physiological response to certain stimuli and the aim of obtaining an 'average' information on the scope for activity of sea bass we pooled the data from the fish used for calibrating swimming speed against EMG. The same procedure was used for testing hard-wired electromyograms against velocity. The possibility of reducing the variability was thus investigated by breaking down the frequency distributions of EMG signals into their gaussian components using the Bhatthacharya (1967) technique after smoothing. This method allowed a more objective cleansing of the data of spurious signals, and thus the identification of unbiased average value for each velocity step. Despite this, standard deviations associated with the mean of each normal component were still rather high (Table 1). Consequently, a precise association between swimming speed and EMG could be unreliable. Instead, a pathway splitting the swimming performance of sea bass in three main phases of activity defined as: (1) resting (0.1–0.3 $msec^{-1}$ speed), (2) moderate (0.4–0.6 $msec^{-1}$), and (3) intense (>0.7 $msec^{-1}$) could be a useful descriptor of swimming ability.

Calibration of the EMG transmitter with hard wire values allowed assessing white and red muscle patterns. Only in this way, with the available technology (EMG radio transmitter), is it possible to associate certain activity levels of red muscle to an intensity of white muscle, and then recognize more accurately a stress-related response (Jain and Farrell, 2003; Chandroo et al., 2005).

Nevertheless, possible improvement of the calibration technique should be further investigated with larger samples and replicates before definitive conclusions are drawn.

The results gained in the study on sea bass indicate that swimming performance tests (U_{crit}, EMGs, 'fine-tuning') represent reliable descriptors of fish condition, and are potentially usable as operational indicators of well-being status/activity of farmed and wild fish. Indeed, reduced U_{crit} may be caused by an interaction of anaerobic (white muscle activity) end product accumulation within the fish musculature and disturbances in ionoregulation (Chandroo et al., 2005). Many stress-related biological factors influence the swimming capacity and, for this reason, behavioural indicators may be able to show the fish response to stressors even when clinical or physiological indicators, such as plasma cortisol, do not (Chandroo et al., 2005). The usefulness of swimming performance tests could be further expanded if recovery tests, where fish performance is analysed in two consecutive sustained exercises, are combined (Farrell et al., 1998; Carbonara et al., 2006). Nonetheless, the description of the fish physiological state based only on a single indicator may not accurately reflect the overall fish state and an integrated diagnostic frame (behavioural tests, plasma cortisol, etc.) would be more suitable.

Overall, our results showed that defining scope for activity, based solely on the red muscle, would overestimate the level of aerobic compensation available to the individual to respond to any external stressor or, in the case of cultured fish, to any handling protocol. Fine-tuning of the scope for activity for the contribution from white muscle would permit an accurate estimate of the aerobic energy available for the fish to expend in response to stressors, such as husbandry protocols. Obtaining the ability to accurately monitor the EMG level, using physiological telemetry, would permit an estimate of the scope or energy remaining. This estimate could be used to monitor the ability of fish to compensate for further stress attributable to handling protocols and the level of compensatory activity, available to the individual, could be viewed as a measure of its present well-being.

Acknowledgement This study was supported by the SFOP EU funds, co-financed by regional government of Puglia (Italy).

References

Beddow, T. A. & R. S. McKinley, 1998. Effect of thermal environment on electromyographical signals obtained from Atlantic salmon (*Salmon salar* L.) during forced swimming. Hydrobiologia 371/372: 225–232.

Beddow, T. A. & R. S. McKinley, 1999. Importance of electrode positioning in biotelemetry studies estimating muscle activity in fish. Journal of Fish Biology 54: 819–831.

Bhattacharya, C. G., 1967. A simple method of resolution of a distribution into Gaussian components. Biometrics 23: 115–135.

Brauner, C. J., G. K. Iwama & D. J. Randall, 1994. The effect of short-duration seawater exposure on the swimming performance of wild and hatchery-reared juvenile coho salmon (*Oncorhyncus kisutch)* during smoltification. Canadian Journal of Fisheries and Aquatic Sciences 51: 2188–2194.

Brett, J. R., 1964. The respiratory metabolism and swimming performance of young sockeye salmon. Journal of Fisheries Research Board of Canada 21: 1183–1226.

Briggs, C. T. & J. R. Post, 1996. In situ activity metabolism of rainbow trout (*Oncorhynchus mykiss*): estimates obtained from telemetry of axial muscle electromyograms. Canadian Journal of Fisheries and Aquatic Sciences 54: 859–866.

Brown, R. S. & D. R. Geist, 2002. Determination of swimming speeds and energetic demands of upriver migrating fall chinook salmon (Oncorhynchus tshawytscha) in the Klickitat river. Bonneville Power Administration, U.S. Department of Energy, Pacific Northwest National Laboratory, 76 pp.

Burgetz, I. J., A. Rojas-Vargas, S. G. Hinch & D. J. Randall, 1998. Initial recruitment of anaerobic metabolism during sub-maximal swimming in rainbow trout (*Oncorhynchus mykiss*). Journal of Experimental Biology 201: 2711–2721.

Carbonara, P., M. Scolamacchia, M. T. Spedicato., G. Lembo & R. S. McKinley, 2006. Swimming performance as a well-being indicator of reared sea-bass. Preliminary results. Biologia Marina Mediterranea 13(1): 488–491.

Chandroo, K. P., S. J. Cooke, R. S. McKinley & R. D. Moccia, 2005. Use of electromyogram telemetry to assess the behavioural and energetic responses of rainbow trout, *Oncorhynchus mykiss* (Walbaum) to transportation stress. Aquaculture Research 36: 1226–1238.

Chatelier, A., D. J. McKenzie & G. Claireaux, 2005. Effects of changes in water salinity upon exercise and cardiac performance in the European seabass (*Dicentrarchus labrax*). Marine Biology 147: 855–862.

Cooke, S. J., K. P. Chandroo, T. A. Beddow, R. D. Moccia & R. S. McKinley, 2000. Swimming activity and energetic expenditure of captive rainbow trout *Oncorhynchus mykiss* (Walbaum) estimated by electromyogram telemetry. Aquatic Research 31: 495–505.

Dewar, H. & J. B. Graham, 1994. Studies of tropical tuna swimming performance in a large water tunnel, I Energetics. Journal of Experimental Biology 192: 13–31.

Demers, E., R. S. McKinley, A. H. Weatherley & D. J. McQueen, 1996. Activity pattern of largemouth and smallmouth bass determined with electromyogram biotelemetry. Transactions of the American Fisheries Society 125: 434–439.

Ellerby, D. J., J. D. Altringham, T. Williams & B. A. Block, 2000. Slow muscle function of Pacific Bonito (*Sarda chiliensis*) during steady swimming. Journal of Experimental Biology 203: 2001–2013.

Farrell, A. P., K. Gamperl & I. K. Birthwell, 1998. Prolonged swimming, recovery and repeat swimming performance of mature sockeye salmon *Onchorhyncus nerka* exposed to moderate hypoxia and pentachlorophenol. Journal of Experimental Biology 201: 2183–2193.

Fry F. E. J., 1971. The effect of environmental factors on the physiology of fish. In Hoar, W. S. & D. J. Randall (eds), Fish Physiology, VI, Academic Press, New York.

FSBI, 2002. Fish Welfare. Briefing Paper 2, Fisheries Society of the British Isles, Granta Information Systems, 82A High Street, Sawston, Cambridge CB2 4H.

Gayanilo, F. C. Jr., P. Sparre & D. Pauly, 1996. FAO-ICLARM Stock Assessment Tools (FISAT) User's Manual. *FAO*–Computerized Information Series–fisheries, 1–126.

Geist, D. R., C. S. Abernethy, S. L. Blanton & V. I. Cullinan, 2000. The use of electromyogram telemetry to estimate energy expenditure of adult Fall Chinook Salmon. Transactions of the American Fisheries Society 129: 126–135.

Geist, D. R., R. S. Brown, K. Lepla & J. Chandler, 2002. Practical application of electromyogramm radiotelemetry: the suitability of applying laboratory-acquired calibration data to field data. North American Journal of Fisheries Management 22: 474–479.

Jain, K. E. & A. P. Farrell, 2003. Influence of seasonal temperature on the repeat swimming performance of rainbow trout *Oncorhynchus mykiss*. The Journal of Experimental Biology 206: 3569–3579.

Massee, K. C., M. B. Rust, R. W. Hardy & R. R. Stickney, 1995. The effectiveness of tricaine, quinaldine sulfate and metomidate as anesthetics for larval fish. Aquaculture 134: 351–359.

McFarlane, W. J., K. F. Cubitt, H. Williams, D. Rowsell, R. Moccia, R. Gosine & R. S. McKinley, 2004. Can feeding status and stress level be assessed by analyzing pattern of muscle activity in free swimming rainbow trout (*Oncorhynchus mykiss* Walbaum)? Aquaculture 239: 467–484.

McKinley, R. S. & G. Power, 1992. Measurement of activity and oxygen consumption for adult lake sturgeon (*Aciepenser fulvescens*) in the wild using radio-trasmitted EMG signals. In Priede, I. G. & S. M. Swift (eds), Wildlife Telemetry. Ellis Horwood series in environmental management, science and technology, 307–318 pp.

Økland, F., B. Finstad, R. S. McKinley, E. B. Thorstad & R. K. Booth, 1997. Radio-trasmitted electromyogram signals as indicators of physical activity in Atlantic salmon. Journal of Fish Biology 51: 476–488.

Økland, F., I. Fleming, E. B. Thorstad, B. Finstad, S. Einum & R. S. McKinley, 2002. EMG radio tag in recording the spawning behaviour of Atlantic salmon: effect, reliability and accuracy. In Moore, A. & I. Russell (eds), Proceeding for the Third Conference on Fish Telemetry held in Europe, 20–25 June, Norwich, UK. Advance in Fish telemetry, The Centre for Environment, Fisheries and Aquaculture Science, Suffolk, UK, 51–58 pp.

Quintella, B. R., N. O. Andrade, A. Koed & P. R. Almeida, 2004. Behavioural patterns of sea lampreys' spawning migration through difficult passage areas, studied by electromyogram telemetry. Journal of Fish Biology 65: 961–972.

Randall, D. J., D. Mense & R. G. Boutilier, 1987. The effects of burst swimming on aerobic swimming in Chinook salmon (*Oncorhynchus tshawytscha*). Marine Behaviour and Physiology 13: 77–88.

Reidy, S. P., S. R. Kerr & J. A. Nelson, 2000. Aerobic and anaerobic swimming performance of individual Atlantic cod. Journal of Experimental Biology 203: 347–357.

Rome, L. C., I. -H. Choi, G. Lutz & A. Sosnicki, 1992. The influence of temperature on muscle function in the fast swimming scup I. Shortening velocity and muscle recruitment during swimming. Journal of Experimental Biology 163: 259–279.

Smit, H., J. M. Amelink-Koutstaal, J. Vijverberg & J. C. Von Vaupel-Klein, 1971. Oxygen consumption and swimming efficiency of swimming goldfish. Comparative Biochemistry and Physiology 39: 1–28.

Thorstad, E. B., B. Finstad, F. Økland, R. S. McKinley & R. K. Booth, 1997. Endurance of farmed and sea-ranched Atlantic salmon (*Salmon salar* L.) at spawning. Aquatic Research 28: 635–640.

Thorstad, E. B., F. Økland, A. Koed & R. S. McKinley, 2000. Radio-transmitted electromyogram signals as indicators of swimming speed in lake trout and brown trout. Journal of Fish Biology 57: 547–561.

Weatherley, A. H., P. A. Daseloo, M. D. Gare, J. M. Gunn & B. Lipicnik, 1996. Field activity of lake trout during the reproductive period monitored by electromyogram radiotelemetry. Journal of Fish Biology 48: 675–685.

Hydrobiologia (2007) 582:281–287
DOI 10.1007/s10750-006-0546-9

Efficacy and accuracy of portable PIT-antennae when locating fish in ice-covered streams

Tommi Linnansaari · Jean-Marc Roussel · Richard A. Cunjak · Jo H. Halleraker

Abstract Active tracking of passive integrated transponder (PIT)-tags using portable antennae is becoming an increasingly common technique in fish habitat studies in shallow rivers. We carried out "blind testing" to test the efficacy (% tags found) and accuracy (distance between predicted and true tag location) of a portable antenna system (Texas Instruments) in winter conditions using 23-mm PIT-tags. Up to 90 cm reading range was achieved and signals penetrated ice, rock, wood and water. In the "blind test" trials, a majority of the hidden tags ($N = 12–30$) were found indicating high tracking efficacy. PIT-tags that were oriented with their cylindrical axis parallel to the plane of the antenna coil inductor loop resulted in a bimodal detection field that had low detection range in the centre of the loop. The utilization of this bimodal detection field proved to be a very accurate method for identifying tag position (mean ± SE distance from predicted to true tag location 10.9 ± 1.4 cm) and thus well suited for microhabitat and activity studies in winter conditions. Aggregations of tags (multiple tags within 1 m^2) and obstacles for the antenna maneuvering (e.g., boulders, logjams) reduced the pinpointing accuracy (mean ± SE 13.3 ± 1.8 cm), but the reduction in accuracy was statistically non-significant between the single and aggregated tags.

Keywords Passive Integrated Transponder · Portable antenna · Accuracy · Efficacy · Winter

Guest editors: P. R. Almeida, B. R. Quintella, M. J. Costa and A. Moore
Developments in Fish Telemetry

T. Linnansaari (✉) · R. A. Cunjak
Canadian Rivers Institute and Department of Biology, University of New Brunswick, P.O. Bag Service 45111, Fredericton NB, Canada E3B 6E1
e-mail: tommi.linnansaari@unb.ca

J.-M. Roussel
Laboratoire d'Ecologie Aquatique, Institut National de la Recherche Agronomique, UMR 985, Rennes cedex, France

J. H. Halleraker
SINTEF Energy Research, NO-7465 Trondheim, Norway

Introduction

Use of passive integrated transponder (PIT)-technology has become increasingly common in various ecological research settings and especially so in aquatic research. The early use of the PIT-technology in aquatic ecology includes monitoring of fish passage using stationary antennae for automatic data storage (Prentice et al., 1990; Brännäs et al., 1994; Castro-Santos et al., 1996) whereas portable PIT technology used handheld scanners that require a tagged individual to be physically recaptured to read a tag code due to the very limited reading range (e.g., Donaldson et al., 1992). Recently, however, effective portable antennae systems have been developed and successfully used to track movements, habitat use and/or survival of crustaceans (e.g., Bubb et al., 2002) and stream-dwelling cottids and salmonids (Roussel et al., 2000, 2004; Barbin Zydlewski et al., 2001; Cucherousset et al., 2005). The portable antennae systems are usually operated by one person carrying the large reader apparatus, but two-person operated portable gate antennae have also been successfully used to detect PIT-tags in adult salmon (*Salmo salar* L.) residing in pools (Connell, 2005).

The potential value of portable PIT-antennae systems to study movements and habitat use of fish in fluvial habitats requires that the efficacy (i.e., % of tags detected) and accuracy (i.e., distance between predicted and true tag location) of the systems are known. The data-at-hand indicates that the efficacy of active PIT-tracking is good (70–80%) for various antennae systems and PIT-tag types when measured in open-water conditions in summer (Roussel et al., 2000; Bubb et al., 2002; Cucherousset et al., 2005). By contrast, it is evident from Barbin-Zydlewski et al. (2001) and Bubb et al. (2002) that high accuracy (i.e., within a 10-cm radius) is associated with low detection ranges (15–50 cm). Roussel et al. (2000) reported a good detection distance (up to 1 m) but a decreased accuracy (1 m^2) further suggesting a trade-off between the two measures.

Increasing interest to study winter microhabitat use and behaviour of juvenile salmonids (e.g., Alfredsen & Tesaker, 2002) provides an exiting opportunity to utilize this improved technology (see Roussel et al., 2004). However, improving detection distance is imperative to enable efficient tracking surveys when ice cover precludes the maneuvering of the antennae in the water column. The concurrent reduction in the pinpointing accuracy of the PIT-tags may further limit the use of this technology in microhabitat, activity and movement studies. Therefore, it is important to find an efficient means to locate the tags accurately, and simultaneously be able to utilize adequate detection distances. Thus, our objective in this study was to test the efficacy and accuracy of the portable antenna system in winter conditions using the knowledge of the shape of the antenna detection field.

Methods

Portable antenna system

We used a portable antenna system (Texas Instruments TIRIS S-2000) similar to that described by Roussel et al. (2000) and Barbin Zydlewski et al. (2001) with a piezoelectric buzzer to indicate tag detection. The following modifications were introduced: (1) we used coil inductor loop of 55 cm in diameter (six loops, either 4 $(mm)^2$ or 12-gauge multistrand plastic coated copper wire) with an adjustable joint for changing the angle between the antenna pole and the coil inductor loop; (2) the system was powered with 18-V DC rechargeable batteries (either 12-V and 6-V connected in series or 24-V with a DC converter to regulate the voltage to continuous 18-V DC) instead of 12-V; (3) the power input was equipped with on-off switch to facilitate operation, and a rectifier and a 2-A fuse helped avoid damage that might result from misconnections or super voltage; (4) the support frame for the coil inductor loop (protected by 2 mm rubber tube or epoxy impregnated paper) was constructed of single piece of solid aluminum (12 mm diameter) or of turbonite (centre of support frame) because PVC construction was not robust during freezing conditions. All the equipment was enclosed in two separate weatherproof cases that were attached to a rigid backpack frame.

Portable antenna performance

The detection field of the portable antenna was determined using 160 × 160 cm area comprised of 5 × 5 cm grid cells to produce a total of 1,024 measurement nodes. A Hewlett Packard DC power supply amplifier (model 6824a) was used to provide constant 18.5 DC voltage while the detection range of the antenna coil inductor loop was measured (to nearest 5 mm) using a metre stick. The detection range was measured separately for a 23-mm Texas Instruments PIT-tag oriented with its cylindrical axis perpendicular (hereafter vertical) and parallel (hereafter horizontal) to the plane of the antenna coil inductor loop. The detection range data from these nodes was further analyzed in Surfer (Golden Software Inc., 1999), where the data were smoothed using the weighted matrix smooth method. Further, the spatial interpolation method of kriging (Cressie, 1991) was used to create detection range estimates for unrecorded locations.

The antenna efficacy and accuracy was tested in the field by carrying out "blind testing" during the winters of 2003–04. In these tests, a number of horizontally-oriented 23-mm PIT-tags were hidden in locations unknown to the person carrying out the tracking. The number of hidden tags was always unknown to the operator.

In March 2003, two field trials were conducted in the South Rusagonis River, New Brunswick, Canada, during conditions of partial ice and snow cover. For field trial 1, 30 PIT-tags (14 single tags and eight paired tags, i.e., two tags ~30 cm from each other) were hidden in the stream. Tag locations were 10–60 cm deep with 5–30 cm of ice and up to 25 cm of snow. The combination of these measures did not exceed the detection distance of the portable antenna. Thus, detection of the tags was always possible. The operator carried out the survey, moving in a downstream direction, scanning the entire study area from bank to bank and recording the tag codes detected.

Field trial 2 was designed to address the problems faced in the first trial, namely to better detect tags in clusters. With our system, it is impossible to simultaneously detect multiple tags in the antenna field (e.g., Castro-Santos et al., 1996). It is, however, possible to maneuver the antenna to approach each tag from a different angle so that the "strongest" tag signal can be detected for each tag in a cluster. However, if the data stream includes multiple tag codes, differentiating between codes is not possible. To solve this problem, a new data logging software ("Lockout", programmed by A. Haro, USGS, Massachusetts, USA) was used. The program "locked out" (for user determined time period) the identification code when a tag was first read and thus enabled the data stream only from unrecovered tags during the lock out period. In field trial 2, only paired tags (N = 12, tags ~30 cm of each other) were used and the tags were located under a 60 cm thick combination of water and ice. The tracking procedure was similar to field trial 1.

In March 2004, field trial 3 was carried out to better discern individual tags in clusters and to test the accuracy of identifying tag location. In this field trial, the person carrying out the tracking was not aware if the tags were clustered (i.e., in aggregations). The tags were hidden so that single tags (N = 23) were placed randomly in a rectangular field ($A = 500\ m^2$). Clusters (i.e., multiple tags within $1\ m^2$) were also set (N = 7; 2 × 2 tag and 1 × 3 tag aggregations). The tracking took place after a heavy snowfall that masked all tracks identifying tag locations. The tracking was carried from side to side with a 3-metre wide section scanned each time. The experimental field had up to 40 cm of snow and obstacles (wood, debris and boulders). When a tag was located, the "blind-spot method" (see results for description) was used to pinpoint the tag and a wooden stake was used to mark the presumed location of the tag. After scanning the entire area, tags were gently uncovered and the shortest distance from the presumed to true tag location was measured to nearest 5 mm.

Results

Maximum detection distances of 90 cm and 65 cm were achieved for vertically and horizontally oriented tags, respectively. The size of the detection field (measured in area) was substantially larger for vertically oriented tags (~150 cm in diameter, round in shape) than horizontal tags (~100 cm in diameter, ellipsoidal in shape) (Fig. 1). Most notably, the horizontal tag produced a bimodal detection distribution; best detection distance was achieved when the tag was above or just outside the inductor coil loop wires whereas the detection was lessened towards

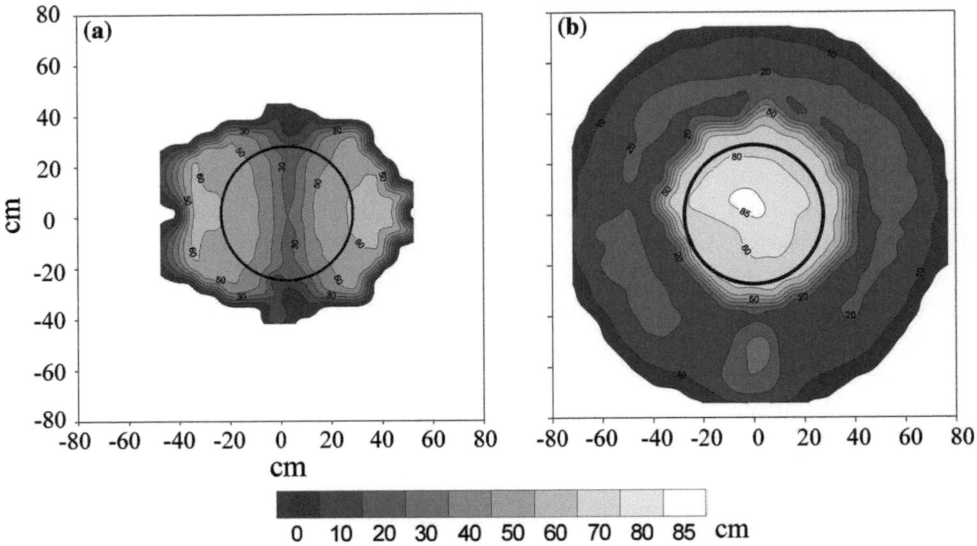

Fig. 1 Representation of a 23-mm PIT-tag detection pattern using a portable antenna. (**a**) horizontal tag, i.e. tag held parallel to x-axis showing a bimodal detection pattern; (**b**) vertical tag, i.e., tag held parallel to z-axis showing an unimodal detection pattern. The black circle shows the open coil inductor loop (55 cm in diameter)

the centre of the loop (up to 40 cm difference, Fig. 1A). Thus, when the centre of the antenna was held ~20 cm directly above the tag, it was undetected (thus, no sound from piezoelectric buzzer) suggesting a "blind spot" in the middle of the antenna field. As the "blind spot" was narrow (Fig. 1A), any movement of the antenna exceeding a few centimeters to any direction resulted in re-detection of the tag.

In field trial 1, 93% of single tags and 62.5% of the double tags were found with a single pass survey in an ice-covered stream (Table 1). Of the eight paired-tag aggregations, both tags in a cluster were recovered three times; one of the double tags was recovered four times; and one cluster was not found. In field trial 2, 100% of the tags were detected in the stream in winter, even though all tags were clustered (paired tags, Table 1).

In field trial 3 (snow covered field), all of the 30 hidden tags were found indicating 100% tracking efficacy (Table 1). The findings from the antenna magnetic field detection range were used to pinpoint the hidden tags. After a PIT-tag was initially located, the antenna was moved across a tag until the sound from piezoelectric buzzer suddenly disappeared, indicating that the tag was in the "blind spot" (i.e., middle of the inductor coil loop). Any further movement of the antenna resulted in a detection sound from the buzzer. The pinpointing accuracy (mean ± SE) using the "blind-spot method" was 10.9 ± 1.4 cm for single tags. One of the hidden tags was in close proximity to an obstacle (boulder) and the

Table 1 The efficacy (% of tags detected) and accuracy (distance between predicted and true tag location) of portable antenna system to detect 23-mm PIT-tags in three field trials

	n (single)	n (clustered)	Efficacy (%, single)	Efficacy (%, clustered)	Mean accuracy ± SE (cm)
Trial 1	14	16	93[c]	62.5[d]	N/A
Trial 2	0	12	N/A	100[c]	N/A
Trial 3	23	7	100[c]	100[c]	A) 10.9 ± 1.4; B) 13.3 ± 1.8

The accuracy in Trial 3 is reported separately for single tags (A) and all tags in the trial (B, i.e., including clusters). N/A = data could not be computed, n = number of tags hidden. The values sharing a common superscript (c or d) within each column do not differ at the 0.05 level of significance (Log-likelihood test G) antennae could not be moved into a position where the blind spot could be precisely identified; consequently, the tag was assumed 31 cm off from the true tag location. Also aggregations of tags (i.e., magnetic field of multiple tags overlapping) reduced the pinpointing accuracy (for aggregated tags, mean ± SE; 20.0 ± 5.0 cm) but the reduction was not statistically significant when compared to single tags (Mann-Whitney *U*-test: $U = 55.0$, $P = 0.121$). When all the tag data were considered (i.e., including aggregations and the tag associated with obstacles), mean pinpointing accuracy was 13.3 ± 1.8 cm (Table 1).

For single tags in the open area, the "blind spot" was easily found within ~20 s, but considerably more time was required for tags in aggregations (~1–2 min) and the identification of the "blind spot" was not always possible.

Discussion

The utilization of the bimodal detection field of a horizontally oriented tag proved to be a very accurate method for identifying tag locations and thus, well suited for microhabitat and activity studies in winter conditions. High accuracy can also be used to differentiate the tags that are left on substrate from those that are still inside a fish because even the smallest movement can be detected and thus associated with alive fish. We have been able to effectively retrieve PIT-tags from the riverbed both in open-water (snorkeling) and surface-ice conditions (by drilling a hole through the ice after which the tag could be observed laying on the bottom). As PIT-tags do not have an internal battery and thus can be re-used after recovery, it may become useful to retrieve tags especially in systems where the tagging takes place in high quantities. If the "blind spot method" is to be used for accurate tag positioning, the antenna dimensions should be round in shape and at least 55 cm in diameter, as the magnetic field produced by smaller antenna dimensions may not have the "blind spot" in the middle of the open coil inductor loop. Also, the detection distance may be greatly decreased with smaller antenna dimensions.

The knowledge of the shape of the antenna field can also be used to predict the orientation of the fish within the streambed under ice. As an example, in field experiments with PIT-tagged Atlantic salmon parr, reading ranges exceeding 80 cm were achieved in areas that were ~65 cm deep (distance from top of substrate to top of ice). As the reading range for our antenna for a horizontally positioned tag did not exceed 65 cm and no "blind spot" could be detected (indicating unimodal detection pattern), it was obvious that the fish was burrowed into the substrate in vertical position (Fig. 2) as was confirmed later with snorkeling observations. Caution has to be exercised, however, when predictions of the fish orientation are done without visual observations; the orientation of the PIT-tag in the body cavity of a fish can vary, especially in large-bodied fish (Baras et al., 2002). Furthermore, we have also been able to recognize fish mortalities by using the knowledge about the detection field of our antenna. This was accomplished by gaining detection of the tag exceeding 70 cm measured from the top of the ice, indicating that the tag was somewhere in the ice column in non-horizontal position (Fig. 2).

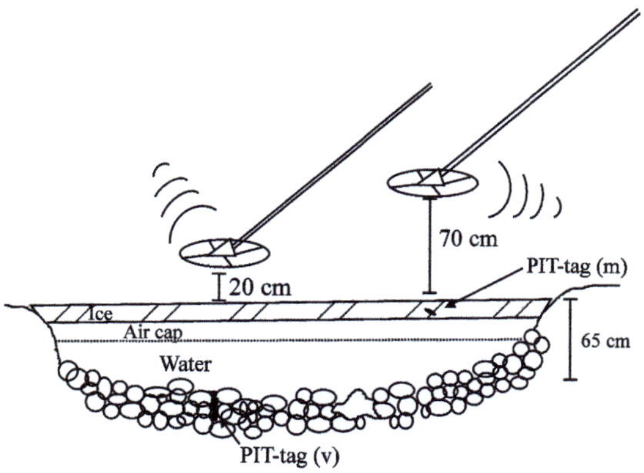

Fig. 2 Interpretation of PIT-tag position and orientation. In the example, the distance from the top of the substrate to the top of the ice is 65 cm, i.e., the upper limit for detection of horizontally oriented tag for the described portable antenna. (A) Antenna measures 20 cm from top of ice with no blind spot, indicating unimodal tag detection pattern with a total reading of over 80 cm and thus, vertically oriented tag (v) within substrate; (B) Antenna measures 70 cm reading from top of ice, indicating tag loss or mortality (m) with non-horizontal tag in the ambient ice column

Based on the "blind-testing", it was demonstrated that the active PIT-tracking was also very efficient in winter conditions. However, it is recommended that either "lock out" feature in datalogging program (as in trial 2) or knowledge about the shape of the antenna field (as in trial 3) is utilized in order to detect tags in aggregations as these methods resulted in significantly better detection efficacy. By contrast to open water-conditions, it was also easy to ascertain that the whole area under study is efficiently scanned as the tracks on snow could be used to guide the path of the person carrying the tracking.

Despite the fact that the PIT-tags used in the efficacy and accuracy testing were "dummy" tags and thus not capable of moving themselves, we believe that our results for the efficacy and accuracy can be applied to represent the situation in stream environments during winter. There is wide evidence that rheophilic salmonids are hiding in crevices in a relatively non-active state during the day (e.g., Rimmer et al., 1983; Cunjak, 1988; but see also e.g., Hiscock et al., 2002). Thus, the fright-bias that can be caused by the tracking can be assumed low. Based on tracking surveys of juvenile Atlantic salmon we have observed no immediate escape responses either during day or night when portable tracking was done from the top of full surface ice. The same also applied for open water conditions during the winter (i.e., water temperatures below 5°C), except if the fish were physically hit by the antenna. We have also observed that fish are more susceptible to the presence of the tracker during nights when frazil- and anchor ice is forming in extensive amounts. The methodology can also be assumed to be very suitable for studying other strongly rheophilic species like cottids. However, the ability to detect fish with high accuracy might not be as useful during summer months for species that are easily frightened due to tracking operations.

It should be noted that care has to be taken, so that the results concerning the efficacy would not produce too good of a picture. Even though the "blind testing" implies that the recovery rate of tags is high, the testing was done under conditions where the tag detection was always possible (i.e., tags were within the detection range). This assumption is not valid in all natural settings and tags may be situated in areas where the detection is not possible. This might be the case during the winter when the antenna cannot be lowered near the substrate due to surface ice and/or snow. Indeed, the knowledge of the study site (i.e., bathymetry, recent ice thickness) has to be taken into account to determine if the detection has been possible in the area scanned with the antenna. Furthermore, as mentioned already above, rheophilic salmonids conceal themselves in the substrate material during daytime in late fall and winter and even 10–30 cm burial depths have been recorded (Erkinaro et al., 1994). Fish can also bury themselves in loose gravel or even fine sediment (Smith & Griffith, 1994; Emmett et al., 1996; Meyer & Griffith, 1997). Thus, fish are capable of placing themselves out of detection range even in relatively shallow areas during winter. This behaviour can affect the tracking efficacy, even if the bathymetry and ice conditions of a particular study site would theoretically allow efficient tracking.

Overall, it was observed that active PIT-tracking can successfully be used to explore fish behaviour in ice-covered, shallow streams.

Acknowledgements We would like to thank C. Breau, H.-P. Fjeldstad, A. Fraser, H. Förster, and J. Karhunen for insightful ideas and help during the fieldwork. We also want to thank A. Haro and D. Bateman for sharing their knowledge about PIT-technology and M. Stickler for comments on an earlier version of this article. The research was funded by Norwegian Research Council, Canada Research Chair Program and T. Linnansaari was further financially supported by Finnish Cultural Foundation and Helsingin Sanomat Centennial Foundation.

References

Alfredsen, K. & E. Tesaker, 2002. Winter habitat assessment strategies and incorporation of winter habitat in the Norwegian habitat assessment tools. Hydrological Processes 16: 927–936.

Baras, E., C. Malbrouck, M. Houbart, P. Kestemont & C. Melard, 2002. The effect of PIT tags on growth and physiology of age-0 cultured Eurasian perch *Perca fluviatilis* of variable size. Aquaculture 185: 159–173.

Barbin Zydlewski, G., A. Haro, K. G. Whalen & S. D. McCormick, 2001. Performance of stationary and portable passive transponder detection systems for monitoring of fish movements. Journal of Fish Biology 58: 1471–1475.

Brännäs, E., H. Lundqvist, E. Prentice, M. Schmitz, K. Brännäs & B.-S. Wiklund, 1994. Use of the Passive Integrated Transponder (PIT) in a fish identification and monitoring system for fish behavioral studies. Transactions of the American Fisheries Society 123: 395–401.

Bubb, D. H., M. C. Lucas, T. J. Thom & P. Rycroft, 2002. The potential use of PIT telemetry for identifying and tracking crayfish in their natural environment. Hydrobiologia 483: 225–230.

Castro-Santos, T., A. Haro & S. Walk, 1996. A passive integrated transponder (PIT) tag system for monitoring fishways. Fisheries Research 28: 253–261.

Connell, C. B., 2005. Fry stocking and adult translocation as strategies to address patchiness in Atlantic salmon (*Salmo salar* L.) spawning distributions within Clearwater Brook, NB. M.Sc. thesis, University of New Brunswick, Canada.

Cressie, N. A. C., 1991. Statistics for spatial data. John Wiley and Sons, New York, USA.

Cucherousset, J., J.-M. Roussel, R. Keeler, R. A. Cunjak & R. Stump, 2005. The use of two new portable 12-mm PIT tag detectors to track small fish in shallow streams. North American Journal of Fisheries Management 25: 270–274.

Cunjak, R. A., 1988. Behaviour and microhabitat of young Atlantic salmon (*Salmo salar*) during winter. Canadian Journal of Fisheries and Aquatic Sciences 45: 2156–2160.

Donaldson, W. E., D. Schmidt, L. Watson & D. Pengilly, 1992. Development of a technique to tag adult red king crab, *Paralithodes camtschaticus* (Tilesius, 1815), with Passive Integrated Transponder tags. Journal of Shellfish Research 11: 91–94.

Emmett, B., G. M. Kruzynski & I. K. Birtwell, 1996. The use of a pulp mill effluent mixing zone during winter by juvenile chinook salmon in the Fraser River at Prince George, British Columbia. Canadian Manuscript Report for Fisheries and Aquatic Sciences 2341: 38 p.

Erkinaro, J., I. L. Shchurov, T. Saari & E. Niemelä, 1994. Occurrence of Atlantic salmon parr in redds at spawning time. Journal of Fish Biology 45: 899–900.

Golden Software Inc, 1999. Surfer 7. User's guide. Contouring and 3D surface mapping for scientists and engineers. Golden Software Inc., Colorado, USA.

Hiscock, M. J., D. A. Scruton, J. A. Brown & C. J. Pennell, 2002. Diel activity pattern of juvenile Atlantic salmon (*Salmo salar*) in early and late winter. Hydrobiologia 483: 161–165.

Meyer, K. A. & J. S. Griffith, 1997. Effects of cobble-boulder substrate configuration on winter residency of juvenile rainbow trout. North American Journal of Fisheries Management 17: 77–84.

Prentice, E. F., T. A. Flagg, C. A. McCutcheon & D. F. Brastow, 1990. PIT-tag monitoring systems for hydroelectric dams and fish hatcheries. American Fisheries Society Symposium 7: 323–334.

Rimmer, D. M., U. Paim & R. L. Saunders, 1983. Autumnal habitat shift of juvenile Atlantic salmon (*Salmo salar*) in a small river. Canadian Journal of Fisheries and Aquatic Sciences 40: 671–680.

Roussel, J.-M., A. Haro & R. A. Cunjak, 2000. Field-test of a new method for tracking small fishes in shallow rivers using passive integrated transponder (PIT) technology. Canadian Journal for Fisheries and Aquatic Sciences 57: 1326–1329.

Roussel, J.-M., R. A. Cunjak, R. Newbury, D. Caissie & A. Haro, 2004. Movements and habitat use by PIT-tagged Atlantic salmon parr in early winter: the influence of anchor ice. Freshwater Biology 49: 1026–1035.

Smith, R. W. & J. S. Griffith, 1994. Survival of rainbow trout during their first winter in the Henrys Fork of the Snake River, Idaho. Transactions of the American Fisheries Society 123: 747–756.

Hydrobiologia (2007) 582:289–299
DOI 10.1007/s10750-006-0553-x

FISH TELEMETRY

Growth, survival, and tag retention of steelhead trout (*O. mykiss*) surgically implanted with dummy acoustic tags

D. W. Welch · S. D. Batten · B. R. Ward

Abstract Recent advances in micro-electronics make the study of the migration of even small marine animals (>12 cm) over many 1000s of kilometres a serious possibility. Important assumptions in long-term studies are that rates of tag loss caused by mortality or tag shedding are low, and that the tagging procedure does not have an unacceptable negative effect on the animal. This paper reports results from a study to examine the retention of relatively large (24 × 8 mm) surgically-implanted dummy acoustic tags over a 7-month period in steelhead pre-smolts (*O. mykiss*), and the effects of implantation on growth and survival. Although there was some influence on growth to week 12, survival was high for animals > 13 cm FL. In the following 16-week period, growth of surgically implanted pre-smolts was the same as the control population and there was little tag loss from mortality or shedding. Currently available acoustic tags can be implanted in salmonid fish ≥12 cm FL, although combined losses from mortality and tag shedding were 33–40% for animals in the 12 and 13 cm FL size classes. By 14 cm FL, combined rates of tag loss (mortality plus tag shedding) for surgically implanted tags dropped to <15% and growth following surgery was close to that of the controls. Our results suggest that studies of ocean migration and survival over periods of many months are now feasible even for animals as small as salmon smolts. Surgically implanted salmon smolts are therefore good candidates for freshwater and coastal ocean-tracking studies on relatively long time scales (months). On such time scales, even relatively small salmon smolts may move thousands of kilometers in the ocean.

Keywords Tagging · Tag shedding · Surgical implantation · Growth · Survival · Telemetry

Guest editors: P. R. Almeida, B. R. Quintella, M. J. Costa and A. Moore
Developments in Fish Telemetry

D. W. Welch (✉) · S. D. Batten
Kintama Research Corp., Building 373, Malaspina University-College, 900-5th St., Nanaimo, BC, Canada V9R 5S5
e-mail: david.welch@kintamaresearch.org

D. W. Welch
Pacific Biological Station, Fisheries & Oceans Canada, Nanaimo, BC, Canada V9R 5K6

B. R. Ward
British Columbia Fisheries Science Section, University of B.C., 2204 Main Mall, Vancouver, BC, Canada V6T 1Z4

Introduction

The potential for sea-floor observatories to dramatically increase our understanding of processes occurring within the ocean has recently been reviewed (National Research Council, 2000); surprisingly, no mention was made of the ability of such observatories to potentially allow the tracking of marine animals during their extensive marine migrations. Recent developments in acoustic technology (Voegeli et al., 1998; Lacroix & Voegeli, 2000) offer the prospect of establishing extensive networks of underwater acoustic listening lines. Arrays formed of a series of such lines might eventually stretch the entire length of continents (e.g., Welch et al., 2003; see also www.postcoml.org). The development of continental-scale arrays would allow tracking the movements of even quite small animals for months or years over vast regions of the continental shelf. However, although important aspects of the technology needed to eventually allow the deployment of long-term monitoring posts on the seabed still need to be established and validated, an equally critical aspect of this work involves the question of how animals of different sizes recover from tagging with acoustic tags, and their rates of tag retention—the development of long-lived acoustic tags with operational lifetimes of months or years is meaningless if animals die from tagging or rates of tag loss are high.

This study was designed to examine the effect of tagging small salmon pre-smolts (steelhead trout; *Oncorhynchus mykiss*) with acoustic tags. The goals were: (1) to determine the appropriate size ranges and protocols for pre-smolt tagging, and (2) to establish rates of tag loss over time.

Materials and methods

Hatchery-reared steelhead trout were selected from part of a large-scale production within the British Columbia Fish Culture program, at the Vancouver Island Trout Hatchery in Duncan, B.C. Approximately 200 pre-smolts (Cowichan River stock) were selected and placed in a single rectangular fibreglass rearing tank several days before the start of the experiments at the end of February, 2001. At the end of the tagging procedure described below, all fish were transferred to a second fibreglass tank, which served to hold all of the control and tagged fish together in freshwater for the duration of the experiment. Feeding and care of pre-smolts followed standard hatchery rearing procedures.

Dummy acoustic tags (8 mm diameter × 24 mm long; 1.4 g) were cast from quick-setting epoxy resin with a 12 mm PIT tag embedded in the body of the tag to allow unique identification of each pre-smolt. The epoxy resin was mixed with a bright yellow colouring agent at the time the hardening agent was added to make it easier to see any tags shed onto the bottom of the hatchery raceway. Control fish also were implanted with a PIT tag, thereby ensuring that all fish used in the experiment were uniquely identified, and allowing the two groups to be placed in the same holding tank and eliminating possible tank effects on the results.

Treatment protocol was similar for the two groups of pre-smolts, but we made some slight modifications to our surgical procedure over the 3-day tagging period (19, 20, and 24 February, 2001). Two pre-smolts were netted from the initial holding tank and placed in a bucket of freshwater, into which had been added clove oil (80 ppm target strength). The clove oil was mixed 50:50 with ethanol and shaken, in order to improve the dispersion of the clove oil at the low temperature of the water (7°C). Once quiescent the first (control) pre-smolt was weighed, fork length recorded, and a PIT tag was injected into the abdominal cavity using a modified hypodermic syringe before placing the pre-smolt in the second fibreglass raceway. This process took approximately one minute to complete, allowing deeper anaesthesia to be induced in the second pre-smolt.

The second pre-smolt was handled identically except for the implantation process. After measuring and recording size, pre-smolts were placed ventral side up in a V-shaped trough formed from a sheet of acrylic plastic. The channel was partially filled with water to ensure that the head and gills remained submerged while the abdomen was exposed above the water line. A small amount of paper towelling was submerged in the trough, cushioning the fish and helping to prevent it from moving during surgery.

A fine-bladed scalpel was used to make an incision just large enough to allow passage of the dummy tag along the ventral midline anterior to the pelvic fins (11~12 mm), and the tag lightly pushed through the incision and then forwards until the body of the tag lay fully within the abdominal cavity. The tag was then moved backwards using either the point of the scalpel or a pair of pointed forceps, so that the posterior end of the tag was located beyond the posterior end of the incision. In this way the two ends of the 24 mm long tag were both located under uncut abdominal musculature and the central section of the tag body lay under the incision. As monofilament suture material is generally viewed as preferable for use in fish (Gilliland, 1994; Wagner et al., 2000), the incision was closed with two simple interrupted sutures tied in Ethicon PDS-II 2-0 monofilament polydioxanone using a FS-1 cutting needle. The two stitches approximately trisected the incision, with the needle and suture passed fully through the abdominal wall on each side of the incision. The position of the tag acted as a shield, facilitating the completion of each suture without puncturing the internal organs or sewing them into the incision.

No major attempt was made to sterilise the working area other than to keep it clean and reasonably hygienic, as might be encountered in a field program. The surgeon (DWW) wore surgical gloves but did not change them between fish. The scalpel blade was dipped in betadine antiseptic at the start of each operation, but the blade was only changed when needed to ensure a sharp edge. The drop of betadine left on the tip of the blade was first placed on the abdomen of the fish and then spread out over the area to be incised with the tip of a gloved finger. Care needed to be used here because betadine in contact with the gills is lethal. (A second reason for the use of paper towelling, in addition to holding the anaesthetised fish in one position was to form a dam between the abdomen and the gills, minimising the chance of antiseptic-contaminated water reaching the gills). No antibiotics were used.

Some modifications to the surgical procedure were made during the course of the 2-day procedure. Oxygen was not initially used during surgery, but we found that the pre-smolts came out of their anaesthesia more rapidly with its use, and accordingly used oxygen bubbled into the water using an airstone placed near the head thereafter. The water in the surgical table was still changed after every 2–3 fish to minimise the chance of the betadine antiseptic reaching the gills and to prevent thermal shock from the water in the surgical tray warming to air temperature.

Handling procedures were identical except for the surgery, since both the control and surgically implanted pre-smolts were held in the same tank. A record was kept during implantation of the relative size distributions for the two groups, and the tagging sequence was occasionally varied to ensure that the size distribution of pre-smolts tagged using the two procedures was similar. Pre-smolts were fed daily to satiation by an automatic feeder. Hatchery staff maintained the water temperature in the rearing trough at 7°C for the first 10 days after tagging, and then increased the temperature to 11°C for the duration of the experiment. We tried to ensure a similar size distribution of pre-smolts chosen for control and surgical implantation (Table 1), except that the smallest available pre-smolts were clearly too small to be implanted. Below ca. 11 cm FL, it was not possible to insert the tag into the body cavity and close the body wall. (There were significant fat bodies lying along the intestinal mesentery of these hatchery-reared fish, occupying space that in a lean wild pre-smolt might make it easier to place the tag in the body cavity).

Pre-smolts were netted in small batches from the indoor holding tank on 16 May, 2001, ~12 weeks after tagging, and lightly anaesthetised with clove oil. The PIT tag number for each animal was recorded and each animal measured for weight and length, and visually examined to assess the extent of healing. The same procedure was repeated again on 8 September 2001, ~16 weeks later (~29 weeks after implantation). During the holding period the tank was checked daily for dead fish or shed tags and the date of recovery was noted. The weight of the dummy tag was subtracted from the fish weights measured during the May and September evaluations. Statistical comparisons were based on an analysis of covariance.

Table 1 Rates of potential acoustic tag loss by size class to week 29

Fork Length (cm)	<11	11–12	12–13	13–14	14–15	15–16	16–17	Total
Mean Weight (g)	11.1	16.8	21.6	26.7	33.3	39.6	53.3	
N								
Dummy	1	6	15	20	15	15	11	83
PIT	7	15	7	16	18	16	6	85
Mortality								
Dummy	100.0%	50.0%	33.3%	10.0%	6.7%	6.7%	9.1%	15.7%
PIT	0.0%	0.0%	0.0%	0.0%	0.0%	0.0%	0.0%	0.0%
Tag shedding								
Dummy	–	16.7%	6.7%	30.0%	20.0%	6.7%	9.1%	13.3%
PIT	0.0%	0.0%	0.0%	6.3%	0.0%	0.0%	0.0%	1.2%
Overall loss								
Dummy	100.0%	66.7%	40.0%	40.0%	26.7%	13.3%	18.2%	28.9%
PIT	0.0%	0.0%	0.0%	6.3%	0.0%	0.0%	0.0%	1.2%

The initial number of fish (*N*) in each 1 cm size category is indicated, as is the mean weight of each length class. Mortality rates for fish surgically implanted with dummy acoustic tags dropped sharply for animals greater than 12 cm fork length. Although the rate of tag shedding (but not mortality) remained fairly high for the 13 cm size class, no dummy tags were lost before 39 days after implantation. Tag weight ranges from 15% of body weight in the <11 cm size class to 3.5% in the 15–16 cm size class

Results

Effect on survival

Some difficulty was initially encountered in the early stages of the surgery. It initially took considerable time to induce anaesthesia in the pre-smolts, so we replaced our standard stock solution of clove oil (which was several years old) with freshly purchased clove oil from a pharmacy, where it was sold as a toothache remedy. Use of the new stock resulted in 6 of the 13 subsequently implanted pre-smolts not recovering from the anaesthesia (including a run of five sequential deaths), and the death of the only PIT tagged pre-smolt associated with the tagging procedure. The problem appeared to be with the clove oil, as surgical times were low (~2 min). After switching back to the original stock solution of clove oil no further immediate mortalities were encountered.

These deaths were thus associated with a specific anaesthetic problem, and the animals surgically implanted under such conditions did not recover from the surgery. Implanted tags in such cases can be recovered and used in a different animal if animals are held for a brief period of observation post-surgery. We therefore excluded these cases from the analyses reported below, as it seems most relevant to examine what fraction of animals released into the environment would be expected to survive and retain their tags.

All but one mortality occurred during the initial 12-week period. Mortality was strongly associated with size, at least for the smallest animals examined (Table 1). For animals less than ca. 11.5 cm FL, it proved impossible to close the body cavity with the dummy acoustic tag inside, and the few animals that we attempted to surgically implant all died. However, although the dummy acoustic tag filled much of the body cavity, surgically implanted animals >11.5 cm FL (20 g) survived the surgery and mortality rates dropped sharply with increasing size over the interval 11–13 cm (Table 1). For animals ≥13 cm mortality remained low and constant at <10%. Mortality in the PIT-tagged control group was zero for the same size classes. Over the subsequent 16-week period (to week 29, 7 months post-implantation), only one additional dummy-tagged pre-smolt died (on 24 July). The surgical incision on this animal (12.1 cm at time of tagging) had not healed and mesenteric fat was protruding from the incision on both the May and September census. This animal had progressively lost weight between each examination.

Effect on growth

To examine the effect of surgical implantation on growth, we compared relative growth over the first (12 week) and second (16 week) study periods for the two groups (Fig. 1). Both controls and

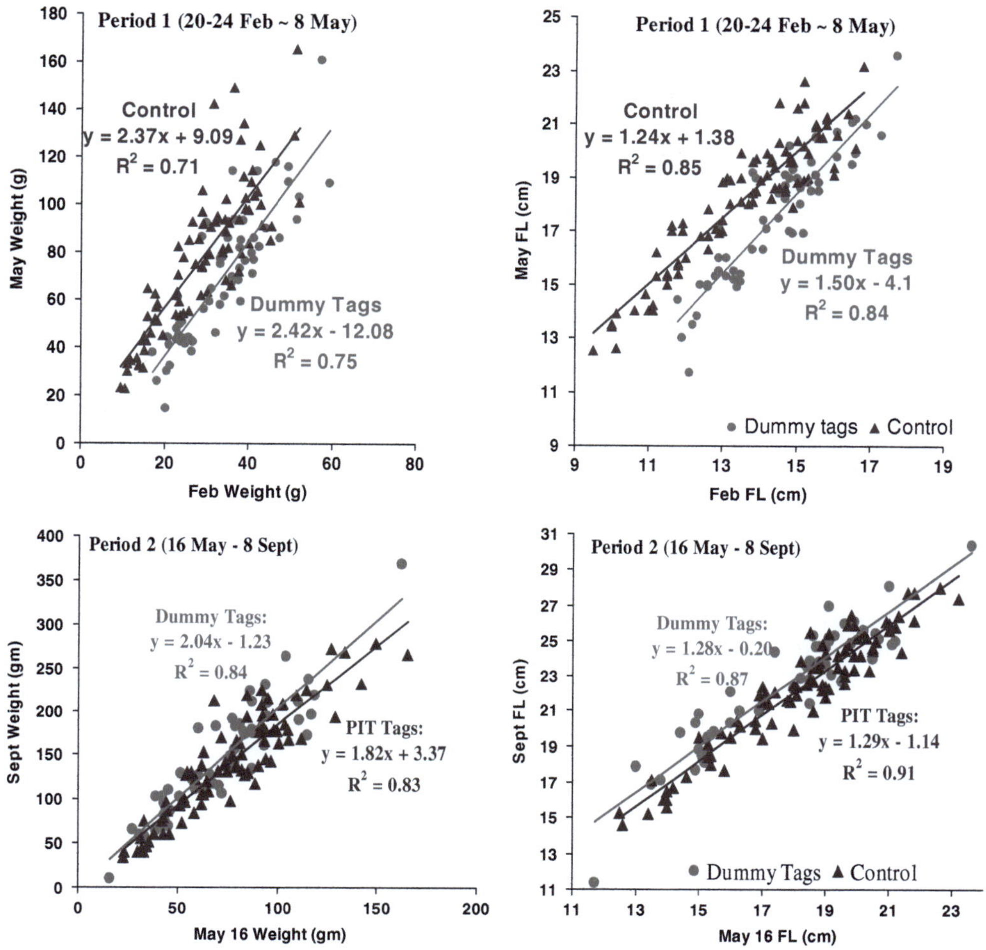

Fig. 1 Comparison of achieved growth for implanted pre-smolts over the first (12 week) and second (16 week) periods. Growth of surgically implanted pre-smolts was reduced relative to the controls in the first period (top row), but not the second (lower row)

implanted pre-smolts grew over the initial 12-week study period, although surgically implanted animals grew less for a given size than the controls (ANCOVA, $P < 0.05$). Linear regressions fitted to the data show that the growth patterns had nearly parallel slopes but different intercepts, suggesting that the effect of the surgical implantation was a reduction in growth that was similar for all size groups. Fitting a second model with a common slope but different intercepts supported this suggestion, as the fit was indistinguishable ($R^2 = 0.729$ for both). The difference in intercepts was 2.4 g, suggesting that animals of all sizes had reduced growth of roughly 2.4 g as a result of surgical implantation. For the smallest animals (~20 g or ~12 cm) this growth difference corresponds to a 35% reduction in achieved growth.

Closer examination of the residuals from the regression lines calculated using the data for the first period indicated that there was some tendency to heteroscedasticity, with the residuals tending to increase with increasing size. Log-transforming the size data and repeating the analysis resulted in a slight increase in R^2 (0.793) and a better residual fit, but showed a statistically significant difference in slopes (ANCOVA; $P < 0.05$), so that the difference in growth increment between implanted and control fish decreased even further at larger sizes. Although surgical implantation reduced growth at all sizes examined, the results indicate that growth differences were progressively smaller and showed greater overlap between control and implanted pre-smolts when initial sizes were greater than 14 cm FL (roughly 30 g body mass; Fig. 1).

Comparison of growth increments for the second period (16 May to 8 Sept, 2001), indicates no statistically significant difference in growth increment for any size group (ANCOVA; $P < 0.05$; Fig. 1, lower row). On average, the surgically implanted animals grew slightly more than the controls.

Rates of tag loss

Tag loss (shedding) was also related to body size (Table 1; Fig. 2). By week 12 a total of 11 dummy acoustic tags (a 13% loss rate) were expelled from the body, apparently without damage to the pre-smolts. In the following 16-week period only two additional tags were shed, both animals in the 14 cm size class (a 16% loss rate). Pre-smolts recovered from the tank at week 12 that were no longer individually identifiable by the presence of a PIT tag in all cases appeared to be healthy and the surgical incision appeared to be fully healed. When shedding rates are compared for different size classes it is clear that tags were retained over the full size range examined, but that shedding rates dropped from 16–30% in pre-smolts ≤13 cm to a low and stable rate of ~7% for animals ≥15 cm at implantation (Table 1). Tag shedding rates during the initial 12-week period were similarly low, but increased to 20% when the two shed tags found on 3 July and 18 August are included in the analysis (weeks 19 and 25; Fig. 3). The time course of tag losses (Fig. 3) shows that most tags were shed in a period around 4–12 weeks post-surgery, with only one tag shed >20 weeks post-surgery. The data thus suggests that the process is self-limiting and does not affect all tagged animals.

Some insight into the process of tag expulsion was obtained from examination of those tagged animals ($N = 18$; 22%) that initially appeared to be in the process of shedding the tag. These fish all had distended abdomens, apparently as a result of the tag being firmly pressed against the abdominal muscles from the inside (Fig. 4). (The remaining pre-smolts did not show evidence of the outline of the tag against the body wall). In each case the surgical incision was well-healed and showed no evidence of being likely to rupture. One pre-smolt was observed in the last stages of tag expulsion (Fig. 4). The abdomen was greatly distended around the tag and a "pore" had formed near the anterior end, through which the tag was completely exposed and clearly

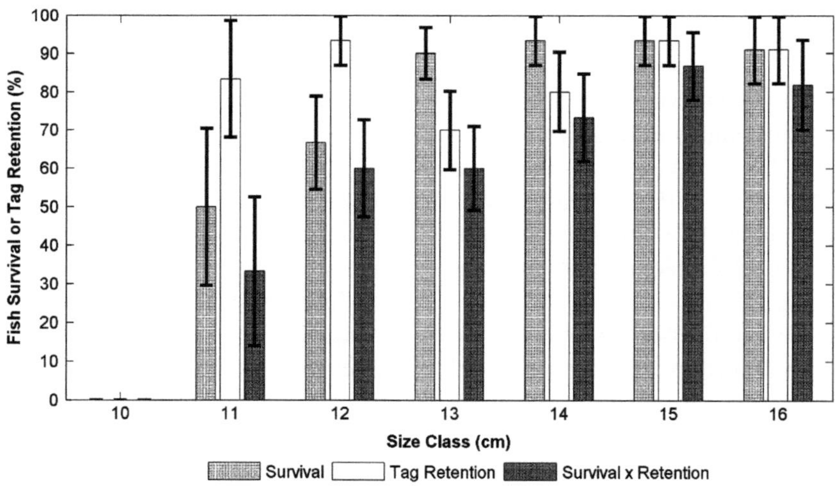

Fig. 2 Comparison of rates of pre-smolt survival, tag retention, and overall tagging success (survival × retention) to 6 months post-implantation. Overall success increases rapidly with initial size at implantation from 33% for 11.5 cm pre-smolts to >80% of the surgically implanted pre-smolts. Error bars indicate ±1 SE on the proportion

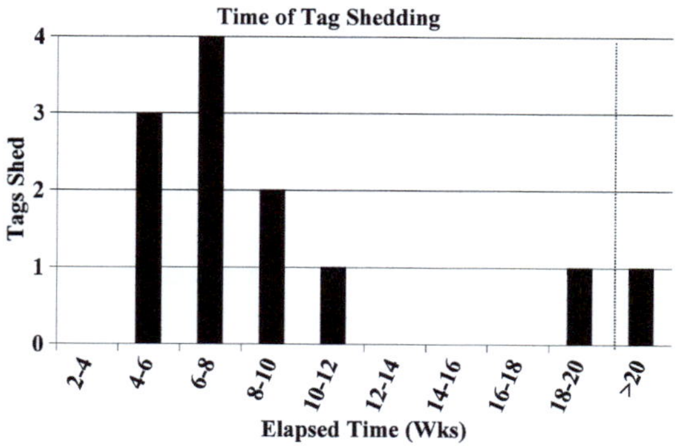

Fig. 3 Time course of tag shedding. Tag losses have been binned into consecutive 2-week periods. Median time to loss was 7.9 weeks, average time 9.8 weeks. A total of 13 tags were shed from a starting population of 90 implanted smolts

visible. The tissue forming the pore was remarkable for the absence of any evidence of trauma—the edge of the opening was smooth and neither the skin nor the underlying abdominal muscles showed any evidence of a ragged edge or infection, and there was no evidence of oozing from the wound. Rather than giving the impression of an ulcerated wound in the abdominal wall, the neatly rounded hole into the abdominal cavity seemed to cause little trouble for the fish and no evidence of significant trauma.

Discussion

Studies of the large scale movement patterns at sea of animals such as salmon pre-smolts will require monitoring their movements over thousands of kilometers and many months (or years) at sea. Newly-developed acoustic tracking technologies (e.g., Klimley et al., 1998; Voegeli et al., 1998; Lacroix & Voegeli, 2000; Lacroix et al., 2004; Welch et al., 2004; Lacroix & Knox, 2005) show great promise in revolutionising the study of the ocean migrations of these animals. An important aspect of such studies will involve the development of sampling protocols that will permit designing studies around animals of an appropriate size and age.

The purpose of the current study was to determine minimum size guidelines for applying the acoustic tags now commercially available to salmon pre-smolts. These tags have active lifespans approaching 4–6 months. Two important aspects were initially identified during the planning phase: (a) to identify those sizes at which mortality is minimised and (b) the sizes at which growth is not seriously disrupted. In addition to these initial issues, we were surprised to discover that a significant proportion of pre-smolts appear to have a well-developed biological mechanism for eliminating these large tags from their body cavities, apparently with little trauma to the body. Tag shedding is a serious issue for long-term studies because either the death of the salmon pre-smolt or the shedding of the tag results in the loss of the tagged animal from the study population.

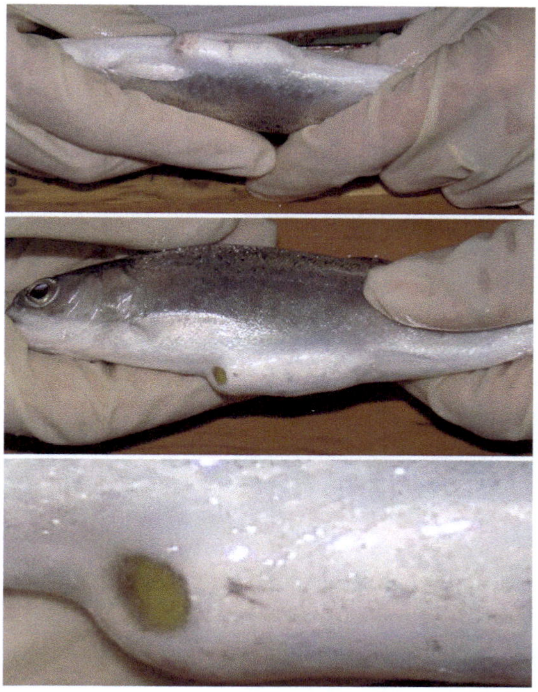

Fig. 4 Example of a pre-smolt in the process of extruding a tag. (Top) The abdomen is distended by the pressure of the tag against the abdominal wall. The right side of the abdomen shows some slight evidence of abrasion of the skin surface at the anterior end on the side facing the camera, presumably from contact with the tank wall. (Middle) The opposite (left) side of the pre-smolt shows a well-developed "pore", through which the head of the dummy tag is clearly visible. (Bottom) Close-up view of the pore. Note the clean unulcerated edge of the pore; the skin and abdominal muscles have neatly drawn back, with no evidence of infection. This tag was shed 7 weeks after the photographs were taken and 19 weeks post-implantation

Although steelhead trout and Atlantic salmon pre-smolts can be quite large (often >16 cm), the other species of Pacific salmon seldom achieve that size in freshwater. As a result, we implanted the tags into animals as small as 10.8 cm. Our results indicate that for animals ≥14 cm fork length, overall rates of tag loss were relatively low and roughly constant (13%) over the initial 12 weeks of the study, and losses from mortality and tag shedding were equivalent contributors (Table 1; Fig. 2). Losses over the following 16 weeks were minor. At smaller sizes mortality and tag shedding rates both increased, but some tags were successfully retained to the end of the study even for animals in the 11 cm size range. At sizes less than 11 cm FL it was not possible to close the body cavity around these large tags. We suspect that mortality in the 11–12 cm size class might be lower for wild pre-smolts, because the hatchery fish used in this study had substantial mesenteric fat bodies that largely filled the abdominal cavity. This limited the amount of space available for accommodating an acoustic tag and would not be as large a problem in lean wild fish.

Several studies have examined the effect of surgical implantation of tags in small salmon. Lucas (1989) and Moore et al. (1990) noted no difference in growth between control and implanted groups of rainbow trout and Atlantic salmon, respectively. In contrast, Adams et al. (1998a, b) found initially reduced growth of surgically implanted chinook relative to control animals during the first 3 weeks post-surgery for animals of similar size (11.4–15.9 cm) to those used in the present study. They also noted that the growth rates and swimming capabilities of surgically implanted fish was superior to that of animals where the tag was placed in the stomach. Lacroix et al. (2004) used Atlantic salmon smolts and found similar results to our study, with an initial period of mortality followed several months later by a period of tag shedding.

Our results indicate that animals of all sizes examined grew after surgical implantation, but that there was some reduction in growth relative to controls over the initial 12 week period. For pre-smolts greater than 14 cm initial size (~30 g), the growth depression was less evident and a substantial proportion of the surgically implanted animals achieved individual growth rates that were greater than the mean rate for the controls (Fig. 1). No difference in growth was evident after week 12.

Various studies have commented on the issue of determining an appropriate ratio of tag weight to fish weight, with several authors advocating a 2% rule. More recent studies suggest that a higher ratio may be reasonable. Zale et al., 2005 (review), suggested 4% was appropriate in at least some studies, while Lacroix et al. (2004) suggested that the ratio should not exceed 8% for Atlantic salmon smolts in the 25–45 g range. In our study, the tag to body weight ratio of the 14 cm size class (33 g) is 4.2%, close to Zale et al.'s suggestion.

Tag loss via extrusion directly through the body wall or passage into the intestine and then out the anus has been noted previously (Summerfelt & Mosier, 1984; Chisholm & Hubert, 1985; Marty & Summerfelt, 1986, 1990; Baras & Westerloppe, 1999; Lacroix et al., 2004). Shedding rates seem to be species-specific, ranging from 59% in a 24-week study of rainbow trout (Chisholm & Hubert, 1985), 49–59% over a 17-week study in channel catfish (*Ictalurus punctatus*; Marty & Summerfelt, 1990), and essentially zero in a 120 week study in tilapia (*Oreochromis aureus*; Thoreau & Baras, 1997). In our study, we observed an overall tag shedding rate of 13% to week 12, with the highest shedding rates in the smallest size classes, but with rates that then dropped quickly and stabilized at about 7% over 12 weeks in pre-smolts ≥14 cm. Two additional shed tags over the following 16-week period raised the shedding rate to 20% for the 14 cm size class (Table 1). This suggests that, as with mortality, the loss of animals from a tracking study are relatively small and occur over a relatively short duration early in the study period. Multi-year studies are likely to be feasible because tag-induced mortality and tag loss are transitory phenomenon, and not processes that persist over the long-term.

Tag shedding rates may be compared with the shedding rate for the PIT-tagged controls used in our study of 1%, similar to rates of tag loss reported in other studies (ranging from 1% to 5% for coded wire tagged salmon (e.g., Eames & Hino, 1983; Blankenship, 1990) and 3% for PIT-tagged brown trout (Ombredane et al., 1998)). Thus rates of tag loss may be expected to be higher than for the simpler CWT or PIT tag technologies. Studies need to balance the additional information gain with the higher expected loss rates of these larger and more expensive tags.

The mechanism by which extrusion occurs appears to be similar to that of mammals (Lucas, 1989), with the tag first being completely encapsulated and adhesion of the capsule to either the coelomic wall or the intestines. Subsequent contraction of the myofibrils within the encapsulation tissue resulting in the tag being pressed in close contact with the intestine or the body wall. Contact necrosis then appears to result in the gradual passage of the tag through the wall of the intestine or abdomen, and then out of the body (Summerfelt & Mosier, 1984; Marty & Summerfelt, 1986).

We did not examine the detailed histological changes occurring within the body cavity. However, the lack of any visual evidence of disruption of the initial incision and the markedly distended shape of the abdominal wall around the tag in a number of cases (Fig. 3) strongly suggested that encapsulation of these tags was initially occurring and that the pressure from capsule contraction would eventually force these tags out of the body. Surprisingly, during the following 16 week period only two tags were shed, and all of the animals initially noted at week 12 to have significantly distended abdomens were categorized as nearly normal by week 29; abnormalities were restricted to minor lumps only slightly detectable by finger tip or eye.

Although this distension presumably has some effect on swimming ability, the lack of any ulceration evident in the tag extrusion process (Fig. 4) strongly suggests that the biological mechanism underlying the extrusion process is well-developed and does not appear to place great stress on the animal. Furthermore, the regression of many animals initially expected to shed their tags at week 12 suggests that most tag loss was over by the seventh month of this study.

The foreign body response (Coleman et al., 1974) is not unique to fish, and is found to frequently occur when implants are placed into humans. For example, encapsulation of artificial breast implants in women has been reported to occur at rates of 27–40% within the first year following surgery (Little & Baker, 1980; Moufarrage et al., 1987) and is thought to represent a rejection mechanism expressed in these patients. As the reported rates of encapsulation in women are similar to those seen for the dummy acoustic tags used in the present study, this suggests that a broader biological mechanism is at work that determines the biological basis of the host response to the implant, since surgical implantation in humans occurs under highly sterile conditions with much more care taken in the choice of biomaterials. It is significant that in the case of breast implants, only a proportion of the patients developed a capsule around the breast implants, as was the case in the present study. In the majority of women, encapsulation of the breast implant seems to be uncommon if not expressed during the first year of implantation and surgical intervention to remove the implant and surrounding capsule was rarely needed past one year.

In tracking studies in fisheries, where tags with multi-year lifespans are now available, issues concerning long-term retention of the tags are just as important as direct effects of tagging on mortality. Tagging studies should be designed recognizing the potential for continued rates of loss of tags over the long-term from mortality or shedding to affect the goals of such studies. Ideally, protocols need to be developed to minimize long-term rates of tag loss.

Conclusions

In both the Atlantic and Pacific oceans there is growing recognition that marine survival has dropped sharply for many stocks of salmon, making formerly viable populations unsustainable even in the absence of exploitation. This phenomenon was unexpected and not consistent with standard fisheries models, which predict that populations recover rapidly when fishing pressure is relaxed. The increased mortality experienced at sea has in many cases been greater than the combined influence of large-scale sport and commercial fisheries on some stocks, such as southern British Columbia coho or steelhead (e.g., Smith & Ward, 2000; Ward, 2000; Welch et al., 2000). In a large number of British Columbia salmon stocks ocean survival has dropped to one-tenth that of only 25 years ago. Such changes make it imperative that a better understanding be developed of where salmon go in the ocean, because other stocks of salmon whose rivers are in reasonable geographic proximity have much higher ocean survival. Large-scale telemetry studies offer great promise for answering a number of fundamental biological questions about the ocean life of salmon.

Acknowledgements This is a contribution to the Census of Marine Life. We thank Ray Billings and Brian Martin of the Vancouver Island Trout Hatchery, British Columbia Fisheries, for their generous support of the experiments described here, Mary Thiess for statistical advice, Erika Welch for field assistance, and Adrian LaDouceur, Marc Trudel, and Jen Zamon for comments on the manuscript. This project was supported as part of the Northwest Power Planning Council's Innovative Proposal process (Proposal 22001), funded under BPA Contract 00003839.

References

Adams, N. S., D. W. Rondorf, S. D. Evans & J. E. Kelly, 1998a. Effects of surgically and gastrically implanted radio transmitters on growth and feeding behavior of juvenile chinook salmon. Transactions of the American Fisheries Society 127: 128–136.

Adams, N. S., D. W. Rondorf, S. D. Evans, J. E. Kelly & R. W. Perry, 1998b. Effects of surgically and gastrically implanted radio transmitters on swimming performance and predator avoidance of juvenile chinook salmon (*Oncorhynchus tshawytscha*). Canadian Journal of Fisheries and Aquatic Sciences 55: 781–787.

Baras, E. & L. Westerloppe, 1999. Transintestinal expulsion of surgically implanted tags by african catfish *Heterobranchus longifilis* of variable size and age. Transactions of the American Fisheries Society 128: 737–746.

Blankenship, H. L., 1990. Effects of time and fish size on coded wire tag loss from chinook and coho salmon. Transactions of the American Fisheries Society 7: 237–243.

Chisholm, I. M. & W. A. Hubert, 1985. Expulsion of dummy transmitters by rainbow trout. Transactions of the American Fisheries Society 114: 766–767.

Coleman, D. L., R. N. King & J. D. Andrade, 1974. The foreign body reaction: a chronic inflammatory response. Journal of Biomedical Materials Research 8: 199–211.

Eames, M. J. & M. K. Hino, 1983. An evaluation of four tags suitable for marking juvenile chinook salmon. Transactions of the American Fisheries Society 112: 464–468.

Gilliland, E. R., 1994. Comparison of absorbable sutures used in largemouth bass liver biopsy surgery. The Progressive-Fish Culturist 56: 60–61.

Klimley, A. P., F. Voegeli, S. C. Beavers & B. J. Le Boeuf, 1998. Automated listening stations for tagged marine fishes. Marine Technology Society Journal 32: 94–101.

Lacroix, G. L. & F. A. Voegeli, 2000. Development of Automated Monitoring Systems for Ultrasonic Transmitters. In Moore, A. & I. Russell (eds), Fish Telemetry: Proceedings of the 3rd Conference on Fish Telemetry in Europe. CEFAS: Lowestoft, UK, 37–50.

Lacroix, G. L. & D. Knox, 2005. Distribution of Atlantic salmon (*Salmo salar*) post smolts of different origins in the Bay of Fundy and Gulf of Maine and evaluation of factors affecting migration, growth, and survival. Canadian Journal of Fisheries and Aquatic Sciences 62: 1363–1376.

Lacroix, G. L., D. Knox & P. McCurdy, 2004. Effects of dummy acoustic transmitters on juvenile Atlantic salmon. Transactions of the American Fisheries Society 133: 211–220.

Little, G. & J. L. Baker, 1980. Results of closed compression capsulotomy for treatment of contracted breast implant capsules. Plastic and Reconstructive Surgery 65: 30–33.

Lucas, M. C., 1989. Effects of implanted dummy transmitters on mortality, growth and tissue reaction in rainbow trout, *Salmo gairdneri* Richardson. Journal of Fish Biology 35: 577–587.

Marty, G. D. & R. C. Summerfelt, 1986. Pathways and mechanisms for expulsion of surgically implanted dummy transmitters from channel catfish. Transactions of the American Fisheries Society 115: 577–589.

Marty, G. D. & R. C. Summerfelt, 1990. Wound healing in channel catfish by epithelialization and contraction of granulation tissue. Transactions of the American Fisheries Society 119: 145–150.

Moore, A., I. C. Russell & E. C. E. Potter, 1990. The effects of intraperitoneally implanted dummy acoustic transmitters on the behaviour and physiology of juvenile Atlantic salmon, *Salmo salar* L. Journal of Fish Biology 37: 713–721.

Moufarrege, R, G. Beauregard, J.-P. Bosse, J. Papillon & C. Perras, 1987. Outcome of mammary capsulotomies. Annals of Plastic Surgery 19: 62–64.

National Research Council, 2000. Illuminating the Hidden Planet: The Future of Seafloor Observatory Science. (Committee on Seafloor Observatories: Challenges and Opportunities). National Academy Press, Washington, DC, 135 pp.

Ombredane, D., J. L. Baglinière & F. Marchand, 1998. The effects of passive integrated transponder tags on survival and growth of juvenile brown trout (*Salmo trutta* L.) and their use for studying movement in a small river. Hydrobiologia 371/372: 99–106.

Smith, B. D. & B. R. Ward, 2000. Trends in wild adult steelhead (*Oncorhynchus mykiss*) abundance for coastal regions of British Columbia support the variable marine survival hypothesis. Canadian Journal of Aquatic Sciences 57: 1–14.

Summerfelt, R. C. & D. Mosier, 1984. Transintestinal expulsion of surgically implanted dummy transmitters by channel catfish. Transactions of the American Fisheries Society 113: 760–766.

Thoreau, X & E. Baras, 1997. Evaluation of surgery procedures for implanting telemetry transmitters into the body cavity of tilapia *Oreochromis aureus*. Aquatic Living Resources 10: 207–211.

Voegeli, F. A., G. L. Lacroix & J. M. Anderson, 1998. Development of miniature pingers for tracking Atlantic salmon smolts at sea. Hydrobiologia 371/372: 35–46.

Wagner, G. N., E. D. Stevens & P. Byrne, 2000. Effects of suture type and patterns on surgical wound healing in Rainbow Trout. Transactions of the American Fisheries Society 129: 1196–1205.

Ward, B. R., 2000. Declivity in steelhead *(Oncorhynchus mykiss)* recruitment at the Keogh River over the past decade. Canadian Journal of Fisheries and Aquatic Sciences 57: 298–306.

Welch, D. W., B. R. Ward, B. D. Smith, & J. P. Eveson, 2000. Temporal and spatial responses of British Columbia steelhead (*Oncorhynchus mykiss*) populations to ocean climate shifts. Fisheries Oceanography 9: 17–32.

Welch, D. W., G. W. Boehlert & B. R. Ward, 2003. "POST—the Pacific Ocean Salmon Tracking Project". Oceanologica Acta 25: 243–253.

Welch, D. W., B. R. Ward & S. D. Batten, 2004. Early ocean survival and marine movements of hatchery and wild steelhead trout (*O. mykiss*) determined by an acoustic array: Queen Charlotte Strait, British Columbia. Deep-Sea Research 51(6–9): 897–909. DOI: 10.1016/j.dsr2.2004.05.010.

Zale, A. V., C. Brooke & W. C. Fraser, 2005. Effects of surgically implanted transmitter weights on growth and swimming stamina of small adult westslope cutthroat trout. Transactions of the American Fisheries Society 134: 653–660.

MIX
Papier aus verantwortungsvollen Quellen
Paper from responsible sources
FSC® C105338

If you have any concerns about our products,
you can contact us on
ProductSafety@springernature.com

In case Publisher is established outside the EU,
the EU authorized representative is:
Springer Nature Customer Service Center GmbH
Europaplatz 3, 69115 Heidelberg, Germany

Printed by Libri Plureos GmbH
in Hamburg, Germany